AF713405

Groundwater in the Arab Middle East

Wolfgang Wagner

Groundwater in the Arab Middle East

Springer

Wolfgang Wagner
Hohenzollernstr. 21
30161 Hannover
Germany
uwwagner@gmx.de

ISBN 978-3-642-19350-7 e-ISBN 978-3-642-19351-4
DOI 10.1007/978-3-642-19351-4
Springer Heidelberg Dordrecht London New York

Library of Congress Control Number: 2011931780

© Springer-Verlag Berlin Heidelberg 2011
This work is subject to copyright. All rights are reserved, whether the whole or part of the material is concerned, specifically the rights of translation, reprinting, reuse of illustrations, recitation, broadcasting, reproduction on microfilm or in any other way, and storage in data banks. Duplication of this publication or parts thereof is permitted only under the provisions of the German Copyright Law of September 9, 1965, in its current version, and permission for use must always be obtained from Springer. Violations are liable to prosecution under the German Copyright Law.
The use of general descriptive names, registered names, trademarks, etc. in this publication does not imply, even in the absence of a specific statement, that such names are exempt from the relevant protective laws and regulations and therefore free for general use.

Cover design: deblik, Berlin

Printed on acid-free paper

Springer is part of Springer Science+Business Media (www.springer.com)

To Ursula and to my colleagues and friends in the Arab countries

If all the waters would be salty and man would have to change it to achieve its sweetness, great efforts would arise to man. The sublime Creator has saved, in his grace, his creatures these efforts through the effect of the sun on the water of the sea and the rise of the vapour from the sea; the winds then disperse these vapours wherever he wants, and it arrives as rain water at these locations. That is then stored in rivulets within the mountains and under the earth. He then brings out part of the water and lets flow wadis and rivers, and springs and wells appear in such a degree that it is sufficient for the creatures for 1 year until rain comes again in the following year.

Al Qazwini Zakariya ibn Muhamad ibn Mahmud abu Yahya, ca. 1262 (AH 665)

Introduction

The region covering the Arabian Peninsula and the adjoining northern Arabian countries coincides approximately with a specific large geologic structure: the Arabian Plate. Politically the region includes the countries of the Arabian Peninsula together with the northern Arab countries: Al Mashreq – the eastern part of the caliphate or of the Arab World. In a geographic political view the region may be denominated Arab Middle East (Ash Sharq al Awsat) or Western Asia.

The Arab Middle East with an area of 3.7 million km^2 forms a small subcontinent between the Mediterranean Sea, Red Sea, Arabian Sea, the Gulf and, in the north, the Zagros–Taurus mountain chains. About 90% of the region are semi-arid to arid steppe or desert areas. As perennial rivers exist only in the northern and western margins of the Arab Middle East, the use of groundwater resources is an essential basis for the economic development and survival of the countries. The region includes 12 Arab countries; water demand/supply previews indicate precarious situations in the near future for most of these countries.

The idea of compiling a book on "Groundwater in the Arab Middle East" arose from the professional activities of the author as hydrogeologist in the services of the German Government between 1965 and 1998, much of which was devoted to groundwater projects in the Middle East. The information presented in the book is based on reviews of a large number of publications, reports and documents as well as on field experience in various Arab countries.

The groundwater projects in the Middle East, in which the author had the chance to participate, were carried out in the framework of Technical Cooperation between national or international institutions of the region and the Federal Institute of Geosciences and Natural Resources, BGR, Hannover, partly in connection with activities of the German Agency for Technical Cooperation, GTZ, Eschborn. The projects were sponsored by the Ministry for Economic Cooperation and Development, Bonn. Regional information on groundwater conditions in the Middle East were obtained, in particular, through long-term assignments of the author to international institutions: The Arab Centre for the Studies of Arid Zones and Dry Lands of the Arab League (ACSAD), Damascus, and the United Nations Economic and Social Commission for Western Asia (ESCWA), Amman and Beirut.

Contents of the Book

The book gives an outline of prevailing hydrogeologic conditions on the Arabian Plate together with the geologic background. Emphasis is given to relationships between the main features influencing the hydrogeologic conditions:

- Regional and sub-regional geologic developments
- Paleogeographic conditions
- Morphology
- Climate and paleo-climate

and the resulting hydrogeologic features:

- Formation of aquifers
- Distribution of major aquifers
- Major groundwater flow systems
- Occurrence of renewable and fossil groundwater

Information on groundwater conditions in the various hydrogeologic sub-regions of the Arabian Plate is presented in a summarized form with comparative evaluations of hydrochemical and isotope data from different hydrogeologic units. Reported data on hydraulic aquifer parameters, recharge rates and groundwater flow volumes are listed with a view to arrive at characteristic values under specific hydrogeologic and climatic conditions.

The review of hydrogeologic conditions in this book is obviously a look back to information accumulated by the scientific–technical community during the past decades. It does not deal, in detail, with main actual problems and techniques for groundwater utilization, such as quantitative assessments of groundwater resources, groundwater management planning, legal aspects of safe groundwater exploitation and groundwater protection, artificial measures for groundwater management.

The presentation of groundwater conditions refers prevailingly to the state of knowledge in the late 1990s but includes information from more recent publications. For some areas, no recently updated information has been available. This will not affect the relevance of the description of the general hydrogeologic conditions, but the state of groundwater exploitation, of water quality and of remaining resources may have changed significantly in some areas during short periods.

The area considered in the outline of groundwater conditions covers approximately the Arabian Plate, as delineated in Chap. 1, with minor deviations: political boundaries are not coinciding with the plate boundaries, e.g. in the north, where the plate extends into Turkish territory. Information on the following countries is included: Bahrain, Iraq, Jordan, Kuwait, Lebanon, Oman, Qatar, Saudi Arabia, Syria, United Arab Emirates, West Bank and Gaza, Yemen. Relevant information from results of investigations and research in Israel is considered, but no attempt is made for including an outline of hydrogeologic conditions in Israel.

Remarks on Graphic Presentations, Transcriptions and References

Much of the information presented in this book is shown in a generalized manner.

Maps are included as rough sketch drawings indicating locations and geographic, geologic or hydrogeologic features. The sketch maps are intended to provide a general orientation of locations and features mentioned in the text. Geographic latitudes and longitudes are indicated at the frame of most sketches.

For discussion of hydrochemical and isotope data, standardized diagrams are used, mainly Piper diagrams and plots of $\delta^{18}O/\delta^2H$ data.

In the trilinear "Piper diagram" (Piper 1944), concentrations of major ions in meq% are plotted in a cation triangle and an anion triangle, and the cation and anion points of one water analysis are combined through projection into a diamond field (Fig. 1.5). Hydrochemical water types are generally defined according to the predominant cation and anion of a water group. Frequently, major ions of groundwaters from a specific aquifer and area vary over considerable ranges and the definition of water types has to include more than one cation and/or anion (e.g. Ca–Na–HCO_3–Cl type water).

The $\delta^{18}O/\delta^2H$ plots are presented in a uniform scale with $\delta^{18}O$ between −9‰ and +3‰ and δ^2H between −70‰ and +15‰. For orientation two meteoric water lines are shown on each $\delta^{18}O/\delta^2H$ diagram (Fig. 1.6): the Global Meteoric Water line (GMWL, $\delta^2H=8\times\delta^{18}O+10$) and the Mediterranean Meteoric Water Line (MMWL, $\delta^2H = 8 \times \delta^{18}O + 22$).

Transcriptions of Arabic location names are made in a somewhat simplified manner:

- Symbols for characters which do not exist in English script are omitted, such as hamza ء or ain ع,
- Duplication of consonants are largely avoided
- The feminine ending with the Arabic consonant h ه is written as "a" (not "ah")
- The consonant dad ض is written as dh

In order to avoid multiple citations of references within the text, evaluated publications are generally listed at the end of each sub-chapter, except for direct quotations. In Chap. 1, relevant publications are, in most cases, cited also at the appropriate passages of the text.

Denomination of Stratigraphic Units

Names of stratigraphic units have been adopted as commonly applied in the hydrogeologic literature of the concerned countries or regions. Generally, litho-stratigraphic units are denominated "formation" or, in relation to hydrogeologic features, "aquifer" or "aquitard", e.g. "Umm er Radhuma formation", "Umm er Radhuma aquifer", "Rus formation", "Rus aquitard". The term "group" is applied, where several formations compose a larger stratigraphic or hydro-stratigraphic complex, but no definite differentiation has been followed between the terms formation, group or super-group.

In the geologic literature of the Arab countries, the Cretaceous has been subdivided into

- Lower Cretaceous (up to Albian) and Upper Cretaceous (Cenomanian to Maastrichtian)

or

- Early Cretaceous (up to Aptian), Middle Cretaceous (Albian–Turonian), Late Cretaceous (Coniacain–Maastrichtian)

In this book, the differentiation into Lower Cretaceous and Upper Cretaceous is applied.

Acknowledgements

The activities in projects, carried out by the author in Arab countries, involved the cooperation with a great number of specialists from national and international institutions. It will not be possible to name all colleagues who directly contributed to the knowledge which is summarized in this book. Especially extensive professional information and stimulations were received from Jean Khouri in Damascus, Bader Hirzalla in Aman, Hans Bender and Mebus A. Geyh in Hannover. A special privilege was the opportunity to participate in discussions with Louis Dubertret and D.J. Burdon, scientists who had established fundamental parts of the hydrogeology of the Middle East.

Particularly valuable information on groundwater investigations or problems in specific countries or sub-regions has been provided by the following colleagues: Abdulla Droubi, Refat Rajab, Watheq Rasoul Agha, Nabil Rofail, Said Zahra in Damascus (ACSAD), Aziz Ghadban and Hisham Abu Ayyash in Damascus, Adnan Al-Baeini in Soueida, Rifaat Haoui in Aleppo, Samir Hijazin, Mohammad Almomani, Raja Gedeon, Jamil Rashdan, Ali Subah in Aman, Bassam Adib Jabir and Hassan Jaafer in Beirut, Amitabha Mukhopadhyay in Kuwait, Mohamed Sager Al-Asam, Muhamed Saeed Abdullah and Muhamed Abdul Haq in Dubai, Waleed K. Al-Zubari in Bahrain, Eduard Boeckh, Manfred Hobler, Armin Margane, Christian Neumann-Redlin, Hellmut Vierhuff in Hannover (BGR), Yucel Yurtsever in Vienna (IAEA), Omar Touqan, Omar Joudeh, Mohamed J. Abdelrazzak in Aman and Beirut (ESCWA).

The missions and investigations in the different Arab countries were supported ex-officio, but in many cases also through personal encouragement, by officials and political representatives of many ministries, governmental authorities and institutions in the concerned countries, in Germany and in organizations of the United Nations and of the Arab League. That support as well as the contribution of information from the above mentioned colleagues and of many scientists, who are not mentioned here explicitly, is gratefully acknowledged.

Several authors whose publications were used rather extensively for compilation of the book should be included in the acknowledgements: V.P. Ponikarov et al. (1967a, b): geology and hydrogeology of Syria; Zuhair Kattan: isotope hydrologic aspects in Syria; Friedrich Bender and collaborators: geology of Jordan; Elias

Salameh: various features of the hydrogeology of Jordan; Al-Sayari and Zötl (1978), Jado and Zötl (1984) and H.S. Edgell (1997): hydrogeology of Saudi Arabia; Alsharhan et al. (2001): hydrogeology of the eastern part of the Arabian Peninsula; John W. Lloyd: hydrogeologic aspects of various areas of the Middle East.

The author is also indebted to a considerable number of specialists of BGR (Hannover) for their essential contributions for the execution of groundwater projects in Arab countries in various geoscientific fields.

Special thanks have to be extended to Springer Verlag and to Mr. E. Balamurugan and his team of SPi Technologies India in Chennai for the efforts in editing and publishing the book.

Contents

List of Figures

List of Tables

Chapter 1
The Arabian Plate: Geology and Hydrogeologic Characteristics

1.1 Geology and Main Orographic Units

1.1.1 Geologic and Structural Features

1.1.1.1 Arabian Plate

The Arabian Plate extends approximately over the region known as Arab Middle East, which includes the Arabian Peninsula together with the adjoining northern Arab countries (Al Mashreq). Until the early Tertiary, the Arabian Plate was connected to the African Plate, one of the huge main plates of the earth crust. During the Tertiary, the sector which became the Arabian Plate was separated from the African Plate with the opening of the Red Sea and Gulf of Aden rift structures.

Descriptions of the regional geological framework of the Arabian Plate can be found in publications by Ponikarov et al. (1967a), Powers et al. (1966), Bender (1974a), Beydoun (1991), Alsharhan and Nairn (1997), Alsharhan et al. (2001), Al-Sayari and Zötl (1978).

The development of the Arabian Plate included the following main stages (Beydoun 1991):

- The basement of the plate was formed during the Precambrian as part of the Nubo-Arabian craton.
- During the Paleozoic until the Permian, the sector comprising the present Arabian Plate occupied a marginal shelf position on the Gondwana supercontinent.
- In the early Mesozoic, plate boundaries developed on the northeast, southeast, north and northwest margins with the opening of the Neo-Tethys and the Indian Ocean.
- During the middle to late Tertiary, collision of the Arabian Plate with the Eurasian continent compressed and modified the north and northeast margins of the plate, and rift tectonics opening the Red Sea–Dead Sea graben and the Gulf of Aden created a new northwest margin and the south and southwest margins.

W. Wagner, *Groundwater in the Arab Middle East*,
DOI 10.1007/978-3-642-19351-4_1, © Springer-Verlag Berlin Heidelberg 2011

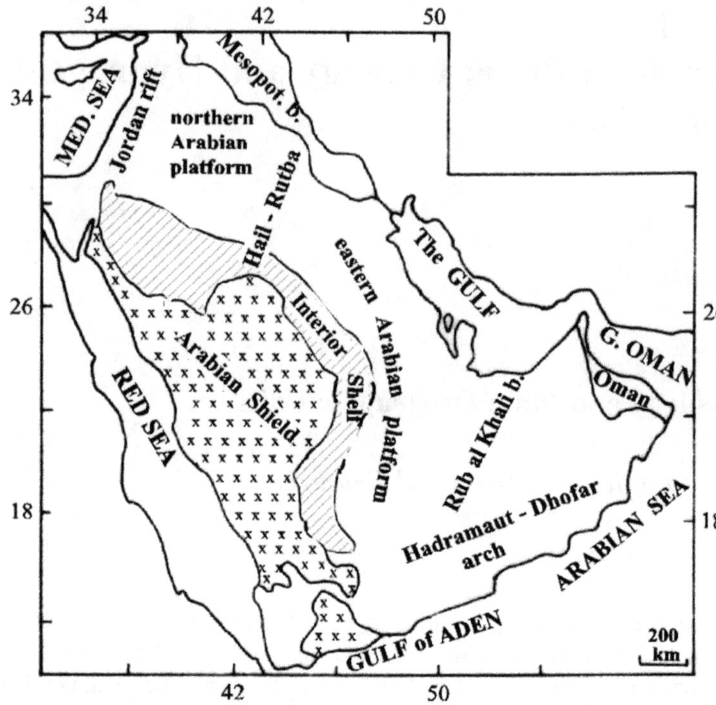

Fig. 1.1 Main geologic-structural units of the Arabian Plate. Modified after Alsharhan et al. (2001), Edgell (1997), ESCWA (1996b)

During most of its geologic history from the Precambrian to the early Tertiary, the Arabian sector of the African Plate was situated at latitudes south of the equator or at tropical latitudes. From the Triassic to the Paleocene, the northern margin of the African–Arabian Plate formed the shore of a vast continent along the Neo-Tethys, the precursor of the Mediterranean Sea, which was connected with the Indian Ocean. Northward movement of the Arabian Plate during the Eocene to Neogene lead to the closure of that sea connection and brought the plate into its present position between around 10°N and 35°N.

The Arabian Plate consists of two major geological units: the Arabian Shield and the Arabian Shelf (Fig. 1.1).

1.1.1.2 Arabian Shield

The Arabian Shield is composed of Precambrian crystalline and metamorphic rocks and occupies the southwestern part of the Arabian Plate, extending over around 770,000 km^2 from southern Jordan across western Saudi Arabia into Yemen.

Until the early Tertiary, the Arabian Shield and the Precambrian complex of northeastern Africa were connected and formed the Nubo-Arabian Shield.

The development of the shield during the Precambrian comprised the following major phases (Issar 1990):

- Sedimentation of several kilometers of sands, clays and flysch deposits in a marine basin, intrusion of basic magmatites
- Folding and metamorphism of the sedimentary–magmatic sequence, intrusion of magmatites with acid composition
- Regional uplifting, eruption of volcanic rocks along major fault lines
- Erosion and deposition of erosion products as sandstones
- Peneplanation at the end of the Precambrian

The main lithologic units of the Precambrian shield comprise:

Schists, amphibolites and gneiss
Metamorphosed basalts, greywacke, chert and andesite
Non-metamorphosed volcanic rocks with clastic intercalations

That sequence is penetrated by pegmatitic and porphyric to diabasic and gabbroic dikes.

By the end of the Precambrian, most of the Arabian Plate constituted a deeply eroded peneplain and the shield has been a rather stable land mass since that time. In the late Precambrian to early Paleozoic, the Arabian Shield was affected by the Nejd orogeny, which created a southeast–northwest trending system of transcurrent faults. The faults extend as tectonic lineaments of several hundred kilometers length over the surface of the shield (Beydoun 1991).

During the Mesozoic, the southern part of the shield began to break up along zones of weakness into uplift and basin structures. Marine ingressions inundated the basin zones during the Mesozoic and early Paleogene as narrow sea-arm invasions. Jurassic–Cretaceous fluviatile, epicontinental and marine sediments were deposited in graben and basin structures on the margins of the shield, such as the Marib–Jawf graben, but also in a belt of graben systems within the shield between Sada, Sanaa and Taiz (Beydoun 1991; Kruck et al. 1996).

Rift movements in the Tertiary caused the break-up of the Nubo-Arabian Shield along the Gulf of Aden–Red Sea structure and the Arabian Shield was uplifted on the shoulders of the rift graben to elevations of presently up to >3,000 m asl. The rift tectonics were accompanied by wide spread volcanism, and significant parts of the shield area are covered by Neogene–Quaternary volcanic rocks.

1.1.1.3 Arabian Shelf

The eastern and northern parts of the Arabian Plate form the Arabian Shelf, which was inundated during various cycles of marine transgression from the Paleozoic to early Cenozoic by epicontinental seas. The transgressions intruded from a great marine basin, which bordered the shelf area in a generally northern position: the Paleo-Tethys during the Paleozoic and the Neo-Tethys during the Mesozoic to

Tertiary. Terrestrial erosion and marine transgressions left a cover of continental and shallow marine sedimentary rocks on the shelf.

The shelf extends from the boundaries with the Arabian Shield in the west and south to margins of the Arabian Plate in the east, north and northwest: the Oman mountains and the Gulf of Oman, the Gulf–Euphrates sedimentary basin, the Zagros–Taurus mountain zone, and the Mediterranean Sea coast. The Dead Sea–Jordan rift zone dissects the shelf near its western margin as an approximately south–north directed major geosuture, extending from the Gulf of Aqaba to the Amiq depression.

The Arabian Shelf constitutes a relatively stable geological province, but is divided by structural features into sub-provinces with different geologic history (Powers et al. 1966; Edgell 1997):

- The Interior Homocline which borders the Arabian Shield in a semi-circle in the east and north
- The Arabian Platform east and north of the Interior Homocline comprising the Interior Platform (Powers et al. 1966) on the Arabian Peninsula and the "North Arabian part of the African Platform" (Ponikarov et al. 1967a: 176), which extends between the Euphrates basin in the east–northeast and the rift zone in the west

Beydoun (1991) and Bender (1974a) distinguish an interior stable shelf and an unstable shelf. The unstable shelf constitutes the northern and eastern parts of the Arabian Plate, which are bordered by the mobile zones of the foredeeps of the alpidic Taurus and Zagros mountains, the Mesopotamian and Gulf basins, and, in the southeast, the nappes of the Oman mountains.

The tectonic development of the Arabian Shelf was characterized by stable platform conditions until the Turonian. During a late Cretaceous (Turonian to Maastrichtian) stage of orogenic activity, a foredeep was formed along the Tethys margin, and ophiolite–radiolite nappes were emplaced on the southeastern margin of the plate. From the late Cretaceous to Miocene the platform was relatively stable; during the late Tertiary the Alpine orogeny created the Taurus–Zagros anticlinal structures along the northern margin of the Arabian Plate.

A simplified sub-division of the Arabian Shelf into *Interior Shelf* and *Arabian Platform* is applied here, considering the present extent of the deposits of the main Upper Cretaceous–Tertiary marine transgressions.

The Interior Shelf is characterized by outcrops of Paleozoic and prevailingly continental Mesozoic formations, which dip away from the shield toward east and north. The platform comprises prevailingly outcrops of Cenozoic and marine Mesozoic formations with flat to undulating layering or anticlinal–synclinal structures. The boundary between Interior Shelf and platform is set roughly along the upper – transgressional or erosional – outcrop boundary of sediments of the marine Upper Cretaceous and Paleogene transgressions.

The Arabian Platform, as defined here, encloses the tectonic unit "Interior Platform" or "unstable shelf", but reaches into the Interior Homocline and extends over the boundary between stable and unstable shelf.

The Interior Shelf and the Arabian Platform are sub-divided into different segments by broad and very long *uplift structures*: the Hail–Rutba arch in the north, the Central Arabian arch (Qatar–south Fars Arch, Murris 1984) extending from central Saudi Arabia to the Gulf, and the Huqf–Dhofar arch in the south.

The south–north trending Hail–Rutba arch (Jauf–Gaara–Mosul arch, Rutba–Gaara–Khleissia high) appears to have been a positive structure in particular in the Cretaceous and Paleogene and constitutes a significant structural element influencing the paleogeographic situation in the northern part of the Arabian Platform.

The central Arabian arch extends in east to northeast direction from central Saudi Arabia to the Gulf where it appears to continue into the north–northeast trending Qatar anticline (Qatar–Fars arch). The arch has been a positive structure in particular during the Devonian–Permian and constitutes a hinge zone between northeast dipping segments of the Arabian Platform in the north and east dipping southern segments. The arch is covered by a thick sequence of Cretaceous–Paleogene sediments. A series of north–northwest trending tectonic depressions, the central Arabian graben and trough system, developed during the Cretaceous– Paleogene along the zone of major tension on the hinge of the uplift.

The northeast–southwest trending Huqf–Haushi arch (Huqf–Jaalan axis) in southern Oman and the WSW–ENE trending Hadramaut arch in southern Yemen accompany the southern–southeastern rim of the Arabian platform. The Huqf–Haushi arch developed since the Precambrian as a positive structure, the Hadramaut arch emerged as an uplift structure during the Paleogene.

The *Interior Shelf* is sub-divided into various segments with different strike direction of the strata (Edgell 1997): Tabuk segment, Widyan basin margin, northern Tuwayq segment, southern Tuwayq segment. The Wajid basin in the south of the Arabian Peninsula, which is covered by Paleozoic sandstones, may be considered a southern continuation of the Interior Shelf. Structural elements separating the different segments are: the Hail–Rutba arch between the Tabuk and Widyan segments, the Wadi el Batin fault between the Widyan and northern Tuwayq segments, and the central Arabian arch between the northern and southern Tuwayq segments.

The Tabuk segment in the north and the Wajid segment (Wajid basin, Wajid sandstone plateau) correspond to Paleozoic intracratonic basins, which have been consolidated as structural highs with mainly Paleozoic cover. The homoclinal structure the Interior Shelf is particularly distinct in the northern and southern Tuwayq segments, which adjoin the Arabian Shield in the east as a 100–200 km wide belt of Paleozoic and Mesozoic strata over a north–south distance of 1,300 km. The homocline shows a gentle but rather constant dip of the strata from the boundary of the shield toward the platform.

On the *Arabian Platform,* the Hail–Rutba arch forms a dominant structural element separating the platform into a segment situated north of the Interior Shelf and a part of the platform adjoining the Interior Shelf to the east. South of the central Arabian arch, the platform is occupied by the Rub al Khali basin and its slope areas. The three segments of the Arabian platform are denominated in the following description northern Arabian platform, eastern Arabian platform and Rub al Khali basin, respectively.

The *northern Arabian platform* ("North Arabian part of the African Platform", Ponikarov et al. 1967a) is covered, over wide parts, by Paleogene deposits. Mesozoic sedimentary rocks are exposed in anticlinal structures, Neogene marine and terrestrial deposits cover structural basins. Part of the northern platform is occupied by Neogene to Quaternary basalt flows of the north Arabian volcanic province.

The northwestern margin of the Arabian Shelf has been strongly affected by Cenozoic uplift and rift movements. The Dead Sea fault zone with the adjoining uplift and graben structures constitutes a predominant tectonic element, which separates the northern platform into a western and eastern block (Lovelock 1984).

The Dead Sea rift is part of a major geosuture which is recognizable from East Africa to southern Turkey over a distance of about 6,000 km. On the northwestern margin of the Arabian Plate, the Dead Sea rift extends from the Gulf of Aqaba through Wadi Araba–Dead Sea–Jordan valley–Lake Tiberias toward north into the graben structures of the Bekaa, Orontes and Kara Su. The graben structures of the rift are accompanied by a system of major faults and flexures. The fault zone crosses the northern Arabian platform and forms, in the Wadi Araba–Red Sea area, the western margin of the Interior Shelf and the Arabian Shield.

The eastern part of the northern Arabian platform is characterized prevailingly by plateau landscapes which are, however, interrupted by mountain belts of the Palmyrean zone and the volcanic province of Jebel el Arab and by sedimentary basins, the most extensive of which comprise the depression zones of Wadi Sirhan–Azraq–Yarmouk, Ad Daw–Sabkhet el Mouh.

The *eastern Arabian platform* is bordered by the Hail–Rutba arch and the Interior Shelf in the west and the Euphrates–Gulf depression in the northeast. In the southeast and south, the allochthonous Oman mountains and the Hadramaut arch mark the boundaries of the platform.

The northern part of the eastern Arabian platform comprises zones of flat to slightly anticlinal structural elements with generally northeast dipping strata. The generally flat structure of the eastern platform is interrupted, in its eastern part, by approximately south–north oriented anticlinal axes and by synclinal zones: the Salman zone and the Dibdiba basin.

The vast *Rub al Khali basin* forms an intra-plate sedimentary basin in the south of the platform. The basin constitutes an elongated synclinal structure plunging gently toward northeast across Abu Dhabi into the Gulf. The synclinal structure of the Rub al Khali apparently originated in the early Paleozoic, but basin development took place prevailingly during the Tertiary, resulting in a relatively great thickness of Paleocene, Eocene and Neogene sediments in the center of the basin (Powers et al. 1966).

In the southwest, the Rub al Khali basin is adjoined by the Wajid segment of the Interior Homocline, in the south to southeast by the Hadramaut and Dhofar segments of the platform, which culminate in anticlinal structures of the Hadramaut arch and the Huqf–Haushi arch. The Hadramaut plateau is a broad, generally east–west trending swell, on which a syncline, corresponding roughly to the channel of Wadi Hadramaut, is superimposed. The Hadramaut plateau and the Dhofar mountains, which are covered in wide parts by Paleogene formations, can be considered to form the southern margin of the Arabian platform. The Arabian Shelf is, on the southern margin of the Rub al Khali basin, not clearly divided into a platform and interior shelf area.

The eastern border of the Rub al Khali basin is formed by the Oman thrust zone and by a fault zone along which the Paleozoic–Mesozoic formations of the Hajar mountains have been uplifted. The Oman thrust zone delimits the allochtonous Semail and Hawasina complexes from the Arabian Platform.

In the southeast of the Arabian Plate, salt basins developed during the Infracambrian to early Cambrian. The deposits of the salt basins are now buried under a thick cover of Paleozoic to Quaternary sediments.

1.1.1.4 Margins of the Arabian Plate

The margins of the Arabian Plate have been created mainly by tectonic events during the Mesozoic to Neogene (Beydoun 1991):

The northwest margin along the present eastern Mediterranean Sea originated from the break-up of the Gondwana continent and the opening of the Neo-Tethys during the early Mesozoic and is now accompanied by a zone of marginal blocks, separated from the Arabian Platform by the Tertiary rift tectonics.

In the north and northeast, the shelf is adjoined by a zone of Neogene marginal troughs composing the synclinal structures of the Euphrates–Gulf depression, which is part of the Meopotamian foredeep of the Alpidic Taurus and Zagros mountain chains on the margin of the Eurasian Plate.

The southeast margin of the plate has been formed mainly by events during the Cretaceous: the separation of the Arabian and Indian Plates with the opening of the Indian Ocean–Arabian Sea and the overthrusting of the plate margin by nappes of ophiolotic and metamorphic sedimentary rock complexes.

The south and southwest margins "evolved during the Neogene as a result of Eocene–Oligocene extension, rifting and break-up of Afro–Arabia" with the opening of the Gulf of Aden and the Red Sea (Beydoun 1991).

1.1.2 Orographic Sub-division of the Arabian Plate

The geologic development of the Arabian Plate has formed mountain and highland belts along its margins and boundaries and vast zones of plateau landscapes and lowlands in the central and northeastern parts of the plate.

The "northwestern mountain and rift zone" in the west of the northern Arabian platform is occupied by two chains of mountain massifs running sub-parallel to the Mediterranean Sea coast, which are separated by the rift depression of Wadi Araba–Dead Sea–Jordan–Bekaa–Ghab. The coastal mountain belt comprises the highlands of Judea (peak altitudes around 1,000 m asl), the Lebanon mountains (peak 3,088 m asl) and the Ansariye mountains (1,552 m asl). The mountain zone east of the rift depression includes the highlands of Jordan (peak 1,592 m asl), the Antilebanon mountains (2,814 m asl at Mount Hermon) and Jebel ez Zaouiye (870 m asl).

The eastern parts of the northern Arabian platform extend over the Aleppo plateau and the steppe and desert areas of "Al Badiye" and the Hamad, which are characterized morphologically by plateau landscapes and local depressions. The plateau area is crossed by the northern and southern Palmyrean mountains (up to 1,390 m asl), which enclose the Ad Daw plain and are adjoined in the east by the Sabkhet el Mouh depression.

The eastern Arabian platform is occupied by plateau landscapes and vast sand seas, such as the Great Nefud and the Rub al Khali which covers most of the southern parts of the eastern platform. Toward northeast and east, the Arabian Platform dips into the lowlands of the Euphrates depression and the Gulf coastal area. In the south and west, the eastern Arabian platform grades into plateau, hill and highland zones including the escarpment landscapes of the Tuwayq mountains (up to 1,500 m asl) and the highlands of Hadramaut (2,225 m asl) and Dhofar (1,210 m asl).

The Arabian Shield in the southwest of the Arabian Plate rises to high mountain chains near the boundary of the plate at the Red Sea rift structure with altitudes of the Asir mountains of up to 3,133 m asl in Saudi Arabia and 3,620 m asl in Yemen.

In the southeast of the Arabian Plate, the Oman mountains constitute an up to 3,009 m high morphologic barrier between the Gulf of Oman and the interior of the Arabian Peninsula (Fig. 1.2).

References. Al-Sayari and Zötl (1978), Alsharhan and Nairn (1997), Alsharhan et al. (2001), Bender (1974a), Beydoun (1991), Edgell (1997), ESCWA (1999b), Ponikarov et al. (1967a), Powers et al. (1966).

1.2 Occurrence of Aquifers

1.2.1 Paleogeographic Environment and the Formation of Aquifers

The Arabian Plate with its extent over around 3.7 million km^2 comprises a multitude of lithostratigraphic units with lateral and vertical variations of lithologic facies, including a great number of aquiferous layers of local to sub-regional extent. Major sedimentation cycles on the Arabian Shelf were related to wide-spread slow epeirogenic movements of the plate with deposition of erosional material and marine sedimentation over long periods. The sequence of sedimentary rocks therefore contains several aquifers which expand over very large areas: The Paleogene Umm er Radhuma aquifer stretches over a 1,200 km long belt on the Arabian Peninsula, Mesozoic aquifers extend on the eastern platform over a length of

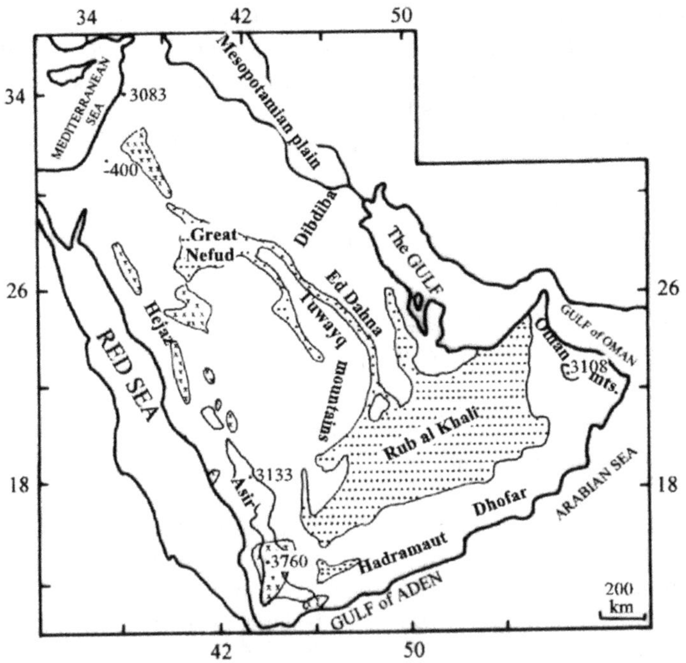

Fig. 1.2 Morphologic features of the Arabian Plate. Compiled after Wohlfahrt (1980) and various other sources. …… sand dunes, sand seas, lava fields, volcanoes; . 3,108 peak elevation (m asl); ____ 2,000 m contour on Asir and Oman mountains, not shown for Lebanon and Antilebanon mountains

1,800 km, the Rub al Khali groundwater basin covers an area of more than 600,000 km^2. Some marl or clay formations, which form continuous aquitards, also extend over wide areas in some parts of the shelf.

1.2.1.1 Geologic Developments

The occurrence of main aquifer units and aquitards on the Arabian Plate is directly related to the geologic developments. The paleogeographic history and structural developments of the shelf are controlling factors for the position of aquifers at the level of active groundwater circulation or at levels where they may provide extensive fresh water or brackish water reservoirs.

Of particular importance for the assessment of groundwater resources is information on lithostratigraphic units, which may constitute porous, fissured or karstic aquifers or extensive aquitards and which are situated at shallow to intermediate depth – down to a few hundred metres below sea level in some areas. Of main interest are, in that connection:

- Paleogeographic developments during the Paleozoic–Mesozoic on the interior parts of the shelf during the Mesozoic–Tertiary on the Arabian Platform and during the Tertiary in the deep basins, such as the Euphrates depression, the Dibdiba basin and the Rub al Khali basin
- Near surface accumulations of Tertiary–Quaternary sedimentary and volcanic rocks
- Occurrence of deep sandstone aquifers, which may contain exploitable fresh to brackish water down to depths of 1,000 m below surface in some of the large groundwater basins

Deep seated salt water bearing formations may generally be neglected in a review of exploitable groundwater resources.

1.2.1.2 Environmental Conditions

An overview of the environmental conditions favouring the *deposition of potential aquifers* on the Arabian Plate was given by D. J. Burdon (1982). Burdon in his lecture discusses global and regional influences on the sedimentary environment and on the deposition of formations which may become primary or secondary aquifers.

Primary aquifers have a primary porosity and permeability; typical primary aquifers are composed mainly of gravel, sand and silt. They are deposited in areas of high erosional and transport energy in continental or marine environments. The permeability of secondary aquifers is controlled by secondary processes: tectonic processes as bending and folding, solution effects, in particular karstification, weathering effects, shrinkage during lithification. The formation of secondary aquifers is favoured especially by tectonic activities and the presence of rocks which deform by fracturing and not by plastic flow. Secondary aquifers are, in particular, carbonate rocks and consolidated sedimentary or volcanic siliceous rocks.

Main global influences on aquifer formation on the Arabian Plate are according to Burdon:

- Eustatic changes in sea level
- Oceanic anoxic events
- Climatic changes

Main regional influences are:

- Tectonic movements of the plate: horizontal movement, tilting, epeirogenetic up and down movements, rifting, subsidiary tectonic influences
- Volcanism
- Morphology
- Cycles of erosion

The influence of *global and regional geologic developments* on the sedimentation history of the Arabian Plate may be illustrated by some examples given by Burdon and other authors: Burdon cites main periods of high global eustatic sea levels during the Cambrian, Ordovician, Triassic, Jurassic and Cretaceous, which lasted between 2 and 40 million years. Eustatic rises in sea level "usually resulted in great transgressions of the sea onto continental margins and deep into the heart of the continents. The seas tended to be shallow, essentially neritic, with much life as sunlight reached the sea floor, and waves and storms kept the water in motion. In such seas, the limestones and dolomites, so much the secondary aquifers of the Arabian Plate, were initially deposited or grew. Here also cherts formed" (Burdon 1982).

Climatic conditions influencing the depositional environment on the Arabian Plate were controlled by global climatic developments, but also by the migration of the Afro–Arabian continent. The northern part of the continent comprising the present Arabian Plate apparently was situated (Burdon 1982; Beydoun 1991)

- In prevailingly arid climate zones during deposition of the Paleozoic to Mesozoic "Nubian Sandstone"
- In the tropical to sub-tropical belt from the Permian to Cretaceous
- At sub-tropical latitudes during the Paleogene

Arid climate conditions favoured the deposition of sandstone aquifers; carbonate deposits, later converted into secondary aquifers, were deposited prevailingly in marine to coastal tropical to sub-tropical environments.

Glacial episodes occurred on parts of the Arabian Plate during the Ordovician, the Permo–Carboniferous and Quaternary. Major glaciations influenced the extent of marine inundations of the shelf areas through the global lowering of ocean levels. On local scale, glacial deposits may form aquitards with, however, rather erratic distribution.

Oceanic anoxic events, during which wide parts of the world ocean became deprived in oxygen, are characterized by

- The deposition of black shales containing pyrites and relatively high organic carbon
- Reduced sedimentation of carbonates because of the relatively high acidity of the sea water

Such conditions apparently prevailed during parts of the Upper Cretaceous in the northwest of the Arabian Shelf and caused the formation of a sub-regional aquitard.

1.2.1.3 Depositional Environment

Global and regional influences together are responsible for the development of the depositional environment. According to Murris (1984) the depositional pattern on the Arabian Shelf results "from the interplay of sea level changes, epeirogenetic movements plus rejuvenation of relief, climatic variations and synsedimentary structural growth". Murris discusses depositional environments and lithologies typical for various environments with respect to hydrocarbon habitat, but these features can also help to define relationships between sedimentation history and

Table 1.1 Depositional environment and main lithologic varieties (after Murris 1984, modified)

Main lithology	Depositional environment
Sandstone	Alluvial – upper coastal plain, lower coastal plain in humid climate
Limestone, dolomitic limestone, dolomite	Shallow shelf
Breccia, conglomerate	Proximal synorogenic basin
Siliceous limestone	Deep marine
Shale, marl, claystone	Lower coastal plain, clastic platform, deeper shelf, deep marine
Anhydrite, rock salt	Evaporite basin, sabkha in arid climate

groundwater occurrence. Main lithological varieties originating from typical depositional environments are listed in Table 1.1.

Carbonate deposition on the shelf is a major factor for the formation of secondary aquifers. Murris distinguishes two basic types of carbonate depositional systems: carbonate ramp and differentiated carbonate shelf. The carbonate ramp is characterized by a significant influx of clastics from the adjoining continental areas and by the cyclic alternation of more or less argillaceous units. Carbonate units comprise pelletoidal–bioclastic varieties, the more argillaceous units contain pyrites and grade into marl and shale. The differentiated carbonate shelf develops in particular during periods with relative high stands of the sea level. Algal and foraminiferal limestones are typical for the shallower parts of the platform, while in the deeper parts limestones and marls were deposited in reduced thickness often under euxinic conditions.

"Ramp sedimentation was contemporaneous with periods of high clastic inflow which occurred during marine low stands. In contrast, when sea level was high, clastic flow diminished and the shallow parts of the differentiated platform contained clean algal–foraminiferal wackestone/packstone and ooidal–peloidal packstone and grainstone. In deeper water depositional rates were low and dominated by the deposition of marl and mudstone. Such carbonate cycles were less consistent in thickness and lithological character than the ramp carbonates".

"The ramp sequence is characterized by cyclic sequences of shale and carbonate units each of which can be correlated over hundreds of kilometers with little change in lithology or thickness" (Alsharhan et al. 2001: 64).

1.2.1.4 Tectonic and Volcanic Activities

The impact of *horizontal movement* of the Arabian Plate is particularly evident in the sedimentary history of Neogene deposits. Northward movement of the plate and collision with Eurasia separated the Tethys Sea into the present-day Indian Ocean and the Mediterranean Sea during the Paleogene. Subsequently, the previous gulf of the Indian Ocean along the Tigris–Euphrates Mesopotamian valley was filled with thick fluviatile, lagoonal and evaporite deposits (Burdon 1982).

The Arabian Plate has been *tilted* during long geological periods from SSW to NNE. The tilting, which is still remarkable in the general morphological structure descending from the highest peak of Arabia at an elevation of 4,300 m in Yemen to and below sea level in the Gulf, had a significant impact on the drainage pattern and sedimentation: Long drainage basins with low slopes extend from the Asir mountain areas in Saudi Arabia and Yemen to the east, northeast and north, while steep escarpments border the highlands of Yemen and Saudi Arabia to the west.

Rift tectonics with the subsidence of the Red Sea–Jordan valley graben produced steep cliffs on the western margin of the Arabian Plate. The western plate rim became mountainous with steep river beds and fast flowing surface waters. Detrital sediments of considerable thickness were deposited on coastal plains and in the rift depressions along the Wadi Araba–Jordan valley, the Beqaa and the Ghab graben.

Volcanism with extrusion of thick sequences of sheet lavas produced volcanic aquifers in particular along a belt east of the Red Sea rift zone: acidic to basaltic volcanics since the early Tertiary in the Yemen highlands, Neogene to Quaternary basalts on the western margin of the Arabian Shield in Saudi Arabia and in the Golan–Jebel el Arab–Al Harra area in Syria, Jordan and Saudi Arabia.

1.2.1.5 Morphology

The morphologic relief and erosional cycles are of major importance for the formation of detrital sediments. The morphology of the source areas of sediments is formed by tectonic and erosional forces and the resulting morphology of these source areas dominantly controls the nature of the deposited material. Tectonic–morphologic cycles with particular importance for the formation of aquifers were

- The erosion and peneplanization of the Precambrian craton producing the wide spread deposition of the Nubian Sandstone
- The tilting of the plate which created uplifted highlands as source areas of erosion and marine and continental sedimentation during much of the Mesozoic to Tertiary
- Uplifting of mountains and highlands on the western margin and in the south and southeast of the Arabian Plate during the Neogene to Quaternary and related sedimentation of wadi and coastal sediments

1.2.1.6 Paleogeographic Environment

From the above cited examples, we may, in a very generalized view, summarize the main *paleogeographic environments*, in which *major aquifers and aquitards* were formed on the Arabian Plate:

- Primary aquifers composed of sandstones and sands and gravels were deposited in continental basins, on coastal and piedmont plains, and in wadi channels

- Secondary aquifers, in particular limestones, dolomites and chalky limestones extend over shallow marine shelf areas
- Sediments with low permeability – shales, marls, argillaceous limestones – characterize sedimentation at deeper parts of the marine shelf and marine or continental basins with low erosional energy
- Evaporites – anhydrite, gypsum, rock salt – occur in arid lagoonal or sabkha areas
- Volcanic aquifers are found along extrusion zones parallel to the rift zone

1.2.2 Stratigraphic Distribution of Main Aquifers

The distribution of aquifers with major economic importance on the Arabian Plate is closely related to the paleogeographic and tectonic-structural development of the plate, which controlled the history of erosional–depositional phases. The main exploitable aquifers comprise:

- Paleozoic–Mesozoic sandstone accumulations on the Interior Shelf
- Mesozoic marine carbonates on parts of the northern platform, in particular in the western mountain and rift zone
- Paleogene carbonate formations on wide parts of the northern and eastern platforms
- Neogene–Quaternary sedimentary and volcanic formations in various areas of the plate

A more detailed assessment of these aquifers and aquifer systems is given in the chapters delineating the hydrogeologic conditions of different sub-regions of the plate (Chaps. 2–9). The following pages provide an overview of main paleogeographic developments and the related formation of aquiferous rock sequences.

1.2.2.1 Precambrian

The Precambrian crystalline rocks of the Arabian Shield provide, in general, aquifers of only local importance in weathered or fractured sections. Groundwater resources on the shield are restricted mainly to wadi and coastal aquifers composed of Neogene–Quaternary sands and gravels. In the south, sedimentary aquifers of mainly Mesozoic age and aquiferous Tertiary–Quaternary volcanics are enclosed within the area of the Arabian Shield

1.2.2.2 Arabian Shelf

During various sedimentation cycles from the Paleozoic to the Tertiary, a thick sequence of continental and marine formations has been deposited on the Arabian

Shelf. The thickness of the sedimentary rocks reaches 4–5 km on the northern platform in Syria and eastern Jordan, more than 8 km in the Euphrates graben and increases to around 13 km in the Gulf area.

1.2.2.3 Paleozoic

During the Paleozoic, the African–Arabian Plate was situated on the shelf area of the Gondwana continent, adjoining the Paleo-Tethys ocean. Throughout the Paleozoic, sedimentation conditions on the Arabian Shelf were prevailingly continental with intermittent coverage by shallow seas. The Paleozoic clastics, deposited in terrestrial basins and in extensive shallow epicontinental seas, were derived from sediment sources in the shield areas and in uplift structures within the shelf.

Outcrops of Paleozoic formations extend over the Interior Shelf, surrounding the shield in the north and east, and again on the eastern margin of the shelf on the border of the Oman mountains. The earliest non-metamorphic deposits in Oman comprise siliciclastics, carbonates and evaporites of the Huqf group (Infra-Cambrian–Cambrian) and clastics of the Haima group (Cambrian–Silurian). The sedimentary sequence of the Huqf group indicates deposition in shallow seas, tidal and terrestrial lowlands, and lagoons with super-saline conditions. Carbonate–anhydrite platforms and salt basins extended during the late Precambrian to early Cambrian along the edge of the Gondwana continent (Beydoun 1991). Deeply buried salt deposits, which occur in Oman and along parts of the Gulf area, gave rise later to halo-kinetic tectonics.

Neglecting occurrences of stagnant saline groundwater in Infracambrian rocks of Oman (Alsharhan et al. 2001), we find the stratigraphically oldest aquiferous formations of the Arabian Shelf in Paleozoic sandstones. Important aquifers are constituted by Cambrian–Ordovician sandstones in southern Jordan, the Interior Shelf of Saudi Arabia and in northern Yemen. These formations, denominated Disi (also Rum or Ram group) sandstones in Jordan, Saq sandstone in northern Saudi Arabia and Wajid sandstone in southern Saudi Arabia and Yemen, were probably deposited under fluviatile to epicontinental conditions.

While sedimentation of the Wajid sandstones continued in the south until the Carboniferous (Kruck et al. 1996), alternations of sandstones with green, red and grey shales occupy the lithologic sequence of the Ordovician to Devonian on the Interior Shelf of northwestern Saudi Arabia and in southern Jordan. Various sandstone aquifers of the Tabuk and Jauf formations, included within that sequence, are separated by shale aquicludes. The deposition of organic-rich shales during the Ordovician–Silurian on the Arabian Shelf "is associated with several cycles of global sea level rise and transgression in a temperate climate" (Beydoun 1991). The succession of aquifers and aquitards in the Ordovocian–Devonian formations in Saudi Arabia and Jordan is shown in Table 5.2.

Sandstones with interbeds of siltstone, shale and limestone of Carboniferous to Middle Permian age form the Berwath and Unayza aquifers on the northern to eastern Interior Shelf.

An extensive period of carbonate deposition occurred in the late Permian. Rift tectonics separated the area of central Iran from the Gondwana continent and opened the Neo-Tethys ocean. The Arabian Shelf, then situated at tropical latitudes, was inundated by a warm water transgression from the Neo-Tethys with deposition of extensive shelf carbonates of the Khuff formation (Beydoun 1991). The Khuff formation is composed mainly of limestones and dolomites, anhydrite interbeds act as aquicludes and divide the formation into four carbonate aquifer units. Outcrops of the Khuff formation extend over a belt of 1,200 km along the Interior Homocline and are found in the subsurface over much of the Arabian Platform until the Rub al Khali basin and the Gulf area (Edgell 1997). In southern Oman, clastic sediments of the Permo-Carboniferous Haushi group provide an important aquifer (Alsharhan et al. 2001).

In the Oman mountains, deposition of limestones in shallow marine waters started in the Upper Permian and continued until the Cretaceous (Hajar aquifer).

1.2.2.4 Triassic–Jurassic

In the late Permian and the Triassic, break-up of the Gondwana continent and the opening of the Neo-Tethys ocean lead to the development of a very wide, shallow marine (epicontinental) shelf in the northeast of the Arabian Plate, extending from northwest Iraq to Oman. Arid to semi-arid conditions prevailed during the Triassic with sedimentation of red beds, shallow-water carbonates, evaporites and shales (Beydoun 1991). Triassic–Lower Jurassic carbonates and sandstones constitute the Jilh and Minjur aquifers in central Saudi Arabia. The early Triassic Sudair shales form an aquiclude above the Paleozoic Khuff and Tabuk aquifers in northern Saudi Arabia.

During the Jurassic, transgression of the Mediterranean Sea created open sea conditions in the areas of the Lebanon, Antilebanon and Ansariye mountains with sedimentation of limestones, dolomites and marls. Further east in Syria, argillaceous carbonates and evaporites were deposited under more lagoonal conditions. Anhydrites also occur in some stages of the Jurassic in the Antilebanon and Ansariye mountains. The Jurassic carbonates constitute important karst aquifers in Lebanon and the western mountain area of Syria.

In the Tuwayq mountains on the Interior Shelf of central Saudi Arabia, outcrops of the Jurassic comprise shallow water limestones, marine shales, interbeds of shale, siltstone, sandstone and anhydrite. The sequence contains aquifers of secondary or local importance, such as the Dhruma and Hanifa aquifers in the Riyadh area in central Saudi Arabia. Shales, argillaceous sediments and anhydrites of Jurassic to Lower Cretaceous age act as aquicludes, aquitards or as aquifers with limited local importance. Summary information of the hydrogeologic properties of the Tuwaiq mountain limestone, Hanifa, Jubaila, Arab formations, Hith anhydrite, Sulay, Yamama, Buwaib formations are given by Edgell (1997) and Alsharhan et al. (2001).

In the Oman mountains, sedimentation of carbonates forming the Hajar aquifer, continued from the Permian through the Triassic and Jurassic into the Cretaceous.

In Yemen, the Jurassic comprises Lower–Middle Jurassic shallow marine to fluviatile sandstones (Kohlan sandstone) and Upper Jurassic mainly shallow marine carbonates (Amran limestone). Together with the overlying Cretaceous Tawila sandstone, the Jurassic sediments constitute an important aquifer system in northern Yemen.

1.2.2.5 Cretaceous

During the early Cretaceous, most of the northern Arabian platform has been under continental or coastal marine conditions after a regional regression of the Tethys Sea, which was initiated by epeirogenic movements in the late Jurassic.

During the Lower Cretaceous, clastic sediments were deposited in wide parts of the Arabian Platform and marginal areas of the Interior Shelf. Lower to Middle Cretaceous sandstones form the extensive Wasia–Biyadh aquifer system in central Saudi Arabia and the Kurnub aquifer in Jordan. In Yemen the Tawila sandstone is a major aquifer in the highlands and on the western escarpment.

A marine ingression of the Tethys intruded into the area of the Lebanon, Antilebanon and Palmyrean mountains during the Albian and spread over wide parts of the northern Arabian platform during the Cenomanian. Limestones and dolomites of Upper Cretaceous age (Cenomanian–Turonian–Santonian–Campanian) constitute a highly productive karst aquifer in the highlands and mountain ranges of western Jordan, the West Bank, Lebanon and western Syria.

Continental conditions prevailed during the early Cenomanian in the vicinity of the Arabian Shield and of the Rutba arch. The sandy facies then retreated during younger stratigraphic stages further southward to the vicinity of the Arabian Shield, in southeastern Jordan sandstones reach up to the Santonian. Limestones and dolomites were deposited on the northern Arabian platform throughout the Cenomanian–Turonian, in some areas up to the Santonian. In central and northern Syria, the sediment facies changed during the Campanian to chalky foraminiferal limestones, sandy, organogeneous and detrital limestones, with intercalations of silicified limestones, cherts and phosphorite.

The marine transgression extended over most of the northern Arabian platform in the Maastrichtian – possibly with the exception of the central part of the Rutba arch. The marine Maastrichtian sediments are characterized by chalky and argillaceous limestones with chert and phosphorite intercalations. East of Wadi Sirhan and of the Jordan–Iraqian border, the marls grade into neritic limestones and silicified limestones and to shallow water calcareous deposits with sandy intercalations in the south. Oil shales occurring in the Maastrichtian in particular in southern Jordan indicate an influence of an anoxic ocean event.

The Maastrichtian marls act as a sub-regional aquitard above the Upper Cretaceous carbonate aquifer. Towards east, the Maastrichtian limestones replacing the marl facies constitute a joint aquifer with overlying Paleogene sediments. On the eastern Arabian platform, the Upper Cretaceous is represented by the Aruma formation, composed typically of chalky and calcarenitic limestones and dolomites, with a shale member at the base. The Aruma formation is exposed in an escarpment along the Interior Shelf and over wide areas of the Widyan plateau. It underlies almost all of the Arabian Platform and forms part of a complex aquifer system together with overlying Paleogene formations. The basal shales act as an aquitard base for the aquifer system. In the southern part of the eastern platform, "the Aruma Formation is developed in sandy facies and forms together with the underlying Wasia and Biyadh sandstones the Cretaceous Sand Aquifer" (Edgell 1997).

On the eastern boundary of the Arabian Shelf, sediments of late Cretaceous age include carbonates, marls and shales, which act as aquitards or aquifers with low permeability.

During the Campanian–Maastrichtian, a complex of sediments and meta-sediments (Hawasina formation) and ophiolites (Semail formation) was emplaced over the eastern margin of the Arabian Shelf in the area of the present Oman mountains. Overthrusting of ophiolites over the shelf margin also occurred on the northern boundary of the shelf, including the Basit area in northwestern Syria. The Hajar super-group (Permian–Jurassic carbonate formations) has, to some extent, been thrust over the Upper Cretaceous Juweiza formation on the western rim of the northern Oman mountains.

1.2.2.6 Paleogene

The Paleogene deposits of the Arabian Shelf comprise prevailingly carbonate rocks: limestone, chalk, dolomite, with intercalations of marls and of chert layers or nodules. The occurrence of sandstones is restricted to limited stratigraphic stages and marginal parts of the basin of Paleogene sedimentation. The Paleogene carbonate rocks are the deposits of the last extensive transgression of epicontinental seas over the Arabian Shelf. Depending on the paleo-environmental conditions, the deposits are composed of platform carbonate, pelagic carbonate, evaporite or clastic rocks. Outcrops and extent of Paleogene deposits on the Arabian Shelf are, to a large extent, controlled by major features of the tectonic structure.

On the eastern Arabian platform, the sequence of Paleogene sedimentary rocks is generally known as Hasa group, which is sub-divided into Umm er Radhuma, Rus and Damam formations.

The Paleogene deposits of the northern Arabian platform constitute deposits of a marine transgression of the Mediterranean Sea and are separated from the Paleogene of the Gulf basin by the Hail–Rutba uplift structure.

Although the lithologic sequence of the Paleogene on the Arabian Platform is rather uniform in general, considerable variations occur in the lithologic composition of rocks of the Paleogene stratigraphic units over the large area of the platform. Major trends of subregional lithologic variations are:

- The base of the Paleogene (Paleocene–Lower Eocene) is represented by prevailingly argillaceous sediments in the northwest (Syria, Jordan) and by carbonate rocks (Umm er Radhuma formation) in most of the Arabian Peninsula.
- The upper part of the Lower Eocene comprises carbonate rocks with abundant chert components in the northwest and gypsiferous sediments in parts of the Arabian Peninsula (Rus formation).

- Limestones with varying portions of chalk and marl extend over most of the Middle Eocene and, in some areas, the upper part of the Lower Eocene.
- Sediments of Upper Eocene and Oligocene age are restricted mainly to the northern part of the platform and to marginal areas in the east (Fars formation).

Main Paleogene aquifers are:

- On the northern Arabian platform: chalky and nummulitic limestones, chalks and cherts of prevailingly Eocene age
- On the eastern Arabian platform (Arabian Peninsula and southern Iraq): limestones and dolomites of the Umm er Radhuma and Damam formations (Paleocene–middle Eocene)

In the mountains and highlands adjoining the rift zone, Paleogene aquifers of limited extent occur in synclinal structures and can be important at least for local water supplies, such as the Jenin aquifer in the West Bank or the Umm Rijam aquifer in northern Jordan.

In northwestern Syria, Paleogene chalks and limestones provide a fissure type aquifer with generally moderate productivity. In southern Syria, eastern Jordan, southwestern Iraq and northwestern Saudi Arabia, Paleogene chalks, limestones and cherts constitute an aquifer with generally low to moderate productivity and fossil brackish groundwater. Fresh water lenses along major wadi systems are sustained by infiltration of wadi runoff.

The Paleogene aquifer system is underlain in the northern Arabian platform by an aquitard composed of Upper Cretaceous–Paleogene marls. Towards structural highs – the margin of the rift zone in the west, the Rutba uplift in the east and the Palmyrean mountains – the aquitard is missing or leaky and the Paleogene aquifer becomes unsaturated or connected with the deeper Upper Cretaceous limestone and dolomite aquifer.

The Euphrates–Gulf basin comprises an outcrop belt of Paleogene carbonate rocks, extending over around 1,500 km from southern Iraq to eastern Saudi Arabia. The carbonate rocks are karstified in some areas and are part of a complex aquifer system, which extends from the outcrop belt eastward and northeastward until the Euphrates and the Gulf. The Paleogene aquifer system constitutes one of the most important groundwater reservoirs in eastern Saudi Arabia and the Gulf area. The groundwater is, however, prevailingly fossil, present-day recharge is very limited according to the arid climate conditions. Fresh water lenses are sustained by recent recharge under karstic surfaces in the Gulf area, e.g. in Qatar.

The Umm er Radhuma formation generally comprises a joint aquifer together with the upper part of the Cretaceous Aruma limestones, while the lower part of Aruma formation acts as aquitard separating the Umm er Radhuma–Aruma aquifer from underlying sandstone aquifers.

In the Rub al Khali sub-basin, the Paleogene is, to a large extent, covered by sand seas and the groundwater in the Paleogene aquifers is prevailingly brackish. On the southern fringes of the Rub al Khali in Hadramaut and Dhofar, some present-day recharge apparently occurs in wadis on outcropping Paleogene rocks.

1.2.2.7 Neogene–Quaternary

At the beginning of the Neogene, the sea had retreated from most of the Arabian Shelf. Fluviatile and lacustrine sedimentation filled several large intra-platform basins during the Neogene with shallow marine sedimentation. Extensive Neogene sediments cover the eastern platform:

- The Lower and Upper Fars formations in the Euphrates area
- Sandstones with marls, chert and some gypsum of the Hadrukh formation, and marls, limestones and clays of the Dam formation on the platform area adjoining the Gulf

The thickness of these sediments reaches around 100 m in the Euphrates basin and more than 500 m in the Dibdiba basin. Sandstones and locally karstified limestones of the Neogene formations provide aquifers in limited areas.

On the northern part of the Arabian Shelf, fluviatile to lacustrine Neogene sediments accumulated:

- In basins within the platform: Ad Daw basin, Matah, Jaboul, Damascus plain, Qaa el Azraq, Wadi Sirhan
- In rift depressions: Wadi Araba, Jordan valley, Beqaa, El Ghab

The basin sediments contain major aquiferous sections in the Damascus plain, Wadi Sirhan and the Aleppo plateau.

Neogene–Quaternary formations contain important aquifers with renewable groundwater in various areas of the Arabian Plate:

- Coastal aquifers in plains adjoining the Mediterranean Sea, Red Sea, Arabian Sea and the Gulf of Oman
- Wadi aquifers within mountain massifs and particularly at the foot of mountain areas: the Asir mountains of western Saudi Arabia and Yemen, the Oman mountains, the Wadi Araba–Dead Sea–Jordan Valley at the foot of the Jordanian and Judean Highlands
- Volcanic aquifers in the area east of the rift zone in Yemen and in the Jebel el Arab basalt field (Golan–Jebel el Arab–Al Harra in Syria, Jordan and northwestern Saudi Arabia)

1.2.3 Major Aquifer Complexes

The following scheme gives a summary of major aquifer complexes within the stratigraphic rock column on the Arabian Shelf and of aquiferous zones on the Arabian Shield and the Oman mountains:

- Paleozoic sandstone aquifers:
  - Wajid sandstone in Saudi Arabia and Yemen
  - Saq–Disi (Rum) sandstone aquifers in Saudi Arabia and Jordan
  - Gaara sandstone in Iraq

- Paleozoic carbonate aquifers:
  - Khuff aquifer in Saudi Arabia
- Mesozoic sandstone aquifers:
  - Wasia–Biyadh aquifer (Saudi Arabia)
  - Tawila (Mukalla) sandstone (Yemen)
  - Kurnub sandstone (Jordan, West Bank)
  - Rutba sandstone (Iraq)
- Mesozoic carbonate aquifers:
  - Hajar aquifer (United Arab Emirates, Oman)
  - Jurassic carbonate aquifers (Syria, Lebanon)
  - Upper Cretaceous limestone and dolomite aquifers (Jordan, West Bank, Syria, Lebanon)
  - Aruma and Tayarat aquifers (Saudi Arabia, Iraq)
  - Amran limestone (Yemen)
- Paleogene carbonate aquifers:
- Chalk and limestone aquifers on the northern Arabian platform
- Umm er Radhuma and Damam aquifers in the Arabian Peninsula
- Neogene sedimentary aquifers:
  - Intra-platform basins on the northern Arabian platform
  - Euphrates–Gulf–Rub al Khali basin
  - Rift depressions
- Pleistocene–Quaternary wadi and coastal aquifers
- Volcanic aquifers
  - North Arabian volcanic province
  - Basalt aquifers on the Arabian Shield
  - Yemen volcanics
- Ophiolite aquifers of the mobile belt

References. Al-Sayari and Zötl (1978), Alsharhan et al. (2001), Burdon (1982), Edgell (1997), ESCWA (1999b), Jado and Zötl (1984), Ponikarov et al. (1967b).

1.3 Main Groundwater Flow Systems

1.3.1 Types of Hydrogeologic Basins

In extensive hydrogeologic basins with considerable topographic relief, a composite groundwater flow pattern can be found with local, intermediate and regional groundwater flow systems. These different types of groundwater flow systems are distinguished by the distance between recharge to discharge zones, by depth of aquifers and time scales of groundwater circulation. Groundwater retention periods of days to years, centuries and millenia may occur simultaneously in different aquifer zones of a composite basin (Tóth 1995). The development of local or regional flow systems depends on the topographic relief and the hydrogeologic structure of the basin, but also climatic factors can have an influence on the scale of prevailing groundwater flow systems. In humid areas, actively circulating local or intermediate flow systems often mask the slower groundwater movement in deeper layers. In arid basins, local groundwater flow systems with short retention periods may be active only seasonally or may not exist. In many areas of the Arabian Shelf, the groundwater surface is situated at several hundred metres depth below the surface. The limited quantities of present-day recharge leak through the unsaturated zone into deeper aquiferous horizons, which drain into distant discharge zones. Under these conditions, regional groundwater flow systems transport the main volumes of groundwater circulation.

As a result of the geometry of extensive sedimentary basins and the climatic conditions, a relatively small number of large hydrogeologic basins with predominant regional groundwater flow has developed on the Arabian Shelf. Most of the basins have a composite groundwater flow regime with multi-aquifer systems and relatively narrow discharge zones.

1.3.2 Major Hydrologic Basins and Groundwater Flow Systems

Major hydrologic basins of the Arabian Plate are (Fig. 1.3):

- Mediterranean Sea basin with the Orontes and Litani sub-basins
- Yarmouk–Jordan–Dead Sea basin
- Internal drainage basins of the northern Arabian platform: Jaboul–Matah, Sabkhet el Mouh, Damascus, Azraq, Wadi Sirhan
- Euphrates basin
- Gulf–Rub el Khali basin
- Wadi and coastal aquifer systems on various parts of the shelf and the shield

1.3.2.1 Mediterranean Sea basin

The boundary of the Mediterranean Sea basin in the northwest of the Arabian Plate is defined by the Mediterranean Sea coast between Lataqiye in the north and the Gaza Strip in the south and by surface water divides on the Judean highlands and the Hermon, Antilebanon, Jebel ez Zaouiye mountains.

The drainage system of the Orontes river, which discharges into the Mediterranean Sea after crossing the swamp area of Amiq north of Antakiya, may be considered a tributary sub-basin of the Mediterranean Sea basin. The Litani river catchment constitutes a sub-basin in the south of the Lebanon mountain zone. Main groundwater discharge areas are located at the Mediterranean Sea coast, in the rift depressions of the Beqaa and El Ghab, and at outcrop boundaries of the aquifers on the mountain slopes.

Major aquifers of the Mediterranean Sea basin are Jurassic and Upper Cretaceous karstic limestones and dolomites, overlain in some areas by aquifers of secondary

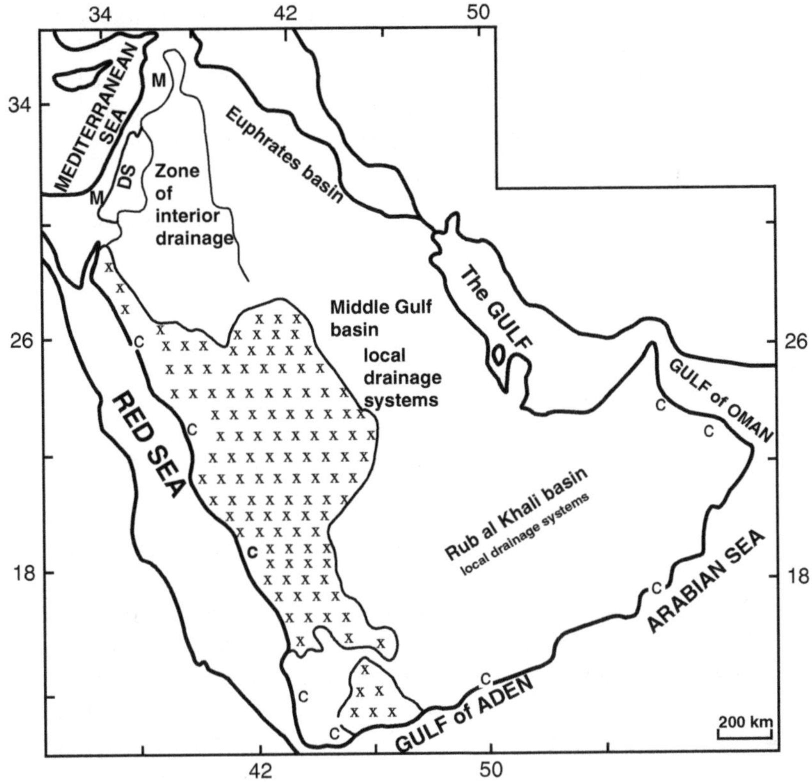

Fig. 1.3 Main hydrologic basins of the Arabian Plate. M Mediterranean Sea basin; DS Dead Sea basin; **c** coastal drainage systems toward the Red Sea, Gulf of Aden, Arabian Sea and Batina plain (schematic); x x x Precambrian shield

importance: Paleogene limestones and chalks, Neogene–Quaternary deposits. Several local to intermediate groundwater flow systems exist in the mountain areas. Deep groundwater flow in the Upper Cretaceous aquifer enters the hydrologic Mediterranean–Orontes basin from the east and the groundwater catchment of the Orontes system possibly extends until the northern Palmyrean mountains and the Taurus mountains in Turkey.

The Mediterranean Sea basin is situated in sub-humid to semi-arid climate zones and receives substantially higher precipitation volumes than other basins of the Arabian Plate. Several aquifers systems with high potential sustain substantial volumes of spring discharge.

1.3.2.2 Yarmouk–Jordan–Dead Sea basin

The drainage system of the Yarmouk–Jordan–Dead Sea basin is oriented to the lowest point of the earth surface at 460 m below sea level. The eastern catchment boundary runs over the western highlands of Jordan and Jebel el Arab, the western boundary over the Judean highlands.

Several larger wadi or stream systems in the eastern catchment of the basin reach lengths of more than 100 km. The western flank of the Yarmouk–Jordan–Dead Sea basin, the "West Bank", comprises catchments of short steep wadis.

Major groundwater flow systems of the basin occur in:

- The Upper Cretaceous carbonate aquifer system of the highlands of Jordan and Judea
- Paleogene chalk aquifers
- Pliocene–Quaternary basalt aquifers

The catchment of deep groundwater flow toward the Jordan–Dead Sea Valley extends far into the adjoining plateau areas in the east. In the north, subsurface inflow occurs from the Damascus basin into the Yarmouk basin.

The Yarmouk–Jordan–Dead Sea basin includes sub-humid to arid zones.

1.3.2.3 Internal Drainage Basins of the Northern Arabian Platform

A zone of internal drainage basins extends over the northern Arabian platform between the river systems of the Orontes, Yarmouk–Jordan, and Euphrates. The zone comprises a number of closed basins, which act as drainage for intermediate and sub-regional groundwater flow systems. The centres of the closed basins are occupied by seasonal lakes or salt flats: Jabboul, El Matah, Sabkhet el Mouh, lakes Ateibe and Hijjane in the Damascus plain, Qaa el Azraq, Wadi Sirhan. Groundwater moves to the basin centres in Mesozoic to Paleogene carbonate aquifers and Neogene–Quaternary basin sediments and volcanics.

Mean annual precipitation in the zone of internal drainage ranges from 500 to 700 mm in the mountain areas to <200 mm in the basin centres.

The western margin of the closed basin zone is formed by the eastern slope of the Hermon–Antilebanon–Jebel ez Zaouiye mountains. Groundwater flow in the Antilebanon mountains, including Mount Hermon, comprises several separate systems with discharge in partly very large springs, from which perennial rivers rise, such as the Aouaj and Barada rivers in the Damascus basin. Some deep groundwater movement appears to be directed from the closed basins to topographically lower discharge areas of the Orontes, Euphrates and Yarmouk–Jordan–Dead Sea drainage systems.

Hydrogeologic details of the internal drainage zone are given in the description of the hydrogeologic provinces of the northwestern mountains and highlands, Jebel el Arab and the Badiye and Hamad (Chaps. 2–4).

1.3.2.4 Euphrates Basin

The hydrologic Euphrates basin comprises the Mesopotamian foredeep on the boundary between Arabian Shelf and the Alpidic mobile zone and extends in the north into the Zagros mountains and over the Taurus mountains far into the Anatolian highlands. The Euphrates runs along the boundary of the Arabian Shelf as an allogenic river, which receives its major flow volume in more humid climate zones on the Anatolian highlands. In the east, the Euphrates river merges with the Tigris river forming the up to two kilometres wide Shatt el Arab which discharges into the Gulf.

The southwestern to southern catchments of the Euphrates river are situated on the Arabian Shelf with groundwater flow systems in mainly Mesozoic–Tertiary aquifers.

The southwestern Euphrates catchment comprises prevailingly plateau landscapes, which are crossed by several southwest–northeast oriented wadi systems with ephemeral flow ending in the Euphrates river plain. Groundwater discharges in a number of large springs and in extensive lakes along the river plain. In the east, the plateau morphology grades into the Dibdiba plain at the lower end of the Wadi Rima–Wadi el Batin wadi system which originates on the Arabian Shield.

The plains on the left bank of the Euphrates in northeastern Syria, the Jezire, are crossed by the Khabour and Balikh rivers with perennial flow.

The Hamad in the southwest of the hydrologic Euphrates basin comprises a plateau area with ephemeral surface runoff into local sabkhas. Groundwater movement in the main aquifer complex, composed of Cretaceous–Tertiary carbonate rocks, is directed generally towards the Euphrates basin in the east.

The climate of the southwestern Euphrates catchment area and in the Hamad is prevailingly arid with mean annual precipitation of <100 mm.

1.3.2.5 Gulf–Rub al Khali Basin

The Gulf–Rub al Khali basin comprises a huge catchment area with prevailingly arid plateaus and plains without any significant perennial or seasonal surface runoff. Groundwater moves in a multi-aquifer system radially to the Gulf coast with main natural discharge zones in oases and sabkhas. The whole catchment extends over an area of around 1.8 million km^2 from the Gulf coast to the Asir mountains in Saudi Arabia and Yemen and to the Hadramaut highlands and Oman mountains. The huge basin consists of several hydrologic–hydrogeologic drainage zones:

- The slopes of the highlands on the southern and southeastern margins of the basin with local groundwater flow systems along wadis in gravel and sand aquifers or fissured Paleogene carbonate aquifers, which discharge partly into small springs or local depressions, but leak to a large extent into intermediate or regional aquifers.
- A belt of Paleozoic to Mesozoic aquifers on the Interior Shelf of Saudi Arabia, which become downstream deeper components of the regional multi-aquifer system.
- An extensive groundwater flow system in Paleogene–Neogene aquifers on the platform of eastern Saudi Arabia and of the Nejd in Oman.

- Groundwater flow systems in Neogene–Quaternary aquifers overlying deep Paleogene aquifers in the Rub al Khali Basin.
- Local groundwater flow systems in karstic or detrital aquifers on the Gulf coast.

Groundwater circulation sustained by present-day recharge occurs on the mountain slope areas, and also the groundwater movement in the upper section of the regional multi-aquifer system is, to some part, related to recharge in extensive sandstone and karst outcrops. Groundwater in the deeper aquifers and in most of the Rub al Khali appears to be prevailingly stagnant and to originate from fossil Pleistocene recharge. Limited amounts of present-day recharge sustain shallow local groundwater systems on the Gulf coast.

1.3.2.6 Fluviatile and Coastal Sand and Gravel Aquifers

Pleistocene to Quaternary fluviatile or coastal sand and gravel aquifers extend over many areas of limited size on the Arabian Plate. Some sand and gravel aquifers provide important reservoirs of renewable groundwater resources, where favourable infiltration conditions exist in main runoff zones.

Sand and gravel aquifers occur, in particular, on mountain slopes and in coastal plains. They are of particular importance

- As aquifers, where they overlie rocks of low to moderate permeability, e.g. on the Arabian Shield or on the Oman ophiolite mountains
- As fresh water aquifers overlying or adjoining zones of brackish to saline water in coastal plains
- As media collecting infiltrating runoff, which then leaks into permeable underlying aquifers, e.g. karstified carbonates

The catchments of some wadi aquifers comprise carbonate or sandstone aquifers with considerable groundwater potential, in particular in the northern Oman mountains (Hajar mountains), the Qara mountains north of Salala, Wadi Hadramaut and the Yemen highlands.

1.3.3 Hydrogeologic Provinces and Sub-regions

Lerner et al. (1990:24) defined areas with specific climatic, geologic and soil conditions as hydrogeologic provinces. The most common provinces are according to Lerner et al. (1990):

- Alluvial fans and riverbeds
- Sand and sandstone
- Limestone and dolostone
- Chalk

- Volcanic
- Plutonic crystalline

A strict application of that classification for the Arabian Plate would lead to a differentiation into a very large number of hydrogeologic provinces. Major hydrogeologic sub-regions of the Arabian Plate may be delineated through a more generalized definition of hydrogeologic provinces considering

- Main lithologic and hydrogeologic characteristics of major aquifers
- Morphologic–climatic zones
- Hydrologic basins and main groundwater flow systems
- Groundwater quality characteristics

The differentiation into hydrogeologic sub-regions may follow, in principle, the megatectonic framework of the Arabian Plate with its main geologic units: Arabian Shield, Arabian Shelf comprising the Interior Shelf and the Arabian Platform, and marginal mobile zones.

The Arabian Platform, as a geologic–hydrogeologic unit, includes the sub-regional carbonate aquifer systems, which are formed by the deposits of the Mesozoic and Tertiary marine transgressions on the Arabian Shelf. The Interior Shelf comprises the main outcrop areas and zones of main groundwater circulation in Paleozoic to Mesozoic sandstone aquifers, but sandstone aquifer systems are exposed also in some structural highs within the Arabian Platform: the Rutba uplift and the Hadramaut arch.

The geologic–hydrogeologic framework of the Arabian Plate, as defined above, comprises the following major units:

- Arabian Shield
- Interior Shelf
- Northern Arabian platform with a western highland and rift zone, an eastern part with prevailing plateau landscapes, and the marginal Mesopotamian plain
- Eastern Arabian platform with the hydrogeologic Euphrates–Gulf–Rub al Khali basin
- The Oman mountains

Sand and gravel aquifers in wadis, morphologic depressions and coastal plains constitute hydrogeologic provinces dispersed over various areas in different parts of the Arabian Plate.

Some of the main hydrogeologic sub-regions of the Arabian Plate and the characteristics used for the definition of these sub-regions are listed schematically in Table 1.2. Considerable variations of climatic and geologic conditions can occur within some of the sub-regions. The mountain and rift zone in the northwest of the Arabian Plate includes, e.g.

- Large climate variations from the sub-humid mountains to the arid Dead Sea area
- Outcropping Mesozoic carbonate aquifers in anticlinal mountain ranges and outcrops of Pleistocene–Quaternary aquifers in synclinal basins
- Occurrence of saline water and brines in the Dead Sea zone within a sub-region characterized mainly by fresh water

Table 1.2 Selected hydrogeologic sub-regions of the Arabian Plate

Hydrogeologic sub-region	Major aquifers	Climate	Groundwater salinity
Western highlands and mountains, Jordan, Syria, Lebanon	Karstic limestones and dolomites	Sub-humid	Mainly fresh
Badyie and Hamad	Fissured carbonate aquifers	Semi-arid	Brackish, fresh in limited areas
Basalt area, Jordan and Syria	Fissured basalt	Semi-arid to sub-humid	Mainly fresh
Sandstone area in Jordan and Saudi Arabia	Sandstone	Prevailingly arid	Fossil fresh groundwater
Carbonate aquifers in Saudi Arabia and the Gulf area	Fissured or karstic carbonate aquifers	Prevailingly arid	Mainly brackish
Arabian Shield	Unconsolidated wadi aquifers	Arid, semi-arid in mountain areas	Fresh to brackish
Western highlands and escarpment in Yemen	Limestone, sandstone, volcanics	Semi-arid to sub-humid	Fresh to slightly brackish
Oman mountains	Ophiolites	Semi-arid	Low yielding fresh water aquifers
Coastal areas	Carbonate rocks		Fresh to brackish
	Unconsolidated coastal aquifers	Semi-arid to arid	Brackish to saline, fresh water lenses

Boundaries between hydrogeologic sub-regions cannot always be delineated precisely along geographic lines, and transition zones may have to be considered in several cases.

References. ESCWA (1999b: 3.2), ESCWA-UNEP (1996), Lerner et al. (1990: 23 ff.).

1.4 Groundwater Regimes: Quantitative Aspects

1.4.1 General Features of Groundwater Regimes on the Arabian Plate

Volumes of groundwater flow through an aquifer may be determined through an analysis of

- Recharge and inflow of groundwater into the aquifer
- Throughflow of groundwater through the aquifer
- Outflow from the aquifer

Inflow components comprise lateral inflow into the aquifer and leakage from adjoining aquifers. Recharge occurs as natural recharge from precipitation and runoff infiltration, irrigation return flow, waste water infiltration and surface water infiltration through special artificial measures, such as recharge dams.

Throughflow of groundwater through an aquifer or section of an aquifer is generally assessed from hydraulic data: hydraulic conductivity, aquifer geometry, transmissivity, hydraulic gradient. Values of aquifer porosity are indirectly related to the hydraulic conductivity and essential for the evaluation of groundwater storage and of changes in storage.

Outflow of groundwater from an aquifer occurs as natural discharge in springs or seepages and through evapotranspiration, as artificial discharge mainly through wells, and as leakage into hydraulically connected adjoining aquifers.

Groundwater regimes on the Arabian Plate are, to a large extent, influenced by specific hydrogeologic conditions of the region:

- The majority of extensive aquifers is composed of fissure-type or karstic carbonate rocks or porous and fissured sandstones.
- Relatively high recharge rates are found in some areas of the sub-humid highlands in the northwest of the Arabian Plate, but in most parts of the region, groundwater replenishment is restricted essentially to indirect recharge, while direct recharge from precipitation is low to negligible.
- Under the recharge conditions in the semi-arid to arid zones, many of the important groundwater resources are contained in aquifers of large extent characterized by very long groundwater flow periods, low hydraulic gradients, predominance of fossil groundwater and partly very large volumes of stored groundwater.
- Groundwater occurrences with significant recharge and short-term circulation are, in the dry areas, restricted mainly to special morphological features, in particular wadi systems or morphologic depressions with limited lateral extent and thickness.

Substantial information on groundwater balances and hydraulic data is available for a number of aquifers of the Arabian Plate. The data are, to some part, representative for certain stages of groundwater development only or to limited areas. A comparative evaluation of the available information of aquifer hydraulics and groundwater flow volumes, which is presented in the following sections, may provide a general overview of the order of magnitude of the groundwater potential of major aquifers of the region.

1.4.2 Hydraulic Characteristics

1.4.2.1 Hydraulic Parameters

Calculations of groundwater flow and groundwater storage can be made from data of hydraulic aquifer parameters and aquifer geometry. Parameters needed for describing the hydraulic aspects of groundwater flow and of storage volumes in extensive aquifers are, in particular, hydraulic conductivity, transmissivity, hydraulic gradient, porosity, storativity of confined aquifers or specific yield of unconfined aquifers, respectively, and extent of the aquifer. A general characterization of the potential of major aquifers may be obtained by mean values or ranges of values of hydraulic conductivity (K) and transmissivity (T). Values of effective porosity may give indications on groundwater storage volumes.

Ranges of hydraulic parameter values in consolidated aquifers are listed in textbooks (Freeze and Cherry 1979; Matthess and Ubell 1983) as:

	K (m/s)	Effective porosity (%)
Karstic limestone and dolomite	10^{-6}–10^{-2}	5–50
Fissured limestone	10^{-9}–10^{-6}	0.7–13
Chalk		1–2
Sandstone	10^{-9}–10^{-4}	5–30
Permeable basalt	10^{-7}–10^{-2}	

Hydraulic conductivities in unconsolidated aquifers range from 10^{-3} to 10^{0} m/s for gravels to 10^{-7}–10^{-3} m/s for silty sand. Effective porosities of gravel and sand aquifers generally range from 10 to 25%.

1.4.2.2 Ranges of Hydraulic Parameter Values of Main Aquifers of the Arabian Plate

Table 1.3 lists reported values of horizontal hydraulic conductivity, transmissivity and specific well capacity for some aquifers of the Arabian Shelf.

High transmissivities are related in particular to karstified carbonate aquifers and to sandstone aquifers with great saturated thickness. Hydraulic conductivities are relatively large in some unconsolidated Quaternary aquifers, but limited thickness restricts the transmissivity and exploitable quantities of these aquifers.

Transmissivities of the Paleogene Umm er Radhuma and Damam aquifers are moderate to high in some areas of the eastern Arabian platform and relatively high transmissivities are found in Paleozoic sandstone aquifers of the Interior Shelf. Low to moderate transmissivities prevail in the Mesozoic aquifers of the Interior Shelf, the Paleogene chalk–limestone aquifers of the northern Arabian platform and most of the Neogene–Quaternary aquifers.

Relatively high transmissivites are found in the karstified limestone and dolomite aquifers of the western mountain and rift zone of the northern Arabian platform. Hydraulic conductivities of these aquifers range from 10^{-5} to 5×10^{-3} m/s. The aquifer thickness of the Jurassic and Upper Cretaceous of the high Lebanon mountains is in the order of 1,000 m each with transmissivities of generally 1,000–90,000 m^2/d. High transmissivities of around 100,000 m^2/d occur in karstified zones, while transmissivities in poorly fissured limestones may be as low as 10 m^2/d. Effective porosity is assumed to be around 5%.

Relatively high transmissivities are indicated for some parts of the Upper Cretaceous aquifers of the highlands of Jordan and of Judea and very high transmissivities occur in Upper Cretaceous aquifers at some locations of the Jordan valley and in Paleogene aquifers in the Jezire on the Syrian–Turkish border area.

Table 1.3 Hydraulic parameter values of major aquifers of the Arabian Plate

Aquifer	Country	K (m/s)	T (m^2/d)	Q/s ($m^3/h/m$)
Mesozoic limestone and dolomite aquifers				
Jurassic-Cretaceous	Lebanon mts.	10^{-5} to 5×10^{-3}	1,000–90,000	
Aman-Wadi Sir	Jordan highlands	1.3×10^{-4} to 10^{-3}	10–10,000	0.02–2,100
Hummar			200–3,000	
Mountain aquifer	West Bank	10^{-5} to 10^{-3}	10,000–100,000	
Paleogene carbonate aquifers				
B4/B5	Jordan	3.5×10^{-6} to 10^{-4}	43–10,000	
Rijam		2×10^{-9} to 2×10^{-5}	0.03–450	0.01–37
Shalala				0.08–10
Umm er Radhuma	Saudi Arabia	3×10^{-7} to 4×10^{-6}	4.3–55,000	
	Bahrain		9.2–40,000	
	Qatar		230–8,400	
Damam	Kuwait	10^{-5} to 10^{-3}	13–4,600	
Khobar	Saudi Arabia		0.3–25,000	
Alat			2.3–25,000	
Damam	Bahrain		47–41,000	
Khobar			0.1–900	
Alat			2–400	
Paleozoic–Mesozoic sandstone aquifers				
Disi	Jordan	10^{-6} to 5×10^{-4}	300–1,000	
Saq	Saudi Arabia	3×10^{-6}	700–3,000	
Tabuk, aquifer members		3×10^{-6}	18–2,600	
Wasia			26–2,400	
Wajid		3×10^{-6} to 2×10^{-5}	50–1,800	
Sandstones Rub al Khali			600	
Berwath			260	
Unayza		1.5×10^{-8} to 10^{-5}		
Kurnub	Jordan	3×10^{-8}		
Basalt aquifers				
Neogene basalt	Jebel el Arab, Syria, Jordan	6×10^{-5}		

1.4.3 Groundwater Recharge

1.4.3.1 Climatic Features

The Arabian Plate is situated predominantly in zones with semi-arid to arid climate. Sub-humid climate conditions with mean annual precipitation of 300 to more than 600 mm are restricted to mountain ranges on the northern margins of the region and at the southwestern and southeastern fringes of the Arabian Peninsula (Fig. 1.4).

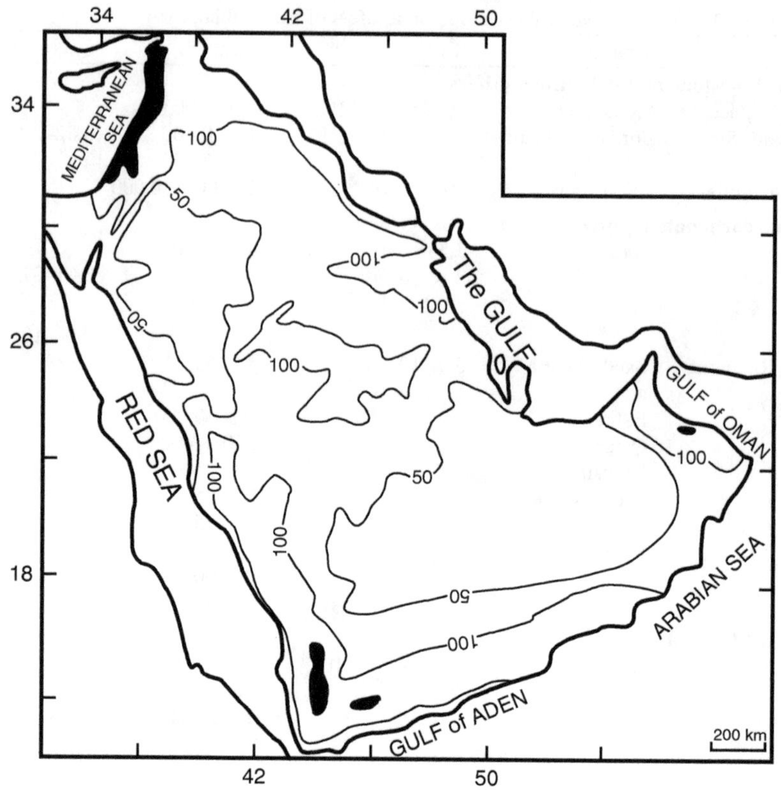

Fig. 1.4 Mean annual precipitation on the Arabian Plate. After ESCWA (1999b). 50 mm and 100 mm isohyetes. black fields >500 mm/a

Precipitation on most of the Arabian Plate is concentrated to the winter months between November and March, when also temperatures and evaporation rates are relatively low. Long dry seasons are common in the whole region. Even in the area with sub-humid climate in the northwest of the Arabian Plate, the dry season without precipitation extends over 5–6 months.

Precipitation events in the Arab Middle East are dominated by the movement of air masses from the eastern Mediterranean Sea. The eastern Mediterranean climate is affected during summer by a subtropical high pressure belt, which produces, together with a semi-permanent surface heat trough centered over Iraq and Iran, stable hot and dry weather. The high pressure belt is related to the position of the Inter-Tropical Convergence Zone (ITCZ) over the deserts of Arabia and the Sahara. Air masses moving northward from the tropics become heated and dry while descending within the convergence zone.

According to the position of the sun, the ITCZ moves southward during winter, and the Mediterranean area and adjoining land masses are then temporarily influenced by low barometric pressures bringing moist air and rain from the Atlantic and the North Sea. Low pressure systems, which form over the western part of the Mediterranean Sea and southern Europe during the winter season, enter into western Asia and produce precipitation over the areas adjoining the Mediterranean Sea. The movement of low pressure belts into the area adjoining the eastern Mediterranean Sea is highly different from year to year.

The precipitation pattern in the northwest of the Arabian Plate is strongly influenced by the mountain chains running sub-parallel to the Mediterranean Sea coast: the Judean highlands–Lebanon–Ansariye mountains and Jordanian highlands–Antilebanon mountains create a pronounced lee effect with rapidly decreasing rainfall towards east.

Part of the rainstorms following barometric lows in the northwest of the Arabian Plate intrude into the eastern and southern desert areas as convective cells with limited extent of a few kilometers to some tens of kilometres. Cold frontal troughs, originating from the Mediterranean Sea, the North Atlantic Ocean or the Red Sea, can reach until the southeastern Arabian Peninsula and cause rainfall in winter and early spring in the United Arab Emirates and Oman. Rainfall is often restricted to the areas of the center of the convective cells and rainstorms in the steppe and desert areas may be of high intensity during a few minutes up to a few hours. In particular in spring, flows of hot air from the desert toward barometric lows can initiate abundant dust storms, occasionally broken by a heavy rainstorm.

Advection of cold air from central Asia can cause occasional heavy rainfall during December to April on the Arabian Peninsula. In the Gulf area, easterly winds from the Indian Ocean together with atmospheric depressions produce sporadic rainfall during winter.

Average annual rainfall on most of the Arabian Peninsula is between 50 and 110 mm with high variability from year to year and from place to place. The mean annual precipitation is reported as 73–96 mm on the Gulf coast of Saudi Arabia, up to 110 mm in Kuwait, 74 mm in Bahrain, 75 mm in Qatar.

In the Rub al Khali desert, precipitation usually ranges from 25 to 50 mm/a, although in some areas rain has not fallen for 20–30 years. In general, rainfall decreases from north to south and from west to east.

In some areas on the margins of the Arabian Plate, the seasonal rainfall pattern is influenced, apart from the regionally predominant Mediterranean meteorologic regime with winter rains, by sources of air moisture from the east to south. In the Asir mountains of southwestern Saudi Arabia and the highlands of Yemen and in the Oman mountains, orographic or convectional rainfall events are experienced during the summer months. High temperatures create convectional cells, into which moisture intrudes from adjoining oceans, such as the Gulf of Oman and the Arabian Sea. Convectional rainfall occurs in the Asir mountains mainly in April to May, in the Oman mountains during the summer months, and generally affects only the mountain chains near the coast.

Tropical cyclones originating over the Arabian Sea or India can bring heavy rainfall on the southeastern coast of the Arabian Peninsula in the months from May to December, but rarely cross the coastal area further inland.

Monsoonal rainfall which occurs in the southern parts of the Arabian Peninsula during summer (southeast monsoon) generally affects only the coastal mountains. Monsoon rains reach the southern coast of Oman and the Asir highlands normally in late July and may extend, after main precipitation events in August, until September–October.

The Mediterranean Sea with its high heat capacity acts as a gigantic *temperature* regulator. The influence of the Red Sea, Dead Sea and the Gulf on air temperatures is limited to their close vicinities.

In the central areas of the Arabian Peninsula summer temperatures range up to 50°C, while winter temperatures average between 10 and 20°C. Differences of air temperature between day and night often reach 15°C to 20°C. A more temperate climate persists along the humid Gulf coast, where temperatures range up to 38°C.

Inland humidities can vary from 5% to 90%, depending on the penetration of saturated air masses (Al-Sayari and Zötl 1978).

Potential *evaporation* in most of the Arabian Plate is in the range of 1.5–3 m/year with very high values of up to 4.5 m in some desert areas. On the Gulf coast, humidity frequently approaches 100% and annual potential evaporation averages between 1.8 and 2 m.

1.4.3.2 Recharge Processes

Present-day recharge on the predominantly semi-arid to arid Arabian Plate is, in general, very limited and is principally restricted to areas with favourable rainfall and/or infiltration conditions. Significant recharge occurs

- In the northwestern mountains and highlands with relatively high precipitation on outcrops of karstified or fissured carbonate rocks
- In the dry zones on karstic surfaces of limestones and dolomites and on outcropping sandstones
- On pervious surfaces of Neogene–Quaternary sediments or volcanics
- In wadi beds and coastal plains where surface runoff infiltrates into sand and gravel deposits or into sandstone or carbonate aquifers

Limited recharge may occur in sand dune areas. A major source of artificial recharge is irrigation return flow in agricultural areas.

The morphologic relief controls the recharge conditions on the Arabian Plate in particular

- Through the altitude effect on precipitation and the barrier effect of mountain ranges to storm trajectories
- Through the regulation of surface runoff and accumulation of runoff in the wadi network and in morphologic depressions

Depending on climatic and infiltration conditions, different modes of recharge are predominant: direct or indirect recharge. These have been defined (Lloyd 1995) as:

"Direct recharge – water that enters the saturated zone directly from precipitation as unsaturated quasi-vertical flow.

Indirect recharge – water that enters the saturated zone indirectly through a runoff or ponding resultant of precipitation, either by means of unsaturated or saturated flow".

From overviews of recharge conditions in dry areas by Lloyd (1986, 1995), the general occurrence of recharge in different aridity zones may be generalized as follows:

Aridity zone	Mean annual rainfall	Recharge
Hyperarid	<50 mm	No recharge except in rare circumstances
Arid	50–200 mm	Only irregular indirect recharge
Semi-arid	200–500 mm	Indirect recharge dominant in lower range of precipitation but direct recharge becoming dominant with increasing precipitation

Lloyd (1995) gives the following conclusions on recharge volumes with respect to exploitable groundwater resources: "no recharge occurs in hyperarid areas in terms of usable groundwater resources except under rare circumstances". "... except for localised runoff, that may be managed by recharging structures, recharge in arid areas is non-significant in resource terms. Even where channel runoff recharge occurs it is unlikely to be large with respect to regional resources". "In semi-arid areas recharge obviously increases with increasing precipitation, but it can be variable and is not necessarily sufficient to sustain the requirements of modern progressively developing societies".

We may conclude from that certainly realistic assessment of the recharge potential in dry areas, that present-day recharge can balance only very limited groundwater extraction in most areas of the Arabian Plate. Even small recharge rates may, however, be important for groundwater exploitation,

- Where groundwater flow, sustained from recharge in large catchment areas, is concentrated in relatively small discharge areas
- Where seasonal or ephemeral recharge from surface runoff can be efficiently managed for local water requirements

In addition to recharge, aquifers can be replenished by subsurface inflow and by leakage from adjoining aquifers.

1.4.3.3 Recharge Rates in the Sub-humid to Semi-arid Northwestern Arabian Shelf

The highest recharge potential of the Arabian Plate area occurs in the Lebanon mountains, where outcropping karstic limestones and dolomites receive a mean annual precipitation of 1,450 mm over some catchments. Average recharge rates in percentage of annual rainfall vary from around 40% or 200–600 mm/a on outcrops of Jurassic and Upper Cretaceous karst aquifers to 27% or 150 mm/a on Quaternary aquifers of the coastal plains. The average total recharge to all outcropping aquifers in Lebanon is estimated at 2.5×10^9 m^3/a corresponding to 25% of the total volume of precipitation (Khair et al. 1992).

For the main groundwater basins of Lebanon, the following ranges of mean annual recharge rates have been calculated:

Northern Bekaa basin	180 mm
Southern Mt. Lebanon and southern Bekaa basins, Antilebanon mountains	210 mm
High Lebanon mountains	350 mm

On the outcrops of the Upper Cretaceous mountain aquifer in the Judean mountains, recharge rates of 25–30% of precipitation are indicated, with an average annual recharge of 100×10^6 m^3 over an area of 2,200 km^2. Recharge on the Upper Cretaceous Amman–Wadi Sir aquifer in the highlands of Jordan varies generally between 14 and 30% of mean annual precipitation.

Recharge rates on outcrops of the Upper Cretaceous limestone and dolomite aquifer of the Ansariye mountains in northwestern Syria reach around 30% of the mean annual precipitation of 1,050 mm.

In the semi-arid to arid basalt areas of Jebel el Arab in southwestern Syria and northern Jordan, recharge rates are probably in the order of 3–6% of mean annual rainfall or 2.4–30 mm/a.

1.4.3.4 Recharge Rates on Arid to Hyper-arid Areas of the Arabian Shelf

The eastern Arabian platform and the Interior Shelf are situated in the arid climate zone with mean annual rainfall below 150 mm. Groundwater replenishment is restricted mainly to irregular indirect recharge from sporadic high intensity rainfall.

Estimates of average annual recharge are reported as

- 4–8 mm of 60 mm average annual rainfall on aquifer outcrops in the eastern province of Saudi Arabia
- 15% of rainfall in the Rus–Umm er Radhuma aquifer of Bahrain
- 7–10% of annual rainfall of Qatar

Calculations of recharge to the Umm er Radhuma aquifer in northeastern Saudi Arabia indicate recharge rates of 17 mm/a in areas, where surface runoff ponded on the margins of sand dunes infiltrates into the aquifer. Recharge rates in areas, where the aquifer is covered by gravel plains, is much lower with around 2.6 mm/a.

Volumes of recent recharge to the main Tertiary aquifers in northeastern Saudi Arabia are by far highest for the Umm er Radhuma aquifer with an average of around 1×10^9 m^3/a. Estimates of average annual recharge to the Damam and Neogene aquifers are in the order of 2×10^6 m^3/a and 20×10^6 m^3/a, respectively.

Rather high recharge rates have been observed on karstified outcrops of the Umm er Radhuma formation on the As Sulb plateau in eastern Saudi Arabia. Locally, 47% of the average precipitation of 90 mm/a infiltrate into karst openings of the aquiferous formation.

In the Tuwayq area of the Interior Shelf in central Saudi Arabia, mean annual precipitation is around 40–80 mm. Estimates of recharge rates on outcrops of sandstone aquifers in that zone vary from 4 to 22% of mean annual precipitation. The weighted average recharge has been calculated at 10% of annual precipitation or 4 mm/a.

For the Dibdiba plain in Iraq, recharge rates of 3.8–14% of the mean annual precipitation or 94–140 mm/a have been estimated.

In the Hajar carbonate aquifer of Wadi Bih in the northern United Arab Emirates, recharge rates of 10% of mean annual precipitation, or 15.5 mm/a, are indicated.

1.4.3.5 Recharge on the Arabian Shield

The main volume of groundwater recharge on the crystalline Precambrian of the Arabian Shield results from infiltration of surface runoff into wadi aquifers. Estimates of recharge rates in the Hejaz–Asir mountains of Saudi Arabia are in the order of 2–5 mm/a over the entire wadi catchments. Recharge rates may be significantly higher in the more humid mountain areas of Al Asir.

1.4.3.6 Recharge Rates on Arid Coastal Plains

For the Jizan coastal plain on the Red Sea coast of Saudi Arabia, a mean recharge of 17% of wadi runoff has been estimated. Assuming a runoff coefficient of 10% of precipitation in the catchment, the recharge may be in the order of 1.7% of precipitation or around 7 mm/a.

From calculations of recharge rates for the Fujayra coastal plain in the United Arab Emirates, recharge from flood flow infiltration and local direct recharge appears to amount to around 9 mm/a.

1.4.4 Groundwater Discharge and Flow Volumes

In wide parts of the Arabian Plate, the limitation of groundwater recharge support only moderate to low volumes of groundwater flow. High quantities of groundwater flow, however, through extensive fissured and karstified aquifers in the more humid areas and move, in the arid zones, through aquifer complexes with very large dimensions. These conditions may be illustrated by a few examples:

In the sub-humid mountain and highland zone in the northwest of the Arabian Plate, large volumes of groundwater flow through karstic Jurasic and Upper Cretaceous aquifers:

Flow volumes in the Upper Cretaceous mountain aquifer of the Judean highlands are estimated for major catchment areas as

360×10^6 m^3/a for the western catchment of 1,800 km^2
140×10^6 m^3/a for the northern catchment of 700 km^2
100×10^6 m^3/a for the eastern catchment of 2,200 km^2

Groundwater flow in Upper Cretaceous carbonate aquifers in the highlands of Jordan toward the Jordan valley is in an order of 100×10^6 m^3/a.

Mean discharges of large springs issuing from Mesozoic aquifers in the Antilebanon mountains amount to 7 m^3/s in Ain el Fije, 3.3 m^3/s in Ain Barada, and 2.0 m^3/s in Anjar spring.

Mean base flow of streams in the Lebanon mountains, fed by spring discharge from Jurassic and Upper Cretaceous karst aquifers, reaches 11–43 m^3/s. Spring discharge and submarine discharge from the Upper Cretaceous Sanin aquifer of the northwestern Lebanon mountains reaches 13 m^3/s.

Mean discharge of three large springs from the Upper Cretaceous aquifer of the middle Orontes area (springs Tell Ayoun, Ain Taqa, Ain el Moudiq) range from 1.5 to 5.8 m^3/s. In Ain el Sinn on the coast at the western slope of the Ansariye mountains, a mean volume of 10.5 m^3/s discharges from the Upper Cretaceous aquifer.

The large volumes of groundwater flowing through the major aquifers of the northwestern mountain and highland zone are fresh water resources with salinities mainly between 400 and 700 mg/l TDS.

A large volume of groundwater flows through an Eocene karstic fresh water aquifer toward the Euphrates plain in the Syrian Jezire, supporting a mean spring discharge at Ras el Ain of 38.7 m^3/s.

On the southern rim of the Euphrates valley, groundwater flow through Tertiary aquifers sustains a total discharge of a number of brackish springs of around 1 m^3/s. That in view of the extensive catchment area rather moderate discharge reflects the groundwater regime in an arid climate.

In the arid zones of the Arabian peninsula, groundwater flows in the order of 500×106 m^3/a have been estimated for sandstone aquifers:

- The Paleozoic Wajid aquifer in Saudi Arabia and Yemen, the Cretaceous Wasia–Biyadh aquifer in Saudi Arabia
- The Mukalla aquifer in southern Yemen

Groundwater flow in the Paleozoic Saq aquifer in Saudi Arabia is around 290×10^6 m^3/a, in the corresponding Disi aquifer in Jordan 50×10^6 m^3/a.

Relatively high groundwater flow volumes have also been estimated for the Paleozoic Tabuk sandstone aquifer with 140×10^6 m^3/a and the Mesozoic Minjur sandstone aquifer with 100×10^6 m^3/a. Groundwater in these sandstone aquifers is prevailingly brackish, becoming saline in the confined aquifer zones at the downstream end of the groundwater flow systems.

The relatively high volumes of groundwater flow correspond to very large aquifer dimensions, thicknesses of several hundred metres and vast lateral extent.

For the multi-aquifer system of the Gulf sub-basin on the eastern Arabian platform, a groundwater flow of $>1 \times 10^9$ m^3/a has been estimated. The groundwater is prevailingly brackish and moves, to a major part, through the Paleogene Umm er Radhuma formation.

Groundwater inflow through the Damam aquifer of Kuwait amounts to 0.9 m^3/s from southwest and 1.2 m^3/s from northwest.

References. FAO (1979), Harhash and Yousif (1985), Issar (1990), Khair et al. (1992), Lloyd (1986, 1995), MAW (1984), Mukhopadhyay (1995), Zubari (1997).

More details and references on hydraulic parameter values, groundwater recharge rates and flow volumes in different parts of the Arabian Plate are given in Chaps. 2–7.

1.5 Groundwater Salinity and Major Hydrochemical Processes

1.5.1 General Distribution of Groundwater Salinity

Under the prevailingly dry climate conditions, present-day precipitation produces significant quantities of fresh groundwater only in limited parts of the Arabian Plate. In wide areas, the aquifers contain brackish or saline groundwater.

On regional scale, the pattern of groundwater salinity on the Arabian Plate appears related mainly to the morphologic–climatic conditions:

- Sub-humid climate conditions on the mountainous margins of the plate favour the formation of fresh groundwater. Water with low salinity occurs, in particular, in the karstic aquifers of the northwestern mountain ranges and in the sedimentary aquifers of the highlands and western escarpment at the southwest of the plate.
- In most of the semi-arid to arid areas covering most of the Arabian Peninsula and the steppe (Badiye) in the northern Arab countries, the groundwater salinity is elevated to levels of more than 1,000 mg/l up to several thousand mg/l TDS. This general increase of groundwater salinity toward the more arid regions is certainly, to a large extent, related to the climatic conditions: increasing aridity is accompanied by decreasing recharge, an increasing impact of evaporative processes and low rates of flushing of the aquifers.

The regional groundwater salinity distribution is modified by local or subregional variations influenced by lithologic and hydrographic features:

- The geochemical reactivity of soil and rock material is a dominant factor for the concentration of dissolved substances in the groundwater:
  - Evaporite layers cause elevated groundwater salinity in various aquifers and affect e.g. the groundwater quality in parts of the sub-humid northwestern highlands.
  - Groundwater salinity can be relatively low in semi-arid and arid areas in non-marine siliceous aquifers, in particular basalts and sandstones, e.g. in the Jebel el Arab basalt aquifer system or the Paleozoic sandstone aquifers of Saudi Arabia and Jordan.

- Extensive wadi systems provide favourable conditions for infiltration of surface runoff and the formation of fresh water lenses in wadi sediments, in basin sediments on the foot of mountains, and under outcrops of fissured aquifers along wadi courses. On the other hand, groundwater salinity is generally elevated in flat plains of closed basins or coastal areas.
- Fresh water or brackish water with relatively moderate salinity can accumulate under open karst surfaces, where fast localised recharge in karst openings and extensive flushing reduce the impact of evaporative and lithogenic salinisation. Examples are found in karstic Tertiary aquifers in some areas of northeastern Saudi Arabia and in Qatar and Bahrain.

Brines with more than 100 g/kg TDS are found, in particular, in deeper formations of sedimentary basins and in arid coastal zones, such as the plain adjoining the Shatt el Arab and northern Gulf coast and the Dead Sea coast.

Groundwater with relatively low salinity, stored in some extensive sandstone or karst aquifers on the Arabian Peninsula, appears to originate from recharge during the Pleistocene when, under relatively wet and cool climate conditions, recharge rates were higher and evaporation less intensive.

The outlined general pattern of groundwater salinity refers to the upper aquifers located near the surface down to several hundred metres depth. In most groundwater basins, salinity increases with depth, although deeper sandstone or karst aquifers can contain water with lower salinity than overlying formations.

1.5.2 Dissolved Substances in Precipitation and in the Unsaturated Zone

1.5.2.1 Precipitation

The primary source of direct and indirect groundwater recharge on the Arabian Plate is precipitation, which provides an initial input of dissolved substances to the hydrochemical composition of the groundwater. Concentrations of dissolved constituents in rain water are relatively low. Precipitation in the northwestern mountains and highlands of the Arabian Plate contains mean concentrations of a few mg/l of Na, Mg, Ca, Cl and SO_4, and mean HCO_3 concentrations around 10–20 mg/l. Average electrical conductivity values are around 30–50 μS/cm. In coastal areas, Cl concentrations are generally elevated from evaporated sea water spray. Mean Cl concentrations of rainfall at the Mediterranean sea coast are in the order of 15 mg/l (Wakshal and Nielsen 1982; Tartous station: Kattan 1996b) and decrease slightly with increasing distance from the coast.

East of the western mountain ranges of Syria and Jordan, concentrations of dissolved substances in rain water obviously increase generally with the decreasing amount of precipitation. Mean values of electrical conductivity of precipitation water rise from 40 μS/cm in the Antilebanon mountains and the Hauran–Jebel

Table 1.4 Anion concentrations and electrical conductivity values in precipitation at different meteorologic stations of the Arab Middle East. (after data from Al-Kharabsheh 1995; Alyamani and Hussein 1995; Kattan 1996b; Rosenthal 1987; Salameh and Rimawi 1987a, b; records of UAE Min. Electricity and Water of 1991–92)

Area	Station	Mean precipitation (mm/a)	Rock outcrop	HCO_3 (mg/l)	Cl (mg/l)	SO_4 (mg/l)	EC (μS/cm)
				Mean value or range			
Antilebanon	Bloudan	700	Carbonate	18	4.5	4	38.4
Judean highlands	Jerusalem	400	Carbonate	40.7	10.2	20.6	
Northern Syria	Aleppo	350	Carbonate	52	4	85	111
Highlands of Jordan	Aman	300	Carbonate	2.4–42	1.1–45	0–70	10–49
Hauran, southern Syria	Izraa	280	Basalt	12	4	4.6	30.4
Damascus plain	Damascus	200	Carbonate	45.8	4.5	19.8	119
Central Syria	Palmyra	150	Carbonate	77.3	5.3	14.3	170
Eastern Jordan	Azraq	75	Basalt, carbonate	75.9	23.7	33.7	273
Arabian Shield, western Saudi Arabia	Wadi Fatima	70–280	Crystalline	11.6–21.4	4.6–9.2	2.9–6.2	62–94
Northern Emirates of the UAE	various stations	60–180	Ophiolite, limestone	12–177	7–56	10–120	85–640

el Arab area and 50 μS/cm at Aman to 165 μS/cm at Damascus, Palmyra and the steppe east of Amman (Muwaqar, Khalidiye) and to 273 μS/cm at Azraq in northeastern Jordan. The elevated values of total salinity are accompanied, in particular, by an increase of Ca, HCO_3 and SO_4 concentrations and, in the eastern steppe of Jordan, also by elevated Cl concentrations.

Relatively elevated salinity and elevated concentrations of dissolved substances in rain water are observed in the more arid zones, but the type of rock outcrops apparently also has an essential influence on the chemical composition of precipitation: rainfall salinity is relatively low on the basalt field of Hauran–Jebel el Arab and over crystalline rock outcrops of the Arabian Shield in the arid Wadi Fatima basin, while elevated salinity and HCO_3 concentrations occur in precipitation in less arid climate over the carbonate outcrops of the West Bank and of northern Syria.

Mean values or ranges of anion concentrations and electrical conductivity values in precipitation are listed in Table 1.4 for different meteorologic stations of the Arab Middle East.

Generally, rain water in the Arab Middle East is Ca-HCO_3 type water with percentages of Ca ranging from around 40 to 82 meq% and of HCO_3 from 38 to 67 meq% (Fig. 1.5). Relatively high contents of Ca and HCO_3 in precipitation, which are found at stations with dry climate such as Damascus, Aleppo, Palmyra and Azraq, are probably related to dust deposition.

The variability of concentrations of dissolved constituents in rain water between different precipitation events and also during individual events is high. Generally, precipitation events start with relatively high contents of dissolved constituents,

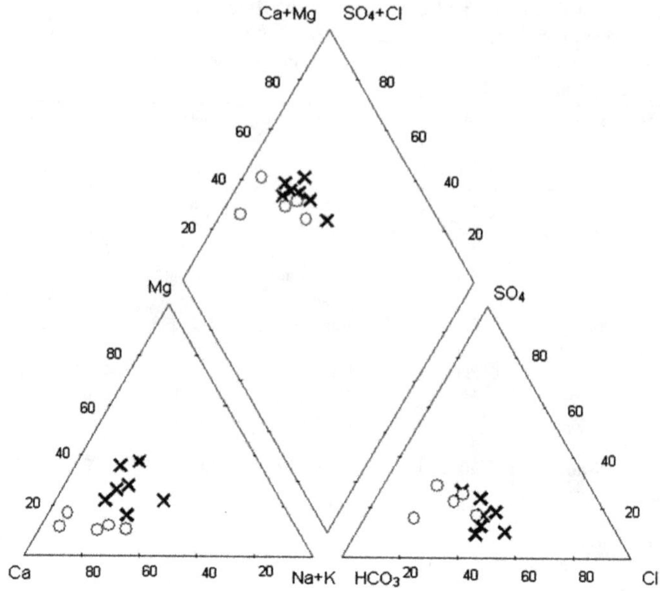

Fig. 1.5 Piper diagram: Rain water samples from meteorologic stations in Jordan and Syria. O northwestern highlands and Hauran; x eastern plateau areas. Data from Kattan (1996b), Salameh and Rimawi (1987a, b), Al-Kharabsheh (1995)

which decrease with higher amounts of precipitation. A decrease of concentrations of Ca, Mg, K, Na, NO_3, Cl and SO_4 has been observed during the course of individual precipitation events (Salameh and Rimawi 1987a, b). In addition to precipitation of dissolved substances, rainfall causes deposition of atmospheric dust, in particular at the beginning of rainfall events. Atmospheric dust over outcrops of calcareous rocks may comprise calcite, silica, clayey material and gypsum particles (Salameh and Rimawi 1987a, b), deposition from washout of dust can therefore contain relatively high amounts of Ca, HCO_3 and SO_4.

In individual samples collected during different rainfall events, ion ratios can deviate significantly from the mean values, and SO_4 or Cl can be predominant anions (Kattan 1996b; König 1994).

The available data on the hydrochemical composition of rainfall in different areas of the Arabian Plate provide very general information on the atmospheric input of soluble substances: the available data generally represent sporadic sampling of individual rain events or arithmetic means of samples collected over one or a few rain seasons. Quantitative calculations of total annual loads of atmospheric deposition can be made from weighted averages in relation of time variability and amount of precipitation. Weighted averages may deviate considerably from arithmetic means. At Aman station, the following values of electrical conductivity have been observed during 1984/85 to 1985/86 (Salameh and Rimawi 1987a):

- Total range: 10–490 μS/cm
- Weighted average: 46.6 μS/cm
- Geometric mean: 67.5 μS/cm
- Arithmetic mean: 97.9 μS/cm

Na/Cl ratios of wet deposition in Jordan are relatively close to the ratio in sea water, reflecting a marine origin of the precipitation. Ion ratios in "wet only" deposition correspond prevailingly to Ca-HCO_3 type water, while in measurements of total deposition, including wet and dry deposition, ion ratios are shifted toward higher SO_4 and/or Cl percentages. An impact of continental conditions is noticed in increases of Ca/Cl and K/Cl ratios in the dry deposition in comparison to ratios of sea water (König 1994).

In Jordan, the proportion of dry Cl deposition ranges between around 50% and 80% of the total annual Cl deposition. These values correspond to observations at several stations with mean annual precipitation between 21 and 400 mm and mean Cl concentrations in precipitation between 13 and 37 mg/l (König 1994).

We may conclude from the above summary of data on atmospheric input of soluble substances:

- Precipitation on the Arabian Plate provides a diluted source of dissolved substances with mean electrical conductivity values between 40 and 400 μS/cm or salinities of 30 to 250 mg/l TDS.
- A substantial proportion of soluble substances is precipitated as dry deposition in particular in semi-arid and arid zones and in areas with outcropping carbonate rocks.
- The water in heavy and long lasting rainstorms is generally more diluted than water at the beginning of rainfall events or in drizzle rains. Weighted averages, rather than arithmetic means, can therefore be representative for the concentration of dissolved substances in precipitation.

The concentration of dissolved substances in precipitation has, in many areas, a minor impact on the hydrochemical composition of the groundwater in comparison to concentrations caused e.g. by dissolution processes. But "even though rainwater has a very low dissolved solids content, this may be highly significant in arid zones where only a very small proportion of precipitation becomes run-off or groundwater recharge. In this case the rainwater salts are concentrated in the soil". "In areas of low or intermittent recharge . . . cyclic salting and the ensuing salt concentration in the unsaturated zone may lead to significant salination of groundwaters when recharge pulses occur. Important cycling salting effects are recorded in a number of arid zone countries" (Lloyd and Heathcote 1985).

1.5.2.2 Infiltration of Rain Water

During infiltration of rain water into the underground, two major processes cause an increase of dissolved substances at the surface and in the soil zone:

- Interaction of biogenic CO_2 with soil material and water
- Evaporative enrichment of the atmospheric input

In the soil zone, CO_2 is generated by the decay of organic matter and the respiration of plant roots. Through the reaction of CO_2 and H_2O, acid H_2CO_3 is produced, and concentrations of HCO_3 and corresponding cations, mainly Ca and Mg, increase in the soil water through the reaction of H_2CO_3 with dissolvable minerals, in particular carbonates.

Evaporation of the precipitation water on the surface or in the soil zone leads to an enrichment of the dissolved substances, in particular Cl, SO_4 and corresponding cations. Salinity in the soil water is therefore, in general, significantly higher than rainwater salinity. Mean Cl concentrations in soil water at various observation fields in Jordan are in a range between 100 and 17,000 mg/l (König 1994). High variations of the Cl concentrations are observed between different measuring points at close distance, reflecting local variations of the soil profile. Distribution of salinity in the groundwater is more homogeneous than the highly varying salinity of water in the soil zone, and concentrations of dissolved substances in shallow groundwater tend to be much lower than concentrations in the soil water. Generally, HCO_3 concentrations in the unsaturated zone are much higher than in the underlying groundwater (Matthess 1990: 271). At soil water observation fields in Jordan, Cl concentrations in groundwater are lower than mean concentrations in the soil water samples by factors of 1:1.8 to 1:280 (Table 1.5).

The diffuse seepage of water through the matrix of the soil zone has, according to local conditions, a varying impact on the groundwater recharge. Under dry climate conditions, the proportion of water percolation through the soil matrix may become insignificant in comparison to other variations of the recharge process:

- By-passing of infiltrating rain water through macro-pores
- Infiltration of rain water on barren rock outcrops
- Indirect recharge which is influenced in its chemical composition by the quality of surface runoff

Computations of the proportion of groundwater recharge through soil matrix and through macro-pores at observation stations in Jordan show, that the by-pass infiltration through macro-pores produces 54% of the total recharge in sub-humid areas (Queen Alia airport near Aman) to more than 99% in arid areas (König 1994).

Table 1.5 Mean Cl concentrations in soil water and groundwater at observation stations in Jordan (from König 1994)

	Mean Cl concentration soil water (mg/l)	Cl concentration ground water (mg/l)
El Jafr	17,185	190
Queen Alia airport	101	56
Azraq	7,120	282
Mafraq	17,768	63
Irbid	234	59

1.5.3 Hydrochemical Composition of Runoff Water and Shallow Groundwater

The hydrochemical composition of shallow groundwater in the Arab Middle East is determined mainly by

- Atmospheric inputs: substances dissolved in infiltrating rainwater
- Pedogenic inputs: substances dissolved through interaction of biogenic CO_2 with soil material and water; salts dissolved from the soil zone
- Lithogenic inputs: substances dissolved from rock material in the saturated and unsaturated zone
- Input of cyclic salts concentrated on the surface
- Intrusion of brackish or saline water from surface water bodies (sea water, sabkhas) or from adjacent aquifers or aquitards

Solution of carbonate, chloride and sulfate minerals at or near the surface and within the aquifers is, in general, a major source of groundwater salinity. Solution processes are of prime importance in determining the chemistry of groundwater, since the composition of rainwater is far from chemical equilibrium with aquifer matrix minerals. Aquifer matrix solution is most pronounced with more soluble lithologies such as carbonates and evaporites, but is significant even with silicates and quartz. In many sandstone aquifers solution of traces of carbonate present either as cement or detrital grains may swamp any chemical contribution from the silicate minerals (Lloyd and Heathcote 1985: 205).

The significance of the above cited processes on the hydrochemical composition of surface and near-surface water in different areas depends largely on the climatic and recharge conditions and the geologic substratum: aridity, direct or indirect recharge, soil cover, vegetation density, lithology of rock outcrops.

The HCO_3 concentrations in runoff water, soil water and groundwater are generally several times higher than in rain water. The HCO_3 regime is predominantly controlled by the biological CO_2 production in the soil zone and reactions of carbonic acid with carbonates and silicates. Biological CO_2 production is generally decreasing with increasing aridity; on the other hand, the relative influence of indirect recharge becomes more important in arid areas. Surface water being in contact with the soil zone over extended periods can have considerable concentrations of HCO_3 even under dry climatic conditions with low CO_2 production. In stream waters of central Syria and in Jordan, mean HCO_3 concentrations of 50–160 mg/l were analyzed (Droubi 1983; Al-Kharabsheh 1995; Salameh 1996).

Salinity of indirect recharge is influenced, in particular, by SO_4 and Cl enrichment. In dry areas, SO_4 and Cl compounds are concentrated on the surface through evaporation processes and occur as dispersed salts on the land surface. Flood waters in Jordan contain Cl concentrations of 5–56 mg/l and SO_4 concentrations of 5–98 mg/l (Salameh 1996). In seasonal runoff of streams in central Syria, Cl concentrations reach up to 135 mg/l and SO_4 concentrations up to 870 mg/l (Droubi 1983).

The chemical composition of indirect recharge is determined by the composition of surface water rather than by the composition of precipitation water.

The chemical composition of infiltrating water in the unsaturated zone can be determined directly through lysimeters or suction devices in shallow boreholes or pits. These techniques require, however, considerable expenditure to achieve representative coverage of wider areas and various depth ranges.

The impact of hydrochemical processes at and near the surface can also be deduced from the composition of

- Very recently recharged groundwater in near surface aquifers from shallow wells and springs
- Flood water
- Water stored above the aquifer zone in cisterns

The unsaturated zone is considered, in such a comparative evaluation, as a black box. Recharge water is released from the unsaturated zone into shallow aquifers through macro-pore pathways, from stream beds and uncovered rock outcrops and, to some percentage, through soil matrix percolation. The impact of soil matrix percolation may be insignificant in particular in arid areas, where no well developed soil cover may exist and where, over long periods, low precipitation rates are not sufficient to cover the soil moisture deficit. Table 1.6 shows examples of the range of values of hydrochemical parameters in flood water, water from cisterns and from shallow groundwater in northern Jordan and in Saudi Arabia.

The water collected in cisterns in northwestern Jordan is similar in its hydrochemical composition to flood water, with slightly higher Ca and HCO_3 concentrations in the flood water. The enrichment in Ca and HCO_3 concentrations is more significant in the spring water, reflecting an increasing lithogenic component through dissolution of carbonate. The higher NO_3 contents in spring water indicate an anthropogenic impact.

The mean values of ion concentrations in freshly recharged water in northwestern Jordan may be representative for conditions characterized by outcropping limestone terrain and an average rainfall of around 300 mm/year. Groundwater with similar hydrochemical composition discharges from small to medium size springs in the Antilebanon mountains in Syria.

In the arid Oman mountains of the northern Emirates of United Arab Emirates, groundwater with low salinity occurs in shallow wadi aquifers after recharge from sporadic flood events. The range of ion concentrations of groundwater recharged from flood flows can be seen from water samples collected from Shaara well field near Fujayra. The well field is situated directly downstream of a recharge dam on Wadi Ham, which receives occasional floods from an extensive catchment in prevailing ophiolite formations with mean annual precipitation of 110 mm. EC values of groundwater in the well field vary from 900 to 1,370 μS/cm, predominant ions are Mg, Na and Cl (Wagner 1995a).

The groundwater in the Shaara well field corresponds in its hydrochemical composition to water collected from falajes, which drain shallow groundwater in the ophiolite mountains in Fujayra and Al Ain areas (Alsharhan et al. 2001).

Table 1.6 Concentration of major ions in flood water, cisterns, spring water and shallow groundwater in Jordan, Syria and western Saudi Arabia (data from Al-Kharabsheh 1995; ESCWA 1996; Klinge 1982; Rimawi and Udluft 1985; Salameh 1996; Salameh and Rimawi 1987a, b; investigations of ACSAD-BGR 1980)

	Flood water			Cisterns	Springs	Shallow groundwater		
	Highlands northwest Jordan Mean values	Jordan, Azraq area	Tihama, western Saudi Arabia Range	Jordan	Jordan	Highlands northwest Jordan Mean values	Antilebanon Syria Range	Jebel Arab Syria Mean values
EC	244	282	475–880	200–335	436–569	437	178–275	250
Ca	30.2	25.7	45–111	18.9–30.1	44.1–78.8	50.7	36–48	27.1
Mg	6.2	4.0	6–26	7.3–13.4	10.5–23.1	15.3	3.5–10	12.6
Na	5.7	29.1	29–144	9.7–21.2	14.0–29.4	17.4	2.5–5.5	15.4
Cl	5.3	8.7	44–144	17.0–41.8	22.0–50.7	26.2	36–74	22
SO_4	4.8	7.3	56–180	4.8–26.9	2.9–18.3	3.8	3.5–14.5	10
HCO_3	46.4	160.2	116–262	77–109	199–250	202	110–183	120
NO_3	0.2	4.8	2.4–66	2.5–5.0	29.0–62.9	37.5	8.0–10.5	15.7

EC electrical conductivity in μS/cm, ions in mg/l

In hyper-arid central Oman, fresh water lenses are created from recharge of floods during sporadic cyclonic storm events (Macumber 1995; Macumber et al. 1995). The storm runoff produces fresh water lenses with a salinity of 123–255 mg/l TDS.

The shallow groundwater from the mentioned areas shows an enrichment in particular in Ca + Mg and HCO_3 concentrations in comparison to precipitation and flood water. Differences in the hydrochemical composition of shallow groundwater reflect different hydrochemical conditions:

- Enrichment of dissolved solids is relatively moderate in the highly karstified Antilebanon mountains
- Relatively high Ca and HCO_3 concentrations occur in the highlands of Jordan, where diffuse infiltration in areas with extensive soil cover may provide a significant contribution to the groundwater recharge
- Relatively high Mg, Na and Cl concentrations in the shallow groundwater of Fujayra can be attributed to the Mg rich aquifer material derived from ophiolite rocks and to evaporative enrichment of flood water within an extensive catchment in an arid climate
- The relatively low enrichment of dissolved substances in the shallow groundwater of hyperarid central Oman indicates a fast infiltration of flood water

1.5.4 *Hydrochemistry of Main Aquifers*

Changes of groundwater hydrochemistry during longer retention periods in the aquifer are related mainly to

- Dissolution and precipitation of minerals
- Oxidation and reduction processes
- Ion exchange and adsorption

Changes in groundwater hydrochemistry along the flow path can include, in particular,

- Increase of Cl and SO_4 concentrations through dissolution of salts and evaporite (gypsum, anhydrite) layers
- Increase of HCO_3 concentrations through interaction between dissolved CO_2, water and carbonate or silicate minerals
- Redox phenomena modify the groundwater hydrochemistry mainly in confined aquifers, where organic matter is available. Redox processes comprise in particular reduction of dissolved oxygen, nitrate and sulfate
- Oxidation of organic matter

Common redox processes influencing the hydrochemical composition of the groundwater can be defined schematically by the formula

$$SO_4{}^{2-} + 2H_2CO \text{ (organic matter) } |2HCO_3{}^{-} + H_2S$$

This reaction contributes dissolved inorganic carbon which accounts for the rise in bicarbonate concentration (Lloyd and Heathcote 1985: 215f.).

The concentrations of lithogenic components are largely controlled by the geochemical composition of the rock formations, with which the groundwater comes into contact. Main lithological types of aquifers and aquitards, which influence the groundwater quality on the Arabian Shelf, are:

- Carbonate aquifers: limestones, dolomites, chalk
- Siliceous aquifers: sandstones, volcanic rocks
- Aquifers with sulfatic layers: carbonate rocks, sandstones or semi-consolidated Pleistocene basin deposits with gypsum or anhydrite intercalations
- Aquitards of marine origin with dispersed salts and pyrite
- Aquitards deposited under anoxic conditions, which contain organic material
- Unconsolidated or semi-consolidated Pleistocene to recent aquifers in wadis, basins and coastal areas with varying geochemical rock types

In many areas, groundwater salinity increases along flow paths and with aquifer residence periods chiefly because of dissolution Cl and SO_4 salts. Simple dissolution may be a major source of groundwater salinity in unconfined karstified evaporite aquifers, such as in Qatar. Where permeability is high and recharge significant, soluble salts are removed by flushing through active groundwater circulation, and dissolution will generally not lead to high groundwater salinity. The rate of flushing or groundwater flow will determine in part the rate of removal of connate groundwater, the strength of diffusion influences and dissolution rates. In the predominant number of cases where salinity increases down the hydraulic gradient it is a function of decrease in flow and, although dissolution may play a part, the origin and certainly the age of the saline water is often very different from that in the active fresh water part of the aquifer. The increase in salinity is not necessarily a simple progression (Lloyd and Heathcote 1985: 154, 155).

"...salts may be concentrated in the shallow subsurface due to evaporation rates that exceed precipitation by up to one order of magnitude. Significant recharge pulses may dissolve this salt and flush it into ground water" (Richter et al. 1993: 18).

High groundwater salinity occurs in shallow to intermediate aquifers in the Arab Middle East in particular in closed basins and in coastal areas. In closed basins, natural recharge along the surrounding highlands flows toward the basin centers. Along the flow path, ground water dissolves mineral matter, resulting in a general increase in TDS content from recharge areas to discharge areas. Evaporation in the basins and especially in salt flats in the center of the basins is the most influential process in the development of the chemical composition of the shallow, saline ground water in these settings. Groundwater evolution in closed basins may be similar to any other evolution along the flow path, from a low TDS, $Na–Ca–HCO_3$ recharge water to a high TDS, Na–Cl water, as a result of reactions such as calcite dissolution and precipitation, cation exchange on clay minerals, and evaporation in discharge areas near the center of the basins (Richter et al. 1993: 18, 26).

Saline groundwater extends, e.g., adjacent to zones where groundwater discharges from extensive multi-aquifer systems in sabkhas and springs in eastern

Saudi Arabia. Saline groundwaters tend to occur, in particular, downstream of major discharge zones. Such conditions are commonplace where there are low recharge mounds and large groundwater throughputs (Lloyd and Heathcote 1985: 155).

1.5.5 Main Hydrochemical Types of Groundwater

Fresh groundwaters on the Arabian Plate are mainly HCO_3 or Cl type waters. In wide parts of the Arab Middle East, aquifers contain brackish or saline groundwater. Predominant anions in brackish groundwaters with a salinity of up to 3,000 mg/l TDS, which extend over wide parts of the Arabian Shelf, are SO_4 or Cl waters. Brackish to saline groundwaters with more than 4,000 mg/l TDS are generally Cl waters. Some waters with high salinities are Na–SO_4 type waters with sulfate concentrations controlled by gypsum precipitation.

Groundwaters of aquifers of the Arabian Plate naturally show wide variations of major ion ratios. A few hydrochemical types of groundwater can, however, be defined, which appear to be characteristic for specific aquifers and particular sub-regional hydrogeologic conditions.

1.5.5.1 HCO_3 Water

Groundwater with predominance of *Ca* and HCO_3 occurs over wide areas in the Jurassic and Upper Cretaceous karst aquifers of the mountains and highlands in the northwest of the Arabian Plate. The Ca–HCO_3 waters form a rather homogeneous hydrochemical group, which is found in the highlands of Jordan and Judea, the Lebanon and Antilebanon mountains, the Ansariye mountains and Jebel ez Zaouiye. The group is characterized by low to moderate salinity of 180–700 mg/l TDS, HCO_3 concentrations between 200 and 300 mg/l and Mg/Ca ratios of 0.4 to 0.7. The hydrochemical composition reflects a regime with active groundwater circulation, in which dissolution of limestone and dolomite is the main hydrochemical process.

Ca–HCO_3 type waters are common in nummulitic limestone aquifers in Syria and Lebanon. Groundwater salinity in the nummulitic limestone aquifers ranges from 250 to 400 mg/l TDS, typical Mg/Ca ratios are 0.03 to 0.1, differing significantly from ratios in the Mesozoic dolomitic limestone aquifers.

Groundwaters with predominance of Ca and HCO_3 are found in recharge areas of Paleogene chalk aquifers in the northern Arabian platform and of basalt aquifers in the Jebel el Arab mountain area. The Ca–HCO_3 waters of the chalk aquifers are distinguished from the Mesozoic karst groundwater by higher Na, Cl and SO_4 concentrations and generally lower Mg percentages.

In the Paleozoic sandstone aquifers of the Interior Shelf and the Paleogene carbonate aquifers of the eastern Arabian platform, typical Ca–HCO_3 waters are found only at few locations in the outcrop belts of the formations.

In the highlands and the western escarpment of Yemen, Ca–HCO_3 water is the most common water type in wadi and basin aquifers as well as in the Tawila sandstone aquifer and the Yemen volcanics. Groundwater salinity in these Ca–HCO_3 waters generally ranges from 300 to 600 mg/l TDS.

Groundwater with predominance of *Na* and HCO_3 occurs over wide areas of the Jebel el Arab basalt aquifer complex in Syria and Jordan and at or near the outcrop belt of the Paleozoic Saq–Disi aquifer in the Interior Shelf. The high percentage of Na in these groundwaters with generally moderate salinity may be related to weathering–dissolution processes of silicate minerals and/or ion exchange.

Groundwater in shallow ophiolite aquifers of Oman and the United Arab Emirates is generally $Mg\text{–}HCO_3$ type water with salinities between 250 and 1,000 mg/l TDS. The hydrochemical composition of these waters results mainly form weathering of silicates of the ophiolite rocks and evaporative processes in a system open to the atmosphere.

1.5.5.2 Na–Cl Water

Waters with predominance of Na and Cl prevail in many of the brackish and saline aquifers, which extend over wide parts of the Arabian Shelf. On the eastern Arabian platform, groundwaters with a salinity of >4,000 mg/l TDS are generally Na–Cl waters. Elevated Na and Cl concentrations are derived from evaporative enrichment, recent or previous sea water intrusions, and dissolution of remnants of marine NaCl in aquitards or poorly flushed aquifers

Alkaline Na–Cl waters with high OH and Ca concentrations issue from deeper groundwater flow in springs with low discharge within the ophiolite areas of the Oman mountains.

1.5.5.3 SO_4 Water

Brackish Ca–SO_4 water is tapped through boreholes in the confined Upper Cretaceous aquifer of the Aleppo plateau in northern Syria. Elevated SO_4 concentrations are probably derived from dissolution of evaporites intercalated in the Upper Cretaceous carbonate–marl sequence. SO_4, HCO_3 and Ca concentrations appear to be maintained at rather homogeneous levels through a saturation equilibrium. Na and Cl concentrations show wide variations; with increasing salinity the hydrochemical character of the groundwater changes to Cl type water with high SO_4 concentration. The water in the confined aquifer is anoxic, strong H_2S odour indicates SO_4 reduction processes caused by oxidation of organic matter originating probably from the overlying Maastrichtian marl aquitard.

Brackish groundwaters in the Paleogene and Neogene aquifers of the eastern Arabian platform are, to a considerable percentage, Ca–SO_4 or Na–SO_4 waters, alternating with the widely distributed Na–Cl waters. SO_4 waters are particularly common in the salinity range of 3,000–4,000 mg/l TDS.

Main sources of the elevated SO_4 concentrations are evaporatic intercalations in particular in the Paleogene formations and Neogene sediments, and sulfate minerals in surfacial layers of sabkhas, wadi beds and windblown dispersed deposits. SO_4 waters occur from outcrop areas of the aquiferous formations until a belt of groundwater discharge in sabkhas near the coastal plains (sabkha discharge line). The coastal zone downstream of the discharge belt is, apart from shallow groundwater lenses with moderate salinity, occupied by Cl waters.

References. Al-Kharabsheh (1995), Almomani (1996), Al-Sayari and Zötl (1978), Alsharhan et al. (2001), Droubi (1983), Kattan (1996b), König (1994), Lloyd and Heathcote (1985), Macumber (1995), Macumber et al. (1995), Matthess (1990), Richter et al. (1993), Rimawi (1985), Salameh (1996), Salameh and Rimawi (1987a), Salameh et al. (1991), Wagner (1995b, 1996c).

1.6 Recent and Fossil Groundwater: Aspects of Isotope Hydrology

1.6.1 Isotope Investigations of Groundwater in the Arab Middle East

Under the climatic–hydrogeologic conditions of the Arab Middle East, the age of groundwater and the distribution of fossil and of renewable groundwater resources are major critical factors to be considered in the planning of water resources exploitation. Isotope data, providing a unique tool for groundwater age assessment, have therefore a particular importance for groundwater studies in the region.

Routine analyses of environmental isotopes for groundwater studies generally include analyses of isotopes of the water molecule – hydrogen and oxygen – and of dissolved carbon, which is present mainly in the HCO_3-ion. Isotope analyses commonly comprise determination of the stable isotopes of hydrogen, oxygen and carbon – ^{2}H, ^{18}O, ^{13}C – and of the radioactive isotopes of hydrogen and carbon – ^{3}H and ^{14}C. For the area of the Arab Middle East, data of around 3,000 analyses were available from isotope hydrologic studies carried out, until 1998, in the different countries of the region, including data of about 1,000 ^{14}C analyses.

The results of isotope investigations of groundwater in different areas of the region are summarized in Chaps. 2–9. This section presents an overview of the regional distribution of isotope values in major aquifers of the Arab Middle East.

General information on the application of isotope hydrologic methods for groundwater studies can be found in various proceedings of IAEA and in textbooks and publications, e.g. Freeze and Cherry (1979), Fritz and Fontes (1980), Wagner and Geyh (1999).

1.6.2 ^{3}H and ^{14}C Data

^{3}H and ^{14}C data indicate, that significant present-day recharge occurs:

- In Jurassic and Upper Cretaceous karst aquifers in wide parts of the northwestern mountain and highland zone of the northern Arabian platform
- Within the Jebel el Arab basalt field on the higher mountain area and at some locations of the Hauran plain
- In shallow wadi aquifers of the Arabian Shield
- In the coastal plains of the Tihama of Saudi Arabia (Red Sea coast), of Al Batina in the United Arab Emirates (Gulf of Oman), and along main wadi courses in Al Batina of Oman

Measurable tritium levels are found in shallow brackish water lenses in the Paleogene aquifer of Qatar and the Neogene aquifer of northern Kuwait.

In the northwestern mountain and rift zone of the northern Arabian platform, ^{14}C values of most groundwater samples are between 52 and 62 pmc; higher values in some springs indicate very recent (post-bomb) recharge after 1955. Retention periods of the groundwater in the Jurassic–Upper Cretaceous karst aquifers range generally from some decades to a few thousand years. Higher groundwater ages are found in some areas of the Jordan valley.

In the Jebel el Arab basalt field, ^{14}C data indicate contemporary groundwater recharge in the Suweida area and the eastern parts of the Hauran plateau. Groundwater ages increase in direction of groundwater flow toward west and south to 1,000–8,000 years in the Yarmouk spring discharge area and 12,000 to 27,000 years in the discharge area of the Azraq plain.

On the Aleppo plateau, ^{3}H values of groundwater in the shallow mainly Paleogene aquifer and in the deeper Upper Cretaceous aquifer are generally below detection level. ^{14}C data indicate ages of 6,000–15,000 years in the shallow aquifer and of 22,000–35,000 years in the deeper Upper Cretaceous aquifer. In the Khabour area of the Syrian Jezire, groundwater ages of 9,000 years have been calculated for the Paleogene aquifer and of 6,700–12,000 years for the deeper Upper Cretaceous aquifer.

In the Palmyrean fold belt, groundwater ages tend to vary from 10,000 to >20,000 years. In the southern Syrian steppe and Hamad as well as in the Paleogene Rijam aquifer of the east Jordanian limestone plateau, the groundwater ages are generally >20,000 years, reaching up to 40,000 years. The high groundwater ages may have to be corrected, to some degree, as ^{14}C values possibly are lowered by secondary effects of redox processes and carbonate dissolution from the aquifer rocks.

Groundwater with ages of a few thousand years occurs within the plateau areas of the southeastern parts of the northern Arabian platform along larger wadi systems, where recent runoff infiltration dilutes the older groundwater.

On the Interior Shelf, detectable tritium contents are found mainly in groundwater of shallow wadi aquifers. ^{3}H of >4TU were analysed in a few wells tapping the Disi aquifer in southern Jordan (data of 1970s).

Groundwater ages of the sandstone aquifers of the Interior Shelf are generally 6,000–20,000 years at or near the outcrop areas, increasing to >30,000 years in the confined aquifer zones below the sedimentary cover of the Arabian Platform.

On the eastern Arabian platform, ^{3}H values of >4 TU were found in a few samples from outcrop areas of Paleogene aquifers in Saudi Arabia and Oman and in shallow fresh water in the gravel plains of the Oman mountain foreland. Local infiltration of flood water into karst openings appears not to be sufficient to create measurable ^{3}H levels in groundwater pumped from the several hundred metres thick Umm er Radhuma aquifer.

Groundwater ages in northeastern Saudi Arabia are 16,000–20,000 years in the Damam aquifer and 20,000–26,000 years in the Umm er Radhuma aquifer. In the multi-aquifer discharge zones of the Hasa and Qatif oases, groundwater ages range from 22,000 to >34,500 years. Groundwater age of the Damam aquifer of Kuwait is, on average, 22,000 years.

In the Nejd plateau on the southern margin of the Rub al Khali sub-basin, groundwater ages in Paleogene aquifers range from 6,000 to 30,000 years, reflecting the groundwater flow regime with limited present-day recharge in shallow aquifer zones in the upstream areas in the south and fossil groundwater in deeper aquifers and in the downstream areas.

The Hajar carbonate aquifer of the Musandam peninsula in the United Arab Emirates appears to contain mainly old brackish to saline groundwater with admixtures of recently recharged fresh to brackish groundwater. From low ^{14}C values of the mixed groundwater, ages of several tens of thousands of years may be assumed for the high salinity component of the water. The recent recharge is proved by ^{3}H values between 2.8 and 8.8 TU in various samples.

^{14}C data of groundwater in the coastal plains of the Arabian Peninsula indicate groundwater turnover periods of a few years to around 2,500 years. In the Batina coastal plain of eastern Oman, shallow aquifers along active wadi channels contain groundwater with ages of a few years. In the interfluvial areas, groundwater retention periods range from some tens to hundreds of years, and confined fresh groundwater at greater depth appears to originate from recharge at distant areas several thousand years ago.

1.6.3 Stable Isotopes of Oxygen and Hydrogen

1.6.3.1 Isotopic Composition of Precipitation

For precipitation, an empirical relationship exists between the monthly mean δ values of oxygen and hydrogen and temperature with a gradient of about −0.7‰ $\delta^{18}O$ and −5.6‰ $\delta^{2}H$ per °C (Dansgaard 1964). An empirical linear relationship between the $\delta^{18}O$ and $\delta^{2}H$ values of precipitation follows the equation:

$$\delta^{2}H = s \times \delta^{18}O + d.$$

A corresponding straight line with a slope s = 8 for continental precipitation is called meteoric water line (MWL). d is the "deuterium excess" defined as

$$d = \delta^2H - 8x\delta^{18}O.$$

d = +10‰ is observed in most continental precipitations and is representative for the "Global Meteoric Water Line" (GMWL). The rain in the eastern Mediterranean region has a d ≈ 22‰ with a corresponding meteoric water line

$$\delta^2H = 8 \times \delta^{18}O + 22,$$

The "Mediterranean Meteoric Water Line" (MMWL, Gat and Carmi 1970).

Long term observations of $\delta^{18}O$ and δ^2H data of precipitation are available from the two stations on the Arabian Plate: Bet Dagan situated on the Mediterranean Sea coast and Bahrain on the Gulf coast.

Long term average values at Bet Dagan are:

$\delta^{18}O$ −5.29‰, δ^2H −22.8‰, d +19.5‰ (Gat and Dansgaard 1972).

The weighted means of delta values of rainfall at Bahrain are

$$\delta^{18}O\ -1.27‰, \quad \delta^2H\ -7.2‰, \quad d\ +11‰,$$

if values of low intensity rainfall are disregarded (Yurtsever and Payne 1979).

Samples of precipitation in the northwestern mountain and rift zone of the northern Arabian platform scatter around the MMWL with $\delta^{18}O$ values between −4‰ and −8‰. The variation in mean $\delta^{18}O$ values at different stations can be attributed to an "altitude gradient": $\delta^{18}O$ values become more negative with increasing altitude of the observation stations. The altitude gradient in the northwestern mountain and rift zone is around −0.25‰ $\delta^{18}O$/100 m of topographic altitude.

d values in precipitation of more arid zones in the east of the northern Arabian platform (stations Aleppo, Palmyra, Azraq, d = +16.4‰ to +19.5‰) are similar to values in precipitation in the western highlands with somewhat lower $\delta^{18}O$ values of −3.65‰ to −7‰, reflecting a dominant influence of Mediterranean meteoric conditions with a slight effect of aridity.

For $\delta^{18}O$ and δ^2H values of precipitation at Bahrain and in the Oman mountains, meteoric water lines with slopes between 4.26 and 6.3 and d values between +8‰ and +11.5‰ have been stated by different authors. Considering only rainfall events with more than 20 mm in northern Oman, a meteoric water line

$$\delta^2H = 7.5x\ \delta^{18}O + 16.1‰$$

has been evaluated.

The $\delta^2H/\delta^{18}O$ relationship considering all samples from rainfall in the Oman mountains is non-linear with a bend to a slope of around 5 toward samples from lower intensity rainfall.

Stable isotope values of monsoon precipitation, which reaches the Arabian Plate in the Dhofar mountains on the Arabian Sea coast, vary only slightly from sea water with $\delta^{18}O$ values around 0 and positive $\delta^{2}H$ values.

1.6.3.2 Isotopic Composition of Groundwater

$\delta^{18}O$ of groundwater samples on the Arabian Plate range from −9‰ to −0.5‰, as far as the samples are not affected by secondary processes, such as evaporative enrichment and sea water intrusion, or by monsoon precipitation. The most depleted (most negative) $\delta^{18}O$ values are found in karst aquifers of the northwestern mountains and highlands (−6.5‰ to −9‰), the most enriched (less negative) values in samples from the Red Sea coastal plains (−0.5‰ to −3.5‰) and the Salala plain on the Arabian Sea coast (around 0‰). $\delta^{2}H$ values are in a general range of −50‰ to around 0‰.

That variation in $\delta^{18}O/\delta^{2}H$ values represents groundwater with recent recharge. In some areas, the isotopic composition of groundwater with prevailing Pleistocene recharge differs significantly from the composition of actually recharged groundwater.

1.6.3.3 Modern Recharge from Mediterranean Precipitation

Groundwaters with high d values around +20‰ prevail in the northwestern mountain and highland zone of the northern Arabian platform. d values between +19‰ and +24‰ in groundwater of Mesozoic karst aquifers in that zone are the highest d values found on the Arabian Plate. $\delta^{2}H/\delta^{18}O$ values scatter near the MMWL and reflect the dominant influence of Mediterranean precipitation on the groundwater recharge.

The $\delta^{18}O$ values of groundwater in the northwestern mountain and highland zone show a pronounced altitude effect of around −0.25‰ $\delta^{18}O$/100 m. The altitude effect is evident in the average $\delta^{18}O$ values of mountain ranges with different altitudes of main recharge areas; averages of $\delta^{18}O$ are around

- −7‰ in the Antilebanon–Hermon mountains
- −6.2‰ in the highlands of Jordan and Judea
- −5.9‰ in the Ansariye mountains

1.6.3.4 Holocene Recharge on the Eastern Plateau Areas of the Northern Arabian Platform

$\delta^{18}O$ values of groundwaters on the eastern plateau areas of the northern Arabian platform, which originate from Holocene recharge, are mainly in a range between −4.5‰ and −6‰.

Shallow groundwater in Tertiary aquifers of the Aleppo plateau have generally $\delta^{18}O$ values of −4.5‰ to −5.5‰ and d values between +6‰ and +9‰. Many samples show an evaporative isotope enrichment resulting in less negative $\delta^{18}O$ values of up to −3.8‰ and lower d values with a minimum of +1.6‰.

In the Ad Daw plain and the Badiye and Hamad, samples of groundwater originating from Holocene recharge are mainly restricted to runoff infiltration along larger wadi systems. $\delta^{18}O$ values are in a range of −4.7‰ to −6‰; d values of samples with detectable tritium levels vary in a wide range between +8.8‰ and +17.3‰. The scatter of the majority of samples above the GMWL suggests an origin of the groundwater from rain storms approaching from the Mediterranean Sea with varying impact of arid continental conditions.

In some areas in the east of the northern Arabian platform, isotope data indicate recharge from Mediterranean type winter rains:

- d values of spring water from the karstic Eocene Ras el Ain aquifer in the Syrian Jezire are around −20‰, $\delta^{18}O$ values −8‰. The spring water with an estimated age of 9,000 years probably originates from Mediterranean type Holocene recharge on the southern slopes of the Taurus mountains.
- Spring waters of the basalt aquifer in the high mountain zone of Jebel el Arab have d values above +20‰ and $\delta^{18}O$ values of, on average, −6.9‰. These values correspond to Mediterranean winter rains on the eastern slope of the Antilebanon mountains.

The $\delta^{18}O$ values in the Jebel el Arab basalt aquifer complex in southwestern Syria and northern Jordan change from values of <-6.5‰ on the high mountain area of Jebel el Arab at >1,000 m asl to less negative values in the adjoining plateaus situated at lower altitudes(−4.5‰ and −6‰). The general distribution of $\delta^2H/\delta^{18}O$ values appears to reflect prevailing recharge from Mediterranean rainfall with varying evaporation effects in the plateau areas. Groundwater in the main spring discharge zones appears to contain components of water with depleted stable isotope values from recharge on higher altitudes and of isotopically more enriched groundwater from the plateau areas.

Particularly pronounced enrichment of stable isotopes is found in groundwater from irrigation areas at Wadi Dhuleil and the Azraq plain with high variations of $\delta^2H/\delta^{18}O$ and d values.

1.6.3.5 Holocene Recharge on the Arabian Peninsula

$\delta^{18}O$ values of groundwater with significant recent recharge on the Arabian Peninsula are generally in a range of −3.5‰ to 0‰, and are, on average, less negative than $\delta^{18}O$ values in aquifers of the semi-arid to arid plateaus of the northern Arabian platform, where values between −6‰ and −4.5‰ prevail.

Values of deuterium excess of modern groundwater on the Arabian Peninsula tend to scatter around the GMWL with various local to sub-regional exceptions.

Ranges of $\delta^{18}O$ values between −2.0‰ and −3.3‰ are found in Holocene groundwaters from various areas of the Arabian Peninsula:

- In wadi aquifers of the Interior Shelf where ^{3}H values indicate recent recharge ($\delta^{18}O$ −2.0‰ to −3.3‰)
- In Wadi Rima on the Arabian Shield in the main recharge zone along the centre of the wadi course (−2.3‰ to −2.7‰)
- In fresh water lenses in northern Kuwait (−2.9‰ to −3.5‰)
- On the eastern coastal plains of the United Arab Emirates (−1.96‰ to −2.98‰)

Present-day recharge in these areas may result from depleted moisture sources from the Gulf or from Mediterranean storms with some evaporative enrichment of the infiltrating water.

Mean delta values of groundwater in shallow fresh to brackish groundwater in Qatar are $\delta^{18}O$ −2.32‰, $\delta^{2}H$ −8.6‰, d +10‰. These values are very close to the weighted mean delta values of rainfall at Bahrain, reflecting recharge from local moisture of the Gulf.

On the western escarpment of the Arabian Shield toward the Red Sea between Makka and Jedda, $\delta^{18}O$ values in wadi aquifers are mainly between −1.1‰ and −1.8‰, on average slightly lower than on the eastward directed catchments.

Differences in ranges of $\delta^{2}H/\delta^{18}O$ values in different areas of the Arabian Peninsula may be caused mainly by

- Evaporative impacts
- Different altitudes of recharge areas
- Varying ratios of winter precipitation and less depleted summer rains

In the gravel plain aquifers west of the Oman mountains, $\delta^{2}H/\delta^{18}O$ values appear to be affected by evaporative enrichment in many areas. The range of $\delta^{2}H$ values of shallow groundwater between −2.5‰ and −0.5‰ in the northern parts of the gravel plain changes to −1.8‰ to 0.3‰ in the Al Ain area further south and to around −2‰ in the Liwa area on the sand covered margin of the Rub al Khali desert.

1.6.3.6 Pleistocene Recharge

^{14}C data show, that groundwater in the semi-arid to arid zones of the Arabian Plate originate, to a large extent, from Pleistocene recharge. In many areas, groundwater from Holocene and from Pleistocene recharge have different isotope signatures.

In the aquifers of the dry eastern parts of the northern Arabian platform, $\delta^{18}O$ values of Pleistocene groundwaters vary between −6‰ and −9‰, d values between +10‰ and +16‰. On the eastern Arabian platform, $\delta^{18}O$ values of Pleistocene recharge range from −1.4‰ to −5.5‰, d values from +1‰ to 0‰. On the Interior Shelf, fossil groundwaters are found in the Saq–Disi aquifer of the Tabuk–Disi segment with $\delta^{18}O$ values between −7‰ and −5.5‰ and d values above +10‰. In the Tuwayq area of central Saudi Arabia, $\delta^{18}O$ values are −7‰ to −3.8‰, d values <+10‰.

From that distribution of $\delta^2H/\delta^{18}O$ values, a general differentiation of meteorologic conditions on the northern Arabian platform and on the Arabian Peninsula may be assumed with isotopically more depleted precipitation in the northwest and isotopically more enriched precipitation in the southeast. The Interior Shelf may have received precipitation with an intermediate signature of stable isotopes.

On the presently semi-arid to arid eastern parts of the northern Arabian platform (Aleppo plateau, Palmyrean belt, Badiye, Hamad, east Jordanian limestone plateau), $\delta^{18}O$ values vary over a rather wide range from −9‰ and −6.8‰, d values scatter mainly around +15‰, but many deviations to lower d values to about +5‰ are found.

$\delta^{18}O$ values of groundwaters with Holocene recharge in these areas appear to be less negative than the values of Pleistocene groundwaters; d values of Pleistocene and Holocene recharge do not differ significantly.

The range of $\delta^2H/\delta^{18}O$ values of fossil groundwater in the eastern parts of the northern Arabian platform is similar to the range of values in the Qalamoun area. The Qalamoun high plateau is situated on the eastern slope of the Antilebanon mountains on the transition between sub-humid and semi-arid climate conditions and comprises a complex of Mesozoic–Tertiary–Quaternary aquifers. $\delta^2H/\delta^{18}O$ values in the Qalamoun area vary over a rather wide range of $\delta^{18}O$ −9.4‰ to −6.8‰ and d +11‰ to +19‰ without obvious differences in values of individual aquifers or of Pleistocene and Holocene recharge.

The similarity of $\delta^2H/\delta^{18}O$ values in fossil groundwaters of the eastern parts of the northern Arabian platform with values of the Qalamoun area may indicate that, during the Pleistocene, a relatively humid climate, comparable to the present conditions on the Qalamoun high plateau, extended far into the presently semi-arid to arid areas.

In the Syrian Jezire, fossil groundwaters from the deeper confined Upper Cretaceous aquifer are characterized by $\delta^{18}O$ values of −8.2‰ to −7.2‰ and relatively high d values of +15.5‰ to +20.3‰, reflecting possibly a recharge at higher altitudes of the southern slope of the Taurus mountains.

On the eastern Arabian platform, stable isotope data of oxygen and hydrogen show a tendency to less depleted values toward the Gulf coast in the east. $\delta^{18}O$ values between −5.5‰ and −4.5‰ prevail in groundwaters from Tertiary aquifers in the Hasa oasis area, about 80 km west of the Gulf coast, and at Haradh, about 200 km west of the coast. More negative $\delta^{18}O$ values of <−5.5‰ are reported from the Ghawar anticlinal structure, where upward leakage of deeper groundwater from the Umm er Radhuma aquifer is assumed, which has been recharged in more western parts of the catchment. At Qatif near the coast, $\delta^{18}O$ values are less negative in a range of −4.4‰ to −3.4‰. The groundwaters from the Hasa, Haradh and Qatif areas have ages of 16,000–26,000 years.

On the Kuwait plain, confined groundwater from the Damam aquifer, which receives subsurface inflow from catchments situated further west in Saudi Arabia and Iraq, $\delta^{18}O$ values vary from −4.5‰ to −3‰. Groundwater in the lower Kuwait group aquifer, replenished mainly from local rainfall, has $\delta^{18}O$ values of −2.2‰ to −1.5‰. Groundwater ages in both aquifers are around 22,000 years, d values scatter around −3‰. $\delta^2H/\delta^{18}O$ values of recent recharge in Kuwait are, with d values around +15‰, distinctly different from values of the fossil groundwater.

Subsurface inflow of fossil groundwater through the Damam aquifer into Bahrain and Qatar is characterized by $\delta^{18}O$ values of, on average, −4.5‰ and d values of +4‰ to −2‰, corresponding approximately to values on the adjoining coastal area in Saudi Arabia. Average $\delta^{18}O$ values of recent recharge in Qatar of −2‰ are less negative than in the inflowing fossil groundwater.

In aquifers with fossil groundwaters on the upper margin of the Gulf–Rub al Khali basin (sandstone aquifers of the Tuwayq segments of the Interior Shelf and Umm er Radhuma aquifer of the Nejd plateau in southern Oman), $\delta^{18}O$ values tend to be, on average, more negative than on the downstream reaches of the basin. The $\delta^{18}O$ values in these more marginal areas with catchments at higher altitudes concentrate in clusters between −6‰ and −4‰, and d values scatter near the GMWL at around +8‰.

In the Tuwayq segments of the Interior Shelf, $\delta^2H/\delta^{18}O$ values of fossil groundwater differ significantly from values of recent recharge.

In various areas of the Arabian Plate, $\delta^{18}O$ and δ^2H values of Pleistocene groundwaters are more negative than values in groundwaters with Holocene recharge, indicating cooler climate conditions during the Pleistocene.

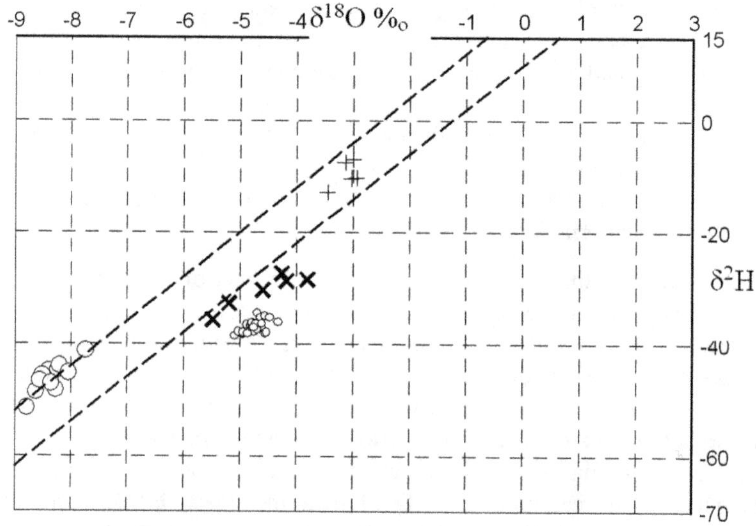

Fig. 1.6 $\delta^{18}O/\delta^2H$ diagram: Examples of Holocene and Pleistocene groundwaters from different areas of the Arabian Plate. O Antilebanon mountains, Holocene groundwater; + northern Kuwait, Holocene groundwater; x Aleppo plateau, Pleistocene groundwater; o Hasa oases, Pleistocene groundwater. Data from Wagner and Geyh (1999), Yurtsever (1992), Wagner (1997), Otkun (1974). *Upper line*: MMWL, $\delta^2H = 8\times\delta^{18}O + 22$; *lower line*: GWML, $\delta^2H = 8\times\delta^{18}O + 10$

The northern Arabian platform and the Arabian Peninsula had, during the Pleistocene, possibly different climatic conditions:

- A semi-arid climate with cooler and more humid conditions than the Holocene climate appears to have prevailed over the eastern parts of the northern Arabian platform with a strong influence of Mediterranean type climate similar to conditions, which are found at present on the leeward side of the Antilebanon mountains.
- On the Arabian Peninsula, $\delta^{18}O$ values of Pleistocene groundwaters are also generally more negative than values of Holocene groundwaters but on average less negative than $\delta^{18}O$ values of Pleistocene groundwaters from the northern Arabian platform. The distribution of $\delta^{18}O$ values probably reflects a less arid climate than at present with a tendency to somewhat more humid conditions toward west, which may be related to an altitude effect and an increasing influence of Mediterranean rainfall events.

Figure 1.6 shows, in a standard $\delta^2H/\delta^{18}O$ plot, examples of the scatter of stable isotope values of recent groundwaters (Antilebanon mountains and northern Kuwait) and of fossil groundwater (Aleppo plateau and Hasa oases on the eastern Arabian platform).

References. Almomani (1996), Gat and Carmi (1970), Gat and Dansgaard (1972), Kattan (1996b), Macumber et al. (1997), Robinson and Al Ruwaih (1985), Wagner and Geyh (1999), Yurtsever (1996), Yurtsever and Gat (1981), Yurtsever and Payne (1979).

Chapter 2
Northwestern Mountain and Rift Zone of the Northern Arabian Platform

2.1 Geographic and Hydrologic Features

The northwestern rim of the Arabian Plate is distinguished from most other parts of the plate by its morphology of elongated mountain chains and deep grabens and by its relatively humid climate, which is dominated by the meteorologic cycles above the adjoining eastern Mediterranean Sea. The mountainous sub-region extending parallel to the Mediterranean Sea coast constitutes, together with vast plateau areas in the east, the northern Arabian platform which occupies the northwestern sector of the Arabian Shelf.

In the west and north, the northern Arabian platform is delimited along boundaries of the Arabian Plate: the Mediterranean Sea coast and the zone of collision of the plate with the Alpidic chains of the Taurus mountains.

2.1.1 Morphology, Climate, Vegetation and Water Supply

2.1.1.1 Morphology

The morphology of the northern Arabian platform is dominated in its western part by chains of mountains and highlands with adjoining intermountain or coastal plains and, in the east, by vast plateau like landscapes. The morphology of the western mountainous part has been formed predominantly by Neogene–Quaternary tectonic events: uplift of the northwestern rim of the Arabian Plate and dissection by graben structures from rift movements. The uplift appears to be related to the northward movement and counter-clockwise rotation of the Arabian Plate. The zone along the northwestern rim of the Arabian Plate, which is characterized by a very pronounced morphological relief, is denominated in the following descriptions as "northwestern mountain and rift zone".

Two main mountain belts run roughly parallel to the Mediterranean Sea in south–north to SSW–NNE direction over the northwestern mountain and rift zone:

W. Wagner, *Groundwater in the Arab Middle East*,
DOI 10.1007/978-3-642-19351-4_2, © Springer-Verlag Berlin Heidelberg 2011

- The coastal mountain chains of the Judean highlands, the Lebanon mountains and the Ansariye mountains
- The eastern mountain belt comprising the highlands of Jordan, the Hermon–Antilebanon mountains and Jebel ez Zaouiye

The two mountain belts are separated by a zone of approximately south–north oriented rift valleys: Wadi Araba as a continuation of the Red Sea graben, the Dead Sea, Jordan valley and Lake Tiberias depressions, the Bekaa valley and El Ghab, ending, in the north, in the Amiq depression and Kara Su valley on Turkish territory (Fig. 2.1).

The rift valleys have their lowest altitude at the Dead Sea shore with 200–460 m below sea level. The valley floor rises to 900 m asl in the Bekaa valley and descends

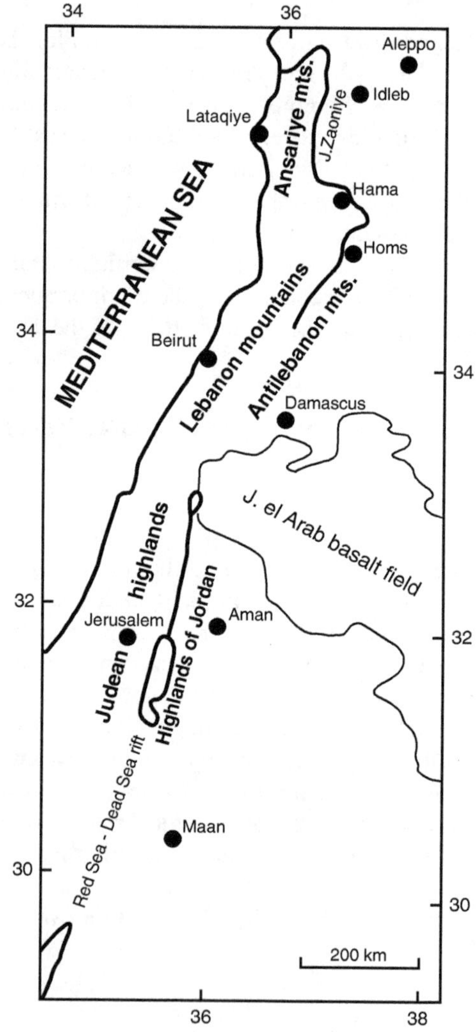

Fig. 2.1 Main orographic–geographic units of the northwestern mountain and rift zone

again to 170 m asl in El Ghab and 80 m in the Amiq depression. The Bekaa plain connects, as an around 10 km wide intermountain valley between Lebanon and Antilebanon mountains, the graben system of the Jordan valley in the south with the Masiaf–El Ghab graben structures in Syria in the north.

The coastal mountain belt is divided by relatively narrow morphologic depressions into four main mountain massifs. The belt comprises, from south to north:

- The Judean highlands, or mountains of Judea and Samaria, which reach their highest elevations in the area around Jerusalem with 800–1,000 m asl
- A zone of valleys between the Jordan river and the sea coast (<300 m asl) and the highlands of Galilea and of southern Lebanon (around 300–500 m asl)
- The Lebanon mountains with a maximum elevation of >3,000 m asl
- The Akar plain including the valley of southern Nahr el Kebir
- The Ansariye mountains rising to >1,500 m asl
- The valley of northern Nahr el Kebir
- The Basit mountains with the 1,700 m high Jebel el Akra

The eastern mountain belt comprises the highlands of Jordan and the Hermon–Antilebanon mountain chain. The mountain belt is adjoined or interrupted by a few superimposed young morphological depressions: the Yarmouk valley and the Damascus, Homs and Acharne plains. In the north, the belt ends in the moderately high Jebel ez Zaouiye and the Idleb plateau.

The *highlands of Jordan and the Judean highlands* rise from the Dead Sea–Jordan valley along steep escarpments with altitude differences of up to more than 1,000 m.

The highlands of Jordan accompany the Dead Sea–Jordan valley to the east in a stretch of 300 km, ranging from the deeply incised Yarmouk valley in the north to the Interior Shelf in southern Jordan. Viewed from the east, the highlands appear rather as a series of hills, to the west the highlands drop in a steep slope over more than 1,000 m down to the Jordan–Dead Sea valley. Topographic elevations of the highland peaks rise from 1,200 m asl in the north to 1,700 m asl at Jebel Mubarak in the south. The highlands are crossed by a number of east–west directed wadis which, as the Yarmouk river on the northern boundary, run in steep incisions and gorges down to the rift valley: wadis Zerqa, Walla, Mujib, al Hasa, which drain different hydrologic tributary basins of the Jordan–Dead Sea valley.

The Judean highlands appear on their eastern face as a reflected image of the highlands of Jordan: a mountain area rising in cliffs and steep slopes from the Jordan–Dead Sea valley at less than 200 m below sea level to altitudes of 800–1,000 m asl near Jerusalem. Toward west, the land surface descends in moderate slopes to the Mediterranean coastal plain. The morphology of the crest zone of the highlands appears as a flat dome, on which partly karstified Upper Cretaceous strata are exposed.

In the north, the highlands are separated from the Lebanon mountain chain by plain and hill areas (hills of Galilea and of southern Lebanon).

The geologic–morphologic structures of the Judean highlands and the highlands of Jordan continue toward north into the *Lebanon and Antilebanon* mountain ranges. Morphologically, the Lebanon and Antilebanon chains are separated from the highlands of Judea and Jordan by a zone of valleys between the Jordan river and the Mediterranean Sea coast and by the Yarmouk valley.

The Lebanon mountains culminate in Qornet as Saouda at 3,088 m asl. The southwest of the Lebanon mountains comprises an area with less pronounced morphologic relief with mountain peaks between around 500 and 983 m asl and a broad zone of hills, which extend toward the Mediterranean Sea in the west. The northwestern boundary of the Lebanon mountains is adjoined by a zone of hill and plain areas, delimited by the Mediterranean Sea coast in the west, the Ansariye mountains in the north, and large faults and flexures in the east, southeast and south. The eastern part of the hill zone is occupied by Neogene basalts of the Tell Kalakh volcanic massif, which extends over the Lebanese–Syrian border area with altitudes between 50 and about 600 m asl. The western part of the hill and plain zone, adjoining the Mediterranean Sea coast, comprises the Akar plain in the north, hills, fluvial and coastal plains around Tripoli, the Koura plateau, and the narrow Chekka coastal plain in the south.

The Antilebanon mountains extend over a length of 165 km from the Houle depression in the south to the Homs plain in the north. Peak altitudes reach 2,814 m asl in the Mount Hermon massif and 2,629 m asl at Talaat Mousa in the northwest of the Antilebanon mountains. The northern prolongation of the Dead Sea–Jordan rift graben structure is formed by the Bekaa valley, which separates the Lebanon and Antilebanon mountain chains. The altitudes of the Bekaa valley floor range from 500 to 1,000 m asl.

The uplift and rift structure of Lebanon mountains–Bekaa valley–Antilebanon mountains continues toward north into the *mountain and rift zone of northwestern Syria* where, however, peak altitudes are more moderate. The mountain and rift zone in northwestern Syria comprises the Ansariye mountains, the Ghab valley and Jebel ez Zaouiye. The Ansariye mountains, the northern section of the coastal mountain belt, extend over 120 km parallel to the coast. The highest peaks of the Ansariye mountains reach altitudes of 1,385 and 1,552 m asl. They are separated from the Lebanon mountains by the valley of Nahr el Kebir el Janoubi and the adjoining coastal plain of Akar. Coastal plains of a few kilometres width extend along the mountain foot in the north and south of the Ansariye mountains.

Topographic elevations range from 120 m asl in the Ghab valley to 500 m on the Homs plain and 870 m and 1,051 m at the peaks of Jebel ez Zaouiye and the Shin plateau, respectively.

2.1.1.2 Climate

The climate of the northwestern mountain and rift zone is dominated by the weather conditions over the eastern Mediterranean Sea. During summer, the position of the Inter-Tropical Convergence Zone over the Arabian and Sahara deserts creates stable hot and dry climate conditions (Sect. 1.4.3.1). During winter, the convergence zone with dry climate and high barometric pressure moves to the south, the Mediterranean area comes under the influence of moist air inflows from the Atlantic, and rain storms enter into the Mediterranean Sea basin from the west to northwest.

Precipitation is concentrated to the winter season between November and April. Even in the relatively humid western parts of the sub-region, a generally continuous dry season extends from May into October.

The morphology of the highland and mountain zone has a dominant influence on the climatic conditions. High precipitation occurs during the rainy winter season on the western slopes and the peak areas of the Lebanon mountains and of Mount Hermon with mean annual precipitation rates of >1,000 mm.

The Lebanon mountains are characterized by relatively high precipitation with an annual average of 1,170 mm and more than 1,500 mm in the high mountain area and bear a thick snow cover during the winter season.

Mean annual precipitation ranges

- From 600 to 1,000 mm on the eastern Lebanon mountain slopes
- From 500 to 800 mm on the Ansariye mountains
- From 700 to 1,000 mm on the coastal plains in Syria and Lebanon, decreasing to around 300 mm toward south in the Gaza Strip
- From 500 to 800 mm in the Judean highlands and the western slopes of the Jordanian highlands and of the Antilebanon–Jebel ez Zaouiye chains, reaching >1,000 mm on the high ranges of Mount Hermon

Rainfall is very low in some parts of the rift valleys, in particular in the Dead Sea valley and at the northern edge of the Bekaa. In most of the Jordan valley–Dead Sea–Wadi Araba depression, rainfall is low with an average of around 50 mm/a. Mean annual precipitation reaches 400 mm around Lake Tiberias on the northern end of the Jordan valley; in the Wadi Araba catchment mean annual precipitation ranges from 250 mm in the highlands in the northeast of the valley to less than 50 mm at Aqaba.

On the leeward eastern slope of the highlands and mountain zone, precipitation decreases rapidly to semi-arid or even arid conditions. Mean precipitation on the Damascus plain does not exceed 250 mm/a and the eastern city boundaries of Greater Damascus and Aman areas reach into the steppe and desert environment of Al Badiye.

2.1.1.3 Vegetation, Water Supply and Historic Developments

The vegetation of the coastal mountain belt and the coastal plains is mediterranean with evergreen shrubs. Forests with pines, oaks and cedars occur on the higher mountain areas. Rainfall sustains rain fed agriculture in many parts of the sub-region, prevailing crops being wheat, barley, grapes and fruit trees, in particular olive trees. Plantations of olive and fruit trees extend over wide areas of the western cultivated zones, poplars, eucalyptus and reed border stream banks. Supplementary winter irrigation and irrigation during summer is used prevailingly for cultivation of wheat, bareley, cotton and vegetables and, in some areas, of sugar cane and tobacco. Along the coast, also bananas and date palms are grown.

Irrigated areas are found in parts of the rift valley area – Jordan valley, Bekaa plain, El Ghab – and in the areas around Homs and Hama, Al Ghouta around Damascus, and the Zebedani valley in the Antilebanon mountains.

The morphologic relief created by the Neogene to Recent tectonic movements, the sub-humid Mediterranean climate with relatively reliable winter rains as well as the extensive outcrops of karstified carbonate formations make the northwestern mountain and rift zone to the area with the most abundant renewable groundwater resources of the Arabian Plate. The easy perennial access to groundwater in springs and shallow wells is certainly one of the reasons that urban and civilization centres have flourished since several thousand years in this area, which was situated between the Mesopotamian and Egyptian high cultures on the Euphrates–Tigris and Nile rivers.

Availability of water from winter rains and from rivers and springs with perennial flow favoured the agricultural and urban developments throughout the past millenia. Archeological and historical sources tell us that

- Settlements at Jericho date back to around 5000 BC
- The towns of Aleppo and Damascus exist continuously since 5,000 years
- Phenician towns flourished along the Mediterranean Sea coast during the second and first millennium BC
- The Israelites met, during their wanderings in the highlands of Jordan and Syria around 1200 BC, with established local kingdoms of Kanaan, Edom, Moab, Amon and Basan
- Around 1000 BC King David established his capital in the already existing town Jerusalem

The civilizations in the northwestern Arabian highlands were mainly local powers between the Mesopotamian and Egyptian empires and later mainly provincial units of the Hellenistic, Roman, Abassid and Ottoman empires. The Umayad Caliphate ruled the Islamic world from Damascus for 88 years.

The naturally available water resources in the northwestern mountain and rift zone were apparently, in general, adequate for the domestic and agricultural water demand of the population during the past millenia, although occasional drought years may certainly have caused temporary supply problems. The relationships between water sources, climatic cycles and development of civilisations is presented in the fascinating book "Water shall flow from the rock" (Issar 1990).

The water supply situation has changed dramatically during the last decades with the increase of population, urbanization and expansion of intensively irrigated agriculture. Recent water supply problems have been caused by the tremendous growth of the major cities. In Jordan, about 90% of the total present population of four million live in cities, towns and villages of the highlands. In spite of the relatively large water resources of the area, the high population density and a high growth rate create imminent problems of water supply. Damascus and the surrounding oasis Al Ghouta have received over millenia adequate water supply from the Barada and Aouaj rivers, which are fed by perennial karst springs. With a population exceeding four million, additional sources for the urban water supply have to be found.

Application of presently available techniques of well drilling and water pumping have lead to a tremendous increase of groundwater exploitation for irrigated agriculture in parts of the plain areas of the northwestern mountain and rift zone.

2.1.2 Main Hydrologic Basins and River Flow

Surface drainage on the northwestern mountain and rift zone is directed to the Mediterranean Sea and to the Dead Sea. The Aouaj and Barada rivers drain from the eastern slopes of the Antilebanon mountains into the closed basin of the Damascus plain (Figs. 2.2 and 2.3).

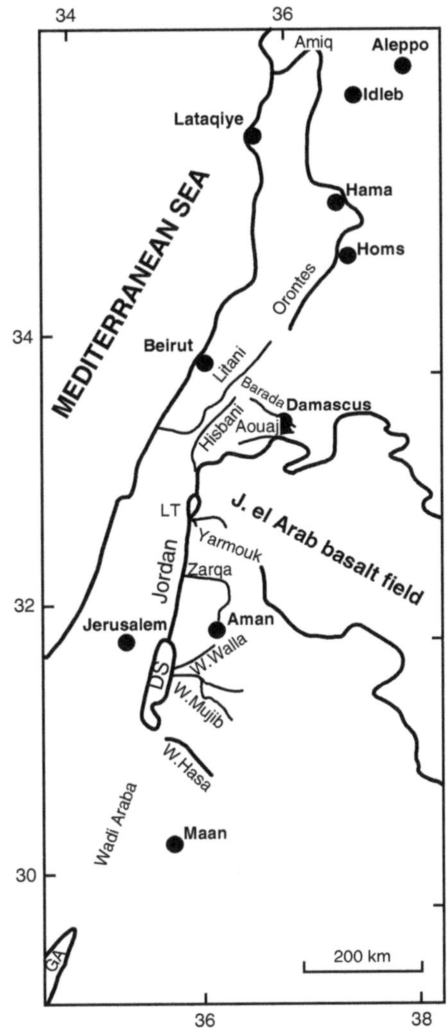

Fig. 2.2 Main rivers and wadis in the northwestern mountain and rift zone. GA Gulf of Aqaba, DS Dead Sea, LT Lake Tiberias

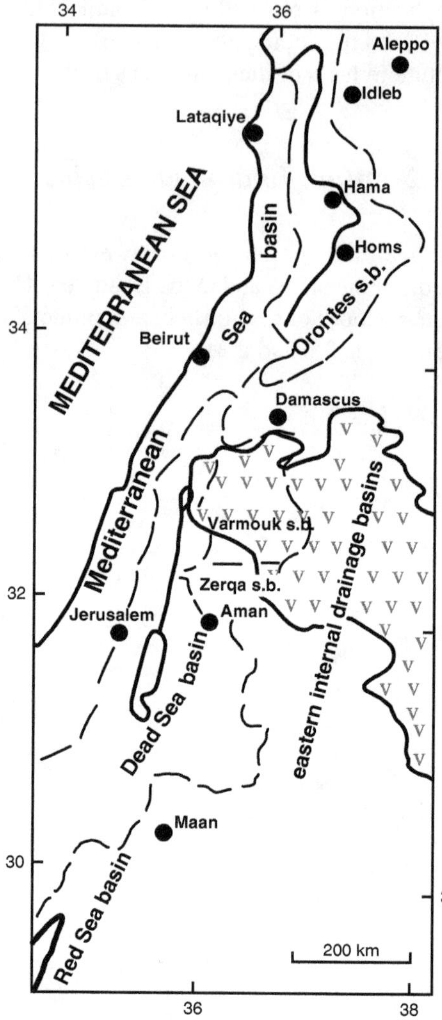

Fig. 2.3 Main hydrologic basins of the northwestern mountain and rift zone. The *Mediterranean* Sea basin with the *Orontes* sub-basin (the Litani sub-basin is not delineated) and the *Dead Sea* basin including the *Zerqa* and *Yarmouk* sub-basins are adjoined in the east by the zone of *interior drainage* of the eastern part of the northern Arabian platform and the *Red Sea* basin within the Interior Shelf

2.1.2.1 Mediterranean Sea basin

The Mediterranean Sea basin includes the sub-basins of the Orontes and Litani rivers and a dense network of coastal rivers and streams on the western slopes of the Lebanon and Ansariye mountains.

The *Orontes river* (Nahr el Aasi) begins in the northern Bekaa plain in Lebanon at Al Labwe spring near Baalbek at an elevation of 400 m asl. With a length of 487 km, the Orontes is the longest perennial river which originates on the Arabian

Plate. Streamflow of the Orontes is fed, to a large extent, by spring discharge on the foot of the mountain ranges adjoining the Orontes valley. A total mean discharge of 23–38 m^3/s from 54 springs is reported (Khouri 1991).

From the Bekaa plain, the Orontes river flows into the Homs plain at around 400 m asl and reaches, after a wide eastward curve between Homs and Hama, the Ghab valley at 120 m asl.

On Syrian territory, the Orontes river is regulated through several reservoirs, such as the reservoirs at Qatine near Homs, and at Rastan and Maharde; the largest reservoir at Rastan between Homs and Hama has a capacity of $228 \times 10^6\ m^3$. River water is used intensively for irrigation, water supply and industry. Irrigation canals divert water from the Orontes river to extensive agricultural areas in the Homs–Hama area, the Acharne plain downstream of Hama and the Ghab valley.

The about 10 km wide Ghab valley, which extends for 50 km in south–north direction between the Ansariye mountains and Jebel az Zaouiye, has been converted through drainage measures and flood regulation of the Orontes river into a large irrigation area.

Mean stream flow of the Orontes increases from 16.1 m^3/s in the Bekaa valley to 75.5 m^3/s near the river mouth at the Mediterranena Sea.

A major tributary of the Orontes is the Aafrin river, joining the Orontes near the northern boundary of Syria.

The 170 km long *Litani river* runs from its headwaters near Baalbek in the Bekaa plain to the Mediterranean Sea coast in southern Lebanon. Flow is regulated by an irrigation and hydroelectric power dam in the Bekaa plain.

The western slopes of the *Lebanon mountains* are dissected by 11 perennial streams and additionally by about 18 wadi systems with seasonal flow. The streams and wadis are generally characterized by steep slopes and relatively short length of the stream courses. The length of the perennial streams ranges from 23 to 50 km, the size of the catchment areas from 90 to 390 km^2, mean annual stream flow from 0.6 to 15 m^3/s. Most of the stream flow during summer originates from groundwater discharge from Mesozoic carbonate aquifers.

The zone of hills and plains between the northwestern margin of the Lebanon mountains and the southern tip of the Ansariye mountains is crossed by a few larger streams: Nahr el Abrache with its headwaters in the Ansariye mountains, Nahr el Kebir el Janoubi, the catchment of which extends over the southern part of the Ansariye mountains and Jebel Akroum in the northern Lebanon mountains, Nahr Oustouene, Nahr Arka, Nahr Abou Mousa–el Bared and Nahr Abou Ali, which originate in the northern Lebanon mountains.

The rivers running toward the Mediterranean Sea on the western slopes of the *Ansariye mountains* are generally characterized by short stream length, steep slopes and deep canyons. Flow on most of these coastal rivers is restricted to the rainy season. Streamflow in Nahr el Kebir ash Shimali and Nahr el Kebir al Janoubi, which have relatively extensive catchments in the north and south of the Anasriye mountains, continues throughout the year.

2.1.2.2 Jordan–Dead Sea basin

The Dead Sea received, prior to installation of intensive irrigation schemes in the Jordan–Dead Sea valley, fresh water from the Jordan river and its tributary stream systems. The headwaters of the Jordan river are formed by the Dan, Banias and Hasbani rivers on the western slopes of Mount Hermon and the Golan heights. Downstream of the confluence of the headwater tributaries, situated at 90 m asl, the Jordan river crosses Lake Houle and Lake Tiberias. The lake and marsh area of Houle has, to a large extent, been drained through a canal system completed in 1958. South of the Houle area, the Jordan river reaches, after crossing a 12 km long gorge, an alluvial plain around Lake Tiberias, which is situated at 212 m below sea level and occupies about 165 km^2.

Quantity and quality of discharge from Lake Tiberias, which receives water from a number of saline springs, is controlled artificially by pumping and diversion of water into canal systems. The water released presently into the Jordan river from Lake Tiberias is brackish.

The Jordan river runs, after leaving Lake Tiberias, in meandering course through the lowlands of El Ghor to the northern end of the Dead Sea 105 km further south at 392 m below sea level.

The from a few kilometres to 20 km wide Jordan valley floor comprises El Ghor, gently sloping fans and terraces east and west of the Jordan river, and El Zor, the narrow flat stream bed of the meandering Jordan river.

Major tributaries of the Jordan river from the east are the Yarmouk river, which rises in the Syrian–Jordanian basalt plateau, and Zerqa river with its headwaters in the Aman area.

The Yarmouk river drains a large catchment area between Mount Hermon and Jebel el Arab in Syria and Jordan. The mean discharge of the Yarmouk into the Jordan river decreased from previously around 500×10^6 m^3/a to presently around 360×10^6 m^3/a, mainly because of increasing groundwater extraction in the catchment area.

The Zerqa river originates from the confluence at Soukhne of Wadi Dhuleil from the east and Sail ez Zerqa from south. The average annual discharge of the Zerqa river into the Jordan river was around 65×10^6 m^3/a and is now regulated through the King Talal dam.

The highlands of Jordan are crossed in general east–west direction by Wadi Mujib and Wadi Walla, which form deep gorges in part of their courses and reach the eastern boundary of the Dead Sea–Jordan valley.

The eastern slopes of the northern part of the Judean highlands are dissected by wadis, incised along tectonic structures, with courses toward the Jordan–Dead Sea valley. The wadis have relatively small catchments and generally steep slopes. In the Nablus–Jenin area, various shallow depressions, related to the Neogene tectonics, constitute closed basins with relatively small dimensions of up to 20 km^2.

The total discharge of the Jordan river into the Dead Sea declined from previous $1{,}370 \times 10^6$ to $250–300 \times 10^6$ m^3/a at present, after diversion of stream flow into large canal systems. The water level of the 85 km long and about 16 km wide Dead Sea is situated, at present, at around 400 m below the Mediteranean Sea level, the deepest point at the bottom of the Dead Sea is 730 m below sea level.

In the eleventh to sixteenth centuries, the water level of the Dead Sea was situated at 375 m below Mediterranean Sea level, and, at the beginning of the twentieth century, at 390 m below sea level.

The depth of the water reaches in the northern part of the Dead Sea more than 300 m. In the south, the Lisan peninsula narrows the width of the Dead Sea to 4 km, separating a southern sector of the sea with merely 5–10 m water depth from the northern Dead Sea basin. The water in the Dead Sea is a brine with a salt content of 340 g/kg.

Diversion of water from Lake Tiberias and the Jordan river has resulted in a drop of the sea water level of 0.5–0.8 m/a. The decrease of the sea water level is expected to continue until a new equilibrium in the water level is reached in about 400 years at 100–150 m below the present level.

Since the Neogene, the rift valley was occupied by a series of lakes with varying degrees of water salinity. The Lisan lake, the ultimate predecessor of the Dead Sea, attained a level of 180 m below sea level around 15,000 years B.P. The salinity of the Lisan lake was probably lower than the salinity of the present the Dead Sea.

South of the Dead Sea, the rift graben extends through the 174 km long Wadi Araba until the Gulf of Aqaba, the northern end of the Red Sea. The valley floor of the 8–25 km wide Wadi Araba rises from 400 m below sea level at the southern shore of the Dead Sea to about 250 m asl at the watershed about 100 km further south and then descends gradually to the Red Sea level.

The highland peaks east of Wadi Araba reach altitudes of around 1,500 m above the valley floor (1,200 m asl) near the south end of the Dead Sea and 1,592 m asl in Jebel Baqir in the south. Elevations of the highlands west of Wadi Araba are lower with altitudes of up to 900 m asl.

Several wadis with relatively small catchment areas between 160 and 500 km^2 enter Wadi Araba from the adjoining highlands in the east, such as Wadi al Hasa, Wadi Feifa, Wadi Fifan, Wadi Musa (Figs. 2.4 and 2.5).

References. Salameh (1996), Sunna (1995), Wolfart (1966), Yechieli et al. (2001).

2.2 Geology

2.2.1 Stratigraphic Sequence in the Mountain Areas

The sequence of sedimentary rocks of the northwestern mountain and rift zone comprises the following major stratigraphic complexes:

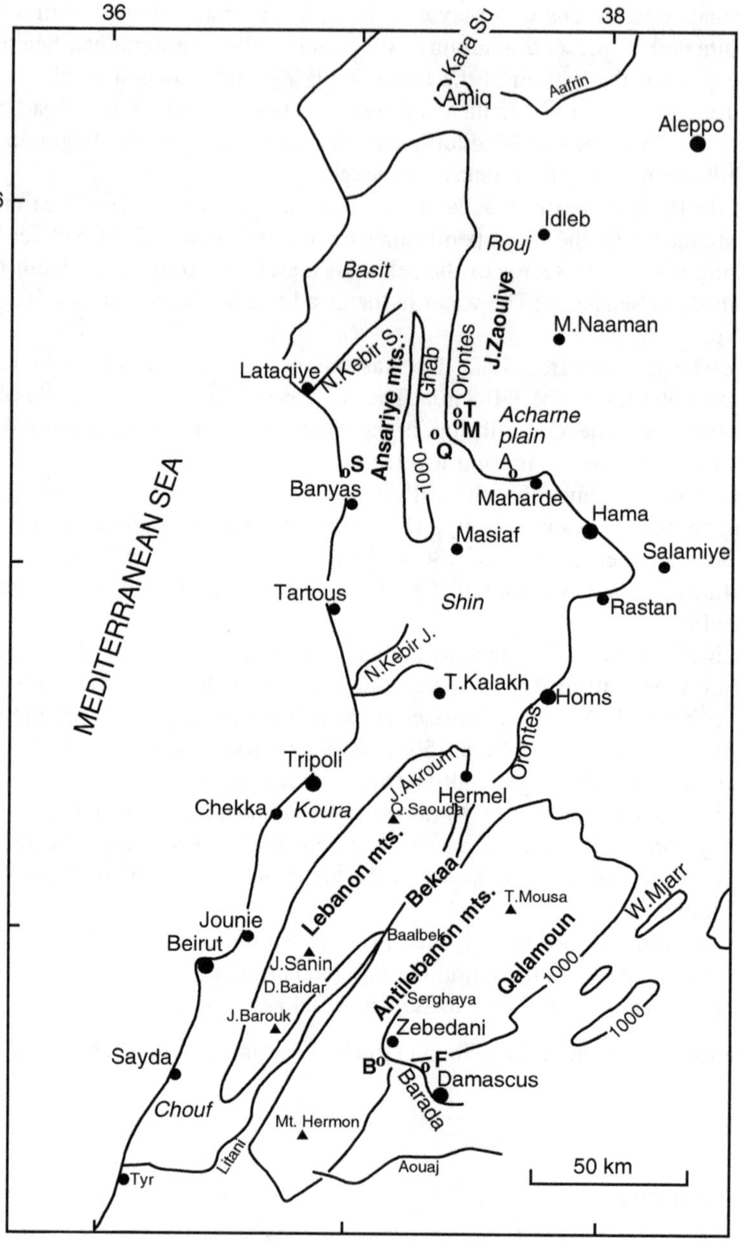

Fig. 2.4 Northwestern mountain and rift zone, location map, northern part. ___1000____ isoline 1,000 m altitude asl; o spring; A Tell Ayoun; F Ain el Fije; M Ain el Moudiq; S As Sinn; T Ain Taqa; Q Abou Qbeis

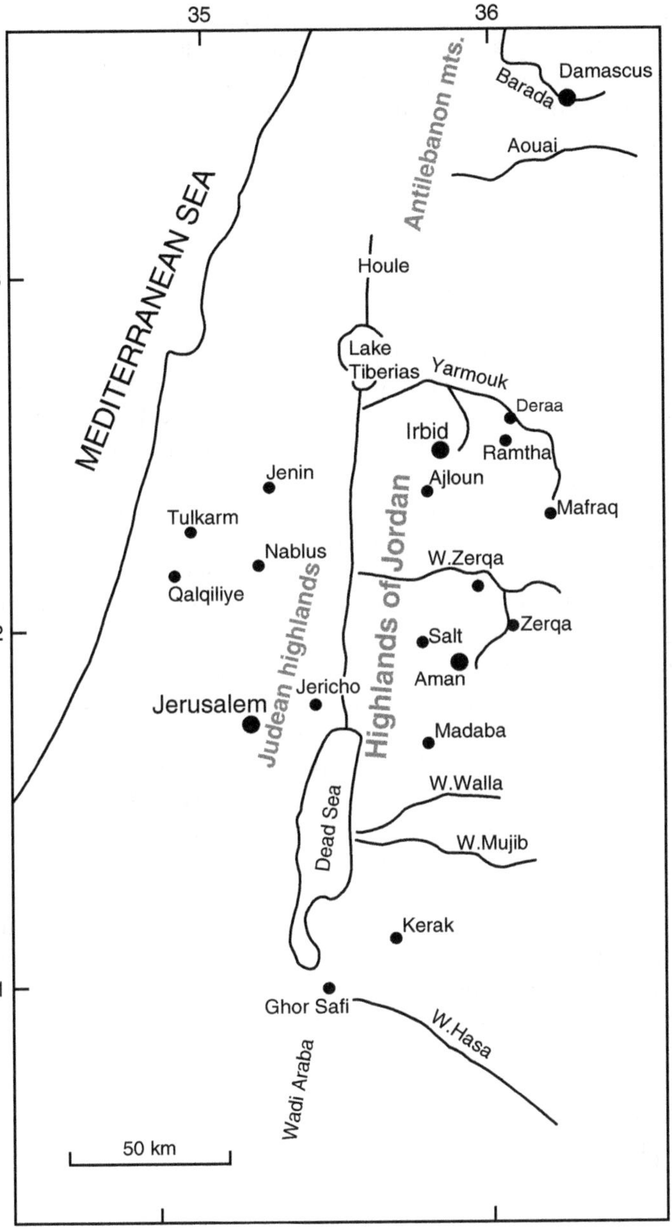

Fig. 2.5 Northwestern mountain and rift zone, location map, southern part

- Cambrian to Lower Cretaceous sandstone formations with shale intercalations
- Jurassic to Upper Cretaceous limestones and dolomites with intercalations of marls
- An Upper Cretaceous–Paleogene bituminous marl formation
- Paleogene limestones and chalks with marl and chert intercalations
- Neogene–Quaternary prevailingly terrestrial sediments in tectonic–morphologic depressions

2.2.1.1 Highlands of Jordan and Judea

The highlands of Jordan and Judea are covered prevailingly by Upper Cretaceous carbonate formations. Rock units of older stratigraphic age are exposed on the Interior Shelf in southern Jordan and in small areas in anticlinal structures and on the escarpments toward the Dead Sea. Paleozoic sandstones (Disi or Rum group) extend in the subsurface of the highlands of Jordan from the outcrop area in the south until the northern edge of the Dead Sea. In southern Jordan, terrestrial to coastal marine sandstone sedimentation continued from the Paleozoic up to the Lower Cretaceous. In the northern part of the highlands of Jordan and in the Judean highlands, Jurassic carbonate formations (Zerqa group in Jordan, Malih formation of the West Bank) indicate a marine ingression.

The Lower Cretaceous Kurnub group in Jordan and the Ramali formation on the West Bank comprise mainly terrestrial sandstones alternating with shale layers. In Palestine, marine marls and limestones related to an Aptian–Albian marine ingression are intercalated, toward west, in the upper part of the sandstone sequence. The Lower Cretaceous sandstones are followed by a thick carbonate complex of mainly Upper Cretaceous (Albian–Campanian) age. The complex, comprising the Ajloun group and the lower part of the Belqa group in Jordan and the Judea group on the West Bank, is built-up by limestones and dolomites with intercalations of marl, chalk and chert. The complex is sub-divided into a number of formations (Table 2.1), which are discussed according to their hydrogeologic properties in Sect. 2.3.3. The Maastrichtian to lower Paleogene is represented by marls, marly limestones and cherts with layers of oil shale (Muwaqar formation in Jordan, Mount Scopus formation of the West Bank). The Paleogene comprises chalk, limestones and marls of the Rijam and Shalala formations in Jordan and the Jenin or Avedat formation on the West Bank. The Maastrichtian–Paleogene deposits overlie the Upper Cretaceous carbonate complex mainly in synclinal structures around Irbid in northern Jordan and in the Nablus–Jenin area in Palestine.

2.2.1.2 Lebanon and Antilebanon Mountains

The stratigraphic sequence of formations outcropping in the Lebanon and Antilebanon mountains contains prevailingly rocks of Mesozoic age. The Upper Cretaceous in the Lebanon and Antilebanon mountains resembles the limestone–dolomite sequence in the highlands of Jordan and Judea. The Jurassic and Lower Cretaceous

Table 2.1 Stratigraphic sequence in the Judean highlands and the highlands of Jordan

		Judean highlands			Highlands of Jordan	
Quaternary						
Neogene						
Paleogene		Jenin (Avedat)		Chalk, limestone	Rijam (B4)	Chalky limestone
Upper Cretaceous	Maastrichtian	Abu Dis (Mt. Scopus)		Chalk, marl	Muwaqar (B3)	Chalky marl
	Campanian				Aman (B2)	Limestone, chert
	Santonian				Umm Ghudran (B1)	Marly limestone
	Coniacian				Wadi Sir (A7)	Limestone, dolomite
	Turonian	Judea group	Jerusalem	Limestone, dolomite	Shuayb (A5/6)	Marl, limestone
	Cenomanian		Bethlehem	Chalky limestone	Humar (A4)	Limestone, dolomite
			Hebron	Limestone	Fuheis (A3)	Marl, limestone
			Yatta	Marl, chalk	Naour (A1/2)	Limestone, dolomite
			Beit Kahil	Limestone, marl		
Lower Cretaceous		Ramali (Kurnub)		Sandstone	Kurnub	Sandstone, shale
Jurassic		Malih		Limestone, marl	Zarqa	Siltstone, sandstone
Triassic						
Paleozoic					Khreim	Siltstone, Sandstone, Mudstone
					Disi (Ram)	Sandstone

Simplified scheme after Haddad et al. (1996), Margane et al. (2002), Sunna (1995), Weinberger et al. (1994)

Table 2.2 Stratigraphic sequence of the Lebanon mountains. Simplified scheme after Khair et al. (1992), UNDP (1970)

Neogene			Marl, conglomerate, limestone
Paleogene			Chalky limestone
Upper Cretaceous	"Senonian"	Chekka marl	Chalky marl, chert
	Turonian	Maameltein limestone	Limestone, marl
	Cenomanian	Sanin limestone	Karstified limestone
Lower Cretaceous	Albian	Hamama marl	Marl, limestone
	Aptian	Mdarej limestone	Limestone, sandstone, volcanics
		Abeih	Argillaceous sandstone
	Neocomian–Barremian	Chouf sandstone	Sandstone
Jurassic	Portlandian	Salima limestone	Marl, limestone, shale
		Bikfaya limestone	Karstified limestone
	Kimmeridgian	Bhannes volcanic complex	Basalt, volcanic tuff, limestone
	Bajocian	Kesrouan limestone	Karstified dolomite, limestone

are represented mainly by thick marine carbonate complexes with intercalations of volcanic rocks in the Upper Jurassic and of detrital deposits in the Lower Cretaceous. The Lower to Middle Jurassic comprises limestones with dolomite beds with a thickness of 1,400 m in Mount Hermon and of around 700 m in the Lebanon mountains (Kesrouan limestone). The Upper Jurassic of the Lebanon mountains includes carbonate, detritic and volcanic formations. In the Antilebanon mountains, the Upper Jurassic consists of a several hundred metres thick succesion of prevailingly limestones and dolomitic limestones with marl intercalations.

Detritic deposits with limestone intercalations of the Lower Cretaceous are followed by a 400–1,000 m thick series of limestones and dolomites of Upper Cretaceous, mainly Cenomanian–Turonian age (Sanin limestone and Maameltein limestone in Lebanon). The top of the Cretaceous consists of a 400–600 m thick series of chalky marl and marly limestone with chert (Chekka marl in Lebanon) of "Senonian" (Maastrichtian–Paleocene) age, which corresponds to the Muwaqar and Mount Scopus formations in Jordan and Palestine. Paleogene chalks or nummulitic limestones occur in synclinal depressions in the Lebanon–Antilebanon zone (Table 2.2).

2.2.1.3 Northwestern Syria

The geologic formations of the Ansariye mountains comprise mainly Mesozoic carbonate formations:

- Jurassic limestones and dolomites with a thickness exceeding 800 m
- Lower Cretaceous (Aptian–Albian) marls and marly limestones, 150 m
- Upper Cretaceous (Cenomanian–Turonian) limestones and dolomites, 350 m
- Maastrichtian marls

Eocene nummulitic limestones occur in some areas on the slopes of the Ansariye mountains.

In the northwest and in the coastal area, the Jurassic to Turonian sequence is covered by formations of Maastrichtian to Quaternary age:

- Maastrichtian marls
- Eocene nummulitic limestone, up to 150 m
- Miocene–Pliocene marly limestones and basalt
- Quaternary limestones and conglomerates

Outcrops in Jebel ez Zaouiye include Cenomanian–Turonian limestones and dolomites, marls of late Cretaceous age in very reduced thickness, and Eocene nummulitic limestones.

2.2.1.4 Neogene–Quaternary Deposits

Neogene–Quaternary deposits extend over tectonic–morphologic depressions within or adjoining the Antilebanon mountains: the Damascus plain and the Qalamoun area on the eastern margin of the Antilebanon mountains, and intermountain basins, e.g. the Zebedani valley.

2.2.2 *Geologic Structure of the Mountain Massifs*

The mountain massifs of the northwestern mountain and rift zone are formed generally by asymetric horst–anticlinal structures with steep fractured flanks toward the rift valleys and more gentle slopes toward the Mediterranean Sea coast and to the eastern part of the northern Arabian platform (Fig. 2.6).

2.2.2.1 Highlands of Judea and Jordan

The mountain area of the Judean highlands shows, as a main structural element, a NNE trending anticlinal dome, the Judean anticlinorium, which was uplifted during the Pliocene. The uplift structure is from 40 to 50 km wide and about 130 km long, and is separated from the Dead Sea depression by the Dead Sea graben fault. North–south to northeast–southwest running faults and flexures accompany the eastern slope of the anticline in a multi-step vertical displacement of, in total, several kilometres. The general anticlinal structure of the Judean highlands is differentiated into tectonic units of local extent by SSW–NNE trending elements – flexures, anticlines, synclines – and, in particular in the northern part of the highlands, by southeast–northwest trending faults.

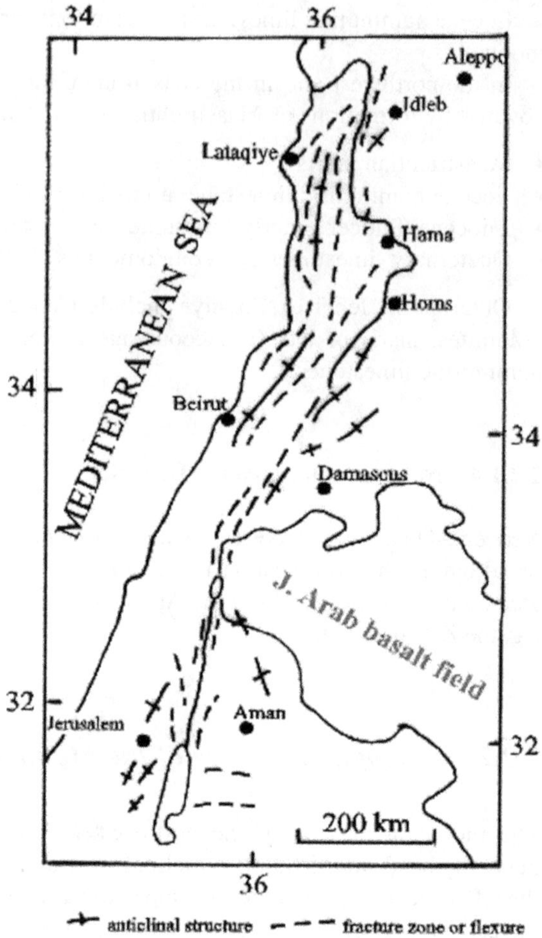

Fig. 2.6 Main geologic structures of the northwestern mountain and rift zone. After Wolfart (1966), Ponikarov et al. (1967a), Lenz (1969), Wiesemann (1969)

Major structural units are:

- A southern SSW–NNE trending anticline, the Hebron anticline, extending over the highland uplift west of the Dead Sea depression
- The Ramalla anticline, which comprises the central part of the highlands and is separated from the Hebron anticline by an east–west shift in the Jerusalem area
- The Fara and Anaba anticlines in the north, surrounding the Nablus syncline on the northwestern slope of the highlands

Main tectonic activities, which formed the structure of the highlands, occurred during the Tertiary until the Pleistocene (Lenz 1969).

The geologic structure of the highlands of Jordan is dominated by the appproximately north–south trending rift fault zone and anticlinal uplifts accompanying the rift zone in the east. The crest of the anticlines runs near the western escarpment of the highlands more or less parallel to the rift graben. Along the anticlinal structures, the limestones and dolomites of the Upper Cretaceous Humar and Aman–Wadi Sir formations are exposed in extensive outcrops. Major uplift structures (Ajloun dome north of the Zerqa river, structural highs at Baqa, Salt and Naour) form the highland crests from northwest to southwest of Aman (Margane et al. 2002: 8). The dip of the strata is generally directed from the centres of the anticlines toward the Dead Sea–Jordan valley in the west, the Yarmouk river in the north and the Jordanian limestone plateau in the east.

The highlands are crossed by several latitudinal, northeast to southeast trending fault systems (Wadi Kerak–Wadi Fayla fracture, the Wadi Zerqa Main–Siwaqa fault, Aman flexure, Wiesemann 1969).

2.2.2.2 Lebanon Mountains

The Lebanon mountains constitute a large anticlinal uplift running in SSW–NNE direction approximately parallel to the Mediterranean Sea coast and comprise, in their central and northern parts, the most intensively uplifted sections of the eastern Mediterranean coastal mountain ranges. Extensive outcrops of Jurassic–Cretaceous carbonate formations occupy the anticlinal crest of the range.

In the west, the mountain slopes rise directly from the Mediterranean Sea coast or from relatively narrow coastal plains. In the east, the anticlinal structure of the Lebanon range is cut off by the sub-regional Yamoune fault system, which extends from Lake Tiberias through Lebanon into northwestern Syria.

The general structure of the Lebanon mountains is sub-divided into three major tectonic–morphologic zones:

- The central and northern mountain area north of Dahr el Beidar with a broad anticlinal zone of high mountains.
- The southern high mountain area of Jabal Barouk centered around a relatively narrow anticlinal uplift, adjoined in the west by the Chouf area with hills and low mountains.
- The hill area of southwestern Lebanon.

The central and northern parts of the Lebanon mountains – north of Beirut and of the main mountain pass Dahr el Baidar – comprise large generally SSW–NNE trending structures with extensive outcrops of Upper Cretaceous and Jurassic carbonate rocks.

In the anticlinal zone of the northern high Lebanon mountain area, Jurassic and Upper Cretaceous karstified carbonate formations are exposed over around 700 km^2. That northern mountain zone includes the highest peaks of the Lebanon chains (Qornet as Saouda, 3,083 m asl).

The central Lebanon mountains are formed by a large about 36 km long SSW–NNE trending anticlinal structure, which is covered by Jurassic and Upper Cretaceous limestone and dolomite formations. The crest zone near the main

Lebanon water divide includes several mountain peaks at around 2,000 m asl and culminates in Jebel Sanin east of Beirut at 2,548 m asl.

The high mountain area of the southern part of the Lebanon mountains south of Dahr el Baidar comprises a 50 km long NNE–SSW trending anticlinal structure with a narrow, 3–6 km wide outcrop belt of Jurassic limestones and dolomites at the anticlinal crest. Altitudes of mountain peaks descend from >1,800 m asl in the north (Jabal Barouk 1,950 m) to 1,200 m asl further south.

An 8–16 km wide outcrop belt of Upper Cretaceous carbonate rocks extends over the eastern slope of the central and northern Lebanon mountains between the mountain peak region and the Bekaa valley boundary. A major structural feature on the eastern mountain slope is the SSW–NNE Yamoune fault, which separates a steep narrow strip of high mountain area from mountain slopes with less pronounced relief east of the fault. The eastern slope of the southern Lebanon mountains consists prevailingly of a narrow, 2–4 km wide strip of outcropping Jurassic carbonate rocks, which is delimited by the Yamoune fault from the Bekaa valley.

In the southern Lebanon mountain zone, two faults branch out from the Yamoune fault system in northwest and northeast direction, respectively: the Ram fault crossing the Lebanon mountains toward the Mediterranean Sea coast and the Hasbaya fault extending from the Bekaa plain into the Hermon–Antilebanon massif.

2.2.2.3 Antilebanon Mountains

The structure of the Antilebanon mountains is dominated by major SSW–NNE directed anticlines with outcrops of Mesozoic formations, the flanks of which dip under Paleogene–Neogene depressions: the Bekaa graben in the west, the Damascus plain and Qalamoun depression in the east. The anticlinal Mount Hermon–Antilebanon chain constitutes the northern prolongation of the uplift structure of the highlands of Jordan on the eastern rift boundary.

The Antilebanon mountains are crossed by major S–N to SSW–NNE directed faults in continuation of the Dead Sea–Jordan fault system (faults of Rachaya, Hasbaya and Serghaya). Along these fault systems, several intermountain basins subsided between the mountain chains, such as the Zebedani basin and the Bekaa plain. The Hasbaya fault delimits the Bekaa valley on the western foot of Mount Hermon, the NNE directed Rachaya fault crosses the western slope of Mount Hermon. The Serghaya fault runs from the eastern slope of Mount Hermon obliquely through the Antilebanon mountains and borders the intermountain basins of Zebedani and Serghaya.

The Barada valley separates the mountain chain into the Mount Hermon or Jebel esh Sheikh massif in the south and the Antilebanon mountains proper in the north. Mount Hermon is formed by a NNE trending anticlinal dome structure with wide outcrops of Jurassic deposits. Cretaceous–Paleogene formations cover the western and eastern foothills of the massif. In the northeast of Mount Hermon, the Jurassic outcrops are adjoined by Neogene conglomerates along a fault with a throw of around 3,000 m.

The Antilebanon mountains north of the Barada valley comprise mainly SSW–NNE directed anticlinal structures with prevailing outcrops of Upper Cretaceous formations. The Jurassic outcrops of the Mount Hermon dome continue into a narrow belt of horst structures near Zebedani–Serghaya.

The Antilebanon mountains are bounded, in the west, by the Bekaa valley, in the east by the volcanic plateaus of Golan and Hauran and by the Damascus plain. In the northeast, the Antilebanon mountains are separated from the SW–NE to WSW–ENE trending Palmyrean chains by the synclinal structure of Qalamoun.

2.2.2.4 Northwestern Syria

The Ansariye mountains constitute an asymetric horst–anticline with the core of the uplift running near the eastern rim of the mountain chain. Jurassic and Cretaceous formations are exposed in the core of the anticline, while Paleogene and Neogene deposits cover most of the western part of the mountains, which descends in a gentle slope toward the Mediterranean Sea coast. To the east, the mountains are bordered by steep cliffs along the rift graben border.

The meridional "Syrian–Lebanon fault" (Ponikarov et al. 1967a), which represents the northern prolongation of the Yamoune fault, marks the eastern boundary of the Ansariye mountains. The fault separates the southern section of the Ansariye mountain area from the Masiaf plateau along a rectilinear scarp with a vertical displacement of more than 500 m. Toward north, the fault splits into two branches, which accompany the Ghab rift graben in the west and east. The vertical displacement reaches around 1,200 m along the western branch of the fault, and 600 m along the eastern branch.

In the north, the Ansariye mountains dip under the narrow Nahr el Kebir plain, which is covered by Paleogene–Miocene formations and separates the Ansariye mountains from the allochthonous ophiolite massif of Al Basit.

East of the Ghab valley, the anticlinal uplift of Jebel az Zaouiye reaches moderate altitudes of up to 870 m asl. Near the northern boundary of Syria, the rift graben diverges into the Orontes valley and, east of the valley, into the depression of the Rouj lake. Further north, the faults disappear under Pliocene sediments in the Amiq lake depression. The western rim of Jebel ez Zaouiye forms a steep cliff, which is cut by a system of faults and flexures, while its eastern slope dips from the anticlinal crest, situated close to the rift boundary, gently to the east.

At the northern tip of Jebel Ansariye, the coastal belt of anticlinal uplifts of Mesozoic formations ends at the highly disturbed margin of the northern Arabian platform. The boundary is marked morphologically by the Nahr el Kabir esh Shimali and Aafrine valleys, which delimit the platform area covered by Mesozoic–Tertiary sedimentary formations from the allochthonous ophiolite massif of Al Basit.

East of the rift graben, a southwest–northeast trending fault system (Aafrine–Kilis fault) separates the Arabian Platform from a zone, which has been intensively disturbed during the Mesozoic–Tertiary by tectonic movements on the margin of the alpidic mountain chains and is covered by allochthonous blocks and molasse sediments in the foreland of the Taurus mountains. Movement along the fault during the Neogene–Quaternary was accompanied by volcanic flows.

2.2.3 Geologic Set-Up of the Rift Valleys

2.2.3.1 Jordan–Dead Sea–Wadi Araba Rift Valley

The Jordan–Dead Sea–Wadi Araba graben cuts in approximately north–south direction as a deep morphologic–tectonic depression through the northern Arabian platform and adjoining areas of the Interior Shelf, separating the highlands of Jordan from the Judean highlands.

During the Mesozoic and until the Eocene, the geologic development of the present Jordan–Dead Sea–Wadi Araba zone corresponded to the events on the adjoining shelf areas of the Arabian Plate. Taphrogenic movements began along an old geosuture in the Oligocene. The movements resulted in the uplift of areas adjoining the graben and in sedimentation of great masses of clastic deposits in the graben zone. Marine influence and fresh water lakes alternate in the graben furrow during the Oligocene to Miocene. A marine connection between the Gulf of Aqaba and the Mediterranean Sea may have existed during the late Miocene–early Pliocene. The subsidence of the graben continued until the Holocene, accompanied by basaltic volcanism in particular during the Peistocene.

The Jordan valley contains a cover of Miocene–Quaternary detrital and lacustrine sediments, the Jordan valley group, above Mesozoic formations.

The sediments of the Jordan valley group comprise:

- Conglomerates of the Al Beida formation (Miocene–Pliocene)
- Clayey deposits of the Lisan formation (Pleistocene–Quaternary)
- Alluvial fans and stream deposits: mainly poorly sorted sandy gravels with silt, clay and boulder intercalations (Pleistocene–Quaternary)

The total thickness of the Jordan valley group sediments reaches 300–400 m near the Dead Sea.

The sediments of the Al Beida formation, the lower part of the Jordan valley group, are composed mainly of coarse detrital, generally cemented deposits. The formation is overlain by a sequence of alternating clays, marls, sand and gravel: the lacustrine Lisan formation and detrital fan deposits.

The mainly Pleistocene Lisan formation, the marl facies of the upper Jordan valley group, is composed of thinly laminated marl, clay, gypsum beds and nodules of sulfur. The formation represents lacustrine deposits of the Lisan lake which occupied, as an ancestor of the Dead Sea, wide parts of the Jordan–Dead Sea valley during the upper Miocene–Pleistocene.

Pleistocene–Quaternary fluviatile deposits form gravel fans, which spread out from the mouths of tributary wadis into the valley floor, where the fan deposits are interfingering with the lacustrine Lisan marls.

Gravel fans of larger extent adjoin the entrances of Wadi Qilt and Wadi Muhallish on the western boundary of the Jordan valley and are found, on the eastern flank, at the confluence of the Yarmouk and Zerqa rivers and of Wadi al Arab with the Jordan valley.

Mesozoic formations occur, within the valley, below the Jordan valley group sediments, in outcrops on the mountain escarpments of the Jordan valley, and insome areas of the valley. The Mesozoic of the Jordan valley corresponds to the sequences of the Upper Cretaceous Ajloun–Belqa groups and Kurnub and Zerqa groups of Lower Cretaceous and Jurassic age in the adjoining highlands (Table 2.1).

The eastern shores and part of the western shores of the Dead Sea are flanked by narrow plains, which, on their outer margins, are bounded by the escarpments of the highlands of Jordan and Judea and geologically by major flexures and fault zones.

The plains are covered mainly by alluvial deposits, which overlie a tectonically disturbed sequence of Precambrian–Lower Cretaceous sandstones, Upper Cretaceous carbonate rocks and Tertiary deposits. Gravel fans of Pleistocene–Quaternary age mark the entrance of the larger tributary wadis from the Jordanian highlands into the Dead Sea valley, e.g. Wadi Walla and Wadi Mujib.

The 174 km long Wadi Araba between the Dead Sea and the Gulf of Aqaba is bordered in the east by a narrow spur of Precambrian rocks of the Arabian Shield and of Paleozoic–Cretaceous sandstone formations of the Interior Shelf. The margins of the NNE trending Wadi Araba graben are demarcated by a complex pattern of NW to NNE trending margin faults.

The highlands east of Wadi Araba and the graben zone itself comprise a sequence of Precambrian to Quaternary formations:

- Precambrian crystalline rocks
- Precambrian–Cambrian quartz porphyry, porphyrite and ignimbrite
- Cambrian, Ordovician and Lower Cretaceous prevailingly sandy sedimentary rocks with minor marine dolomite–limestone–shale intercalations
- Upper Cretaceous–Paleogene carbonate rocks
- Oligocene–Miocene predominantly clastic deposits
- Quaternary terrestrial and lacustrine sediments in the graben
- Pleistocene basalts

The tectonic Wadi Araba depression is filled with marine, lacustrine and fluviatile Neogene of about 2,000 m thickness. Lacustrine sediments of the Lisan lake – Lisan formation – extend over the northern part of Wadi Araba until about 80 km south of the present Dead Sea shore with a thickness of up to 800 m. The Lisan formation is composed of shale, marl with gypsum and sulfur inclusions. Coarse clastics and sands are intercalated on the graben margins. Quaternary wadi sediments, comprising sand, marl and clay, are found in the gorges and valleys in the highlands east of Wadi Araba and in extensive alluvial fans at the mouth of wadis on the margins of the valley. Mud flat deposits and aeolian sand dunes cover part of the central rift valley. The total thickness of Holocene sediments reaches up to 150 m.

2.2.3.2 Bekaa and Midle Orontes Area

The boundary of the SSW–NNE trending Bekaa depression is, along some stretches, demarcated by major faults: the Yamoune fault on the southern part of the western flank of the valley, the Rachaya and Serghaya faults along the eastern mountain foothills. Outcrops in the Bekaa valley comprise:

- Upper Cretaceous (mainly Cenomanian) limestones and dolomites
- Neogene lacustrine deposits and conglomerates
- Alluvial deposits

The southern part of the Bekaa graben comprises along its axis a synclinal structure, which is filled by continental Neogene to Quaternary deposits of several hundred metres thickness. On the southern flank of the syncline, Upper Cretaceous and Paleogene deposits appear on the surface of the Bekaa valley. The flat valley floor changes into a landscape with more accentuated relief, into which the Litani river is incised in gorges until it changes its course toward west and crosses the southern tip of the Lebanon mountain massif in direction to the Mediterranean Sea coast. The graben structure of the Bekaa valley merges in the north into the Homs plain and in the south into the morphologic depression of Houle.

In the middle Orontes area, the Lebanon–Syrian fault system separates the Antilebanon mountains in northwestern Syria from a mosaic of different geologic–morphologic structures:

- Shin basalt plateau
- Homs and Selemiye plains
- Masiaf and Hama plateaus
- Acharne plain and Ghab valley
- Jebel ez Zaouiye
- Idleb plateau

These units constitute, together with the eastern slope of the Ansariye mountains, the middle section of the Orontes river catchment.

The eastern border of the rift graben is accompanied in the middle Orontes basin by moderate tectonic uplifts, the "monoclinal horst anticlines" (Ponikarov et al. 1967a: 190) of the Masiaf hills and Jebel ez Zaouiye. Toward east, the uplifted zones descend to the plateaus of the eastern part of the northern Arabian platform. On the Masiaf hills, the strata dip toward east grading into the nearly horizontal structure of the Hama plateau.

Tectonic depressions of the rift graben system, the Acharne plain and the Ghab valley, separate Jebel ez Zaouiye from the Masiaf plateau and from the Ansariye mountains. The Shin basalt plateau hides the transition from the Lebanon mountains to the Masiaf plateau.

The Homs plain constitutes a shallow depression at the northern margin of the Bekaa plain.

The Idleb plateau north of Jebel ez Zaouiye forms a zone of low hills between the Rouj branch of the rift graben and the Aleppo plateau in the east. The present Idleb plateau constituted, during the Paleogene to Miocene the southern rim of the Aafrin depression, a foredeep of the alpidic Taurus mountains.

The following *stratigraphic units* are exposed in the various tectonic depression zones of the middle Orontes basin:

- Cenomanian–Turonian limestones and dolomites
- Campanian chalks, limestones and marls
- Maastrichtian marls with chert and limestone layers
- Pliocene–Quaternary lacustrine and alluvial deposits in the rift depressions of the Ghab valley and the Acharne plain

Paleogene and Miocene carbonate rocks extend over the Idleb plateau.

References. Al Ejel and Abderahim (1974), Batayneh (2006), Bender (1974a, 1982), Dubertret (1955), Haddad et al. (1996), Khair et al. (1992), Lenz (1969), Margane et al. (2002), Ponikarov et al. (1967a), Salameh and Shaqur (1981), Salameh and Udluft (1985), Shahab (1997), Sunna (1995), UNDP (1970), Weinberger et al. (1994), Wiesemann (1969), Wolfart (1966).

2.3 Main Aquifers

2.3.1 Stratigraphic Distribution of Main Aquifers

The most important aquifers of the northwestern mountain and rift zone are constituted by Jurassic and Upper Cretaceous limestone and dolomite formations. These Mesozoic carbonate aquifers are exposed over wide parts of the mountain chains and have been intensively karstified. Relatively high precipitation on the mountains makes the carbonate formations to highly productive karst aquifers with major national importance for domestic supply and irrigation.

The high degree of karstification is shown in numerous caves and in the occurrence of several large karst springs discharging high water volumes from extensive catchment areas. Some of the caves extend over considerable stretches and show phantastic stalactite formations, such as the caves of Jeita, a major tourist attraction in Lebanon.

The Mesozoic carbonate formations comprise the following aquifers of national importance for the countries of the sub-region (Fig. 2.7):

- Upper Cretaceous Aman–Wadi Sir (B2/A7) aquifer in Jordan
- "Mountain aquifer" (Cenomanian–Turonian) in the West Bank

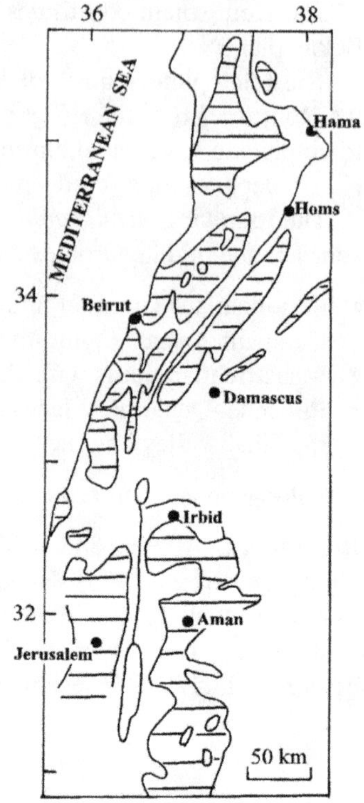

Fig. 2.7 Outcrops of Mesozoic carbonate formations in the northwestern mountain and rift zone generalized, after Dubertret (1966), ESCWA (1999a), Khair and Haddad (1993), Salameh (1996) hatched fields Outcrop of Jurassic and Cretaceous limestone and dolomite formations

- Jurassic and Cretaceous "Water Towers" of Lebanon (Kesrouan and Sanin aquifers)
- Jurassic and Cenomanian–Turonian limestones and dolomites on the Hermon, Antilebanon and Ansariye mountains in Syria

In addition to the Mesozoic karst aquifers, the northwestern mountain and rift zone comprises:

- Paleogene limestone and chalk aquifers in synclinal structures within the mountain and highland areas
- Pleistocene–Quaternary sand and gravel aquifers in basins and coastal areas
- Perched aquifers along the eastern and western shoulders of the rift graben
- Miocene reef limestones in the coastal area of Lebanon

Karstified Jurassic limestones provide important aquifers where they are uplifted in anticlinal structures to levels of present groundwater circulation: in the Lebanon mountains, Mount Hermon, the crests of the Antilebanon and Ansariye mountains. Lower Cretaceous sedimentary rocks contain aquiferous horizons which are generally connected hydraulically to the main Upper Cretaceous limestone and dolomite aquifer. On the flanks of anticlines in the Antilebanon mountains and in the Lebanon mountains, Lower Cretaceous sandstones and limestones form separate aquifers in areas of limited extent.

2.3.2 *Highlands of Judea and Jordan*

In the highlands of Judea and Jordan, the main aquifer system is formed by fractured and karstified Upper Cretaceous limestones and dolomites, the "mountain aquifer" or Judea group aquifer on the West bank and Aman–Wadi Sir and Humar aquifers in Jordan.

2.3.2.1 West Bank

Aquiferous sections of the mountain aquifer of the West Bank are found in the Hebron and Jerusalem formations (Table 2.3). The base of the mountain aquifer, the Lower Cretaceous Qatan formation, is composed of shale, chalk and marl (upper part of the Kurnub group). In synclinal structures the mountain aquifer is confined below the Mount Scopus chalk–marl formation.

The mountain aquifer complex is divided, in part of its extent, into sub-units by argillaceous aquitards:

- Marls of the Bethlehem formation between the aquiferous Jerusalem and Hebron formations
- Marl sequences in the Yatta formation, separating the Hebron aquifer from aquiferous layers in the upper Beit Kahil aquifer
- Argillaceous layers between the upper and lower Beit Kahil formations

The Hebron and Jerusalem formations form the most productive aquifer section of the mountain aquifer complex. Transmissivity in the Beit Kahil formation is generally low.

The total thickness of the mountain aquifer is 600–700 m.

Limestones of the Paleogene Jenin formation constitute an important aquifer in the Nablus synclinal structure around Nablus–Jenin.

2.3.2.2 Highlands of Jordan

In the highlands of Jordan, the about 700 m thick Upper Cretaceous aquifer complex consists of a sequence of limestones and dolomites with marl, chalk and chert intercalations. Considerable volumes of the precipitation, which reaches mean annual rates of >500 mm in the Ajloun–West Aman areas and around 350 mm at Kerak, infiltrate on the karst outcrops and make the crest of the highlands to the main recharge area of the Upper Cretaceous aquifers.

The fractured and karstified Upper Cretaceous limestones and dolomites constitute, in a sub-regional view, a coherent aquifer. In detail, the sequence of Upper Cretaceous carbonate formations of the highlands of Jordan is sub-divided, from bottom to top, into:

Table 2.3 Scheme of main aquifers in the northwestern mountain and rift zone

		Judean highlands	Highl. of Jordan	Lebanon mountains	Antilebanon mts.	Northwestern Syria
Quaternary Neogene				Aquifers in morphologic depressions		
Paleogene		Jenin aquifer in Nablus-Jenin area	Rijam (B4) aquifer in Irbid area		Nummulitic limestone, local aquifer	
Upper Cretaceous	Maastrichtian	Major Aquitard Mount Scopus form	Muwaqqar marl	Chekka marl	marl	marl
	Campanian Santonian Coniacian		Aman (B2) Wadi Sir (A7)			
			major karst aquifer			
	Turonian Cenomanian	Judea group "mountain aquifer"	Humar (A4)	Sanin limestone	limestone and dolomite	
Lower Cretaceous			Kurnub sandstone, deep brackish water aquifer			
Jurassic				Kesrouan limestone	limestone	limestone
					major karst aquifer	
Triassic Paleozoic			Disi sandstone, deep brackish water aquifer			

- The lower Ajloun group (A1–A6, Cenomanian), which is composed prevailingly of marls, shales and marly limestones and includes dolomitic limestone of the Humar formation (A4)
- Limestones and dolomites of the Wadi Sir formation (A7, Turonian)
- Marls of the Wadi Ghudran formation (B1, Santonian)
- Limestones, cherts and chalky limestones of the Aman formation (B2, Campanian)

The Upper Cretaceous aquifer system is underlain by a sandstone aquifer complex of Cambrian to Lower Cretaceous formations (Disi and Kurnub aquifers). The top of the Upper Cretaceous aquifer system is followed by the Maastrichtian Muwaqar marl aquitard, which overlies the Aman formation.

The Aman–Wadi Sir (B2/A7) and the Humar formations provide the major aquifers of the Upper Cretaceous sequence. The *Aman–Wadi Sir aquifer*, consisting of limestones, cherts and sandy, marly and phosphatic limestones, is exposed in extensive outcrops in the highlands of Jordan and on the southern parts of the escarpment toward the Jordan valley. The total thickness of the Aman–Wadi Sir aquifer is 90–350 m. In some areas, in particular in the northern highlands of Jordan, the marly and chalky Wadi Ghudran formation constitutes an aquitard between the aquiferous Aman and Wadi Sir formations.

In the highlands of the Yarmouk basin in northwestern Jordan, the Upper Cretaceous carbonate sequence comprises:

- A highly productive upper aquifer (B2, Campanian)
- A deeper aquifer with lower well yields (A7, Turonian)

The Aman–Wadi Sir aquifer is here generally situated at depths of less than 200 m below surface.

The geologic structure with dips toward the Yarmouk and Jordan valleys creates, along the main mountain slopes and in the Jordan valley, confining conditions in the Aman aquifer and the Wadi Sir aquifer. Piezometric heads in the aquifers, confined below overlying marl aquicludes, reach tens of metres above ground surface.

On the southern part of the Jordan valley escarpment, the Upper Cretaceous carbonate aquifer is, according to the geological structure and morphology, cut into several blocks, which extend for a few kilometres in length and width and are separated by faults or other unconformities.

The *Humar formation* (A4) constitutes a significant aquifer mainly in the Aman–Zerqa and Salt areas and is exposed in a rather narrow outcrop in the Aman area. The aquifer is underlain by the Fuheis marl aquitard (A1/A3) and is, east and west of the outcrop area, confined below the Shueib marl aquitard (A5). In the Sukhne area, the aquifer is situated near the surface below gravels of Wadi Zerqa. The saturated thickness of the Humar aquifer is 40–45 m.

In the Yarmouk basin in northwestern Jordan, the Humar formation occurs at greater depth and has a relatively low permeability, as the carbonate rocks of the formation grade into marl and siltstone beds. South of Aman, the thickness of the Humar formation decreases to about 10 m in Wadi Mujib and grades into marl and siltstone beds.

Around Naour southwest of Aman, the Ajloun group comprises near its base an aquiferous limestone and dolomite sequence with marl, sandstone and siltstone intercalations and a total thickness of 220 m, the *Naour aquifer* (A1/A2, Cenomanian). The Naour aquifer is separated from the overlying Humar aquifer by the Fuheis marl aquitard (A3).

In part of the northwestern highlands of Jordan, the Upper Cretaceous aquifer complex is underlain by a thick sandstone sequence, the Lower Cretaceous *Kurnub sandstone* and, at greater depth, the *Disi sandstone*. The Kurnub group consists prevailingly of sandstones interbedded with siltstone and clay, with a mainly silty upper part and an arenaceous lower part. Together with the Zerqa group, a Triassic–Jurassic sequence of limestones and dolomites with interbedded shale and sandstone, the Kurnub group constitutes a generally deep aquifer system in the highlands. The Kurnub sandstones are exposed at some locations west and northwest of Aman, on the southern part of the Jordan valley escarpment and in deeply incised tributary valleys of the Dead Sea and the Jordan river. The thickness of the sandstones varies, in general, from 185 to 300 m. The Zerqa group disappears toward south. In boreholes in the northern part of the Jordan valley, the Zerqa group has been tapped in a thickness of up to 1,700 m.

The deep Disi sandstone aquifer is reviewed in Chap. 5.

In the sub-regional groundwater regime, various hydraulic connections exist between the Upper Cretaceous aquifer and the underlying Kurnub sandstone aquifer:

- On the eastern slopes of the highland anticline, groundwater leaks generally from the Upper Cretaceous aquifer into the lower sandstone aquifer
- In some parts of the highlands, piezometric levels in the Kurnub sandstone are higher than in the overlying carbonate aquifer, producing some upward leakage of the sandstone water

In the area south of Aman, the Kurnub sandstone aquifer contains confined groundwater with salinities of up to a few thousand mg/l TDS, with artesian conditions in deep boreholes. In the Baqaa valley north of Aman, several springs discharge from the outcropping Kurnub aquifer.

In the Irbid area in northern Jordan, chalks and limestones of Paleogene age form a *shallow aquifer* (Umm Rijam or B4 aquifer) above the Muwaqar aquitard. The Umm Rijam formation, composed of limestones, chalks and chert intercalations, is exposed in a 4–15 km wide strip in the northwestern border area of Jordan between Ramtha and the Yarmouk river. In the north and northeast, the formation disappears under the cover of Neogene–Quaternary basalts. Groundwater flow in the shallow Rijam aquifer follows a differentiated pattern with flow directions toward the Yarmouk river and its tributaries with numerous emergences of springs (Margane et al. 2002: 22).

An alluvial aquifer is formed by gravels of Wadi Zerqa at Sukhne. The shallow Sukhne gravel aquifer with a saturated thickness of 20 m is hydraulically connected with the underlying Humar aquifer.

2.3.3 Jordan–Dead Sea–Wadi Araba Rift Valley

Main fresh water resources in the Jordan valley are found in the Pleistocene–Quaternary gravel fans and, in limited areas, in Upper Cretaceous carbonate formations. The thickness of the aquiferous gravel fan deposits ranges between a few metres to around 50 m.

Zones with higher aquifer productivity are restricted to the near surface gravel fans of larger wadis, which receive recharge from seasonal surface runoff. The groundwater in the gravel fans is generally unconfined; confined conditions can occur at the escarpment boundary and on the fringes of the fans, where the Lisan marls intercalate with the gravel deposits. Fresh water occurrences are generally limited to the gravel fan aquifers, where groundwater with low to moderate salinity extends in lenses above brackish to saline deeper water. Along the Zor area in the graben center, thin layers of shallow fresh water overlie saline groundwater or brines.

The Upper Cretaceous B2/A7 and the A4 formations provide deeper aquifers within the valley with, in many areas, elevated groundwater salinity. At various locations in the northern Jordan valley, wells were drilled into the B2/A7 composite aquifer are artesian and produce thermal water with temperatures of 28–56°C.

Groundwater in the deep Kurnub sandstone aquifer is saline.

Neogene conglomerates provide a fresh water aquifer in limited areas in the north of the Jordan valley. The productivity of the about 100 m thick conglomerate aquifer is generally low.

The extent of the Paleogene Jenin (B4) aquifer in the Jordan valley is restricted to the areas around Jericho and some locations in the northern part of the valley. Groundwater salinity in the Paleogene aquifer is generally high with 5,000–7,000 mg/l TDS. In the Jericho area, water with moderate salinity has been tapped in some boreholes drilled into the about 300 m thick Paleogene chalky limestones. The productivity of the aquifer is generally low and the groundwater in the deeper parts of the aquifer is brackish.

Southeast of the Dead Sea, a 1 km to >5 km wide plain extends between the graben fault structure and the sea shore. The plain is covered mainly by Quaternary sediments with alluvial fans, which consist of sands and gravels with thin marl intercalations (fans of Wadi Kerak and Wadi Isal), and interfinger with lacustrine sediments (clay, gypsum, aragonite) of the Pleistocene Lisan formation.

The sedimentary sequence under the plain contains a complex aquifer system with:

- The Paleozoic–Lower Cretaceous Disi and Kurnub sandstone aquifers
- Aquiferous limestones and dolomites of the Upper Cretaceous Ajloun–Belqa group
- The alluvial Quaternary aquifer

Main freshwater resources are contained in the alluvial fan deposits, which are replenished from flood flow infiltration and lateral flow from the Ajloun and Belqa aquifer across the eastern margin of the Dead Sea valley.

2.3.4 Lebanon and Antilebanon Mountains and Bekaa

2.3.4.1 Lebanon Mountains

The Lebanon mountains certainly contain the most productive aquifer system of the Arabian Plate. According to Khair et al. (1992), the mean volume of annually replenished groundwater circulation in the Lebanon mountains and its foreland on the coast and the Beqaa Valley is more than 2×10^9 m^3, corresponding to about 20% of the total annual precipitation in Lebanon. The high groundwater potential is created by:

- Abundant precipitation in the winter season with a mean of 930 mm/a in Lebanon in general and 1,400–1,500 mm on the upper slopes of the Lebanon mountains.
- High infiltration rates on widely exposed karst surfaces characterized by numerous sinkholes.
- The occurrence of thick karstified sections in Jurassic and Upper Cretaceous limestone and dolomite formations.

In particular the Kesrouan limestone of Jurassic (Bathonian–Kimmeridgian) age and the Sanin limestone of Upper Cretaceous (Albian–Cenomanian–Turonian, mainly Cenomanian) age provide highly productive aquifers and have been denominated the first and second "water towers of Lebanon".

The limestones and dolomitic limestones of the Kesrouan and the Sanin formations compose highly karstified carbonate complexes. The high degree of karstification has been favoured by the wide extent of outcrops of the carbonate rocks, intensive fracturing, high percentage of dolomite and dolomitic limestone which are particularly susceptible to karstification, and the relatively high precipitation rates in the mountain areas. Large interconnected cavern systems penetrate the thick Jurassic and Upper Cretaceous carbonate complexes. The mature stage of karstification in Jurassic and Upper Cretaceous carbonate rocks has significantly increased their porosity (more than 24%) and infiltration rate (40–44% of precipitation).

The Kesrouan aquifer is overlain by the Bhannes volcanic complex, which acts as an aquitard, separating the Kesrouan aquifer from the overlying Bikfaya aquifer. The Jurassic outcrop belt is delimited in the west by outcrops of Lower Cretaceous sandstones and marls, the Chouf sandstone and Aptian–Albian marl aquitards and, in the east, by the Yamoune fault.

The Sanin aquifer is composed of limestones and dolomitic limestones with thin marl intervals and is connected, in some areas, with aquiferous sections in the overlying Maameltein limestone. The 800–1,000 m thick Sanin limestone aquifer extends in the Lebanon mountains over a surface of 4,290 km^2.

The Mastrichtian Chekka marls constitute an aquiclude or aquitard where they extend over the Upper Cretaceous aquifer complex on mountain slopes, foothills and in coastal plain areas. The Neocomian–middle Albian sequence comprises, together with the late Jurassic, prevailingly aquitards and aquicludes with some aquiferous horizons in sandstone and limestone units.

2.3.4.2 Coastal Area of Lebanon

On long stretches of the Lebanon coast, outcrops of the Upper Cretaceous reach directly to the sea shore.

On the northern coast of Lebanon, a zone of plains and hill areas extends in front of the mountain range: the Akar plain in the Lebanon–Syrian border area adjoined to the east by basalt massifs, a coastal plain and foothills of the Lebanon mountains near Tripolis, a narrow coastal plain at Chekka, and the Koura plateau between the coastal plains and the Lebanon mountain chain.

The Akar plain extends over 720 km^2 in the Lebanese–Syrian border area along the Mediterranean Sea coast. The plain is crossed by Nahr Abrache, Nahr el Kebir el Janoubi and Nahr Oustouene. Mean annual rainfall is around 800 mm.

The plain is underlain by a multi-aquifer system comprising as main aquifers:

- Upper Cretaceous limestones and dolomites which are exposed in wide parts of the mountainous catchment area in the Ansariye and Lebanon mountains
- Neogene basalts
- Neogene–Quaternary sediments: sandstones, sand, limestones, conglomerates, alternating with clays, silt and marls

Productivity of the shallow basalt and sedimentary aquifers is generally low to moderate, higher well yields are found in the Upper Cretaceous aquifer, the top of which is situated in the plain area at depths of a few metres to more than 200 m below surface.

In the fractured Pliocene basalts of the Tell Kalakh massif east of the Akar plain, groundwater issues in generally seasonal springs with low discharge and has been tapped by shallow wells and a few deeper wells with low to moderate yield.

The hill and plain area along the coast of Tripoli comprises shallow aquifers in Miocene reef limestones and fluviatile to coastal sediments. South of Tripoli, the Abou Ali river drains, in northwest direction, a local basin with sedimentary aquifers of Miocene–Quaternary age. Deeper groundwater flow in the Upper Cretaceous aquifer is obviously directed westward below the Koura plateau toward the coast at Chekka, independent from the surface drainage.

The coastal plain at Chekka extends for 1–2 km from the shore and is bordered to the east by the Koura plateau, a hill area with topographic altitudes of 300–450 m asl. The Cenomanian–Turonian aquifer is overlain in the Koura plateau by Senonian marls. Part of the Koura plateau is covered by Miocene to Quaternary sediments: Miocene limestones, Miocene–Pliocene marls and clays, alluvial deposits, which contain a shallow aquifer system in the limestones and alluvial deposits.

At Chekka, submarine springs discharge from the confined Upper Cretaceous Sanin aquifer, which is covered by "Senonian" Chekka marls. The submarine springs of Chekka constitute one of the largest known submarine discharges of groundwater into the Mediterranean Sea. The simultaneous occurrence of perennial and seasonal springs at almost the same locations indicates the existence of various karst channels with independent groundwater circulation.

The thickness of the confining layer is about 115 m in the coastal area of Chekka. The springs are supposed to rise at locations, where the thickness of the marls is reduced, or on local anticlinal structures, where the marls are cut by fissures. The springs represent the lowest discharge area of the Upper Cretaceous carbonate aquifer of the northern Lebanon mountains. The Upper Cretaceous aquifer is underlain by low permeability marls of Lower Cretaceous age.

The total thickness of the Upper Cretaceous Sanin limestone formation in the Chekka catchment reaches up to 900 m.

In west central Lebanon, outcrops of the Upper Cretaceous limestone and dolomite formations reach over large strips until the sea coast. A narrow belt of Miocene limestones and Quaternary deposits adjoins the sea coast between Beirut and Jounie.

Main aquifers in the southwestern Lebanon mountains and foothills are Upper Cretaceous limestones and dolomites (Sanin aquifer) and Paleogene limestones. A coastal plain of about 1–2 km width extends over a long stretch of the coast between Saida and Tyr (Sur). Unconsolidated deposits of the coastal plain contain aquiferous layers which are, to some extent, connected with the underlying Paleogene or Upper Cretaceous aquifers. South of Tyr, the perennial springs Rachichiye and Ras el Ain discharge from the carbonate aquifer through the overlying coastal sediments.

2.3.4.3 Antilebanon Mountains

As in the Lebanon mountains, the main aquifers of the Antilebanon are formed by karstic Mesozoic carbonate formations: Jurassic limestones and Upper Cretaceous (Cenomanian–Turonian) limestones and dolomites. Groundwater from the Mesozoic karst aquifers issues in a considerable number of perennial spings, some of them with high discharge volumes.

Chalky limestones and fractured cherts of Eocene age provide aquifers of local importance on synclinal structures within the Antilebanon mountains west of Damascus and in the Qalamoun area between the Antilebanon and Palmyrean mountains north of Damascus. Groundwater on the periphery of the synclines issues in springs with low to moderate discharge (1–10 l/s) and is extracted through shallow wells. In the central parts of the synclines, confined groundwater is extracted from Eocene limestones and cherts.

2.3.4.4 Bekaa Valley

Major aquifers in the Bekaa valley are:

- More than 800 m thick Upper Cretaceous limestones and dolomites
- Eocene karstic limestones extending in particular over the southern end of the valley and the adjoining hill and mountain area of Jebel el Aarbi

- Neogene–Quaternary deposits in the valley plain, composed of conglomerates and alluvial and fluvial sediments

The Upper Cretaceous limestones and dolomites are exposed on the mountain flanks west and east of the valley and in the southern part of the valley and are situated at a depth of several hundred metres to more than 1,000 m below the northern and central valley floor.

The groundwater regime in the Bekaa valley is interrelated with the aquifer systems of the adjoining Lebanon and Antilebanon mountains through:

- Surface inflow of spring discharge on the boundaries of the valley
- Subsurface inflow into the valley mainly in the Upper Cretaceous aquifer

Major fault systems, in particular the Yamoune fault on the eastern slope of the Lebanon mountains, act along some stretches as barriers to groundwater flow, causing groundwater discharge in large perennial springs. Subsurface groundwater inflow from the Upper Cretaceous aquifer of the Lebanon mountains into the Neogene–Quaternary aquifer of the plain occurs in the northern part of the Bekaa, feeding the spring Ras el Aasi.

On the southern section of the Bekaa graben, karstified reef limestones of Eocene age constitute a productive aquifer, which feeds the springs Ain Hasbani, one of the sources of the Jordan river, and Ain et Tine. At its southern tip, the basin branches into the Litani river drainage in the west and the Hasbani river drainage in the east.

Groundwater with moderate salinity of 500–1,000 mg/l TDS is found in alluvial deposits, in particular in gravel fans of the larger wadi systems. The shallow aquifer system in the alluvial deposits is replenished from infiltration of surface runoff and base flow from the tributary wadis and by lateral inflow of groundwater from the Upper Cretaceous carbonate aquifer complex.

Groundwater from the shallow Neogene–Quaternary aquifer is exploited in particular in the central part of the Bekaa valley in the area around Baalbek. The Eocene limestone aquifer is intensively exploited in its outcrop area in the southern Bekaa.

2.3.5 Ansariye Mountains and Middle Orontes Area

In the Ansariye mountains, fractured and karstified Jurassic and Cenomanian–Turonian limestones and dolomites act as major aquifers. The two aquifers are generally separated by a 150 m thick aquitard of Aptian–Albian marls, but faults and main fracture zones create hydraulic interconnections. Claystones interbedded in the middle Jurassic probably act as aquitards and as base of the main groundwater circulation system.

In parts of the high mountain area, the Jurassic and Cenomanian–Turonian aquifers are unsaturated or contain locally perched groundwater bodies. On the lower mountain slopes and in the coastal area, groundwater in the Jurassic and Cenomanian–Turonian aquifers is confined below overlying formations with low permeability: Lower Cretaceous marls above the Jurassic aquifer, Maastrichtian or Pliocene aquitards above the Cenomanian–Turonian.

Several submarine springs with significant flow discharge at the Mediterranean Sea coast in the area of Banias–Tartous–Amrit. Most of the submarine springs issue from the confined Upper Cretaceous limestone and dolomite aquifer.

Several shallow – partly perched – aquifers are developed above the main Mesozoic aquifer system:

- In Mesozoic carbonate rocks, where the Mesozoic formations are uplifted in the mountain areas above the zone of saturation of the main aquifer system
- In Albian limestones, which are separated from the Jurassic aquifer by Aptian clay layers, and from which small springs rise in the crest area of the mountains with discharges of 0.3–2.5 l/s
- In Eocene nummulitic limestones, which feed small springs with discharges of about 3 l/s on the northwestern flank of the Ansariye mountains
- In marine or terrestrial Quaternary deposits in coastal areas south of Tartous and between Nahr el Sinn and Lataquie, and in fluviatile sediments of Nahr el Kebir ash Shimali near Lataquiye and of Nahr el Kish
- In Helvetian–Quaternary limestones in the coastal plain of Lataquiye, in chalky and calcareous marls of the lower Tertiary northwest of Nahr el Kebir esh Shimali

In the Jebel Basit area, shallow groundwater is found in the ophiolite rocks and in Triassic and Upper Cretaceous sedimentary rocks, tectonically intercalated into the ophiolites. Groundwater occurrences have a limited extent and discharge in numerous intermittent springs.

Cenomanian–Turonian limestones and dolomites constitute a main aquifer system on the Masiaf and Hama plateaus, the Acharne plain and Jebel ez Zaouiye. The Upper Cretaceous carbonate aquifer is unconfined in an around 35 km wide zone east of the rift graben, which comprises the main outcrop areas. Further east, the aquifer is confined below Upper Cretaceous aquitards: Campanian chalks and marls, Maastrichtian–Paleogene marls.

A particular characteristic of the areas adjoining the rift zone is the occurrence of discontinuous aquifers and of perched aquifers. Groundwater, which accumulates from local recharge in aquifers underlain by aquitards at levels above the main groundwater system, leaks into the deeper aquifer along fault systems or in zones where the aquitard is thinning or disappearing. Aquifers with water levels above the hydraulic head of the Cenomanian–Turonian aquifer system are found in several areas adjoining El Ghab in the east:

- Pliocene sediments in the Acharne plain
- Eocene nummulitic limestones on Jebel ez Zaouiye
- Paleogene–Neogene carbonate rocks on the Idleb plateau

The aquitards below these aquifers are marl and clay layers in the Pliocene of the Acharne plain and Maastrichtian marls in Jebel ez Zaouiye and the Idleb plateau.

The Maastrichtian marls disappear toward the anticlinal structures adjoining the rift graben.

In the Homs plain and the Shin volcanic plateau, aquifers of prevailingly discontinuous nature are found. In the Homs plain, lacustrine Pliocene sediments contain aquiferous lenses of conglomerates and sandstones between marls with low permeability. Groundwater from the non-persistent aquifer lenses is extracted through 10–80 m deep wells.

In the Shin plateau west of Homs, groundwater has been tapped in fractured basalts through 100–300 m deep boreholes. Well yields are generally low.

Toward east, the Homs plain and the Hama plateau grade into the Selemiye plain, where Paleogene chalks provide a more extensive shallow aquifer. Sporadic surface runoff and groundwater flow in the Selemiye plain are directed toward the Orontes river in the west. The groundwater in the shallow aquifer is intensively exploited by wells and leaks probably, to some extent, into the deeper Upper Cretaceous aquifer.

References. Agrocomplect (1984–85), Al Charideh (2007), Boeckh et al. (1970), Davidson and Hirzallah (1966), GTZ and NRA (1977), Hirzalla (1973), Hobler et al. (1991), Kareh (1967, 1968), Kozlov et al. (1966), Kroitorou et al. (1985), Margane et al. (2002), Mijatovic and Bakic (1966), Nations Unies (1967), Ponikarov et al. (1967b), Salameh (1996, 2004), Salameh and Shaqur (1981), Shahab (1997), Sunna (1995), UNDP (1970), Wagner (1996b), Wolfart (1966).

2.4 Groundwater Regimes

2.4.1 Hydraulic Parameters

Transmissivities are relatively high in many parts of the karstified Mesozoic limestone and dolomite aquifers of the northwestern mountain and rift zone. In the Lebanon mountains, considerable hydraulic conductivities and high aquifer thickness in the order of 1,000 m yield transmissivities of generally between 1,000 and 90,000 m^2/d in the Jurassic as well as in the Upper Cretaceous aquifer complexes. High transmissivities of around 100,000 m^2/d are found in karstified zones, while transmissivities in poorly fissured limestones may be as low as 10 m^2/d. Karstification is assumed to be restricted, in general, to shallow depth, e.g. 18–33 m below land surface in the Rachin area in the northern Lebanon mountains, but may reach several hundred metres in localized zones.

Transmissivities in the confined parts of the Upper Cretaceous Judea group aquifer of the West Bank (mountain aquifer) are in the order of 10,000–100,000 m^2/d; for the phreatic part of the aquifer, transmissivities of several hundred m^2/d are reported. In the western highlands of Jordan, transmissivities of the Upper Cretaceous carbonate aquifers may range from 10 to 3,000 m^2/d.

Very high transmissivities of the Upper Cretaceous karst aquifers are found at some locations: in the Mukheibe well field in the Jordan valley, one well yielded an artesian flow of 1.6 m^3/s.

Values of hydraulic conductivity of Mesozoic karst aquifers in the Lebanon mountains as well as in the highlands of Jordan and of Judea are generally in the order of 10^{-5} to 10^{-3} m/s, effective porosities are in the order of 1–5%.

2.4.2 Groundwater Recharge

In the eastern basin of the Judean mountains, average annual precipitation decreases from 800–1,000 mm on the peak areas to 600–150 mm on the mountain slopes situated in the rain shadow and about 100 mm in the Jordan valley. Balance estimates indicate groundwater recharge rates of 25–30% of precipitation and runoff rates of 5% of the precipitation. Corresponding to the spatial distribution of precipitation, recharge rates are relatively high in the north and west of the Judean highlands and decrease toward east and southeast. An average recharge rate of 144 mm/a has been computed for the mountain aquifer of the West Bank.

Estimates of annual groundwater recharge for various areas of the West Bank are listed in Table 2.4.

Recharge to the Aman–Wadi Sir aquifer in the highlands of Jordan varies generally from 14 to 30% of mean annual precipitation "depending on rainfall distribution, topographic situation, soil cover, karstification, etc." (Margane et al. 2002).

For the Aman–Zerqa basin with an area of 4,586 km^2, a groundwater recharge of 88×10^6 m^3/a has been estimated, corresponding to an average recharge rate of 19 mm/a. The main volume of recharge (42×10^6 m^3/a) is received in the Upper Cretaceous limestone aquifers, while a minor part of replenishment is contributed through inflow from the semi-arid basalt area in the east.

Groundwater recharge rates in the Upper Cretaceous aquifers of the Aman–Zerqa basin are the highest rates in Jordan and sustain 30% of the national renewable groundwater resources.

Table 2.4 Estimates of annual groundwater recharge for various catchment areas of the West Bank after Gvirtzman (1994), Kroitorou et al. (1985, 1992), Sunna (1995)

Catchment area		Estimated groundwater recharge	
	km^2	10^6 m^3/a	mm/a
Highlands north of Jerusalem (eastern catchment)	135	49	363
West of Hebron (northern catchment)	696	77.3	111
Dead Sea	1,045	60.4	58
Jordan valley	805	153	190
Tulkarm–Qalqiliye (western catchment)	1,005	246	245
Jenin–Nablus (northern catchment)	1,050	124	118

In Lebanon with its relatively favourable climatic and infiltration conditions, the average total recharge to all outcropping aquifers is estimated at around 2.5×10^9 m^3/a. That recharge volume corresponds to approximately 25% of the mean annual precipitation of 930 mm over Lebanon's total area of 10,452 km^2.

Estimates of groundwater recharge in individual sub-basins in Lebanon, which extend over catchment areas of 40–851 km^2, indicate the following recharge rates:

- On outcrops of Jurassic and Upper Cretaceous karst aquifers: 37–43% recharge of mean annual rainfall of 550–1,450 mm or 200–600 mm/a
- On outcrops of Eocene limestones and chalks: 27% recharge of mean annual rainfall of 750–900 mm or around 240 mm/a
- On coastal plains covered with Quaternary sediments or basalts: 15% recharge of mean annual rainfall of around 1,000 mm or 150 mm/a

The areas receiving significant recharge comprise 50–90% of the catchments in most sub-basins and 20–35% in some coastal plain catchments.

For the main groundwater basins of Lebanon, the following ranges of mean annual recharge rates have been calculated:

- High Lebanon mountains 350 mm
- Southern Lebanon mountains, southern Bekaa, Antilebanon mountains 210 mm
- Northern Bekaa basin 180 mm

A general groundwater balance of the Upper Cretaceous Sanin aquifer of the northern Lebanon mountains assumes a mean recharge of 585 mm/a or 42% of precipitation, sustaining a total mean annual discharge of 13 m^3/s at springs in the mountain area and, near Chekka, on the coastal plain and under the Mediterranean Sea. Submarine discharge, estimated at $120–350 \times 10^6$ m^3/a, may account for about 40–50% of the total net discharge. The submarine discharge at Chekka comprises 17 springs, 7 of which are perennial. Total spring discharge is reported as:

- 4 m^3/s fresh water base flow
- 60 m^3/s fresh water discharge in winter

The submarine springs show high seasonal discharge fluctuations, some springs cease in summer or show even reverse circulation of salt water into the outlet points.

An evaluation of satellite images and aerial photographs covering the Lebanon mountain area and its foreland, showed that 57% of the area have a high to very high recharge potential with recharge rates of 30–50% of mean annual precipitation. The areas with high to very high recharge potential correspond to the outcrops of fractured and karstified Jurassic and Cenomanian limestones and dolomites on the elevated parts of the Lebanon mountains. Over the total area, around 24% of the precipitation contributes to groundwater recharge.

High recharge rates of around 350 mm/a are indicated for outcrop areas of the Upper Cretaceous limestone and dolomite aquifer of the Ansariye mountains, corresponding to around 30% of the mean annual precipitation of 1,050 mm. A significant part of the recharged groundwater discharges in submarine springs.

2.4.3 *Groundwater Flow Systems and Flow Volumes*

Sub-regional groundwater flow systems in the extensive Mesozoic carbonate aquifers of the northwestern mountain and rift zone are directed toward discharge zones on the Mediterranean Sea coast and along the rift graben. Within the mountain and highland areas, groundwater discharge zones of various hydrogeologic catchments, some of them with large dimensions, are found on aquifer outcrop boundaries at particular structural and/or topographic situations.

Each of the six mountain and highland chains of the northwestern mountain and rift zone comprises several groundwater flow systems with local to sub-regional extent.

2.4.3.1 Judean Highlands

The main aquifer complex in the Judean highlands is recharged mainly by precipitation on the outcrops of the Upper Cretaceous limestones and dolomites at altitudes above 500 m asl. The higher mountain ranges contain phreatic zones of the aquifer complex, the main volumes of groundwater storage are found in confined zones of the aquifer at intermediate to lower ranges of the highlands.

The Judean highlands comprise three main groundwater flow systems with recharge areas on top of the highlands and discharge areas in the adjoining lowlands:

- The eastern basin extending between a water divide, which runs approximately over the crest of the Judean highlands, and the Jordan–Dead Sea valley
- The western basin between the water divide on the highlands and the Mediterranean Sea coast
- The northern basin or Nablus basin, which covers a large synclinal structure in the northern part of the highlands

The groundwater divides separating the basins coincide, in general, approximately with anticlinal structures (Hebron, Ramalla, Anabta, Fara anticlines).

The *eastern basin* extends over an area of around 4,200 km^2 on the eastern flank of the Judean highlands and comprises several separate groundwater catchments with general eastward groundwater flow from the mountain zone to the Jordan valley and Dead Sea lowlands. The eastern boundary of the Upper Cretaceous limestones and dolomites of the mountain aquifer, which constitute the main aquifer complex of the eastern basin, is formed by a large fault system along the rift graben. Groundwater levels in the mountain aquifer system descend from around 450 m asl on the highlands to 400 m below sea level on the Dead Sea shore.

Annual groundwater recharge volumes of 70–172 $\times$ 10^6 m^3 have been estimated for the eastern catchment of the Judean highlands, which covers an area of 2,200 km^2, with an average of around 100 $\times$ 10^6 m^3.

Natural groundwater discharge occurs in several springs and groups of springs on the eastern slope of the highlands and in the Jordan–Dead Sea valley. Springs on the mountain slopes appear to be fed by two major systems of karst conduits:

- An older system related to a former higher level of the Lisan lake, the predecessor of the Dead Sea, around 180 m below sea level
- A younger system which developed after recession of the Lisan lake

The upper system carries relatively fast flowing recent groundwater e.g. to the Elisha springs. The deeper systems contains older groundwater which is mixing through interconnections with the upper karst system. The deeper karstified sections are extensively exploited by well fields. The location of springs on the highland slopes is mainly controlled by the structural–morphological features of the escarpment. Major springs issuing from the Upper Cretaceous mountain aquifer on the mountain slopes of the eastern basin are Ain Auja west of Jericho and several springs in Wadi Qilt: Ein Fara, Ein Fawar, Ein el Qilt situated at 325, 80 and 10 m altitude asl, respectively. Ain Auja has a high seasonal discharge in years with heavy rainfall (e.g. >70 m^3/s in 1991–1992) and runs dry frequently in summer.

In the Jordan valley, numerous springs issue along the rift fault zone: Ain Elisha, Ain Sultan and the springs Duyuk, Sosha and Noeima. The salinity of these springs is relatively low with Cl concentrations around 30 mg/l and around 400 mg/l TDS.

Ain Elisha, located at the rim of the rift valley, is fed from water recharged in the Judean mountains and has a relatively constant discharge of 180 l/s. The total annual fresh water discharge in springs in the eastern basin is in the order of 40×10^6 m^3.

The brackish Feshkha springs on the northwestern tip of the Dead Sea coast constitute a major discharge zone for most of the deeper sections of the aquifer complex. The springs with a discharge of some thousand m^3/h irrigate the largest oasis in the Dead Sea basin. Over twenty springs, streamlets, ponds and drainage canals are spread along a 4 km long strip of the western Dead Sea shore which is here only around 600 m wide (Table 2.5 and Fig. 2.8).

The *northern catchment* (Nablus–Gilboa basin) extends over a large synclinal structure in the north of the highlands of Judea and Samaria and is covered, to a large extent, by Maastrichtian–Paleogene sedimentary rocks: the Mount Scopus marl formation and the Jenin (Avedat) chalk formation. The Jenin formation provides a shallow aquifer, separated from the deeper Upper Cretaceous mountain aquifer by the Mount Scopus aquitard.

Table 2.5 Mean annual discharge of springs on the eastern slope of the Judean highlands after Kroitorou et al. (1985, 1992), Sunna (1995)

Spring	Mean discharge $\times$ 10^6 m^3/a
Auja	10
Feshkha	40
Wadi Qilt	5
Ein Qilt + Sultan	9
Jericho springs	13
Ain Fawar	11–36
Ain Fara	1.4

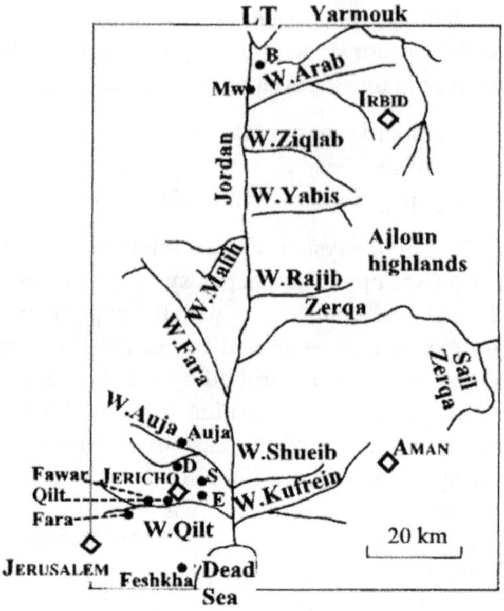

Fig. 2.8 Wadis and springs in the Jordan river catchment.; B Al Himme springs (Balsam and Maqla); D springs Duyuk, Sosha, Noeima; E Ein Elisha; S Ain Sultan; LT Lake Tiberias; Mw Mukheibe well field

Groundwater movement is directed mainly toward north. Groups of springs with small to moderate flow issue from the Jenin aquifer in particular in the Beisan area, where total spring discharge was nearly 100×10^6 m^3/a in 1952–1960. The main volume of groundwater flow emerges from springs or is extracted from wells in the Harod and Beit Shean valleys west of the northern Jordan valley.

The *western catchment* (Yarqon–Taninim basin) occupies the western slope of the Judean highlands between the crest area and the Mediterranean Sea coast. Groundwater moves in the main Judea group aquifer generally toward west. In the pre-development stage, a large percentage of the groundwater flow discharged in springs in the foothills and the coastal plain. In the areas of Tulkarm and Qalqiliye, groundwater is extracted through wells from the Upper Cretaceous aquifer and the Paleogene Jenin aquifer.

Groundwater balance estimates indicate a flow volume of 140×10^6 m^3/a for the northern catchment and of 360×10^6 m^3/a for the western catchment. These estimates do not differentiate between Plaestinian and Israeli territories.

2.4.3.2 Highlands of Jordan

The groundwater flow regime in the Upper Cretaceous carbonate aquifer complex of the highlands of Jordan is differentiated by major groundwater divides into flow systems of four main hydrogeologic sub-basins of Wadi Mujib, the Zerqa river, Yarmouk river and Azraq.

Regional groundwater movement in northwestern Jordan is directed to the Jordan–Dead Sea valley either immediately through the Upper Cretaceous carbonate aquifers or through leakage into the Kurnub sandstones. The general westward oriented groundwater flow system is, however, overlain in wide areas by a system of eastward groundwater flow in the A7–B2 aquifer (Margane et al. 2002: Fig. 2.8). The "configuration of the groundwater table indicates that the groundwater in the Upper Cretaceous aquifer system flows to the east and partly infiltrates to the sandstone aquifer system. In this aquifer it takes a westerly course and is discharged along the western slopes bordering the Dead Sea" (Salameh and Udluft 1985: 46).

In the *Wadi Mujib* sub-basin, groundwater flow is generally directed to the Dead Sea valley. In the *Aman–Zerqa* sub-basin, groundwater moves from recharge mounds around Aman to the Jordan valley in the west and the to Zerqa valley in the east and northeast. Along the south–north course of Wadi Zerqa, groundwater flow direction changes along a barrier of uplifted Muwaqar marls toward north and discharges partly in springs. The groundwater regime in the Aman–Zerqa area is now highly disturbed by pumping.

In the *Yarmouk* sub-basin, groundwater flow is, in general, directed to the Yarmouk and Jordan valleys.

Main groundwater discharge occurs in springs and seepages along the Jordan valley and in tributary wadis on the western escarpment, such as Wadi Arab and Wadi Ziglab, or through lateral outflow into the deposits of the Jordan valley floor.

In the *Azraq* sub-basin, groundwater flows eastward from the highlands into the east Jordanian limestone plateau with significant leakage into the underlying sandstone aquifer complex, in which groundwater movement is oriented to the Dead Sea–Jordan valley in the west.

2.4.3.3 Jordan–Dead Sea Valley

The Jordan–Dead Sea valley receives groundwater from various aquifers of the adjoining highland escarpments. Groundwater reaches the valley:

- As surface base flow from spring discharge in tributary wadis
- As spring discharge on barriers formed by the rift fracture and flexure system
- As subsurface flow through conduits within the rift faults

Direct recharge from precipitation within the valley, with mean annual rainfall of around 100 mm, is negligible, but the catchment areas of the main Mesozoic aquifers extend over wide areas of the adjoining escarpments and highlands, which receive significantly higher precipitation. The aquifers within the valley – mainly the Upper Cretaceous carbonate aquifers and Pleistocene–Quaternary gravel aquifers – are replenished prevailingly by percolation of runoff in tributary wadis, irrigation water return flow, canal losses, infiltration of occasional flood flow and subsurface inflow.

Spring discharge zones are located in several areas on the western tectonic boundary of the Jordan–Dead Sea valley and along the foothills on the northern end of the Jordan valley. A zone of thermal springs extends over the eastern escarpment between the southern end of the Dead Sea and Lake Tiberias. The thermal springs issue from Lower Cretaceous sandstones in the south and, in the north, from Upper Cretaceous carbonate aquifers. Topographic elevation of discharge points of the thermal springs range from 570 m asl in the south to around 100 m below sea level at the sites of the springs Balsam and Maqla in the northern Jordan valley.

Spring discharge in the eastern Jordan valley–Dead Sea catchment amounts to 160×10^6 m^3/a. Main discharge volumes issue from the Upper Cretaceous Humar (A4) and Aman–Wadi Sir (A7/B2) aquifers and from alluvial aquifers in the wadis of the northern highlands and the Dead Sea area. Around 24×10^6 m^3/a are contributed from spring discharge of the deeper sandstone aquifer system, mainly the Lower Cretaceous Kurnub aquifer (data of 1983–1993, Margane et al. 2002). The groundwater discharge supports a baseflow in wadis of the Jordan–Yarmouk and Dead Sea catchments of presently 155×10^6 m^3/a. Total discharge from the sandstone aquifer complex east of the Dead Sea is around 90×10^6 m^3/a.

Artificial groundwater extraction from wells in the Yarmouk–Jordan river–Dead Sea catchment in Jordan reached, between 1993 and 1998, an average of 416×10^6 m^3/a, creating a high deficit of the groundwater balance between recharge/inflow and natural + artificial discharge.

Groundwater movement in the Pleistocene–Quaternary aquifer of the Jordan valley is generally directed from the escarpment to the central Zor area, groundwater streamlines running from the eastern and western escarpment toward the central part of the valley with a slight bend to the south. Depth to groundwater ranges from about 100 m in Mesozoic aquifers at the foothills of the escarpment to around 5 m in the central part of the Zor area. Natural groundwater discharge occurs through evaporation in the Zor area and small seepages into the Jordan river. At present, groundwater discharge is primarily controlled by artificial extraction in wells mainly from the Upper Cretaceous aquifers (B2/A7, A4, A1/A2, Judea group aquifers).

Groundwater extraction in the eastern part of the Jordan valley–Dead Sea area is estimated at 140×10^6 m^3/a. In the *north of the Jordan valley*, wells at the confluence of Wadi el Arab and the Yarmouk valley at Mukheibe discharge artesian water from the B2/A7 aquifer, which is overlain by the Muwaqar marl aquitard. The exploited aquifer is recharged in the highlands and receives some inflow through upward leakage from the deeper A4 and Kurnub aquifers.

Groundwater recharge to the Upper Cretaceous aquifers in the northeastern Jordan valley and tributary catchments (Yarmouk, Wadi el Arab, Wadi Jarim, Wadi Yabis) is estimated at 127×10^6 m^3/a, of which 100×10^6 m^3/a discharge in spring flow and base flow in the tributary wadis. Groundwater extraction through wells of 73×10^6 m^3/a is considered to produce heavy overexploitation of the groundwater resources.

In the *southeastern Jordan valley* catchment, fresh water in the Upper Cretaceous carbonate aquifer system and brackish water in the deeper Lower Cretaceous

Kurnub sandstone aquifer flows from the highlands and the escarpment toward the Jordan–Dead Sea valley. Main discharge zones are found in wadi sediments in the lower reaches of the tributary wadis. The groundwater of the Upper Cretaceous carbonate aquifer system is recharged in the highlands of Aman – As Salt and around Wadi Mujib. The deeper sandstone aquifer complex receives groundwater mainly from inflow from sandstone aquifers of the Interior Shelf, supplemented by leakage from the overlying carbonate aquifers and very limited recharge on outcrops on the escarpment.

Renewable groundwater in the southeastern Jordan catchment is estimated at 10×10^6 m^3/a.

2.4.3.4 Wadi Araba

Wadi Araba may be divided into a northern segment which drains to the Dead Sea, and a southern segment south of the water divide at Jebel er Risha (about 250 m above sea level in the central Wadi Araba) draining to the Gulf of Aqaba. Permanent surface run-off exists neither in the northern nor in the southern regime. In northeastern Wadi Araba, the groundwater flows in a westerly direction with a component toward north to the Dead Sea. In the southeastern part of Wadi Araba, groundwater flow is directed toward the Red Sea in the south.

Recharge in Wadi Araba comes mainly from precipitation on the adjoining highlands in the east and lateral inflow into the fluviatile and alluvial deposits on the wadi floor. A part of the recharge takes place along the wadi courses of the tributary wadis and Wadi Araba itself. The considerable intermittent surface runoff and partly permanent baseflow from the mountains east of the Wadi Araba rift valley, which receive more than 300 mm of precipitation annually, infiltrates completely into the unconsolidated Quaternary sediments within the depression. Baseflow in wadis on the southern escarpment of the highlands of Jordan, the catchment of Ghor Safi–Wadi Araba, is around 76×10^6 m^3/a.

2.4.3.5 Lebanon Mountains

The Lebanon mountains comprise, according to geomorphologic features and geologic structure, several extensive groundwater flow systems in karst aquifers with catchments of several hundreds of km^2. A large number of groundwater flow systems with smaller dimensions are superimposed above the main karst flow systems or extend over catchments limited by outcrops of low permeability formations.

The karstified Jurassic and Upper Cretaceous carbonate formations constitute important aquifers with numerous groundwater flow systems of local extent and a few major flow systems with sub-regional extent. Groundwater discharges in the mountain area from several perennial springs and from many seasonal springs.

Spring discharge from the Mesozoic karst aquifers in the northern Lebanon mountains feeds the headwaters of Nahr Oustouene and Nahr Arka, which run into the Akar plain, and sustain base flow in the river systems of Nahr Abou Mousa–el Bared and Nahr Abou Ali. To some extent, spring discharge from the Jurassic aquifer reinfiltrates into outcrops of the Upper Cretaceous Sanin limestones.

Hydrologic–hydrogeologic regimes in the Lebanon mountains are, in general, rather complex with various interconnections between surface and subsurface flow:

- Streamflow sustained by groundwater discharging at high elevations may recharge aquifers at lower mountain slopes
- Interconnections between different aquifers and between adjoining sub-basins are created by major faults with high throw and locally by abundant minor faults

Khair et al. (1992) defined hydrogeologic units on the territory of Lebanon according to relief, river channels, water divides, faults, anticlinal axes, dip direction of rocks, direction of groundwater flow, synclinal structure, river catchments. The Lebanon mountain area includes around 35 of these hydrogeologic sub-basins.

The main groundwater divide between the catchments of the Mediterranean Sea and the Bekaa valley coincides approximately with the surface water divide connecting the main peaks of the Lebanon mountains and is situated generally at distances of 26–36 km from the Mediterranean Sea coast and 4–16 km from the Bekaa valley.

Groundwater movement in the *northern Lebanon mountain range* is largely controlled by structural and morphologic features. Groundwater discharges in several large perennial springs from the Jurassic and Upper Cretaceous karst aquifers at altitudes between 750 and 1,700 m asl. The northern Lebanon mountain groundwater sub-basin extends in the east to the main water divide on the mountain crest and to the Yamoune fault, which cuts off the Jabal Akroum massif. In the west and northwest, the sub-basin is delimited by a large flexure from the Tell Kalakh volcanic massif, the Akar plain and the Koura plateau. In the south, an approximately west–east directed fault system separates the northern Lebanon mountains from the central Lebanon massif.

Groundwater flow in the mountain area is generally directed toward SW to SSW with numerous springs issuing in particular at the outcrop boundary of the Upper Cretaceous or Lower Cretaceous aquitards.

In the Upper Cretaceous Sanin aquifer, a sub-regional groundwater flow system is directed toward large perennial springs at Rachin, which are situated on the boundary of the Upper Cretaceous aquifer and the overlying Chekka marl aquitard at around 300 m asl. A significant percentage of circulating groundwater of the northern Lebanon basin mountains reaches, however, the Mediterranean Sea coast at Chekka through a confined section of the Sanin aquifer.

A general groundwater balance of the Upper Cretaceous Sanin aquifer of the northern Lebanon mountains assumes a mean recharge of 585 mm/a or 42% of precipitation, sustaining a total mean annual discharge of 13 m^3/s in springs within the mountain area, under the Mediterranean Sea and on the coastal plain near Chekka. Submarine discharge may account for about 40–50% of the total net discharge.

Table 2.6 Mean discharge of major springs at the west slope of the Lebanon mountains after Khair et al. (1992), Shatilah, UNDP (1970)

Spring	River catchment	Altitude (m asl)	Discharge m^3/s	Aquifer
Dalle	Nahr Jaouz	635	1.93	Jurassic
Afka	Nahr Ibrahim	1,150	4.62	Upper Cretaceous
Ar Rueis	Nahr Ibrahim	1,260	3.2	Upper Cretaceous
Al Asal	Nahr el Kelb	1,350–1,660	0.8	Upper Cretaceous
Laban, As Saqia, Sanin	Nahr el Kelb		2.7	Upper Cretaceous
Jeita	Nahr el Kelb		4.48	Jurassic
Antelias			0.56	
Ain ed Delbe	Nahr Beirut		2.2	Jurassic
Safa	Nahr Damour	1,100	1.43	Jurassic
Barouk	Nahr Awali	1,240	1.03	Jurassic

Several large perennial springs sustain the base flow in major rivers, which descend on the western mountain slope to the Mediterranean Sea: Nahr Jaouz, Nahr Ibrahim, Nahr el Kelb, Nahr Beirut (Table 2.6 and Fig. 2.9).

Large springs are located

- At >1,000 m asl along the lower boundary of the outcrop belt of the Upper Cretaceous aquifer of the high mountain area in the catchment of:
  - Nahr Ibrahim: springs Afka and Rueis
  - Nahr el Kelb: springs Al Asal, Laban, As Saqie, Sanin
- Along the lower boundary of the outcrop belt of the Jurassic aquifer, occupying the western part of the central Lebanon mountains, at elevations between 200 and 635 m asl in the catchments of:
- Nahr Jaouz: spring Ad Dalle
- Nahr el Kelb: Jeita
- Nahr Beirut: Ain ed Delbe

Jebel Akroum – on the northeastern tip of the Lebanon mountains with a maximum peak altitude of 1,089 m asl – comprises a separate groundwater flow system between the Yamoune fault in the west and the main Lebanon water divide in the east. Springs of intermediate size discharge in the Jebel Akroum area from the Upper Cretaceous aquifer into the Nahr el Kebir el Janoubi stream system.

Groundwater flow in the *Akar plain* is generally directed toward the sea coast, in the southern part of the plain toward Nahr el Kebir el Janoubi. The shallow aquifers and the deeper Upper Cretaceous aquifer are probably hydraulically connected.

Estimates of groundwater recharge in the catchment area of the Akar plain, extending over around 1,050 km^2, indicate a total volume of groundwater flow of 170×10^6 m^3/a, part of which is extracted by irrigation schemes. It has been assumed that a substantial percentage of the groundwater flow volume discharges into the Mediterranean Sea.

On the western flank of the anticline forming the core of the *southern high Lebanon mountains*, large springs rise at the lower outcrop boundaries between the Jurassic and Upper Cretaceous aquifers and the Lower Cretaceous aquitard. The springs feed the base flow in Nahr Damour and Nahr Awali.

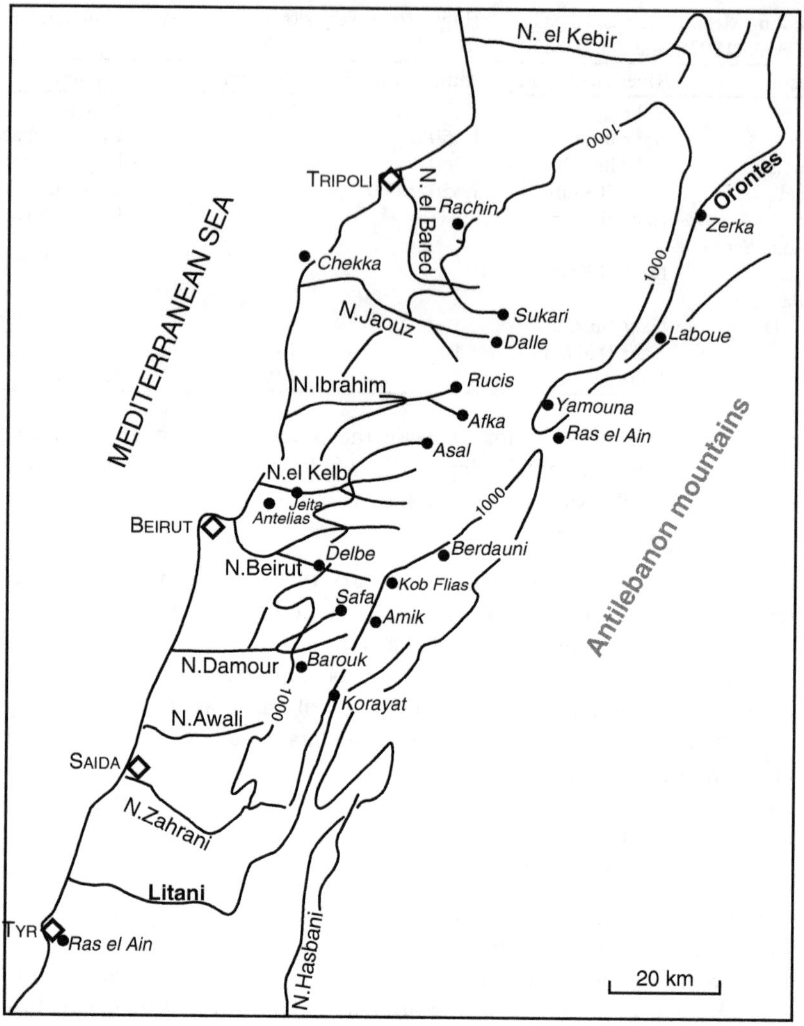

Fig. 2.9 Main wadis and springs in the Lebanon mountain area and the Bekaa. After Khair et al. (1992), Nations Unies (1967). • spring (ain); N. nahr (stream); ___1000___ topographic contour line 1,000 m asl

On the lower western slopes of the *southern Lebanon mountains*, extensive outcrops of the Lower Cretaceous marl aquitard and of the Chouf sandstones divide the karst aquifer system, comprising here prevailingly the Upper Cretaceous Sanin aquifer, into numerous catchments with generally limited extent.

A number of SW–NE to WSW–ENE directed faults and the large SSE–NNW oriented Rum fault dissect the *southwestern hill zone of the Lebanon range* into a complex pattern of tectonic units. Groundwater flow toward the Mediterranean

Sea in the west and to the Houle graben in the east appears to be, to a considerable extent, influenced by the tectonic structure.

The karstic Upper Cretaceous and Jurassic aquifers of the *eastern slope of the Lebanon mountains* are separated into a number of groundwater flow compartments. On the narrow steep slope sections west of the Yamoune fault, groundwater flows generally in an eastward direction. The Yamoune fault acts as barrier at least along part of its extent, causing groundwater discharge from the Upper Cretaceous aquifer section in several springs, in particular Ain Yamoune in the Orontes river catchment with an average discharge of 2.8 m^3/s.

Further important springs issuing along the Yamoune fault are the spring Berdauni from the Upper Cretaceous aquifer in the northern part, springs Kob Elias, Amik, Korayzat from the Jurassic aquifer in the southern part. To some extent, subsurface flow from the eastern mountain slopes reaches the Bekaa plain. On the northeastern mountain sector east of the Yamoune fault, groundwater systems with northeast and south direction feed large springs within the Bekaa valley.

2.4.3.6 Antilebanon Mountains

Groundwater flow in the karst aquifers of the Hermon–Antilebanon mountains is mainly directed to spring discharge areas located on tectonic or stratigraphic boundaries. Springs with high perennial discharge are:

- On the western slope of Mount Hermon: springs Hasbani (Upper Cretaceous aquifer), Chebaa, Sreid and Wazzani (Jurassic)
- On the eastern slope of Mount Hermon: Aouaj spring (Jurassic)
- On the western slope of the northern Antilebanon mountains: springs Ras el Ain at Baalbek, Laboue (Upper Cretaceous)
- On the eastern slope of the northern Antilebanon mountains: Ain Barada (Jurassic) in the Zebedani intermountain basin, Ain el Fije (Upper Cretaceous) in the Barada valley

To some extent, groundwater of the Upper Cretaceous aquifer moves from the mountain area into the adjoining Bekaa valley and Damascus plain.

Ain el Fije, the largest spring in the Antilebanon mountains, issues on the left bank of the Barada river at an altitude of 860 m asl. The spring drains a large catchment of fissured and karstified limestones and dolomites of Cenomanian–Turonian age. Mean discharge is 7.7 m^3/s. Base flow from the spring during the dry season is around 2.5 m^3/s sustained from groundwater storage of about 3.9×10^9 m^3. Ain Fije is the major source of water supply of Damascus City.

Ain Barada issues in the Zebedani valley at 1,100 m asl from Jurassic limestones, with a mean discharge of 3.3 m^3/s. Apart from Ain Fije and Ain Barada, the Barada river receives various minor tributaries, which are fed from springs at altitudes of 1,200–1,750 m asl. The 82 km long Barada river drains a catchment of 1,450 km^2 and has a mean annual discharge of 350×10^6 m^3.

The headwaters of the Aouaj river rise on the eastern flank of Mount Hermon at altitudes of 1,300–1,400 m asl. The Aouaj river has a length of 91 km; mean annual flow is 100×10^6 m^3.

Barada and Aouaj feed the Ghouta oasis in the Damascus plain.

Many small springs issue from shallow aquifers on the mountain slopes and in intermountain basins (Ponikarov et al. 1967b):

- From Bajocian limestones at Arne on the foot of Mount Hermon
- From Aptian and Albian sandstones and limestones of Mount Hermon
- From Lower Cretaceous limestones (Aptian) and sandstones (Albian) in the Zebedani intermountain basin
- From Jurassic (Oxfordian) limestones on the southeast slope of Mount Hermon
- From terrigenous formations (loam, sandy loam, pebbles, lacustrine marls and limestones) at several structural–morphologic depressions between the Antilebanon mountain chains and in the Qalamoun area

Groundwater flow in aquifers of the Damascus plain is directed mainly from the western and northwestern margins toward discharge zones around Lake Ateibe and Lake Hijane in the east. Some subsurface outflow appears to leave the Damascus plain toward east (southern Syrian steppe) and south (Yarmouk groundwater basin).

On the Qalamoun high plain, groundwater flow probably follows mainly the NNE direction of Wadi Mjarr.

2.4.3.7 Bekaa Valley

Groundwater movement within the Bekaa valley generally follows the topographic slope from the foothills toward the axis of the valley and, along the valley axis, toward NNE and SSW, respectively. The groundwater divide between northward flow in the Orontes sub-basin and southward flow in the Litani sub-basin is located approximately on the topographic vertex of the valley near Baalbek with a culmination of the groundwater surface at around 1,010 m asl. The groundwater surface descends to around 700 m asl near Hermel at the northern end of the valley and to 800 m asl at Qaraoun in the south. In the north, the valley grades into the Homs depression.

East of the Yamoune fault and north of the Bekaa water divide between Orontes and Litani river catchments, groundwater flow in the Upper Cretaceous aquifer is directed toward northeast to spring discharge areas in the Bekaa plain.

2.4.3.8 Ansariye Mountains

The main volume of groundwater recharged on the karstic surfaces of the Ansariye mountains circulates in the Mesozoic carbonate aquifers with general flow direction toward the Mediterranean Sea coast.

The groundwater divide between the Mediterranean Sea and the Orontes sub-basins is located along the anticlinal crest at a distance of 2–5 km from the Ghab valley and of around 30 km from the sea coast.

Cenomanian–Turonian limestones and dolomites form an extensive main aquifer on the western slope of the Ansariye mountains. Groundwater flow in the Cenomanian–Turonian aquifer of the northern part of the mountains is directed toward south–southwest to the large springs Banias and Nahr el Sinn, with mean discharges of around 1.5 and 10.5 m^3/s, respectively. Groundwater storage in the catchment of Nahr el Sinn is about 700×10^6 m^3.

In the southern part of the mountains, groundwater in the Jurassic and Cenomanian–Turonian aquifers moves prevailingly toward west and southwest to the coastal area around Tartous. Significant volumes of groundwater may discharge from the Cenomanian–Turonian aquifer system directly into the Mediterranean Sea.

At the foot of the steep slope of the Ansariye mountains on the boundary of the Ghab valley, various springs with discharge in the order of 100 l/s drain catchments of generally limited extent in the Jurassic carbonate aquifer. At the southwestern edge of the Ghab valley, the spring of Abou Qbeis drains a relatively large catchment with a mean discharge of 0.7 m^3/s.

2.4.3.9 Eastern Catchment of the Middle Orontes Area

Groundwater movement in the Cenomanian–Turonian aquifer in the eastern catchment of the middle Orontes sub-basin is directed to discharge points in large springs located at altitudes of 172–180 m asl within the Orontes valley and along the border of the Ghab graben. Major discharge points are the springs at Tell Ayoun near Sheizar, where a mean volume of 5.8 m^3/s flows directly into the Orontes river, and Ain et Taqa and Ain el Moudiq on the edge of the Acharne plain, Jebel ez Zaouiye and Ghab valley, with mean discharges of 3.8–5.7 and 1.3–1.8 m^3/s, respectively.

The subsurface catchment of the Cenomanian–Turonian aquifer system draining into the Orontes–El Ghab depression probably extends far to the east until the northern Palmyrean mountains. Main recharge to the Cenomanian–Turonian aquifer certainly occurs on the fissured and partly karstified outcrops of the limestones and dolomites between Masiaf, Hama and Jebel ez Zaouiye. Additionally, some subsurface inflow comes from the vast area of extent of the aquifer east of the outcrop areas.

Groundwater flow in shallow aquifers is partly directed to the closed basins further east on the Aleppo plateau and, to some part, to local discharge areas, e.g. small seasonal springs on Jebel ez Zaouiye or to areas like Maaret en Naamane where groundwater is extracted from shallow wells. A large amount of the seasonally recharged groundwater leaks, however, into the main Upper Cretaceous aquifer system discharging along the rift valley.

Various springs with low to medium discharge volumes issue along the boundary of the rift valley with Jebel ez Zaouiye and with the Idleb plateau. These springs are alimented mainly from Paleogene nummulitic limestone and chalk aquifers.

In the Acharne plain, Pliocene deposits composed of marls and sandstones provide a shallow aquifer, which is exploited by wells of a few tens of metres depth with generally low yields. The Pliocene aquifer feeds several small springs with mean discharges of 10–19 l/s.

References. Agrocomplect (1984–85), Al Charideh (2007), Bajjali (2006), Chebaane et al. (2004), ESCWA (1999a: 108 ff.), Gat and Dansgaard (1972), GTZ and NRA (1977), Gvirtzman (1994), Hobler et al. (1991), Hughes et al. (2008), Issar (1990), Kareh (1967, 1968), Kattan (1996a), Kozlov (1966), Kroitorou et al. (1985, 1992), Margane et al. (2002), Mijatovic and Bakic (1966), Ponikarov et al. (1967b), Salameh (1996, 2004), Salameh and Udluft (1985), Scarpa (1994), Shaban et al. (2006), Sunna (1995), Wagner (1996a).

2.5 Groundwater Salinity and Hydrochemistry

2.5.1 Mesozoic Karst Aquifers

2.5.1.1 General Hydrochemical Features

The mountain and highland ranges in the northwest of the Arabian Plate – the highlands of Jordan and Judea, the Lebanon, Antilebanon and Ansariye mountains and Jebel ez Zaouiye – comprise karstic Mesozoic limestone and dolomite aquifers, which receive significant recharge. These aquifers contain, in the unconfined zones, a group of waters with rather uniform hydrochemical composition characterized by predominant Ca, Mg and HCO_3 ions and relatively low salinity, a composition which is dominated by limestone and dolomite dissolution, typical for karst environments. Representative for the karst waters of the northwestern mountains and highlands is e.g. groundwater issuing in the large spring Ain el Fije in the Barada Valley upstream of Damascus. Hydrochemical parameter values of Ain el Fije are:

Salinity 230 mg/l TDS
HCO_3 170 mg/l, SO_4 5 mg/l, Cl 6 mg/l
Mg/Ca ratio 0.6

Ain el Fije drains a catchment area of several hundred km^2 in the Antilebanon mountains with a mean annual precipitation of 800 mm. The mean residence time of the spring water in the aquifer is 20–50 years.

HCO_3 and Ca concentrations of Ain el Fije are slightly higher than in very young groundwater recharge (Table 1.6), reflecting a stronger impact of carbonate dissolution during a somewhat longer retention period in the Ain el Fije spring water.

Groundwaters with a hydrochemical composition similar to the Antilebanon karst groundwater occur in wide areas of the northwestern mountain and highland sub-region in:

- The highlands of Jordan: Humar (A4) and Aman–Wadi Sir aquifers
- The Upper Cretaceous mountain aquifer in the Jerusalem–Jericho area of the West Bank
- Jurassic and Cenomanian aquifers in the Lebanon mountains
- Cenomanian–Turonian aquifers in the Ansariye mountains

Mean hydrochemical parameter values of these waters are in the following ranges:

- Salinity 180–710 mg/l TDS
- HCO_3 200–300 mg/l
- Cl 15–45 mg/l
- SO_4 9–45 mg/l
- Mg/Ca ratio generally 0.4–0.7

The hydrochemical composition is controlled primarily by dissolution of limestone and dolomite, which can be represented schematically as

$$CaMg(CO_3)_2 + 2CO_2 + 2H_2O \rightarrow Ca + Mg + 4HCO_3.$$

2.5.1.2 West Bank: The Judean Highlands

The main aquifer of the West Bank is constituted by karstified limestones and dolomites of Upper Cretaceous age (Hebron and Jerusalem formations, Cenomanian–Turonian, "mountain aquifer"). Groundwater with the typical hydrochemical composition of the karst water group of the northwestern highlands are found in the main recharge zone of the Judea aquifer – the outcrop areas along the anticlinal mountain crest. The groundwater in the phreatic part of the Judea aquifer is Ca–Mg–HCO_3 type water with low salinity of around 150–500 mg/l TDS, Cl concentrations in the range of 30–48 mg/l and Mg/Ca ratios of generally 0.7–0.9. These groundwaters, characterized by recent recharge and limestone–dolomite dissolution, discharge in springs, which are fed by phreatic or perched groundwater that flows in the upper part of the Judea group sequence through dolomites and limestones. Groundwater with similar hydrochemical composition is found in wells situated close to the recharge areas (Fig. 2.10).

Downstream of the recharge area, the Judea aquifer is mostly confined under the Mount Scopus aquiclude. Springs with the typical karst groundwater are situated on the northern mountain slope in the Jericho area at elevations between 300 m asl and 200 m below sea level at distances of up to 22 km from the recharge area: the springs Qilt, Fara and Fawar along Wadi Qilt, springs Duyuk, Nueima and Shosa, the spring group of Ain Sultan, and Ain Elisha near Jericho. The springs form outlets of an old karst system developed in the upper part of the Judea group aquifer.

Wells situated in the Jericho area extract groundwater with salinities of around 1,000 mg/l TDS and somewhat elevated Mg, Na and Cl concentrations. "These trends can be explained by intermixing of ground water originating from the Rift

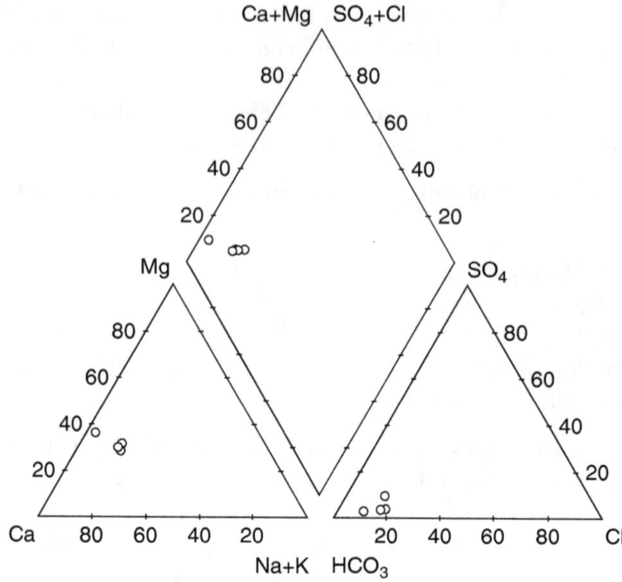

Fig. 2.10 Piper diagram: Spring water samples from the Judea aquifer of the West Bank. Springs Fara, Fawar, Qilt and Elisha, mean values, data from Kroitorou et al. (1985)

Valley Fill, and interaction with the aquifer rocks, which change in composition from limestone in the west to dolomite in the east" (Kroitorou et al. 1992).

The main springs discharge water from an upper fast-flow karstic system, where an impact of water–rock interaction after the recharge process is rather limited. The springs constitute one hydrochemically homogeneous group. The wells tap deeper confined parts of the aquifer with longer retention periods and increasing groundwater salinity in direction of groundwater flow.

A major discharge zone of the Judea aquifer of the West Bank is situated at the Dead Sea shore at Ain Feshka with brackish water of around 3,000 mg/l TDS. In the Jordan valley, groundwater from the Judea aquifer mixes with brackish water in the overlying Quaternary aquifer.

Groundwater salinity in the Upper Cretaceous Judea group aquifer of the Tulkarm–Qalqiliya area on the western foreland of the Judean mountains is generally moderate with electrical conductivity values between 500 and 1,200 μS/cm and Ca and HCO_3 as predominant ions. The major ion composition varies in a relatively wide range. Mean concentrations of 80 samples are 87 mg/l for Cl, 23 mg/l for SO_4 and 320 mg/l for HCO_3. The Mg/Ca ratio is, on average, 0.8 (data from Shahab 1997).

Deviation of ion ratios from the typical groundwater of the Judea limestone–dolomite aquifer observed in many samples may be related to mixture with water from the overlying Avedat chalk aquifer; contamination from the surface may also have a locally significant impact.

2.5.1.3 Highlands of Jordan

Outcrops of Upper Cretaceous limestone and dolomite aquifers provide the main recharge zones of the highlands of Jordan in the Irbid, Ajloun, Wadi Sir areas and the northern Mujib sub-basin. The Upper Cretaceous karst aquifers of these areas contain prevailingly Ca–HCO_3 type groundwater with moderate salinity, typical for the karst groundwater group of the northwestern highlands. The characteristic composition of karst carbonate water of Ca–HCO_3 type with a total salinity around 400–700 mg/l TDS is found in springs, base flow and in wells in the aquifer outcrop area (Fig. 2.11).

Water of base flow fed by groundwater discharge from Upper Cretaceous aquifers in the Yarmouk river and in wadis on the northern part of the western escarpment of the highlands of Jordan is Ca–HCO_3 to Ca–Na–HCO_3 type water with salinities between 320 and 550 mg/l TDS.

Lloyd (1965) characterizes the limestone waters of the highlands of Jordan: "These waters belong almost completely to the Bicarbonate Group of waters and as would be expected contain a predominant amount of the Ca ion. The movement of water in limestones is essentially along joint and fracture plains with the result that the amount of rock surface in contact with the water is small compared to the volume of water transmitted. … the inability of the water to penetrate the crystalline limestones such as those in A4 and A7 formations allows little

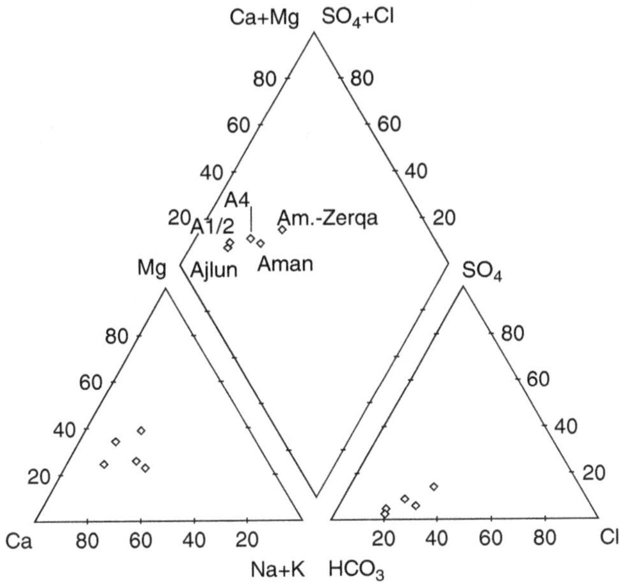

Fig. 2.11 Piper diagram: Groundwater samples from Upper Cretaceous carbonate aquifers of the highlands of Jordan. Mean values of analyses from the A1/2 (Naour) and A4 (Humar) aquifers, from the A7 (Wadi Sir) aquifer in the Ajlun–Irbid and Aman–Zerqa areas, and from the B2 (Aman) aquifer in the Aman–Zerqa area, data from Rimawi (1985)

opportunity for the extraction of the soluble chlorides and sulphates. This results in relatively pure waters".

Variations in the hydrochemical composition of groundwater in the outcrop areas of the Upper Cretaceous aquifer complex appear related mainly to the aquifer lithology. Cl and SO_4 concentrations are particularly low in karstic limestones of the Humar (A4) aquifer in the Irbid, Ajloun, Aman areas and somewhat higher in chalky limestones and dolomites of the Aman–Wadi Sir (B2/A7) aquifer. HCO_3 concentrations in the outcrop areas are generally in a range of 250–300 mg/l. Mg/Ca ratios vary in wide ranges between 0.2 and 1.1 with an average of 0.6.

Groundwater salinity increases, in general, along the path of groundwater flow from the recharge areas in the highlands to the areas with natural groundwater discharge or groundwater extraction downstream. Low salinities are characteristic for the outcrop area of the aquifer, in which tritium values of 4.5–10 T.U. indicate the occurrence of present-day recharge. Salinity in samples from the Upper Cretaceous Aman–Wadi Sir (B2/A7) aquifer ranges from less than 400 to 1,660 mg/l TDS.

The increase of salinity is accompanied by various hydrochemical processes resulting in changes of the hydrochemical composition of the groundwater in the Upper Cretaceous aquifer complex. Groundwaters from the outcrop area of the Aman–Wadi Sir (B2/A7) aquifer system have a geochemical facies typical of karst environments indicating limestone and dolomite dissolution during infiltration. Carbonate material of the aquifer is dissolved until the soil CO_2 introduced during recharge is consumed and a saturation equilibrium is reached at concentrations of around 400 mg/l HCO_3. In the confined parts of the aquifer, dissolution of sulfate and chloride from rock material leads to an increase in Cl and SO_4 concentrations; sulfate reduction and additional carbonate dissolution leads to an increase of HCO_3 concentrations. The redox processes are assumed to involve oxidation of organic carbon mainly in the confining aquitard of the Belqa formation (B3), reduction of SO_4 to HS, and reaction of the resulting CO_2 with aquifer carbonate to increase the HCO_3 concentration up to around 430 mg/l.

In the Ajlun highlands at altitudes between 620 and 1,000 m asl, groundwater in the outcropping Wadi Sir (A7) aquifer has a salinity of 340–435 mg/l TDS with very uniform hydrochemical composition of the water from different wells. The water infiltrating into the karstic formation attains a rather stable chemical composition with HCO_3 concentrations of 240–290 mg/l and Mg/Ca ratios of 0.5–0.6. Cl and SO_4 concentrations are relatively low with 32–43 mg/l Cl and <1–41 mg/l SO_4. Sources of Cl and SO_4 are probably atmospheric precipitation, small amounts of dissolution of salts and evaporite minerals from the marine carbonate aquifer and anthropogenic sources, such as wastes and fertilizers.

In the northern Wadi Mujib sub-basin, groundwater in the Aman–Wadi Sir (A7/B2) aquifer is prevailingly Ca–HCO_3 type water with salinities of around 350–600 mg/l. Groundwater with higher salinity of up to 1,600 mg/l TDS and Cl or SO_4 predominance is found at some locations.

In the Aman City area, the unconfined highly fractured Aman–Wadi Sir (A7/B2) dolomitic limestone aquifer is recharged from precipitation in the winter months and through leakage from the drinking water system and the waste water drainage system. Leakage from cesspools also contributes to the recharge. The groundwater is generally Ca–HCO_3 type water with electrical conductivity values between 450 and 1,350 μS/cm. Anthropogenic impacts are indicated by Cl concentrations of up to 125 mg/l and NO_3 concentrations exceeding 100 mg/l. Nitrate concentrations are particularly high in street runoff with 158–250 mg/l NO_3 in the city centre. A significant correlation between Ca and NO_3 concentrations indicates that dissolution of carbonate rocks is enhanced by the slightly acidic conditions created by the reaction of NO_3 with rain water.

Groundwater in the Upper Cretaceous aquifer in the Aman–Zerqa sub-basin downstream of Aman City is prevailingly Ca–HCO_3 type water with low to moderate salinity (electrical conductivity 295–1,170 μS/cm). NO_3 concentrations between 37 and 112 mg/l indicate wide-spread anthropogenic contamination. Groundwater tapped in deeper parts of the Upper Cretaceous aquifer complex appears to be characterized by Mg/Ca ratios close to 1, indicating an equilibrium with a dolomitic aquifer lithology. Cl and NO_3 concentrations are relatively low in these deeper groundwaters, which appear not to be affected heavily by contamination from the surface.

In deep confined parts of the aquifer complex, brackish groundwater with an electrical conductivity of 1,450 μS/cm is found under reducing conditions.

Ca–HCO_3 type groundwater with moderate salinity reaches within the Upper Cretaceous aquifer into the Jordan valley, where it discharges in thermal springs and is tapped in boreholes.

2.5.1.4 Lebanon Mountains

Groundwaters in the Jurassic and Upper Cretaceous aquifers of the Lebanon mountains are generally typical representatives of the hydrochemical group of karst waters of the northwestern mountains and highlands. Ion ratios are very similar to ratios of the Ain el Fije spring water. The groundwater is Ca–HCO_3 type water with salinities between 215 and 520 mg/l TDS, percentages of HCO_3 are generally in a range of 82–96 meq%, of Ca in a range of 54–82 meq%. HCO_3 concentrations are 230–570 mg/l in the Cenomanian–Turonian aquifer, 215–510 mg/l in the Jurassic aquifer. Mg/Ca ratios range between 0.3 and 0.4 in the Cenomanian–Turonian aquifer and vary from 0.1 to around 1 in the Jurassic aquifer.

Along the Mediterranean Sea coast, Ca–Cl type water is found in the Cenomanian–Turonian aquifer with a salinity of around 750 mg/l TDS, Cl concentration around 250 mg/l and a Mg/Ca ratio of 0.7.

2.5.1.5 Antilebanon Mountains

The characteristics of karst groundwater from the Upper Cretaceous aquifer of the Antilebanon mountains, as represented by spring water of Ain el Fije, have been mentioned in Sect. 2.5.1.1. Variations in the hydrochemical composition of groundwater in the Cenomanian–Turonian aquifer of the Antilebanon mountains in Syria involve mainly variations in HCO_3 concentration, ranging from 110 to 220 mg/l, and of Mg/Ca ratios between 0.14 and 0.66 (Fig. 2.11).

Seasonal variations of the Mg/Ca ratio in the Ain el Fije spring water from 0.76 in the dry season to 0.6 in winter appear to be related to higher Ca dissolution in fresh recharge and a somewhat higher percentage of Mg in the groundwater storage in the dolomitic aquifer.

SO_4 concentrations in the typical groundwaters of the Cenomanian–Turonian aquifer are less than 20 mg/l. Higher SO_4 concentrations in some springs (30–130 mg/l) indicate a contribution from the Jurassic aquifer, which contains layers of gypsum and anhydrite.

Springs issuing from the Jurassic aquifer on the slopes of Mount Hermon have low salinities of 125–300 mg/l TDS. Mg/Ca ratios of 0.17–0.30 and low HCO_3 concentrations of 92–200 mg/l of these Ca–HCO_3 type waters may be related to very short retention periods, during which a hydrochemical equilibrium with the aquifer rocks has not yet been attained (Fig. 2.12).

Reported values of ion concentrations of groundwater from Jurassic and Upper Cretaceous aquifers of the western slope of the Antilebanon mountains show a homogeneous hydrochemical composition of Ca–HCO_3 type water with salinities of about 315 mg/l TDS and HCO_3 concentrations around 200 mg/l. Cl and

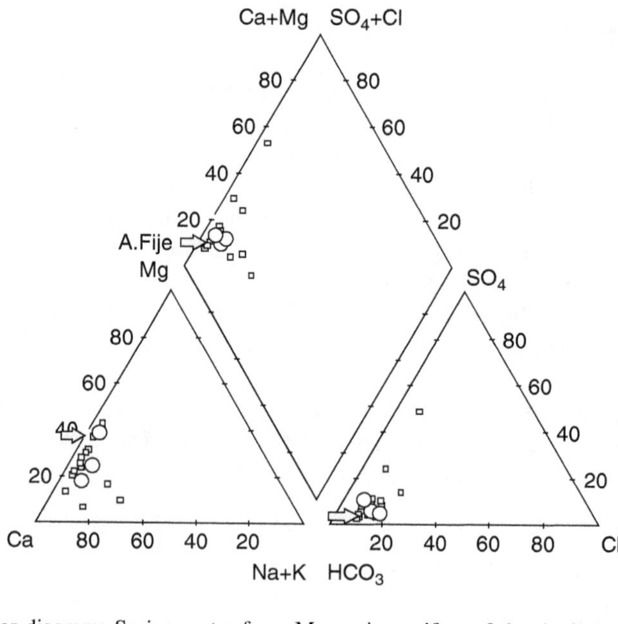

Fig. 2.12 Piper diagram: Spring water from Mesozoic aquifers of the Antilebanon mountains. □ Data from Syria: Ponikarov et al. (1967a), Kattan (1996a); ○ data from Lebanon: UNDP (1970)

SO_4 concentrations are generally low, Mg/Ca ratios range from 0.24 to 0.52. The hydrochemical composition corresponds closely to the composition of most groundwaters on the eastern slope of the Antilebanon mountains.

2.5.1.6 Northwestern Syria

Groundwaters of the Upper Cretaceous and Jurassic aquifers of the Ansariye mountains show, in general, the typical hydrochemical composition of the karst groundwaters of the northwestern mountains and highlands. They are represented by Ca–HCO_3 type waters with salinities of 330–460 mg/l TDS and HCO_3 concentrations of 180–340 mg/l. Mg/Ca ratios of most samples are in a range between 0.15 and 0.9. Cl and SO_4 concentrations are generally low. Slightly elevated Cl concentrations, up to 28 mg/l Cl, in the Upper Cretaceous aquifer of the southwestern part of the Ansariye mountains may be attributed to an impact of marly intercalations within the aquifer.

Water in several springs issuing on the eastern border of the Ansariye mountains has elevated Ca and SO_4 concentrations, up to 250 mg/l Ca and 570 mg/l SO_4. These elevated concentrations probably originate from gypsiferous layers in the Jurassic aquifer, which is tectonically uplifted in the core of the Ansariye anticline. The SO_4 concentrations show seasonal variations with lower values during the rainy season, attributed to an dilution effect from fresh recharge. Seasonal variations are also observed in Mg and HCO_3 concentrations in some spring waters with an increase of concentrations during the dry season. Probably the groundwater stored in the Jurassic aquifer over longer periods is characterized by somewhat higher Mg, HCO_3 and SO_4 concentrations in comparison to the groundwater recharge during the rainy season.

On Jebel ez Zaouiye and in the plateau and plain areas of the middle Orontes catchment (Masiaf plateau, Hama plateau, Acharne plain), waters of the hydrochemical group of karst waters of the northwestern highlands have been tapped in the Cenomanian–Turonian aquifer through several boreholes. These waters extracted from wells are Ca–HCO_3 type waters with a salinity of 375 to around 550 mg/l TDS, HCO_3 concentrations of 218–310 mg/l and Mg/Ca ratios between 0.5 and 0.9.

In the outcrop area of the Cenomanian–Turonian aquifer on the Masiaf plateau, alterations of the hydrochemical groundwater composition through anthropogenic contamination are indicated by NO_3 concentrations of 48–75 mg/l and elevated HCO_3 concentrations of up to 550 mg/l.

Groundwater with moderate salinity and low SO_4 concentration is found in Cenomanian–Turonian aquifer of the middle Orontes catchment only in areas, where the surface of the aquifer is exposed or situated under a cover of Pliocene–Quaternary terrestrial deposits. In the area east of the Hama–Acharne, where the aquifer is overlain by Campanian chalks and marls, the groundwater of this aquifer has elevated SO_4 concentrations of up to 530 mg/l, accompanied by relatively high Ca concentrations (up to 109 mg/l) and Cl concentrations (up to 140 mg/l).

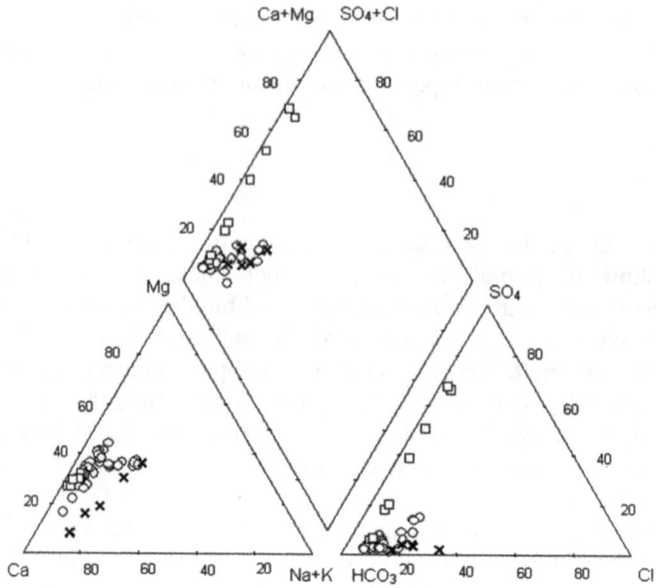

Fig. 2.13 Piper diagram: Groundwater samples from Mesozoic carbonate aquifers of northwestern Syria. ○ Cenomanian–Turonian, Jurassic, Ansariye mountains and Masiaf–Hama–Jebel ez Zaouiye area; □ Jurassic, eastern border of Ansariye mountains; **x** southwestern Ansariye mountains. Data from Boeckh et al. (1970)

The composition of that group of Ca–HCO_3 to Ca–SO_4 type waters may be interpreted as a mixture of local low salinity groundwater recharge with brackish sulfate rich groundwater, which prevails in the Upper Cretaceous aquifer in the area east of the Hama–Aleppo (Chap. 3). In the water discharging from the Upper Cretaceous aquifer in large springs on the Orontes river and on the boundary of the Ghab depression (springs Tel Ayoun, Ain et Taqa, Ain el Moudiq), relatively high SO_4 concentrations of 68–180 mg/l indicate components of the sulfate rich water in the sub-regional groundwater discharge (Fig. 2.13).

2.5.2 Nummulitic Limestone Aquifers

Eocene nummulitic limestone aquifers occur, in not very extensive areas, in Jebel Ansariye, Jebel ez Zaouiye and the southern Bekaa in Lebanon. Nummulitic limestones are composed chiefly of calcite from shells of foraminifera (nummulites) with low Mg content.

Waters from nummulitic limestone aquifers are typically Ca–HCO_3 type waters, characterized by high percentages of Ca and HCO_3 and low concentrations of Mg, Na, Cl and SO_4. Salinity of the waters from nummulitic limestone aquifers in Syria and Lebanon varies between around 250 and 400 mg/l TDS. HCO_3 percentages of

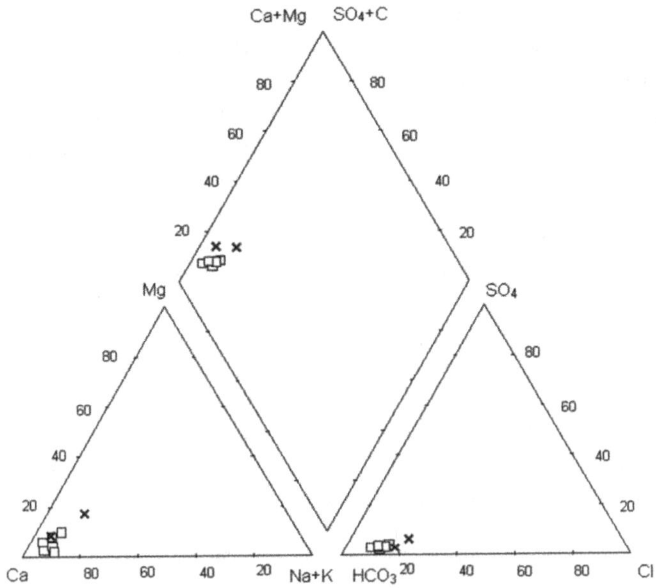

Fig. 2.14 Piper diagram: Groundwater samples from nummulitic limestone aquifers. □ north-western Syria, data from Boeckh et al. (1970); • Bekaa plain, Lebanon, data from UNDP (1970)

the water samples from wells in Syria range from 80 to 88 meq%. Typical Mg/Ca ratios are 0.03–0.1 (Fig. 2.14).

2.5.3 *Aquifers in Intermountain and Foreland Depressions*

The rift valleys and plains on the foot of the northwestern mountains are main recipients and mixing zones of groundwater and surface water inflow from the adjoining mountain slopes. Salinity and hydrochemical composition of groundwater is influenced by subsurface inflow from the mountains, local mainly indirect recharge, salt water bodies in the centres of some plains and, in many areas, also by intensive agriculture.

2.5.3.1 Jordan–Dead Sea valley

Natural water occurrences in the Jordan–Dead Sea valley range from fresh water, issuing in springs on the fringes of the valley plain, to brines of the Dead Sea with a salinity of 300 g/kg TDS. The waters in that wide salinity range include a variety of hydrochemical types. The general salinity distribution in the shallow sand and gravel aquifer (Quaternary Jordan valley group) shows fresh water belts on the western and eastern fringes of the valley and a zone of saline water along the valley floor, the Zor, and around the Dead Sea. The fresh groundwater overlies the saline groundwater in a thin body and is restricted to the sand and gravel deposits on the valley margins and in fans protruding toward the valley floor. The groundwater with low salinity approaches, as lateral flow from the western and eastern margins of the valley, the central valley floor, which is occupied by the salt water body. The position of the interface between fresh water and salt water depends on the quantity of fresh water inflow and the lithologic composition of the Jordan valley deposits and is influenced by seasonal climatic variations and rates of artificial groundwater extraction.

Relatively stable interface conditions are maintained in the central Zor area of the valley, which is covered by the Lisan marls. Horizontal groundwater flow is very low within the marls and evaporation and solution of salts provide permanent sources of groundwater salinization. In some areas, the fresh groundwater extends in shallow lenses relatively far toward the valley centre, but a fresh water layer does not exist in the Zor area and in areas, where the Lisan marl formation replaces the sandy facies of the Quaternary Jordan valley deposits.

Some 15,000 years ago, most of the Jordan–Dead Sea valley was covered by the Lisan salt water lake with a lake water level at 180 m below sea level. The entire alluvial aquifer was then saturated with brine. Since the drying up of the Lisan lake, which left as remnant the present Dead Sea, the fresh water–salt water interface has been pushed from the fringes of the Jordan valley toward its central part by fresh water inflow. The aquifer was continuously flushed by lateral groundwater inflow and local fresh water recharge. "The flushing process of the brine from the aquifer following the base (lake) level drop is relatively rapid. In most cases, a few decades are sufficient for a complete flushing and only locally are brine relicts detected at higher levels" (Yechieli et al. 2001).

On the western side of the Jordan valley in the area around Jericho, fresh groundwater issues from the Upper Cretaceous aquifer in springs while water extracted from boreholes is generally brackish. The spring water represents discharges of the Ca–HCO_3 type water of the karst water group of the northwestern highlands. The well water shows a trend of increasing salinity and increase of major ion concentrations in direction of groundwater from west to east toward the valley floor. The hydrochemical data of the waters indicate a trend between two distinctive end members: Ca–(Mg)–HCO_3 type fresh water of the springs in the west to brackish Na–(Mg)–Cl water in the east. Mg percentages remain rather constant around 35–40 meq%, SO_4 percentages are generally below 20 meq%. The source of the elevated salinity in wells in the east downstream of Jericho, reaching up to 3,660 mg/l TDS and Cl concentrations of 1,860 mg/l, may be a deeper brine and/or leachate of the Lisan marls. In some wells, the groundwater quality appears to be influenced by contamination from agricultural activities.

Aquifers below the shallow Jordan valley aquifer are found in Upper Cretaceous limestones and dolomites (Aman–Wadi Sir aquifer, Judea aquifer) and, in some areas, in Paleogene chalks (Jenin aquifer).

The Upper Cretaceous aquifer system in the northern Jordan valley contains various types of fresh to brackish groundwater:

- Water with hydrochemical composition similar to groundwater in the recharge areas in the highlands with slightly to significantly higher salinity:
  – Ca–HCO_3 type groundwater with a salinity of around 500 mg/l TDS from the B2 (Aman) aquifer in the Mukheibe well field in the northern Jordan valley
  – Ca–Mg–HCO_3 type groundwater with salinities of 700–1,000 mg/l extracted from the A7 (Wadi Sir) aquifer in several boreholes in the northern Jordan valley
- Na–Cl type groundwater with salinities of 900–1,370 mg/l TDS discharging in springs and extracted from boreholes

The groundwaters are partly thermal with temperatures between 39°C and 54°C. The thermal groundwaters have a similar hydrochemical facies as modern groundwater from the outcrop areas, although Mg/Ca ratios and SO_4 concentrations are higher, indicating that dolomite and gypsum dissolution is continuing downstream of the recharge area. The carbonate equilibrium is maintained along the flow path by calcite precipitation.

Variations of Mg/Ca ratios between 0.4 and 0.9 may be related to different degrees of dissolution processes of dolomitic carbonates and mixing processes. Elevated Na and Cl concentrations (up to 300 mg/l Cl) found in some waters may originate from saline water in deeper sections of the Upper Cretaceous aquifer. Significant HS^- concentrations indicate an impact of sulfate reduction in the thermal groundwaters.

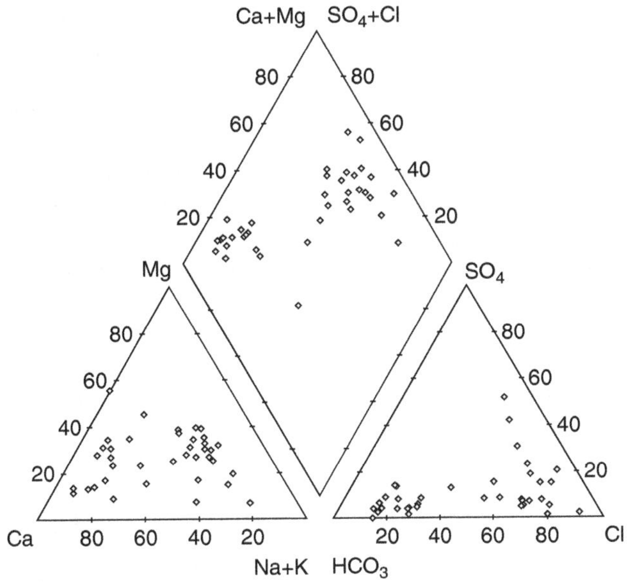

Fig. 2.15 Piper diagram: Groundwater samples from various aquifers of the Jordan valley. Data from Hirzalla (1973)

Thermal brackish water discharges in springs at various points on the eastern rim of the Jordan valley from sandstone aquifers of the Zerqa or Kurnub formations. The water of the thermal springs is mainly Na–Cl type water with salinities between 1,100 and 2,800 mg/l TDS.

Brackish Na–Cl type waters discharging from the Judea aquifer in the Feshkha springs on the eastern shore of the Dead Sea are assumed to constitute a mixture of trapped Mediterranean Sea water and inflowing ground water with no remarkable contribution of Dead Sea brines. The spring water has a salinity of 3,000 mg/l TDS and Cl concentrations of 1,660 mg/l. "...the Mediterranean Sea advanced into the Rift Valley, via the Escherlon Valley, in the geologically recent past, infiltrated into the tectonically shattered terrain and mixed with various amounts of the then prevailing groundwaters. Isolated pockets of these mixed waters became trapped and stayed there although the sea as a whole retreates" (Mazor and Mero 1969). The hydrochemical composition of the trapped sea water has been modified by reactions with the aquiferous rocks. Groundwater with similar hydrochemical characteristics occurs in springs and wells at various locations of the Jordan–Dead Sea valley from Lake Tiberias in the north to the southwestern end of the Dead Sea ("Tiberias–Noit water association" Mazor and Mero 1969). The age of the Mediterranean Sea intrusion into the rift valley is estimated to at least 18,000 years B.P.

In some springs on the eastern Dead Sea shore, saline springs with a Cl concentration of 24,000–40,000 mg/l TDS apparently represent diluted Dead Sea water.

Several springs with fresh to saline water issue in the Lake Tiberias (Buhaira Tabiriye) area at around 200 m below sea level on the northern end of the Jordan valley. In boreholes, fresh to brackish groundwater has been tapped in the area on the eastern shore of Lake Tiberias.

A chain of thermal springs rises along fault zones; the largest spring discharge is concentrated in the Al Himme (Hammat Gader) springs near the right bank of the Yarmouk river. Spring temperatures vary from 26° to 49°C, salinity from 600 to 1,400 mg/l TDS. The spring water is Ca–HCO_3 type water with nearly equal percentages of Ca, Na, HCO_3 and Cl in the higher mineralized waters. Springs on the western shore of Lake Tiberias are mainly brackish Na–Cl waters with salinities between 2,700 and 7,500 mg/l TDS and temperatures of 26–29°C. Salinity reaches 30,000 mg/l TDS in a thermal spring with a temperature of 64°C.

The brackish to saline waters on the western Lake Tiberias shore may be mixtures of three end members: fresh water, remnants of Mediterranean Sea water and rift valley brines. The waters on the eastern shore of Lake Tiberias "could have been formed through dissolution of slightly altered marine waters entrapped in the Judea Group aquifer mixed with diluted Rift Valley brines" (Arad 1988).

In northern Wadi Araba, tongues of brackish or fresh groundwater extend from the edges of the valley to the central zone with highly saline groundwater or overlie the saline groundwater. In the pre-development stage, the Quaternary aquifer contained fresh water with salinities rarely exceeding 800 mg/l TDS. Salinity levels of groundwater in production wells increased to >5,000 mg/l TDS. New wells tapped brackish water with TDS of 4,000–5,500 mg/l. Near the Dead Sea residual brines and salt deposits occur within the Quaternary sediments.

2.5.3.2 Bekaa and Middle Orontes Valley

Some plain zones of the Bekaa and the middle Orontes catchment are covered by young basin sediments, while in hill and plateau zones Mesozoic to Tertiary carbonate formations are exposed.

Main aquifers in the Bekaa plain are Quaternary terrestrial deposits and the underlying Cenomanian–Turonian carbonate aquifer. Groundwater in the Quaternary aquifer is generally Ca–HCO_3 type water with salinities of 500–1,000 mg/l TDS. In the southern part of the plain, pockets of brackish Ca–Cl water are found, probably related to impacts of intensive agricultural irrigation.

The Homs plain is covered widely by lacustrine Pliocene deposits, which provide a shallow fresh water aquifer containing Ca–HCO_3 type water with salinities of 300–600 mg/l TDS. Shallow fresh water also occurs in Pliocene–Quaternary terrestrial deposits of the Acharne plain. The Ghab valley, which is covered by Pliocene clayey marls and calcareous clays, contains no productive aquifer. Groundwater from the Mesozoic karst aquifers discharges along the boundaries of the valley.

2.5.3.3 Depressions of the Eastern Forelands of the Antilebanon Mountains

The Antilebanon mountains are adjoined in the east by the tectonic Damascus depression and the synclinal Qalamoun high plateau.

The Damascus plain comprises an important shallow aquifer composed of Quaternary fluviatile–terrestrial deposits above a complex tectonic basin structure with aquiferous Upper Cretaceous to Paleogene carbonate rocks and Pleistocene–Quaternary volcanics. Groundwater in the Quaternary aquifer of the western – upstream – part of the Damascus plain is generally fresh water of Ca–HCO_3 type with salinities between 300 and 550 mg/l TDS. Groundwater salinity increases in the middle and eastern parts of the plain to levels of generally 1,000–4,000 mg/l TDS with an increase of Cl and SO_4 concentrations. The highest groundwater salinities of around 5,000 mg/l TDS with prevailing Na–Cl waters are found in a wide belt in the east of the plain. The main zone of elevated groundwater salinization encloses the area of an ancient lake without outlet, the remnants of which were the now dried up lakes Ateibe and Hijane.

The main source of the increase of groundwater salinity in the downstream parts of the Damascus plain can be seen in evaporative enrichment in a closed basin environment. At present, the groundwater and hydrochemical regime in the Damascus plain is significantly influenced by urban and agricultural activities: water import from the Antilebanon mountains, deviation of surface water in channels, waste water irrigation return flow, groundwater extraction from wells.

The Qalamoun high plateau comprises a complex system of Upper Cretaceous and Paleogene carbonate aquifers covered by aquiferous terrestrial deposits of Neogene age. Groundwater in the shallow Neogene is generally fresh with EC values of 200 to 700 μS/cm. Elevated SO_4 concentrations are found in some wells tapping the deeper Upper Cretaceous aquifer, in particular the upper sections of the aquifer (Campanian–Maastrichtian), where SO_4 concentrations of up to 650 mg/l have been observed. High HCO_3 concentrations of up to 650 mg/l, occurring at some locations in the shallow aquifer, may indicate an impact of agricultural contamination.

Water extracted through shallow wells or qanats (falaj systems, Sect. 8.3) from Paleogene and Quaternary aquifers in the Qalamoun plain is prevailingly Ca–HCO_3 type fresh water.

References. Abumaizer (1996), Al Charideh (2007), Almasamir and Sarcis (1992), Arad (1988), Bajjali (2006, 2008), Bajjali et al. (1997), Batayneh (2006), Boeckh et al. (1970), Clark et al. (1995), GTZ and NRA (1977), Hirzalla (1973), Hobler et al. (1991), Kattan (1996a, b), Khayat et al. (2006), Kroitorou et al. (1992), Lloyd (1965), Mazor et al. (1973), Mazor and Mero (1969), Mazor and Molcho (1972), Nations Unies (1967), Parker (1970), Ponikarov et al. (1967b), Rosenthal (1987, 1988), Salameh and Rimawi (1984, 1987b), Salameh and Khdier (1985), Salameh et al. (2002), Salameh and Shaqur (1981), Shahab (1997), Shatsky et al. (1966), Sunna (1995), UNDP (1970), Wilson and Wozab (1954), Wolfart (1966), Yechieli et al. (2001).

2.6 Isotopic Composition of the Groundwater

2.6.1 Groundwater Age

^{14}C values of most groundwater samples from the higher reaches of the northwestern highlands vary between 55 and 62 pmc, reflecting a modern age of the actively circulating groundwater in the karst aquifers. In some springs, higher ^{14}C values were analysed (91 pmc in Elisha springs on the West Bank, 103 pmc in spring Al Faouar in the Ansariye mountains) indicating very recent (post-bomb) recharge after 1955.

^{3}H values of groundwater from the karst aquifers in the highland areas are reported in the following ranges:

Antilebanon 18–29 TU (data of 1979/80), 9–24 TU (data of 1989/90)
Ajlun–Nuaime area in the highlands of Jordan 4–10.5 TU (data of 1987–1995)
Springs in the Jericho area of the West Bank 7–22 TU (data of 1980s)

^{3}H values of 10–33 TU and a ^{14}C value of 58 pmc reflect recent groundwater recharge in groundwater of the Ansariye mountains.

All these karst groundwaters obviously contain significant components of recent recharge. For spring water of Ain el Fije with a ^{14}C value of 62 pmc, a mean retention period of the groundwater of around 50 years has been calculated. Spring discharge from the large reservoir of the Ain el Fije aquifer comprises a mixture of fast flowing recent groundwater and base flow of older groundwater. For the spring As Sinn – on the foot of the Ansariye mountains at the Mediterranean Sea coast – with a ^{14}C value of 58 pmc, a mean retention period of 60 years is estimated. For the karst groundwaters with ^{14}C values between 55 and 64 pmc, retention periods of some decades to a few thousand years can be assumed. The comparison of corresponding ^{14}C and ^{3}H values of aquifers on the Qalamoun high plain yields an initial ^{14}C value of 68 pmc, corresponding to an adjustment of the ^{14}C water ages by 3,000 years.

In the wadis running from the highlands of Jordan toward the Jordan valley, ^{3}H values range from below detection level to 10 TU. ^{14}C values of 42–64 pmc have been found at some locations in Wadi Ziglab and Wadi Yabis and in water of a spring on the rim of the Jordan valley.

In the carbonate aquifer of the highlands in Jordan south of Aman, detectable tritium levels occur predominantly in valleys and indicate the importance of indirect recharge. In many groundwater samples from recharge areas, ^{3}H is below detection level, implying that the large volumes of direct recharge can only move through a thin high permeability upper section of the aquifer over the top of the underlying slower moving older waters. Tritium levels decline with increasing well yield.

In the area around Lake Tiberias and the lower Yarmouk valley and in the lower reaches of Wadi el Arab, ^{14}C values in springs and boreholes vary from 3 to 18 pmc. Groundwater samples from these confined parts of the aquifers is generally free of detectable tritium. ^{14}C values appear to have been influenced by secondary hydrochemical processes in the aquifer: SO_4 reduction through oxidation of fossil organic matter and subsequent $CaCO_3$ dissolution. Corrections considering these hydrochemical processes lead to water ages of about 4,200–7,000 years at Mukheibe well field, corresponding to flow velocities of 10–15 m/a between the recharge area and the well field. Retention periods in the deeper less permeable aquifers are calculated as 7,000–15,000 years.

Groundwater ages in the carbonate aquifers of that area range up to 32,000 years, many groundwater samples have ^{14}C activities of less than 10 pmc and old groundwater is present even in the recharge mound areas.

Groundwater in some recharge areas on the top of the highlands is modern with ages of less than 4,000 years. High groundwater ages of up to 26,000 years are found in the southeast of the highlands north of Maan.

In the springs on the slope of the Judean highlands in the Jericho area, an unexpected trend of increasing ^{14}C values with distance from the recharge zone is observed. That trend can be explained by a mixing of recent fast flowing groundwater in an upper karst system with older deeper groundwater. For the relatively older groundwater, which has been tapped by boreholes, retention periods of 1,800–4,000 years were calculated. Springs on the mountain slope appear to contain admixtures of 20–70% of older groundwater. In the spring Ain Elisha downstream of Jericho at 200 m below sea level, only groundwater from the upper flow system is discharging. The end members of the mixed groundwater are represented by Elisha spring with a ^{14}C value of 90 pmc and by older water from boreholes with about 40 pmc.

2.6.2 *Stable Isotopes of Oxygen and Hydrogen*

2.6.2.1 Isotopic Composition of Precipitation

The precipitation throughout the eastern Mediterranean Sea area is characterized by a relatively large deuterium excess

$$d = \delta^2H - 8 \times \delta^{18}O > +15‰.$$

The scatter in the isotope content of precipitation with a high deuterium excess is inherited from processes over the sea. The Mediterranean Meteoric Water Line (MMWL), typical for rain in the eastern Mediterranean area, has been defines as

$$d = \delta^2H - 8 \times \delta^{18}O + 22‰.$$

Rain in observation stations at lower topographic altitudes is more enriched in heavy isotopes and the d value is lower than at mountain stations (Gat and Carmi 1970; Gat and Dansgaard 1972).

Mean stable isotope values of precipitation at stations in the highlands of Jordan, the Antilebanon and Ansariye mountains and adjoining plain areas and on the Mediterranean Sea coast are listed in Table 2.7 and Fig. 2.16.

Table 2.7 Mean $\delta^{18}O$ and δ^2H values of precipitation in different areas of the northwestern mountain and rift zone (data from Abumaizer 1996; Al Charideh 2007; Bajjali 2006; Kattan 1996c; Salameh 2004)

Area	$\delta^{18}O$ ‰	δ^2H ‰	d ‰	Altitude m asl
Highlands of Jordan	−6.63 to −7.25	−29.7 to −33.3	+23.3 to +26.1	555–1,475
Antilebanon mountains (Bloudan)	−8.55	−48.8	+19.6	1,540
Plains and valleys within the highlands of Jordan	−5.42 to −6.06	−23.2 to −26.5	+17.5 to +24.2	350–715
Homs plain	−6.42	−32.2	+19.2	490
Ansariye mountains	−7.3			1,000
Mediterranean Sea coast				
Tartous	−4.96	−22.9	+16.8	5
Bet Dagan	−5.29	−22.8	+19.5	
Jordan valley (Deir Alla)	−3.52	−13.2	+14.9	−224

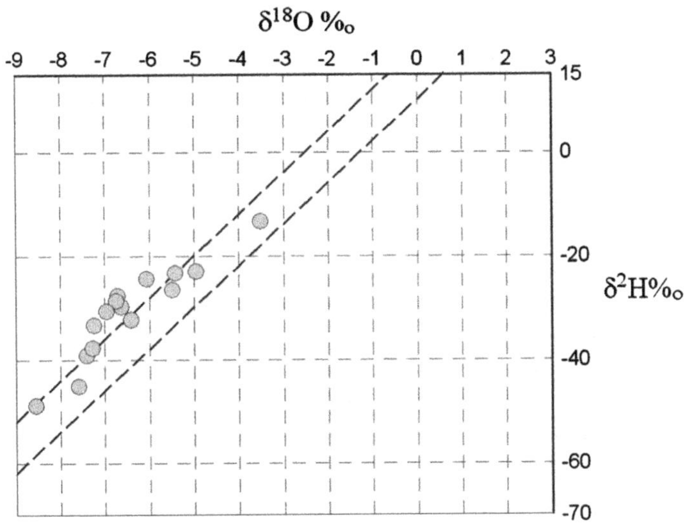

Fig. 2.16 $\delta^{18}O/\delta^{2}H$ diagram: Rain water samples of stations in the northwestern mountain and rift zone. Data from Almomani (1996), Kattan (1996c)

The following trends can be deduced for values of stable isotopes of oxygen and hydrogen in rain water samples:

- In the highlands of Jordan, $\delta^{18}O$ values are between −6.6 and −7.3‰ and d values > +23‰
- On the highest parts of the Antilebanon mountains, $\delta^{18}O$ is −8.55‰ and d +19.6‰
- Generally less negative $\delta^{18}O$ and lower d values are observed on the plain areas adjoining the highlands and, in particular, on the Mediterranean Sea coast and in the Jordan valley

In general, $\delta^{18}O$ values reflect the topographic altitude of the rainfall station. Altitude gradients of $\delta^{18}O$ values in different areas of the northwestern mountain and rift zone range between −0.2‰/100 m and −0.29‰/100 m.

2.6.2.2 Karst Aquifers of the Northwestern Mountain and Rift Zone

Most groundwater samples from the Jurassic–Upper Cretaceous karst aquifers of the northwestern highland zone scatter around the Mediterranean Meteoric Water Line

$$d = \delta^{2}H - 8 \times \delta^{18}O + 22‰.$$

with d values varying between around +19 and +24‰.

Ranges of $\delta^{18}O$ values in different areas of the highlands are

−5.7 to −6.7‰ in the highlands of Judea
−6.0 to −6.5‰ in the recharge areas of the highlands of Jordan
−7.0 to −8.2‰ on Mount Hermon
−7.7 to −9.0‰ in the Antilebanon mountains north of the Barada river
−5.0 to −6.8‰ in the Ansariye mountains

The differing ranges of $\delta^{18}O$ values in different areas of the northwestern highlands apparently reflect the altitudes of the recharge areas with most negative values in the up to more than 2,000 m high Antilebanon–Mount Hermon range. According to unpublished data, $\delta^{18}O$ values may also be more negative than −8‰ in aquifers of the Lebanon mountains, but no published data are available to the author.

d values of $< +15$‰ and deviations from general trends to less negative $\delta^{18}O$ values are observed in some wells in recharge areas, in aquifers influenced by local recharge at lower altitudes and in plain areas adjoining the mountain ranges, such as the Damascus plain, the Qalamoun high plateau and the Jordan valley.

2.6.2.3 Judean Highlands

$\delta^{18}O$ and $\delta^{2}H$ values of water from boreholes and springs on the eastern slopes of the Judean highlands are in the same range as those in the recharge area of the Upper Cretaceous aquifer on the mountain crest. There is no trace of paleowater recharged under a different climatic regime.

$\delta^{18}O$ and $\delta^{2}H$ values in the recharge area around Jerusalem as well as in the main springs in the Jericho area and in springs on the western shore of the Dead Sea are −5.6 to −6.7‰ and −21.1 to −25‰, respectively, with d between +20.5 and +28.6‰.

2.6.2.4 Highlands of Jordan

In the recharge areas of the Upper Cretaceous aquifer complex in the highlands of Jordan around Irbid–Ajlun, $\delta^{18}O$ values are generally between −5.9 and −7.0‰ and d values between +15.7 and +20.8‰.

In some wells tapping the outcropping A7 (Wadi Sir) aquifer in irrigation areas, significant seasonal variations of the isotope values are observed with less negative $\delta^{18}O$ and $\delta^{2}H$ values during the dry season. Seasonal variations of $\delta^{18}O$ values reach 1‰ and of $\delta^{2}H$ values 4.5‰. The range of d values of groundwater samples collected in the dry season extends to relatively low values of +10.7‰.

The variations reflect recharge of the groundwater from two sources. The main source of recharge is infiltration from precipitation on the outcropping karst aquifer, indicated by low groundwater salinity and remarkable ^{3}H contents in the groundwater. A second source of recharge is irrigation return flow. The irrigation water is enriched in heavier stable isotopes through evaporation and affects the isotopic composition of the groundwater in particular in the dry summer months.

From $\delta^{18}O/\delta^{2}H$ values of groundwaters on the slope of the highlands of Jordan to the Jordan valley, altitudes of recharge areas and interconnections between sub-aquifers can be deduced:

$\delta^{18}O$ and $\delta^{2}H$ values of groundwater in the recharge area of the Upper Cretaceous karst aquifer of the Yarmouk catchment in the highlands of Jordan range from −6 to −6.5‰ and −29 to −34‰, respectively, with d values around +18‰.

On the northeastern slope of the highlands, $\delta^{18}O$ values are generally in a range between −6.9 and −5.7‰ with d values varying mainly from +16 to +17.5‰. The values indicate an origin from recharge areas above 1,000 m altitude.

In the intermediate area between the highlands and the Jordan valley at altitudes of 100–400 m asl, wells produce locally recharged water from the B2 aquifer, which is slightly enriched in heavier isotopes ($\delta^{18}O$ −4.5 to −5.7‰, $\delta^{2}H$ −20 to −27‰). In the Jordan valley, where the head in the A7 (Wadi Sir) aquifer is higher than in the overlying B2 (Amman) aquifer and groundwater leakes upward into the B2 aquifer, the isotopic composition resembles that of the recharge area in the highlands ($\delta^{18}O$ −5 to −6‰, $\delta^{2}H$ −5 to −29‰).

Parts of the water from the confined A7 aquifer enters laterally into the Jordan valley deposits. The mixed water shows isotopic compositions, which lie between those of the A7 aquifer and the Jordan valley deposits of local recharge.

The isotopic composition of water from the fault springs Balsam and Maqla on the foothills resembles that of groundwater in the recharge areas with $\delta^{18}O$ values of around −6‰ and d values around +19‰. The isotope data of these springs, which are situated at 200 m below sea level, indicate recharge in the high mountain areas.

Boreholes tapping the A7 aquifer along the foothills of the Jordan valley, e.g. in the Mukheibe well field, generally produce a mixture of water with a major amount of groundwater flowing from the high mountain area and small proportions of water from the B4 aquifer recharged at lower altitudes. $\delta^{18}O$ values of these mixed waters range from −5 to −6‰, d around +20‰. The values fluctuate depending on pumping rates and well interferences.

Brackish groundwater recharged locally within the Jordan valley has $\delta^{18}O$ values from −3.2 to −4‰, ^{3}H values of 2.5–10.7 TU and EC values of 2,600–13,800 μS/cm. Mixtures of lateral groundwater inflow from the B2/A7 aquifer and of locally recharged groundwater in the Adasiye area in the Jordan valley foothills are characterized by low ^{3}H values and $\delta^{18}O$ values of −4.85 to 5.58‰ (Fig. 2.17).

Groundwater from wells tapping the Paleogene chalk and limestone (Rijam) aquifer around and east of Irbid have generally $\delta^{18}O$ values of −4.1 to −5.2‰, and d values around +12‰. ^{3}H values of 1.0–2.9 TU in most samples from these wells indicate a regime with some recent recharge in the shallow aquifer.

2.6.2.5 Antilebanon Mountains and Eastern Foreland

The $\delta^{18}O$ and $\delta^{2}H$ values of groundwater in the Antilebanon mountains and Mount Hermon reflect present-day recharge from Mediterranean precipitation

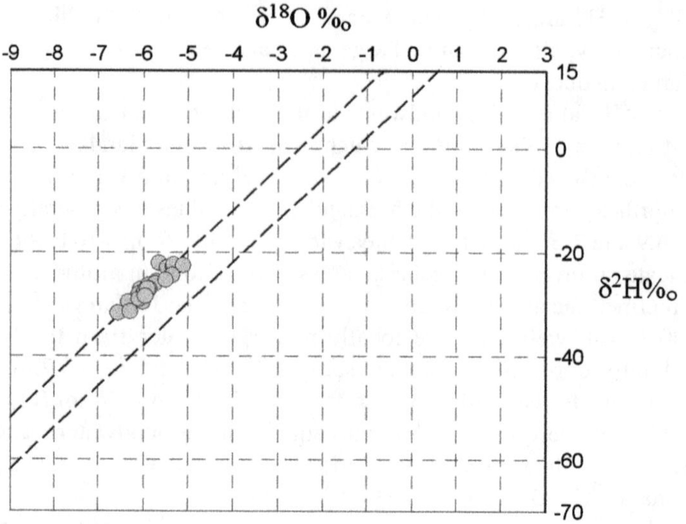

Fig. 2.17 $\delta^{18}O/\delta^2H$ diagram: Groundwater samples from Mesozoic aquifers in northwestern Jordan between the highlands of Irbid – Ajlun and the Jordan valley. Data from Abumaizer (1996)

with a deuterium excess of > +20‰. The $\delta^{18}O$ and δ^2H values are related to the mean altitude of the catchment area: $\delta^{18}O$ values of Ain el Fije, issuing at 860 m asl from a high mountain catchment of the Cenomanian–Turonian aquifer, are around −8.5 to −9‰, $\delta^{18}O$ values of Barada spring, situated at 1,100 m asl and draining a more locally restricted catchment in Jurassic limestones, are around −8.0‰. From the $\delta^{18}O$/altitude relationship the mean recharge altitudes of the groundwater discharged from the springs at Ain el Fije and at Barada are estimated to be 2,100–2,600 m asl and 1,700 m asl, respectively.

Relatively enriched $\delta^{18}O$ values in some wells in the Zebedani plain may indicate an impact of irrigation return flow ($\delta^{18}O$ −6.6‰, d +13.3‰).

$\delta^{18}O$ values in many spring waters of the Antilebanon mountains and probably in the Lebanon mountains are more negativ than −8‰, the most negative values of recent groundwater of the Arabian Plate (Fig. 2.18).

The $\delta^{18}O$ and δ^2H values of samples from springs and wells on the *Qalamoun* high plateau vary from a composition characteristic for present-day rainfall of Mediterranean origin (d values of +17.4 to +22.5‰) to a more enriched stable isotope signature (d values of +12.3 to +14.4‰). The less negative isotope signature in many groundwater samples from wells is attributed to the pronounced change of the local climate along the leeward slope of the Antilebanon mountains with an increasing evaporative isotope enrichment in a more continental climate.

Isotopic compositions are rather homogeneous over the different aquifer units of the Qalamoun groundwater system: the Mesozoic (Jurassic–Upper Cretaceous) carbonate aquifer, the Paleogene chalk and nummulitic limestone aquifer, and the shallow aquifer composed of Paleogene–Neogene detrital deposits.

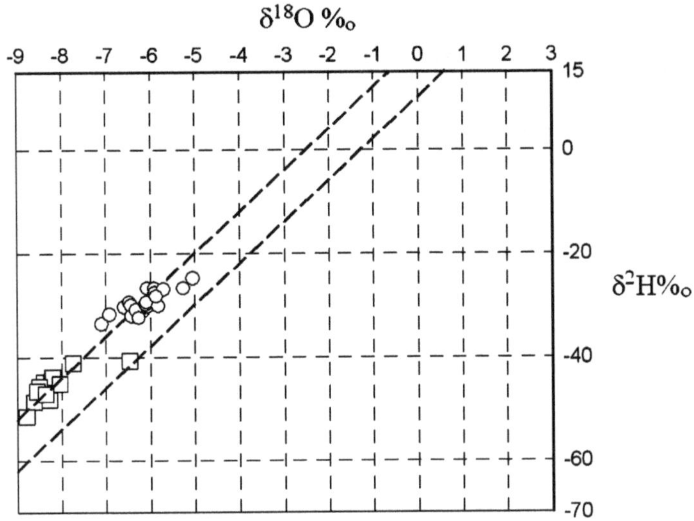

Fig. 2.18 $\delta^{18}O/\delta^{2}H$ diagram: Water samples from springs and wells in the Antilebanon and Ansariye mountains in Syria. □ Antilebanon mountains; ○ Ansariye mountains. After Kattan (1996a, c), Wagner and Geyh (1999)

The wide range of the altitude of the recharge areas between 800 m asl on the Qalamoun plateau and >2,000 m asl on the top of the Antilebanon Mountains is reflected in the wide scatter of the delta values along the Mediterranean Meteoric Water Line. A mean altitude gradient of −0.25‰ $\delta^{18}O$/100 m was estimated corresponding to the altitudes of the sampling points and of the hydrogeologically derived catchment areas as well as the corresponding $\delta^{18}O$ values (Fig. 2.19).

The sand and gravel aquifer of the *Damascus plain* in the foreland of the Antilebanon mountains is replenished from inflowing surface water through infiltration in streambeds, canals, and on irrigated fields, and, to some extent, through subsurface inflow and recharge from local precipitation. The recharge conditions in the plain are reflected in the pattern of $\delta^{18}O$ and ^{3}H values of the shallow groundwater (Table 2.8).

The wide distribution of $\delta^{18}O$ values of less than −7.5‰ in the Damascus plain shows the dominant influence of water originating in the higher Antilebanon mountain range on the groundwater regime of the plain. The water from the mountains reaches the plain through streamflow in the Barada and Aouaj rivers, the water supply network from Ain el Fije, and subsurface inflow in the Upper Cretaceous aquifer. d values in wide parts of the plain are around +20‰.

Groundwater from deeper boreholes in the Damascus plain has prevailingly $\delta^{18}O$ values of −8 to −9‰, d values around +20‰ and, in many cases, detectable tritium. The deep groundwater is probably recharged by relatively fast subsurface inflow from the Antilebanon mountains. The distribution of d, tritium and ^{14}C values indicates differing flow conditions in different parts of the plain: relatively

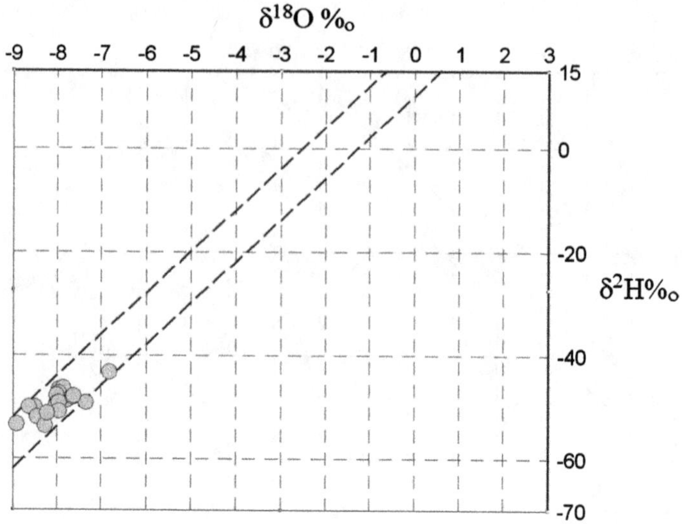

Fig. 2.19 $\delta^{18}O/\delta^{2}H$ diagram: Groundwater samples from the Qalamoun area, Syria. After Wagner and Geyh (1999)

Table 2.8 $\delta^{18}O$ and ^{3}H values in different parts of the Damascus plain (after Selkhozpromexport 1986; Kattan 1996c)

	^{18}O ‰	^{3}H TU
Infiltration of stream water along Barada river	−7.3 to −8.1	17–24
Infiltration within Damascus City of river water and of domestic waste water, originating prevailingly from Ain el Fije	−8.3	>17
Plain north of Damascus City (subsurface inflow of groundwater recharged in the Antilebanon–Qalamoun Mountains)	−8.2	>17
Areas with only local recharge	Less negative than −7	
Area around salt lake	−5.5 to −1.3	

fast groundwater movement from the west, inflow of older groundwater from the north or northwest.

2.6.2.6 Northwestern Syria

The stable isotope values of the water from springs in the *Ansariye mountains* prove its present Mediterranean origin with d values of around +20‰ (+18.4 to +23.8‰, Fig. 2.18). $\delta^{18}O$ values are less negative than those in the Antilebanon mountains according to the lower altitude of the mountain chain with peak elevations of 1,300–1,560 m asl.

$\delta^{18}O$ values reflect an altitude effect and allow an estimate of recharge areas of springs and boreholes situated in the coastal zone. Estimated altitudes of recharge areas of the groundwater in the coastal area range from 400–700 m asl on the northern mountain slopes ($\delta^{18}O$ −6.24‰) to 100–300 m asl for the southwestern mountain slopes ($\delta^{18}O$ −5.56‰).

Groundwater with d values of +15 to +18‰ and no detectable tritium is found in springs issuing from the Cenomanian–Turonian karst aquifer at the eastern border of *El Ghab*. The catchment of these springs extends into the more arid zone of the Syrian steppe.

References. Abumaizer (1996), Al Charideh (2007), Almomani (1996), Bajjali (2006, 2008), Bajjali et al. (1997), Clark et al. (1995), Gat and Carmi (1970), Gat and Dansgaard (1972), Gat et al. (1969), Kattan (1996a, b, c), Kroitorou et al. (1992), Lloyd (1981), Mazor and Molcho (1972), Prizgognow et al. (1988), Salameh (2004), Selkhozpromexport (1986), Wagner and Geyh (1999).

Chapter 3
Eastern Part of the Northern Arabian Platform

3.1 Al Badiye, Hamad and the Transitional Zone

3.1.1 Morphology, Hydrography, Climate and Land Use

3.1.1.1 Morphology

The mountain and highland chains of the northwestern mountain and rift zone grade, in the east, into plateau landscapes of the steppe and desert areas of Syria and Jordan. Semi-arid to arid plateaus extend, over the eastern parts of the northern Arabian platform, from the foreland of the Taurus mountains in the north into the steppe areas of Syria, Jordan and southwestern Iraq. The dry zones of central and eastern Syria, eastern Jordan and southwestern Iraq are generally known as Al Badiye, the land of the bedouins. Toward southeast, the steppe areas of Al Badiye change into a stony desert landscape, the Hamad.

In the east, the northern Arabian platform is separated from the eastern Arabian platform by a geologic uplift structure (Hail–Rutba arch). The distance from the western mountain zone to the eastern boundary of the northern Arabian platform at the Euphrates valley and the Rutba uplift ranges from around 140 km in northern Syria to 350 km in Jordan–Iraq. In the northeast, a zone of plains marks the margin between the Arabian Platform and the alpidic Taurus and Zagros mountain chains.

Most of the Badiye and Hamad is occupied by undulating to flat morphology with moderately incised wadi systems. Two types of mountain landscapes rise above the plateaus: the southern and northern Palmyrean mountain chains and volcanic mountains or hills, the most prominent of which are Jebel el Arab in southern Syria and northeastern Jordan and Jebel el Hass southeast of Aleppo.

The eastern part of the northern Arabian platform comprises – from north to south – the following main geographic units (Fig. 3.1):

- The northern Syrian steppe or Aleppo plateau and, northwest of the Euphrates river, Al Jezire
- The Palmyrean mountain zone
- The southern Syrian steppe grading into the Hamad desert

W. Wagner, *Groundwater in the Arab Middle East*,
DOI 10.1007/978-3-642-19351-4_3, © Springer-Verlag Berlin Heidelberg 2011

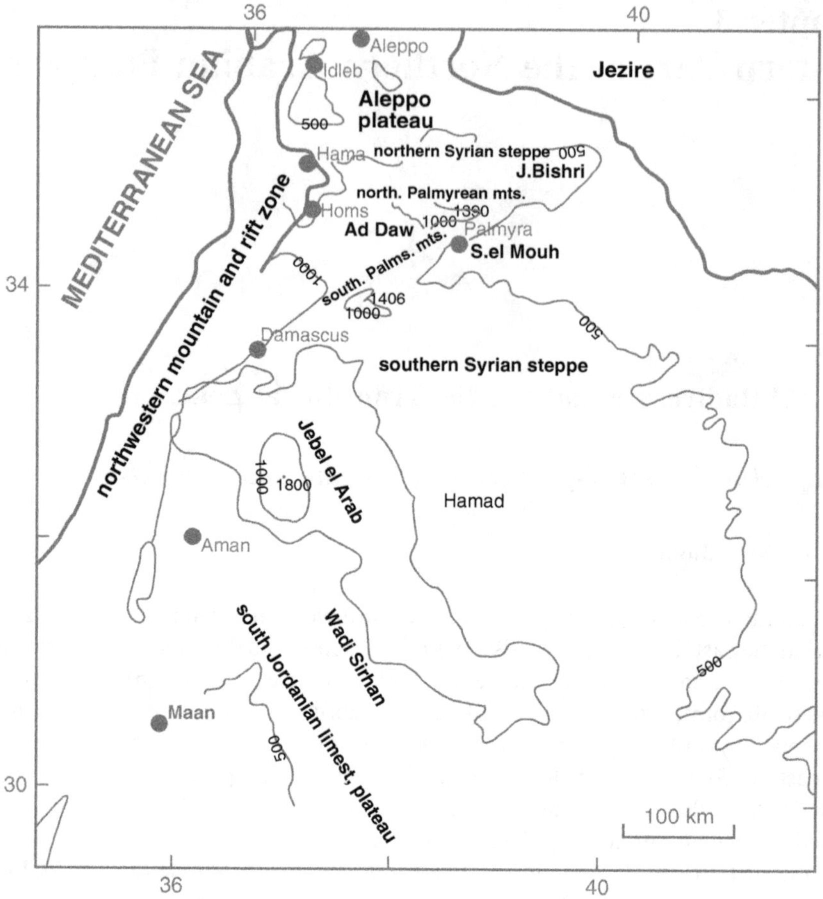

Fig. 3.1 Main orographic–geographic units of the eastern part of the northern Arabian platform. __600__ topographic contour line 600 m asl

- The east Jordanian limestone plateau, separated from the Hamad by the Jebel el Arab volcanic massive

The southwest–northeast to west–east trending southern and northern Palmyrean mountains interrupt the plateau landscapes of the northern Arabian platform. The Palmyrean mountains branch out, in the Damascus area, from the S–N to SSW–NNE directed Antilebanon mountains in three generally NNE to NE trending chains:

- The western mountain zone of Jebel Abu Ata–Jebel Nebek–Jebel Deir Atiye with peak altitudes around 1,400–500 m asl, which is separated from the Antilebanon mountains by the Qalamoun high plain.
- An up to 1,218 m high mountain zone further east, separated from Jebel Nebek by the morphologic depression of Jayroud.
- The southern Palmyrean mountains extending as a 160-km long chain from the Damascus plain at Dmeir until Palmyra.

The main mountain massifs of the southern Palmyrean chains – Jebel Dmeir, Jebel Mghara, Jebel Manqoura, Jebel Ghantous, Jebel Basiri – reach peak altitudes of 1,100–1,390 m asl.

The northern Palmyrean mountains extend in general west–east direction between the Homs depression and the Palmyra area, where they converge with the southern Palmyrean mountains into one chain, which continues toward east into Jebel el Bishri (851 m). Individual mountain massifs of the northern Palmyrean chain – Jebel Shoumariye, Jebel Bilas (1,098 m), Jebel Chaar, Jebel Mraa, Jebel Bouaida (1,390 m) – have a general southwest–northeast trend.

The southern and northern Palmyrean mountains enclose a vast plain, the Ad Daw or Dawwa plain, situated at an altitude of around 600 m asl. To the southeast, the Palmyrean mountain chains are adjoined by the Sabkhet el Mouh salt flat at 370 m asl.

The plateau area north of the Palmyrean mountain zone is occupied by the northern Syrian steppe or Aleppo plateau. The plateau comprises mainly flat to hilly landscapes at topographic elevations around 400 m asl, descending gently to salt flats in the morphologic depressions of Al Matah and Jaboul.

In the north and south, the Aleppo plateau is bordered by mountain uplifts: the eastern Taurus mountains north of Adiyaman and the northern Palmyrean mountains with elevations of up to 2,550 m and 1,060 m asl, respectively. In the west, the plateau is adjoined by the hydrologic Orontes sub-basin in the northwestern mountain and rift zone with elevations between 400 m in the Orontes valley and on the Homs plain, 100 m in the Karasu graben and up to 870 m in Jebel ez Zaouiye. In the northwest, the Aleppo plateau continues into the Idleb plateau and, in the northeast, the plateau is adjoined by the Euphrates valley, which is situated at an elevation of around 300 m asl.

The Euphrates river separates Al Jezire from the Syrian Badiye. Al Jezire, the "island" between Euphrates and Tigris rivers, occupies the northeastern edge of Syria and extends into western Iraq. The topography of the Jezire is dominated by plains and flat valleys at altitudes between 340 and 400 m asl. An approximately west–east oriented series of morphologic elevations interrupts the flat plain landscape and rises up to 950 m asl in Jebel Abd el Aziz and 1,640 m in Jebel Sinjar.

The plateau areas south of the Palmyrean mountain zones extends from the southern Syrian steppe into the Hamad desert. The centre of the Hamad comprises a high plateau with internal surface drainage toward local sabkhas. The plateau with a gently undulating morphology is situated at 700–900 m asl with highest elevations of 940 m at Jebel Aneiza (border triangle of Jordan, Iraq, Saudi Arabia) and more than 1,000 m near Khawr Um Wual in Saudi Arabia. The surface of the plateau is a generally flat, stony semi-desert. To the west, the Hamad plateau is bordered by hill to mountain chains of the Jebel el Arab–Al Harra basalt field.

The southern Syrian steppe and the Rutba and Al Widyan areas north and east of the Hamad comprise undulating plains, which are dissected by mainly shallow wadis with courses toward the Euphrates valley in the northeast.

The east Jordanian limestone plateau occupies a southeastern segment of the northern Arabian platform. The plateau descends from altitudes of 600–900 m asl in the south, on its boundaries with the highlands of Jordan and the Interior Shelf, to about 500 m asl in the morphologic depressions of Wadi Sirhan and Qaa el Azraq in the east. In the northeast, the limestone plateau disappears under the basalt cover of Jebel el Arab field, which separates the Jordanian limestone plateau from the Hamad.

3.1.1.2 Hydrography

The eastern part of the northern Arabian platform comprises a vast zone of internal drainage and, in the northeast, a network of wadis directed toward the Euphrates valley. The by far largest volumes of river flow of the Arabian Plate are concentrated in the Euphrates–Tigris river system. The river system receives its major flow volume in the Anatolian highlands and the mountain chains of Taurus and Zagros, and discharges through the Shatt el Arab into the Gulf. Average flow in the Euphrates river in Syria is in the order of 800 m^3/s; the river flow is, at present, controlled by several large reservoirs.

The Euphrates river enters Syrian territory at Jerablus in the eastern part of the Aleppo plateau. After an about 100 km long north–south stretch the river course turns toward east into the Neogene Euphrates depression. Main tributaries of the Euphrates river in the Syrian Jezire are the Balikh and Khabour rivers. Nahr el Khabour is the main internal river of the Syrian Jezire with a length of 486 km and a mean discharge of 42 m^3/s. The river takes most of its flow from karst springs emerging at Ras el Ain.

Most of the eastern part of the northern Arabian platform is situated in the zone of internal drainage, which extends between the hydrologic basins of the Mediterranean Sea and of the Euphrates river. Wadis in the inland drainage zone are directed to closed basins, generally flat morphologic depressions with seasonal lakes or salt flats.

The large depressions act as groundwater discharge zones: Jaboul and Matah in northern Syria, Sabkhet el Mouh near Palmyra in central Syria, Manqaa ar Rahba and Qaa al Azraq on the fringes of the Jebel el Arab basalt field in Syria and Jordan, Wadi Sirhan on the eastern boundary of the Jordanian limestone plateau. Various smaller depressions without surface outflow are dispersed over the Badiye and Hamad at elevations of up to some hundred meters above the present level of the groundwater surface of the upper aquifer system.

Flow in the wadis of the internal drainage zone is only sporadic after major rain events. In the Queiq river on the Aleppo plateau, seasonal flow drains into the Matah depression.

The southern Syrian steppe and the adjoining Widyan area comprise distinctly incised wadi systems, which are directed toward the Euphrates valley: Wadi el Murabaa and Wadi Miyah in Syria, Wadi Sawab in the Syrian–Iraqian border area, Wadi Akash and Wadi Hauran in Iraq. The hydrography of the Hamad plateau is characterized by mostly short wadis, which end in shallow sabkha depressions. A relatively extensive wadi system, Wadi Ruweishid, is developed on the plateau in northeastern Jordan. Larger sabkhas are Khabra et Tenf and Khabra Hraith in Syria, an extended system of sabkhas at the western border of the plateau in northeastern Jordan and the sabkha in the Khawr Um Wual graben in Saudi Arabia.

The east Jordanian plateau is dissected by a network of wadis, which drain toward closed basins in morphological depressions or to the Dead Sea valley. The catchments of the Dead Sea tributaries Wadi Mujib and Wadi Hasa reach more than 80 km toward east into the limestone plateau. Major closed basins are the Wadi Sirhan and Azraq basins on the eastern border of the plateau, with lowest topographic levels at around 500 m asl, and the Jafr basin, the centre of which is formed by a depression within the limestone plateau situated at around 800 m asl.

Wadi Sirhan comprises a morphologic depression elongated in southeast–northwest direction with a low lying zone of salt pans, lakes and mud flats in northwestern Saudi Arabia; Al Hadhawdha, the largest salt lake, extends over an area of approximately 400 km^2.

The major wadi courses draining toward the basin centres are relatively well defined, have wide channels (50–100 m) and an average slope of 2–5 m/km.

3.1.1.3 Climate

The eastern part of the north Arabian platform has semi-arid to arid climate conditions. The relatively humid climate of the western highlands and mountains of Jordan and Syria changes rapidly to dry climate conditions toward the steppe and desert areas of Al Badiye in the east. The change of the sub-humid climate of the highlands to dry climate conditions of the Badiye takes place in a transitional zone of 50–100 km width. Within the transitional zone, mean precipitation is decreasing toward east and south, the rain season becomes shorter and rainfall more erratic. In the steppe and desert areas of southern Syria and eastern Jordan, mean annual precipitation is less than 50 mm.

Slightly more humid climate conditions in some parts of the sub-region may be related to morphological features: relatively high altitudes (northern Palmyrean mountains) and gaps in the mountain barriers (Homs depression). In northern and northeastern Syria, the dry climate is modified by inflow of moist air masses from the north. Precipitation during the winter season on the Aleppo plateau and the Jezire is associated with Mediterranean fronts or intrusions of cold fronts reaching the area from the continental Anatolian highlands. Mean annual precipitation in the Jezire varies from 300 mm in the northwest to 200 mm in the southeast.

3.1.1.4 Land Use

Al Badiye with mean annual precipitation of <200 mm has been traditionally an area of nomadic tribes. Most of Al Badiye and the adjoining Hamad desert, which extend from the Great Nefud desert in Saudi Arabia to the Euphrates river in the north, is located on Jordanian and Syrian territory. Living conditions have changed considerably over the past few decades. In the Badiye of Jordan, only 5% of the population are still nomadic, the majority has been settled, though livestock production is still a major activity of the area.

Local occurrences of fresh groundwater have been used since antiquity for water supply of towns like Palmyra and Resafe or of isolated settlements like the Umayad desert castles.

A few oases form green spots in the vast steppe and desert landscape of Al Badiye: Palmyra in central Syria, Azraq in Jordan, Qurayat in Wadi Sirhan in northwestern Saudi Arabia.

Palmyra was the centre of a large state, which extended between the Roman and Persian Empires under its rulers Odeinat and Zenobia in the third century AD. The ruins of the splendid Syrian–Hellenistic Palmyra are now a major tourist attraction at the modern small town Tudmor. Other architectonic remains of the Hellenistic and medieval periods are found at isolated locations of Al Badiye: Resafe and Qasr Ibn Wardan in northern Syria, and Umayad desert castles dispersed over the Badiye of Syria and Jordan.

Brackish water was used for irrigation of oases. Through modern drilling techniques, groundwater from deeper aquifers has been exploited in the Badiye in recent decades for watering points of nomadic herds, for irrigation developments, for phosphate mining and domestic supply.

3.1.1.5 Transitional Zone

On the margins between the western highlands and mountains of Jordan and Syria and the dry plateaus in the east, the seasonally or year-round cultivated lands grade into the steppe areas of Al Badiye. In a transitional zone, sufficient precipitation for rain fed agriculture is received, in average and wet years, from winter storms advancing from the western mountain chains and from the Taurus mountains in the north. The high variability of rainfall with frequent occurrence of dry years creates, however, high hazards of agricultural cultivation.

In the transitional zone, the vegetation changes from the highlands covered by perennial trees and plantations to rainfed cereal cultivation in limited areas and to steppe areas with prevailing seasonal pastoral grazing. Production of winter crops relies in the transitional belt on supplementary irrigation, as far as surface or subsurface water resources are available.

The transitional zone has been classified in Syria into agricultural stability zones from:

- Zone 1 with a mean annual precipitation of >350 mm and probability of at least 350 mm in 2 out of 3 years to
- Zone 4 with a mean annual precipitation of 200–250 mm and probability of not less than 200 mm in 1 out of 2 years and
- Zone 5 a mean annual precipitation of <200 mm

The cropping pattern on the irrigated areas depends on the salinity of the irrigation water. Winter legumes are generally not salt tolerant and accept a maximum salinity equal to an electrical conductivity of 1,200 μS/cm. Salt accumulation in the soil can build up rapidly under summer irrigation. Brackish water is therefore applied prevailingly for supplementary irrigation of cereals with a tendency of mono-cropping of wheat.

Supplementary irrigation in the transitional belt is practised, in particular, along perennial rivers: Euphrates, Khabur, and Tigris rivers. Groundwater is extracted for supplementary irrigation in several areas in the eastern part of the northern Arabian platform: in the areas of Aleppo–Idleb and Selemiye in northern Syria, Hauran in southwestern Syria and Mafraq – northeastern desert in Jordan, and in the transitional zone between the highlands and the east Jordanian limestone plateau.

References. ACSAD (1981), Al-Homoud et al. (1995), Khouri (1982), Ponikarov et al. (1967a), Wakil (1994).

3.2 Geology

3.2.1 General Geologic Structure

The eastern part of the northern platform constitutes a relatively mobile zone of the Arabian Shelf, but was less intensively affected by the Neogene rift tectonics than the western mountain and rift zone. The sub-region comprises prevailingly plateau areas with a nearly horizontal or gently dipping sedimentary cover. The anticlinal southern and northern Palmyrean mountains form a major structural element with more intensive tectonic displacements. Basin and graben structures, accompanied by major flexure and fault zones, extend along the northeastern boundary of the platform – the boundary to the Mesopotamian–Euphrates basin – and in the southwest in the Wadi Sirhan–Azraq depression. Toward east–southeast, the platform grades into the paleogeographic uplift zone of the Rutba arch. The western border area corresponds to the zone, where the south–north oriented anticlinal uplifts of the Jordan–Antilebanon highland chains change eastward into flat plateau areas and the mainly southwest–northeast directed Palmyrean mountains (Figs. 3.2–3.4).

The plateaus of the eastern part of the platform are subdivided by the massifs of the Palmyrean mountain and Jebel el Arab into a few major geologic blocks:

- The Aleppo plateau including the northern steppe of Syria
- The Palmyrean fold belt with the basins of Ad Daw and Sabkhet el Mouh
- The southern Syrian steppe and the Hamad
- The Jordanian limestone plateau with the adjoining Wadi Sirhan and surrounding areas in northwestern Saudi Arabia

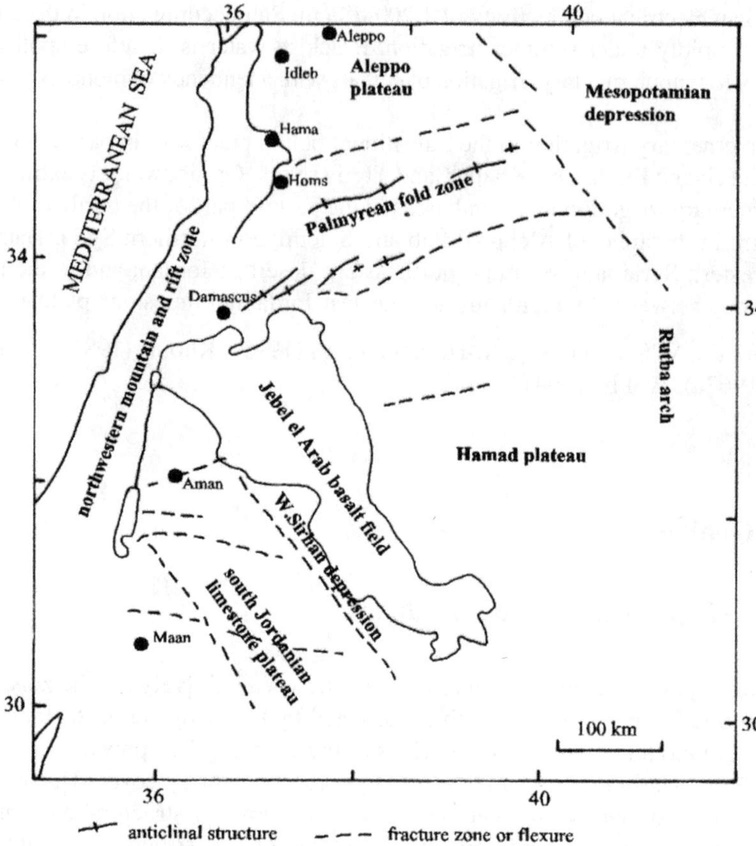

Fig. 3.2 Main structural geologic units of the eastern part of the northern Arabian platform. After Lovelock (1984), Sunna (1995), Wiesemann (1969), Wolfart (1966)

The Arabian Plate dips in the northeast, on the boundary between the Arabian Shelf and the alpidic mountain chains, under the Mesopotamian foredeep. The foredeep, which was formed during the final phase of the alpidic tectonic movements mainly in the Pliocene–Quaternary, extends as a vast plain area in a wide belt along the Taurus–Zagros folded zone and is filled with thick molasse sediments. An intermediate zone between the Arabian platform and the Mesopotamian foredeep is occupied by the Euphrates plain, which covers a structural depression (Al Furatian depression, Ponikarov et al. 1967a) on the northeastern margin of the plateau areas of the northern Arabian platform. Mesopotamian foredeep and Euphrates plain are separated along an approximately west–east trending system of faults and flexures and a series of anticlinal structures: Tuwal el Aba – Jebel Abd el Aziz – Jebel Sinjar (Figs. 3.2–3.4).

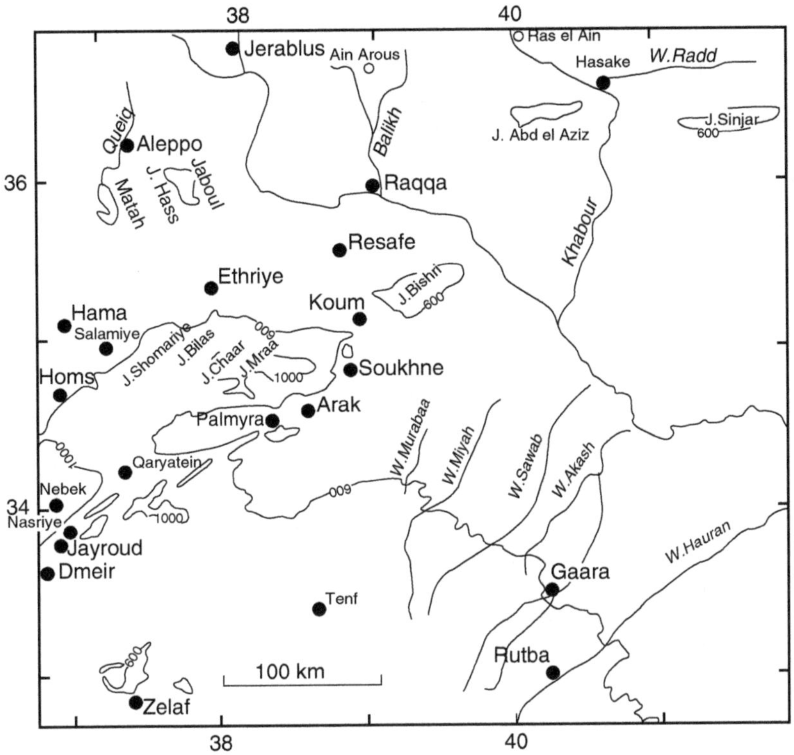

Fig. 3.3 Location map of the northeastern part of the north Arabian platform. __600__ topographic contour line 600 m asl, ○ spring

3.2.2 *Litho-Stratigraphic Sequence*

The eastern parts of the northern Arabian platform are covered, to a large extent, by Paleogene–Neogene deposits. Mesozoic formations are exposed in uplift structures, such as the Palmyrean mountain chains. Accumulations of terrigenous Neogene–Quaternary sediments are found in major structural depressions: the Euphrates–Mesopotamian basin, Ad Daw plain, Wadi Sirhan (Table 3.1).

3.2.2.1 Aleppo Plateau and Mesopotamian Foredeep

The Aleppo plateau is underlain by a several kilometres thick sequence of sedimentary rocks. The main units in the upper few hundred metres of the sequence are (from bottom to top):

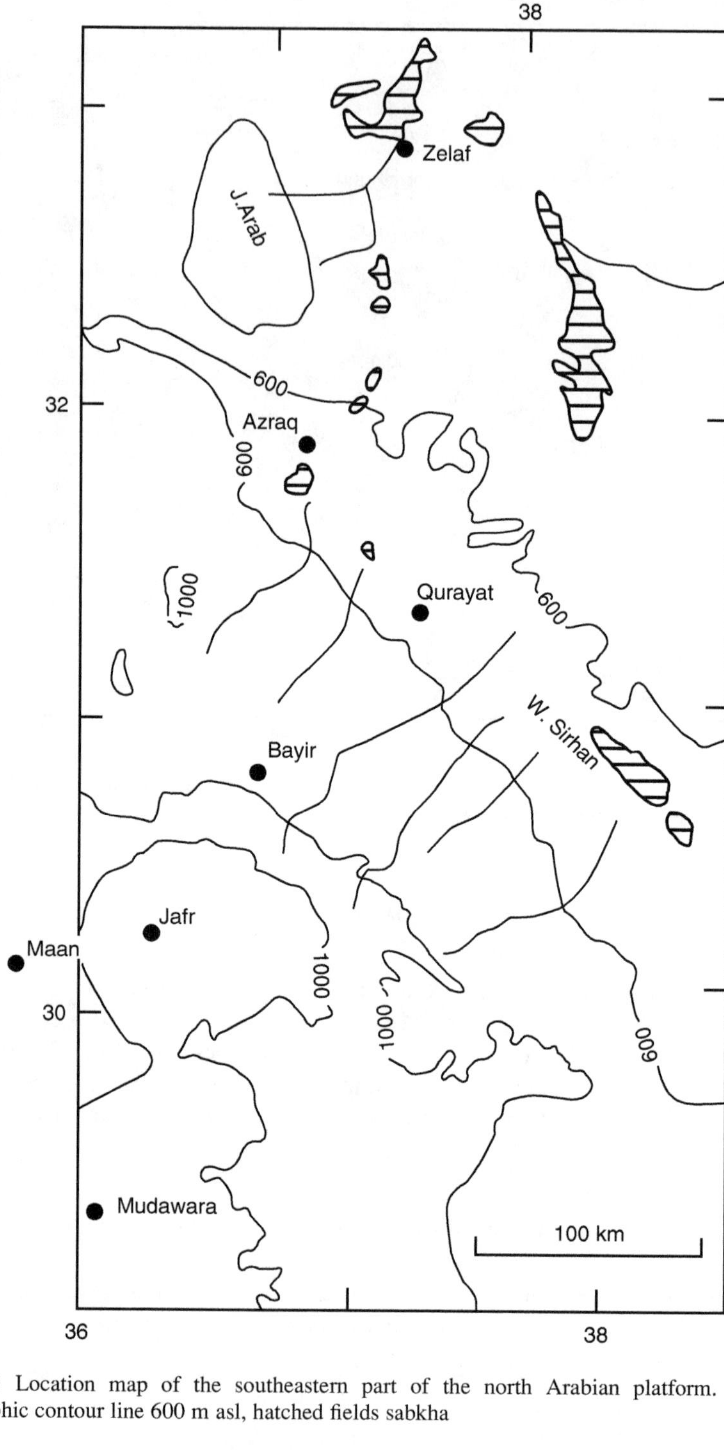

Fig. 3.4 Location map of the southeastern part of the north Arabian platform. __600__ topographic contour line 600 m asl, hatched fields sabkha

Upper Cretaceous:

Limestones and dolomites (Cenomanian–Turonian, in some areas possibly reaching into higher stages of the Upper Cretaceous).

Chalky and marly limestones with chert and phosphoritic intercalations, partly bituminous (Coniacian–Santonian–Campanian)

Upper Cretaceous to Paleogene:

Marls and argillaceous limestones (Maastrichtian–Paleocene)

Paleogene:

Eocene chalks and nummulitic limestones

Neogene:

Helvetian limestones

Tortonian limestones, marls, conglomerates, sandstones

Miocene basalts

Pliocene continental deposits

Pliocene basalts

Quaternary:

Flood plain, terrace and lacustrine deposits

Data of deep drillings near Aleppo and Khanaser indicate the following thickness of Paleozoic–Mesozoic formations:

- Upper Cretaceous carbonate rocks and marls: around 900 m
- Lower Cretaceous–Triassic: 140–300 m
- Paleozoic: >2,900 m

The Aleppo plateau is covered prevailingly by Paleogene to Neogene deposits with some outcrops of Miocene–Pliocene basalts. Exposures of Upper Cretaceous (Campanian–Maastrichtian) formations are limited to Jebel Shbith near Esriye and the southwestern rim of the plateau.

In the depressions of Al Jaboul and Matah–Harayeq, Pliocene lacustrine deposits – sandstones, conglomerates and gypsum layers – overlie Helvetian or Paleogene carbonate formations.

In the Selemiye plain in the southwest of the northern Syrian steppe, the Paleogene is partly covered by Pliocene terrigenous sandstones, clays and conglomerates.

The Mesopotamian foredeep and the Euphrates depression comprise a large Neogene basin, which was temporarily flooded by a marine invasion from the east. During the Miocene, the northern shelf area of the Arabian Platform was divided into two basins with different type of sedimentation. The Mediterranean basin in the west was connected to the ocean and sedimentation occurred under normal saline sea water conditions. The Mesopotamian basin in the east was isolated from the open ocean over considerable periods and the deposits include sediments of varying environments: salt, gypsum, marine, lagoonal and fresh water carbonates, marls and clays. The western and southwestern boundaries of the Euphrates depression follow approximately the shore line of the Neogene basin on the fringes of the Aleppo plateau, the Palmyrean zone and the Hamad plateau.

The Mesopotamian and Euphrates plains are covered prevailingly by Neogene–Quaternary detrital deposits. The thickness of these molasse sediments reaches

500 m in northern Syria, increasing to several thousand metres in the adjoining areas of Turkey and Iraq.

The Jezire includes, on its western boundary, a narrow strip of Paleogene outcrops of the Aleppo plateau on the left bank of the Euphrates river. Cretaceous–Paleogene and Miocene formations are exposed on the hill chain, which separates the Euphrates and Mesopotamian plains.

The underground of the Euphrates depression comprises, above the crystalline rocks of the basement, the following geological sequence:

- About 1,200 m Paleozoic deposits (Ordovician–Permian): mainly mudstones with interbedded sandstone, shale, limestone, sandy shale and clay
- Triassic–Jurassic dolomitic sandstone, shale, dolomite, anhydrite with a total thickness of around 700 m
- 400–600 m Cretaceous sandstones, limestones, marls and chalks
- Paleocene prevailingly argillaceous limestone and bituminous shale
- Eocene argillaceous limestone with chert, nummulitic limestone with interbedded marl, 200–300 m
- Oligocene marls, limestone, dolomite, 20–160 m
- Neogene marl, lacustrine and terrestrial deposits, up to 1,000 m
- Quaternary proluvial, alluvial, lacustrine formations, basalt, gravel, loam, clay, sand

In western Iraq, the Miocene sediments are subdivided into three stratigraphic units:

- Upper Fars (Upper Miocene): continental deposits, clays, sandstones, conglomerates
- Lower Fars (Middle Miocene): clays, marls, shales, limestones, gypsum
- Lower Miocene: reef limestones (Euphrates limestone)

3.2.2.2 Palmyrean Fold Belt

The Palmyrean fold belt constituted a marine sedimentary basin from the Upper Jurassic to Campanian with an up to more than 400 m thick accumulation of Cenomanian–Turonian carbonate rocks. Limestones and marls of up to 300 m thickness were deposited in the basin during the Coniacian, Santonian and Campanian. The Campanian contains abundant chert beds and, in the central part of the Palmyrean zone, thick beds of phosphate rocks.

The Mesozoic stratigraphic sequence of the Palmyrean zone comprises:

- Triassic–Middle Jurassic littoral – lagoonal carbonate and clay sediments with evaporite intercalations
- Upper Jurassic limestones, clayey limestones and marls
- Lower Cretaceous sandstones with beds of marls, dolomite and gypsum
- Cenomanian–Santonian limestones and dolomites
- Campanian limestones and marls with chert bands and thick layers of phosphate rocks
- Maastrichtian marls and marly limestones

Paleogene chalks cover wide areas on the outer parts of the Ad Daw plain, of the surroundings of Sabkhet el Mouh and of the anticline of Jebel Bishri. Jebel Bishri, the northeastern continuation of the Palmyrean chains, dips as a broad anticline toward east to northeast into the Euphrates depression.

The Ad Daw plain is filled by an up to more than 250 m thick sequence of Pliocene–Pleistocene lacustrine and fluviatile deposits: alternations of conglomerates, sandstones, marl, clay, gypsum. Coarse detrital deposits occur mainly at the periphery of the plain.

The Sabkhet el Mouh plain east of Palmyra is underlain by varying Neogene deposits: marl, clay, conglomerates, limestones, sand and sandstones.

In the plain around Jayroud lake, the Paleogene outcrops are partly covered by gypsum sands, which form low dunes.

3.2.2.3 Southern Syrian Steppe, Hamad and East Jordanian Limestone Plateau

The vast plateau areas of the southern Syrian steppe and the Hamad are mainly covered by Paleogene carbonate and chert formations.

In the subsurface, the following sequence of sedimentary formations is found:

- Paleozoic sandstones, which crop out in the Rutba arch east of the Hamad (Gaara sandstone) and are situated at greater depth in the plateau areas.
- Triassic–Jurassic carbonate rocks, in some areas with sandstone intercalations, which are exposed on the Rutba uplift (Upper Triassic Mulussa formation) and have been reached in boreholes in the At Tenf area (Syria) and H5 area (Jordan).
- Lower Cretaceous sandstones (Kurnub sandstone in Jordan, Rutba sandstone in the Rutba area in Iraq).
- Upper Cretaceous limestones, dolomites and cherts (Cenomanian to Campanian), which extend, in general, at a few hundred metres depth below the plateau areas.
- Marls and marly limestones of Maastrichtian–Paleocene–lower Eocene age, grading into limestones and dolomites on the flanks of the Rutba uplift.
- Paleogene chalks and limestones with marl and chert intercalations.

Neogene–Quaternary terrestrial deposits occur in structural and morphologic depressions. Veils of wind blown sand cover some parts of the desert plateaus in the south of the Hamad. Mesozoic sedimentary rocks crop out in a belt on the southeastern margin of the Hamad along the transition to the Widyan plateau as well as in the southern Palmyrean mountains north of the southern Syrian steppe.

The east Jordanian limestone plateau lies in a paleogeographic situation between the western highlands of Jordan and the Hamad. The geologic development during the Mesozoic to Tertiary is dominated by the Upper Cretaceous and Paleogene marine transgressions over older prevailingly clastic shelf deposits. The transition from the limestone plateau to the Hamad is masked by the Neogene–Quaternary basalt field of Jebel el Arab.

The limestone plateau is, to a wide extent, covered by sedimentary rocks of the Paleogene Rijam and Shalala formations:

- Chalky limestone, chalk, chert with phosphatic and bituminous layers of the lower Eocene "Umm Rijam chert–limestone formation"
- Chalk, chalky marl, marly limestone of the middle Eocene "Wadi Shalala formation"

The Paleogene carbonate formations are exposed, in particular, on the western slope of the Sirhan depression and the central part of the Jafr syncline. In the central and southern parts of Wadi Sirhan and on the Hamza graben structure southeast of Azraq, the Paleogene is covered by Neogene to Quaternary sediments. On the eastern slope of Wadi Sirhan and in the northern and eastern parts of the Azraq basin, Neogene–Quaternary basalts overlie the Paleogene.

The thickness of the Rijam formation is generally 100 m and increases to 300 m in tectonic depressions at the eastern margin of the plateau: at Qurayat in Wadi Sirhan and in the Hamza graben structure southeast of Azraq. The Shalala formation has a general thickness of 70 m, increasing to more than 500 m at Qurayat and in the Hamza graben.

On the western slopes of Wadi Sirhan, the Shalala formation comprises an upper unit, consisting of chalky limestone with chert, and a lower unit, composed mainly of bituminous marl with layers of marly limestone.

The Paleogene is underlain by:

Upper Cretaceous marls (Muwaqar formation, B3)
Upper Cretaceous limestones and dolomites (Cenomanian–Campanian, B2/A7)
Paleozoic–Cretaceous sandstones (Disi sandstone, Kurnub sandstone)

The Upper Cretaceous comprises, in the Jordanian limestone plateau, the Lower Ajloun (A1–A6), Aman–Wadi Sir (A7–B2) and Muwaqar marl (B3) formations. The Lower Ajloun formation is composed mainly of marls and marly limestones. To the south, the deposits of the formation become increasingly sandy. The Aman–Wadi Sir (B2/A7) formation consists of limestones and dolomites with intercalated beds of sandy limestones, chalk, marl, gypsum, chert and phosphorite. The Muwaqar formation comprises prevailingly marls and, in some areas, chalks and chalky limestones.

In the tectonic depressions of the Hamza graben and the northern Wadi Sirhan, the Wadi Sir and Aman formations are separated by a sequence of limestones, dolomites, claystones and sandstones. In the southern part of Wadi Sirhan, the Aman formation rests directly on Lower Cretaceous sandstones or on shales of the Paleozoic Khreim group.

3.2.3 *Lithologic Features of the Upper Cretaceous - Paleogene Sequence*

During some stages of the late Upper Cretaceous - Paleogene, differentiated sedimentation conditions caused considerable lithologic variations. The sediments deposited during that period comprise, apart from limestones, dolomites and marl, considerable amounts of chalk, siliceous and phosphatic rocks. A particular feature of the lithologic sequence from the Campanian to Eocene is the abundance of chalk deposits, which act as important aquifers with particular hydrogeologic characteristics in various areas of the platform (Sect. 3.4.2).

Chalk is a sediment composed mainly of skeletal calcite protozoa: coccolithophorida with minor amounts of foraminifera and other biogenic fragments and originates from "planktonic rain" settling as carbonate ooze on the sea bottom. Extensive accumulations of chalk are normally limited to the open ocean but can develop on shelf areas, if sea levels are exceptionally high and erosion on adjoining land areas is very limited. The depth of deposition of chalk on shelf areas below sea level is estimated at 100–250 m. Low erosional activity can be related, in particular, to a non-seasonal arid climate and to times of exceptional tectonic stability. In areas or periods with increasing transport of detrital material, pure chalk deposits generally pass into an alternation of chalk, marl, clays or silty clays.

In addition to the abundance of chalky sediments, the lithology of the late Upper Cretaceous–Paleogene formations is characterized in the eastern parts of the northern Arabian platform by a relatively high percentage of siliceous rocks and the occurrence of phosphates. Siliceous oozes can be deposited in an environment controlled significantly by upwelling ocean water, which is cold and rich in nutrients and silica and has a low oxygen content. Zones of mixing of upwelling ocean water with oxygen rich near-surface water are characterized by rapid organic growth. Decaying organic material, sinking down in the sea water of these zones, consumes the available oxygen and produces CO_2, which dissolves the calcareous plankton, leaving siliceous components of the "planktonic rain" (radiolarians, diatoms) to settle on the sea floor.

3.2.4 Tectonic Structure of the Geologic Sub-units

3.2.4.1 Aleppo Plateau

The Aleppo plateau constitutes a relatively undisturbed block in the northeast of the northern Arabian platform. The plateau is delimited by flexure and fault zones against the Palmyrean fold belt in the south, the allochthonous Basit complex in the northwest and the Euphrates–Mesopotamian depression in the east. Most of the platform is covered by flat to gently sloping Paleogene to Neogene deposits. Outcrops of Maastrichtian rocks are found in the central part of the plateau (Jebel Shbith).

The internal structure of the plateau comprises gentle anticlinal and synclinal structures. Major synclinal zones correspond to the morphologic depressions of Al Jaboul and Matah–Harayeq.

Major fracture zones appear to cross the plateau mainly in southwest–northeast, southeast–northwest and SSW–NNE directions. The plateau area includes several Neogene–Quaternary mainly basaltic volcanic complexes, which may be related to

Pliocene–Pleistocene tensional reactivation of deep seated lineaments within the craton area.

3.2.4.2 Mesopotamian Foredeep

The Arabian Plate dips in the northeast, on the boundary between the Arabian Shelf and the alpidic mountain chains, under the Mesopotamian foredeep. The foredeep, which was formed during the final phase of the alpidic tectonic movements mainly in the Pliocene–Quaternary, extends as a vast plain area in a wide belt along the Taurus–Zagros folded zone and is filled with thick molasse sediments.

An intermediate zone between the Arabian platform and the Mesopotamian foredeep is occupied by the Euphrates plain, which covers a structural depression (Al Furatian depression, Ponikarov et al. 1967a) on the northeastern margin of the plateau areas of the northern Arabian platform. Mesopotamian foredeep and Euphrates plain are separated along an approximately west–east trending system of faults and flexures and a series of anticlinal structures: Tuwal el Aba – Jebel Abd el Aziz – Jebel Sinjar.

The Euphrates plain is filled mainly with marine, lagoonal and continental rocks of Neogene age.

3.2.4.3 Palmyrean Fold Belt

The Palmyrean fold belt (Palmyrean aulacogen, Ponikarov et al. 1967a) comprises a zone of intensively folded dislocations. The zone represents an early Mesozoic rift basin, that was later folded and faulted, and is seen as an intracratonic fold belt isolated within relatively undeformed sectors of the platform. The belt constituted a marine sedimentary basin from the Upper Jurassic to Campanian with accumulations of limestones and dolomites with abundant chert beds and phosphate intercalations in the upper part of the sequence.

Most of the tectonic deformations in the Palmyride belt were produced during the Neogene. The general southwest–northeast trending tectonic structure of the Palmyrean fold belt is cut off, in the southwest, by the longitudinal Dead Sea–Jordan rift system and, in the northeast, by the northwest–southeast oriented Euphrates graben. In a regional view, the Palmyride belt appears as a northeast plunging anticlinorium, upon which complex folds are superimposed. The zone of anticlinal and synclinal deformations of Mesozoic and Cenozoic strata can be traced to a depth of at least 5 km below surface.

The tectonic forces causing the fold and fault structure of the Palmyrean belt are attributed to the Cenozoic movements of the Dead Sea–Jordan rift system and to the shortening of a zone of intraplate weakness between more rigid blocks of the platform.

The fold belt is demarcated in the south and north by systems of deep faults against the southern and northern sectors of the northern Arabian platform. The south Palmyrean flexure between the southern Palmyrean chains and the southern Syrian steppe divides the northern Arabian platform into a stable southern part and a more mobile northern sector, in which displacements surpass 1,000 m. In the northwest, the Palmyrides are also delimited by a series of large fractures, separating the fold belt from the Aleppo plateau and from the Homs–Beqaa depression zone, in which the fault system is covered by Neogene–Quaternary sediments.

The overall structure of the Palmyride belt is a complex composite of uplifted blocks and depressions.

The individual mountain massifs of the southern and northern Palmyrean chains correspond to anticlinal structures with outcrops of Upper Cretaceous limestones and dolomites in the cores of the anticlines and, in some areas, of Lower Cretaceous to Jurassic rocks. The mountain slopes are generally covered by Maastrichtian marls and argillaceous limestones and Paleogene chalks, chalky limestones and marls.

The southern Palmyrean mountain chain comprises prevailingly narrow asymmetrical anticlines of 10–25-km length with steep faulted southern flanks and more gentle slopes in the north. The northern Palmyrean chains are characterized by generally SSW–NNE trending fold structures. In the Pliocene–Quaternary basins of Ad Daw and Sabkhet el Mouh, the Upper Cretaceous formations are situated at depths of several hundred metres below surface.

3.2.4.4 Southern Syrian Steppe and Hamad

The plateau areas of the Hamad and the southern Syrian steppe are subdivided by tectonic movements of prevailingly block-faulting type into structural depressions and arch-like uplifts. The sub-region is bordered by major fault zones:

- The southwest–northeast oriented south Palmyrean fault in the north
- The southeast–northwest trending Faidat fault in the northeast on the boundary of the Euphrates basin

The Hail–Rutba arch is a dominating structural element on the eastern border of the Hamad, which appears as a broad swell in Cretaceous to Paleogene formations. The structural high passes from the Rutba area in SSE direction to Jebel Unaiza (Iraq–Jordan–Saudi Arabia border triangle) and into Khawr Um Wual, a narrow zone of grabens and synclinal depressions along the crest of the arch.

The plateau area west of the Rutba arch is tectonically subdivided into mainly SSE–NNW oriented anticlinal structures and graben or synclinal structures which are occupied by small morphologic depressions (swells of At Tenf and Al Owered, Khabret et Tenf, Kahbret Mashqouqa, Khabret et Tnefat, Ponikarov et al. 1967a). A belt of generally WSW–ENE directed structures marks the northern rim of the southern Syrian steppe; the belt merges in the east into the Euphrates basin and in the southwest into the structural depression of Al Juwef.

South of Rutba, the Hamad is adjoined in the east by the Widyan sector of the eastern Arabian platform. The uplift structure of the Rutba arch, which forms the southeastern border of the northern Arabian platform, is here hidden under a cover of Upper Cretaceous sedimentary rocks.

3.2.4.5 East Jordanian Limestone Plateau

The strata of the Jordanian limestone plateau dip from the western highlands gently to the east; the monoclinal structure is modified by local anticlines, synclines and basin-like depressions.

The base of the Rijam (B4) formation is located at around 1,200 m asl in the western highlands, 950–1,000 m asl in the south of Jordan (south of Bayir) and descends to 766 m below sea level in Wadi Sirhan in Saudi Arabia.

The Paleogene outcrops in the Jordanian limestone plateau are crossed by major fault systems, which extend in east–west, northwest–southeast and NNW–SSE direction over lengths of 50 km to more than 300 km (Kerak–Wadi Fiha fault, Hasa fault zone, Siwaqa fault, Wiesemann 1969; Sunna 1995).

The Jafr depression corresponds to a WNW–ESE oriented tectonic trough within the limestone plateau. The eastern margin of the limestone plateau is formed by the southeast–northwest trending Wadi Sirhan fault zone (Ramtha–Wadi Sirhan trend), which is marked morphologically by the Wadi Sirhan and Azraq depressions. The Hamza graben southeast of Azraq constitutes a particularly deep structural depression, in which Upper Cretaceous sediments reach a thickness of more than 3,000 m.

References. Ahmed and Kraft (1972), Al Ejel and Abderahim (1974), Bender (1974a), Buday (1980), Downing et al. (1993), ESCWA (1999b), GITEC and HSI (1995), GTZ and NRA (1977), Hobler et al. (1991: 49), Kemper (1980), Krasheninikov et al. (1996), Kruck et al. (1981), Litak et al. (1998), Lovelock (1984), McBride et al. (1990), Ponikarov et al. (1967a), Powers et al. (1966a: 104), Razvalayev (1966), Soulidi-Kondratiev (1966), Sunna (1995), Wagner and Kruck (1982), Wiesemann (1969), Wolfart (1966).

3.3 Aquifers and Groundwater Regimes

3.3.1 Main Aquifers

3.3.1.1 Aquifers and Groundwater Flow Systems

A continuous aquifer system at shallow to intermediate depth extends in the Paleogene chalk formations over wide parts of the plateau landscapes of the eastern part of the northern Arabian platform. Zones with increased erosion and with relatively high secondary permeability of aquiferous carbonate rocks have developed along fracture systems. Wadi courses on the rigid blocks largely follow the pattern of the fracture systems and zones with relatively favourable aquifer potential are found preferentially along the fracture and wadi systems.

The Paleogene aquifers are underlain, in general, by a deeper aquifer system in Upper Cretaceous (Cenomanian–Coniacian) carbonate rocks and are separated from that deeper aquifer by an aquitard composed of Upper Cretaceous–Paleogene (Maastrichtian–Lower Eocene) marls, chalks and argillaceous limestones (Table 3.1).

Toward structural uplifts of the Rutba area and the Palmyrean chains, the continuous upper aquifer disappears since:

- The Paleogene sediments are located above the level of saturated aquifers.
- The Paleogene chalk formation and/or the Maastrichtian–Paleogene marl aquiclude are missing because of paleogeographic or erosional gaps.

In the uplift areas, the upper aquifer system is either connected to the deeper Mesozoic carbonate aquifers or becomes unsaturated or discontinuous.

In structural basins and some plateau areas, the upper aquifer system includes Oligocene or Neogene–Quaternary aquiferous sediments, which are generally in hydraulic connection with the Paleogene chalk aquifer.

The Paleogene aquifer system on the eastern part of the northern Arabian platform is separated by structural geologic features into several groundwater flow systems. Major flow systems enclose:

- The northeastern part of the northern platform around Aleppo
- The southeastern part of the northern platform between the southern Palmyrean mountains and the Rutba high
- The east Jordanian limestone plateau

3.3.1.2 Aleppo Plateau

The Paleogene chalk aquifer system and the deeper Upper Cretaceous aquifer system extend continuously over most of the Aleppo plateau. The upper aquifer system is formed by Middle to Upper Eocene chalks and limestones in wide areas around Aleppo and east of Hama and Homs. In some parts of the area, the Eocene rocks form a connected aquifer system with overlying rocks: Helvetian limestones, Pliocene–Quaternary unconsolidated sediments, Neogene basalts. The deeper aquifer comprises Upper Cretaceous (Cenomanian–Turonian) limestone and dolomites.

The upper aquifer system is separated from the deeper aquifer by the Maastrichtian–Paleogene aquitard, which is composed of several hundred metres thick marls and argillaceous or chalky limestones with chert intercalations.

The base of the Upper Cretaceous carbonate aquifer is probably formed by marls of Lower Cretaceous (Aptian–Albian) age. Groundwater contained in deeper sedimentary rocks below the Upper Cretaceous can be expected to be more or less stagnant and brackish to saline.

In the eastern parts of the Aleppo plateau, the hydrogeologic conditions of the upper aquifer system are relatively uniform: Eocene chalks and marly limestones

Table 3.1 Scheme of main aquifers in the eastern part of the northern Arabian platform

		Aleppo plateau	Jezire	Palmyrean fold zone	Southern Syrian steppe and Hamad	East Jordanian limestone plateau
Quaternary		Aquiferous deposits in morphologic depressions	Rad aquifer: unconsolidated deposits, aquiferous in part of the plain areas			
Neogene		Miocene limestone, locally part of upper aquifer system		Aquifers in morphologic depressions		Qirma sandstone: local aquifer in Wadi Sirhan
Paleogene		Chalk, limestone, chert **Main upper aquifer system**	Ras el Ain/Tel Abiad aquifer **Main upper aquifer system**	Chalk, chert, limestone: brackish water aquifer at shallow to intermediate depth	Chalk, chert, limestone: **upper aquifer system**, mainly brackish	Shalala (B5): local aquifer Rijam (B4): fresh water aquifer in Jafr area
Upper Cretaceous	Maastrichtian	marl **major aquitard**	marl **major aquitard**	marl **discontinuous aquitard**	marl **discontinuous aquitard**	Muwaqqar marl
	Campanian to Cenomanian	limestone, dolomite, chert, chalk **main deeper aquifer, brackish water**	limestone, dolomite, chert, chalk **main deeper aquifer, brackish water**	limestone, dolomite, chert, chalk **main deeper aquifer, brackish water**	limestone, dolomite, chert, chalk **main deeper aquifer, brackish water**	limestone, dolomite, chert, chalk **main deeper aquifer, brackish water**
Lower Cretaceous						Kurnub sandstone, deep brackish water aquifer
Paleozoic						Disi sandstone, deep brackish water aquifer

constitute an extensive aquifer, which is underlain by the 100–400 m thick Upper Cretaceous–Paleogene marl aquitard. The productivity of the chalk aquifer is highly varying: its permeability depends, to a large extent, on fracturing of the rock, and productive zones are restricted mainly to areas where tectonic movements have created a relatively high number of fractures and fissures. Such zones are found, in particular, along larger wadis, while aquifer productivity on hilltops and along surface water divides is generally very low. In the morphologic depressions of El Matah and Al Jaboul, the upper aquifer system extends into Pliocene–Quaternary basin sediments overlying the chalk aquifer. In the Jebel al Hass area, basalts constitute locally an upper part of the aquifer system. Leakage through the Upper Cretaceous–Paleocene aquitard to the deeper aquifer system may occur, in particular, along major fracture zones.

The groundwater in the upper aquifer system is generally unconfined, depth to groundwater ranges from a few metres to around 100 m below land surface.

The plateau area in the southeast of the northern Syrian steppe, between the Palmyrean fold zone (Jebel Bishri) and the Euphrates river, is covered mainly by Miocene limestones, sandstones and evaporites, which constitute a shallow brackish water aquifer.

In the western parts of the plateau – the area between Hama, the eastern rim of Jebel ez Zaouiye and the Idleb plateau – the hydrogeologic conditions of the upper aquifer system are complex.

Aquiferous rocks comprise Eocene chalks and nummulitic limestones, and limestones of Miocene age; permeability is relatively high in karstified limestones. The marl aquitard underlying the upper aquifer system is wedging out toward the eastern margin of the rift zone. According to these conditions, the area comprises various discontinuous groundwater occurrences and perched water levels are found in topographically higher areas. The Upper Cretaceous–Paleocene chalky marl and chert formation, which forms part of the main aquitard further east, acts locally as aquifer with low productivity. To a considerable extent, groundwater from the upper aquifer system leaks into the deeper Upper Cretaceous carbonate aquifer. Fractured and karstified Miocene limestones are an important component of the upper aquifer system in the area north of Aleppo. Several perennial springs discharge from the Miocene limestones, such as the springs of Hailan, which provided the main source of water supply for Aleppo until the beginning of the twentieth century.

3.3.1.3 Jezire

In the Mesopotamian–Euphrates plain of the Jezire, main aquifers are found in:

- Quaternary deposits
- Miocene carbonate rocks and sandstones
- Paleogene carbonate rocks

Aquifers of pre-Paleogene age (Permian–Upper Cretaceous) are situated at depths of 200–3,500 m below the plains and contain brackish to saline water.

The most important aquifers of the Jezire have been denominated, according to their main geographic distribution: Ras el Ain, Tel Abiad and Radd aquifers.

The highly productive Ras el Ain aquifer is composed of karstified nummulitic limestone of middle to upper Eocene and Oligocene age. The base of the aquifer is formed by a marl aquiclude of Upper Cretaceous–Lower Eocene age; the aquifer is generally confined by clay, marl, clayey limestone and gypsum of lower to middle Miocene age. In some areas, the overlying Miocene aquiclude is missing and the aquifer is unconfined. In the upper Syrian Jezire, the top of the confined Ras el Ain aquifer is situated at shallow depth of several metres below surface.

The catchment of the Ras el Ain aquifer reaches in the north far into Turkish territory. The aquifer thickness varies between 200 and 300 m in Turkey and increases to around 1,000 m at Qamishly on the northeastern border of Syria. In the south, the aquifer extends until Tell Tamer on the Khabour river. Groundwater discharges from the aquifer in the large springs of Ras el Ain on the Syrian–Turkish border, which is one of the largest karst springs of the world.

The Tel Abiad aquifer is also a karst aquifer in Paleogene limestones with similar hydrogeologic characteristics as the Ras el Ain aquifer. The main discharge of the Tel Abiad aquifer is concentrated in springs on the Syrian–Turkish border: Ain el Arous on the Balikh river and Ain el Arab.

The Rad aquifer constitutes a generally shallow aquifer in unconsolidated to semi-consolidated deposits of Upper Miocene to Quaternary age and extends over parts of the Mesopotamian plains in the Syrian Jezire between Qamishly, Karatchok and the Radd marshes near Hasake. Permeable zones with considerable thickness and moderate to high permeability are developed in gravel, conglomerate and sand layers of the Miocene–Quaternary formations.

Apart from the main aquifers, the Jezire comprises several aquifers in Paleogene–Quaternary sediments, which are of minor importance because of their limited extent and/or relatively low transmissivities.

Eocene marly limestones, exposed along the eastern margin of the geologic Aleppo plateau, are drained along the Euphrates valley and provide village water supplies through shallow wells and springs in the Jerablus area.

Minor aquifers in the Jezire of Syria and Iraq are:

- Limestones and sands of the Oligocene–middle Miocene Lower Fars formation, which yield limited quantities of brackish groundwater
- Clayey limestones and gypsum of the Upper Fars formation
- Pliocene conglomerates, sandstones and marly limestones

Along the anticlinal zone of Jebel Abd el Aziz – Jebel Sinjar, Paleogene–Lower Miocene limestones provide fissure type aquifers with local importance for water supply and irrigation.

Quaternary sediments – sands, sandstones, loam with gravels, gypsiferous conglomerates – form shallow aquifers in the valleys of the Euphrates, Khabour and Balikh rivers and their tributaries.

3.3.1.4 Palmyrean Fold Belt

The Palmyrean fold zone comprises a major sub-regional aquifer system in Upper Cretaceous limestones and dolomites. In the Ad Daw and Sabkhet el Mouh plains, the Upper Cretaceous aquifer system is overlain by aquifers in Paleogene chalks and limestones and in Pliocene–Quaternary sediments.

The Upper Cretaceous carbonate rocks provide a fissure type aquifer. Karstification appears to be much less intensively developed than in the limestone and dolomite aquifers of the northwestern mountain and rift zone. The aquiferous zones of the Upper Cretaceous in the Palmyrean fold zone are formed, in particular, by Cenomanian–Turonian limestones and dolomites, but include also stratigraphically younger members of the Upper Cretaceous, such as layers of chert, limestone and silicified limestone in the Campanian and Maastrichtian. Outcrops of the aquiferous Upper Cretaceous are restricted to the anticlinal cores of the southern and northern Palmyrean mountains and the mountain chains east of Nebek (Jebel Sharq en Nebek). These outcrop areas are assumed to be the main recharge zones of the Upper Cretaceous aquifer. Deep confined sections of the Upper Cretaceous aquifer system are found in the Ad Daw and Sabkhet el Mouh depressions, the confining layers being formed by the Maastrichtian–Paleogene marl aquitard.

Fresh groundwater has been tapped in the Upper Cretaceous aquifer through shallow wells and boreholes in particular in wadis within the southern Palmyrean mountains and the mountains east of Nebek at Nasriye, Qaryatein, Al Barde and around Sawane. The Paleogene sequence of chalk, limestone, chert and marl provides an aquifer with low to moderate productivity at shallow to intermediate depth in several areas of the Palmyrean fold belt: in some parts of the Ad Daw plain and its western surroundings, in the low hills adjoining Sabkhet el Mouh in the north and south.

Pliocene–Pleistocene deposits of the Ad Daw basin are aquiferous in the more permeable layers with water levels at 15–60 m below surface. Aquiferous sections are found, in particular, in coarse detrital layers at the periphery of the plain. Groundwater from deeper aquiferous Campanian siliceous limestones appears to leak, at some points, into the overlying Neogene aquifer.

In the Sabkhet el Mouh plain, sandy layers of the Neogene sequence comprise a brackish water aquifer with salinities of 1,100–6,000 mg/l TDS.

In the eastern prolongation of the Palmyrean fold zone, occurrences of groundwater support water supply and limited irrigation at several locations of the steppe area. The exploited aquifers comprise:

- Eocene chalks at Arak and Hafne.
- Upper Cretaceous, mainly Campanian, limestones and cherts at Soukhne, situated at a depth of 250–340 m below surface.
- Campanian cherts tapped at Al Koum in the Jebel Bishri area at a depth of 250–340 m below surface and Pliocene deposits around Al Koum, which receive groundwater through leakage from the deeper Campanian aquifer.

Groundwater in the aquifers of the eastern outliers of the Palmyrean fold zone is prevailingly brackish (Sect. 3.4.4.1).

3.3.1.5 Southern Syrian Steppe, Hamad and East Jordanian Limestone Plateau

In the southern Syrian steppe and the Hamad, groundwater is found mainly in Paleogene chalks, limestones and cherts, and in fractured and fissured Upper Cretaceous limestones and dolomites. The Upper Cretaceous and Paleogene aquiferous layers are, in some areas, separated by the Maastrichtian–Paleogene marl aquitard (Muwaqar aquitard in Jordan); in structurally uplifted zones, the marls grade into carbonate deposits and the aquitard becomes leaky or is missing.

The Paleogene chalks and limestones form an aquifer at shallow to intermediate depth in wide parts of the plateaus of the southern Syrian steppe and the Hamad, which is, in many parts of the area, connected through leakage with underlying Mesozoic carbonate aquifers or with overlying Neogene–Quaternary sedimentary or volcanic rocks. The main aquiferous sections are found in lower Eocene limestones and cherts and in middle Eocene chalky limestones. In Jordan, these aquiferous members are denominated B4/B5 aquifer or Umm Rijam and Wadi Shalala aquifers. Toward west, the Paleogene plunges under the extensive cover of Neogene–Quaternary basalts of Jebel el Arab.

The stratigraphically lower member of the Eocene aquifer system (lower Eocene chert – limestone aquifer, B4 or Umm Rijam aquifer) extends over wide areas of the southern Syrian steppe and the Hamad and has been tapped by numerous wells in southern Syria, e.g. around Saba Biar, and in northeastern Jordan. The aquifer is used through dug wells of 6–25 m depth in the Wadi el Miyah – Wadi al Murabaa area in Syria, by a group of drilled wells at Humeime (Syria) and by boreholes of 35–100 m depth in the Sawab area (Iraq–Syria) and in northeastern Jordan. The aquifer is, in part of its extent, confined under marly sediments of the middle Eocene. The productivity of the lower Eocene aquifer is generally low.

The stratigraphically higher part of the Eocene contains a rather discontinuous aquifer (middle Eocene chalky limestone aquifer, Wadi Shalala or B5 aquifer), which is locally hydraulically connected with the underlying lower Eocene chert – limestone aquifer. The middle Eocene aquifer has been tapped mainly by dug wells of 50 to nearly 100 m depth in Wadi el Heil and in the Saba Biar–Juwef area in southern Syria, and in northeastern Jordan. Well yields are generally low.

The Paleogene aquifer system becomes unsaturated in the eastern parts of its outcrops near the Rutba uplift: the Paleogene is here situated above the regional groundwater level, the underlying Upper Cretaceous marls are missing or grade into limestone – dolomite facies near the uplift. In the Jordanian part of the Hamad area, the eastern limit of the Rijam aquifer is defined by the limit of saturation in a north–south trending zone around 50–60 km west of the border with Iraq. Toward northeast, the Paleogene aquifer dips under a thick cover of Pliocene sediments of the Euphrates basin.

The aquiferous sections in the Upper Cretaceous of the southern Syrian steppe and the Hamad correspond generally to rock units of Cenomanian–Turonian–Campanian age. On structurally uplifted areas in the east – the Rutba arch and At Tenf and Owered swells – aquiferous layers are found in carbonate rocks of Maastrichtian and possibly Campanian and Jurassic age.

The Upper Cretaceous carbonate aquifer complex has been tapped in the Wadi al Miyah–Wadi Murabaa area in Syria, in the H4–Rishe area in northeastern Jordan, and in southwestern Iraq (Tayarat aquifer).

On most of the western slope of Wadi Sirhan, the outcropping Rijam formation is unsaturated. The chalky and cherty limestones of the Rijam formation constitute a phreatic aquifer in the Jafr and Azraq area and in a 5–10 km wide belt over the western slope of Wadi Sirhan. Confined conditions occur in the Hamza graben and in Wadi Sirhan, where the top of the Rijam formation descends to a depth of 80 m below surface. The saturated thickness of the aquifer is 75–100 m in the water table areas and 100–200 m in the confined sections.

In the central part of the morphologic Jafr depression, the Rijam formation constitutes an aquifer which extends over around 1,250 km^2. The aquiferous layers consist of chalky limestones and cherts with marl and clay layers of the lower part of the B4 (Rijam) formation. The upper part of the B4 formation and the B5 formation are missing. The aquifer reaches a maximum saturated thickness of 30 m. Depth to aquifer is in the range of 15–35 m below surface.

The Shalala limestone member, representing the upper part of the Shalala formation, provides a saturated aquifer in a limited area on the border between Jordan and Saudi Arabia. The aquifer has a saturated thickness of 20–47 m and is situated at a depth of around 60 m below surface. Permeabilities of the Shalala aquifer are generally low.

Moderately high well yields are found in areas, where the Shalala aquifer is overlain by sandstones and sandy limestones of the Miocene Qirma formation. The Shalala and Qirma formations constitute a hydraulically connected unconfined aquifer in the Hamza area and the northern Wadi Sirhan. The aquifer is intensively exploited for irrigation in the Qurayat area in Wadi Sirhan, where groundwater levels are situated at 13–30 m below surface.

On most of the east Jordanian limestone plateau, the Upper Cretaceous aquifer is confined below the Muwaqar aquitard. The limestones and cherts of the Aman formation generally form a connected aquifer with the limestones of the Wadi Sir formation (Aman–Wadi Sir, B4/A7 aquifer). Argillaceous and marly sediments of part of the Ajloun group (A1–A6 formations) probably constitute an aquitard, which separates the Aman–Wadi Sir aquifer from the deeper Kurnub sandstone aquifer.

The Paleozoic–Lower Cretaceous Disi–Kurnub sandstone aquifer complex provides a large groundwater reservoir under the east Jordanian limestone plateau. The hydrogeologic features of the deep sandstone aquifer complex are considered in (Sect. 5.2). Granitic rocks of the basement complex constitute, in general, the impermeable base of the aquifer systems of the Jordanian limestone plateau.

3.3.2 Groundwater Regimes

3.3.2.1 Hydraulic Properties of Main Aquifers

Transmissivities of most aquifers of the eastern part of the northern Arabian platform are low to moderate. Table 3.2 presents some information on transmissivity

Table 3.2 Hydraulic parameter values of aquifers in eastern Jordan

Aquifer	Area	T (m^2/d)	K (m/s)	References
Rijam	El Jafr	40–7,700	10^{-5}–3×10^{-3}	GITEC and HSI (1995)
	Azraq	6–230 (max. 2,250)		Margane et al. (2002)
Aman-W.Sir	El Jafr	40–1,750		Hobler et al. (1991)
Lower Ajlun	El Jafr	320		Hobler et al. (1991)
B4/B5	Hamad		3×10^{-5}	GITEC and HSI (1995)

and hydraulic conductivity values, which have been reported for Upper Cretaceous and Paleogene aquifers of the sub-region. Fresh water aquifers, which occur at shallow depth on the Aleppo plateau, the Jezire, the Jafr area and along some wadi systems, have generally a limited thickness and, accordingly, relatively low transmissivities.

Exceptionally high transmissivities are found in the karstified Paleogene Ras el Ain and Tel Abiad aquifers on the northern boundary of the Jezire, which support high spring discharges.

3.3.2.2 General Groundwater Flow

Groundwater flow on the plateaus and morphologic depressions of the eastern part of the northern platform is directed to:

- Closed basins of the internal drainage zone, which is situated between the Mediterranean Sea, Dead Sea and Euphrates basins: Jaboul, Matah, Sabkhet el Mouh, Damascus and Azraq plains and Wadi Sirhan.
- The rift valleys in the west and the Euphrates river system in the east.

The partly complex flow systems can include different flow directions in upper and deeper aquifers.

3.3.2.3 Aleppo Plateau

Recharge rates in the upper aquifer system of the Aleppo plateau depend on rainfall, rock outcrops and morphology. In the northwestern part of the plateau, outcropping karstified limestones (Eocene nummulitic limestones, Miocene limestones) and relatively high rainfall provide favourable recharge conditions. Recharge rates of 15–20% of annual rainfall may be assumed for the western, more humid zones of the Aleppo plateau. Average recharge rates of 11 mm/a have been estimated for areas covered by Helvetian limestones south of Aleppo, and of 12 mm/a for the Jebel al Hass area covered by Miocene basalts.

In the southeastern dry parts of the plateau, recharge is generally low. Infiltration of rainfall from events with low intensity is not sufficient to exceed the field capacity of the soil and most of the soil moisture is lost by evapotranspiration.

Groundwater recharge may occur along wadi courses, where accumulation of surface runoff can infiltrate, and through percolation in irrigation areas.

Groundwater flow in the upper aquifer system (Paleogene aquifer system) of the Aleppo plateau follows, in general, the topographic–hydrographic slope to closed basins and, in the west and east, toward the Orontes and Euphrates valleys in the west and east, respectively. The groundwater divide enclosing the zone of internal drainage coincides approximately with the surface water divide. The groundwater flow regime in the upper aquifer system is, however, significantly influenced by downward leakage into the deeper Upper Cretaceous aquifer in several zones. Lekage into the deeper aquifer occurs, in particular, in the west of the plateau, where the underlying marl aquiclude is thinning toward the anticlinal structure of the rift zone, and in fracture zones along the lower course of the Queiq river and the western rim of the Matah depression. Groundwater flow toward the closed basins probably reverses its course in some areas after leakage into the deeper aquifer, which drains toward topographically lower river valleys. Main groundwater discharge zones are the salt flats and seasonal lakes of Al Jaboul and Matah, where major volumes of groundwater discharge naturally by evaporation. Seasonal springs issue on the margins of the flats of closed basins and in the Queiq valley upstream and downstream of Aleppo. At present, the groundwater regime of the upper aquifer is significantly affected by groundwater withdrawal for irrigation.

The Upper Cretaceous carbonate rocks underlying the Aleppo plateau constitute a productive brackish water aquifer in an area extending from the border of Syria and Turkey to the northern Palmyrean mountains and, in the west, approximately the Aleppo–Hama highway, an area covering around 20,000 km^2. Well yields of boreholes tapping the Upper Cretaceous aquifer are in a range of 30–140 m^3/h with minor drawdown.

Isotope data indicate that the groundwater stored in the Upper Cretaceous carbonate aquifer of the Aleppo plateau originates prevailingly from recharge in past periods with wetter climate several thousand years ago and may be related to groundwater movement from distant recharge areas. Recharge areas may be presumed at outcrop areas of the Upper Cretaceous limestones and dolomites in the northern Palmyrean mountains or in the eastern Taurus mountains, where present mean annual precipitation is around 400 mm and >600 mm, respectively.

No points of discharge of the deeper groundwater are known on the Aleppo plateau. Upward leakage of deeper groundwater into the upper aquifer system or into sabkhas possibly occurs in the structural and morphologic depressions of Matah–Harayeq. Upward leakage from the Upper Cretaceous aquifer into the overlying aquifers may occur also in the lower ranges of the Euphrates valley in the east of Aleppo, where the Upper Cretaceous is covered by a thick sequence of younger deposits.

Groundwater movement in the Upper Cretaceous of the Aleppo plateau is probably directed, with very low flow velocities, westward to the rift graben, where the old brackish ground water may provide a minor component of spring discharge, and, to some extent, also to the Euphrates valley in the west.

Topographic elevations of the probable main discharge zones are: Jaboul 308 m asl, Matah 249 m, El Rhab 170 m, Er Rouj 218 m, Orontes at the Syrian–Turkish border 100 m, Euphrates 276 m.

3.3.2.4 Jezire

The main recharge areas of the Ras el Ain and Tel Abiad aquifers in the Syrian Jezire extend into the southern ranges of the Taurus mountains, where mean annual precipitation reaches up to 1,000 mm. The recharge areas are covered mainly by Eocene limestones and Pliocene–Quaternary basalts. The catchment area is estimated to extend over 7,500 km^2, and has a mean annual precipitation of 450 mm. The average recharge is estimated at 1.6×10^9 m^3/a.

Groundwater flow is directed toward south to the main spring discharge areas along the Syrian–Turkish border. The total discharge of springs from the Ras el Ain and Tel Abiad aquifers may be around 52 m^3/s. Mean discharge of the spring Ras el Ain is 38.7 m^3/s from an estimated stored groundwater volume of 7.4×10^9 m^3.

In the semi-arid Mesopotamian–Euphrates plain, direct recharge from precipitation is relatively low. Recharge to the shallow aquifers in the plains occurs mainly from surface runoff in the Khabour river and its tributaries. Recharge from precipitation and runoff infiltration to the Upper Miocene–Quaternary Radd aquifer amounts to around 350×10^6 m^3/a. Groundwater flows within the aquifer generally to the extensive Radd marshes in the south, where an estimated volume of 270×10^6 m^3/a discharges through evapotranspiration.

In the eastern Jezire in Iraq, groundwater flow is directed to discharge areas in the Euphrates valley, Wadi Tharthar, the artificial Tharthar reservoir, and to local salt flats.

3.3.2.5 Palmyrean Fold Belt

Present day groundwater recharge in the Palmyrean zone is restricted mainly to the outcrops of the aquiferous Upper Cretaceous and Paleogene formations in the anticlinal mountain chains and to infiltration of sporadic surface runoff. Recharge rates under the present climate conditions with mean annual rainfall of 100–200 mm are certainly low. Most of the groundwater in the Upper Cretaceous and Paleogene aquifers is fossil, originating from Pleistocene recharge.

Groundwater flow in the main aquifers of the Palmyrean zone is directed toward Sabkhet el Mouh, which is situated at an elevation of around 370 m asl and constitutes a sub-regional groundwater discharge zone. In the east of the Ad Daw plain, groundwater moves into the Sabkhet el Mouh plain under the morphologic rise separating the two plains on the surface. The subsurface catchment of Sabkhet el Mouh extends, south of the southern Palmyrean mountains, into the plateau area of the southern Syrian steppe.

Some groundwater discharge occurs in springs on the margins of the Ad Daw plain, in particular at Qaryatein on the southwestern edge of the plain.

In the Maqsam–El Baida area in the southeast of the Ad Daw plain, groundwater apparently leaks from the Upper Cretaceous aquifer probably through a deep fault zone into the overlying Neogene–Quaternary aquifer.

Evaporation in the Sabkhet el Mouh salt flat constitutes an important component of groundwater discharge. To some extent, deep groundwater flow may move from the Sabkhet el Mouh basin eastward to the Euphrates river.

3.3.2.6 Southern Syrian Steppe, Hamad and East Jordanian Limestone Plateau

The climate in the southern Syrian steppe and Hamad and on the east Jordanian limestone plateau is arid with mean annual precipitation between 120 and 50 mm. Present-day recharge is, accordingly, very limited. Most of the groundwater is fossil.

An impact of recent indirect recharge through infiltration of surface runoff is, however, indicated by isotope data (Sect. 3.5) and by the occurrence of shallow groundwater bodies along some wadi stretches. Shallow groundwater sustained from recharge is e.g. found in Wadi al Miyah, Wadi Murabaa and Wadi Sawab in southeastern Syria and southwestern Iraq and along Wadi Ruweishid in northeastern Jordan.

Groundwater moves in the southern Syrian steppe and the Hamad in a general east-northeast direction to the Euphrates valley. A belt of the southern Syrian steppe adjoining the Palmyrean zone is included in the hydrogeologic Sabkhet el Mouh catchment.

Separate groundwater regimes exist in some parts of the southern Syrian steppe in the upper Paleogene aquifer and the lower Mesozoic carbonate aquifer. In the east near the Rutba uplift zone, the marls of the Maastrichtian–Paleogene aquitard grade into limestone facies and the Paleogene aquifer is either unsaturated or hydraulically connected with the deeper aquifer.

Subsurface outflow toward the Euphrates valley appears to occur through a multi-aquifer system in Mesozoic to Neogene deposits.

In the Upper Cretaceous and Paleogene aquifers of the east Jordanian limestone plateau, groundwater flows generally toward east-northeast to the morphologic depressions of Wadi Sirhan and Azraq. The Paleogene constitutes only in parts of the plateau a saturated aquifer. Groundwater flow in the Upper Cretaceous aquifer system is, in some areas, controlled by major faults, as indicated by the groundwater salinity distribution (Sect. 3.4.4.3).

In the deep sandstone (Disi–Kurnub) aquifer system, groundwater movement follows a sub-regional flow path from outcrop areas in the Interior Homocline in the south in a wide semi-circle to the Dead Sea basin (Chap. 5).

References. Abu-Ajamiieh (1967), ACSAD (1983), Ahmed and Kraft (1972), Boeckh et al. (1970), Burdon and Safadi (1963), ESCWA (1999b), GITEC and

HSI (1995), GTZ and NRA (1977), Hobler et al. (1991), Kattan (2002), Khouri (1982), Luijendijk and Bruggemann (2008), Margane et al. (2002), Medvedev (1966), Mikhailov (1964), Oufland (1966a, b), Ponikarov et al. (1967b), Sunna (1995), United Nations (1982: 158), Wagner (1997), Wagner et al. (1982), Ward and Smith (1994), Wolfart (1966).

3.4 Groundwater Salinity and Hydrochemistry

3.4.1 Groundwater Salinity Distribution

On the plateau areas in the east of the northern Arabian platform, fresh groundwater occurs, in particular, in the upper aquifer system of the semi-arid zones of the Aleppo plateau and the western fringes of the east Jordanian limestone plateau. Large quantities of fresh groundwater discharge from Paleogene karstified carbonates on the northern boundary of the Syrian Jezire.

In the deeper aquifer system and, in the more arid areas in the east and southeast, also in the upper aquifer system, the groundwater is, in general, brackish.

In the upper aquifer system of the Aleppo plateau, groundwater salinity is generally low to moderate in the recharge areas and increases toward the local discharge areas in the closed basins of Jaboul, Matah and Harayeq. In the sabkha areas the groundwater is generally brackish with total salinity up to 7,000 mg/l TDS.

In the Palmyrean fold zone, the extent of fresh water aquifers is very limited. Groundwater with moderate salinity is found in some of the hill zones on the peripheries of the Ad Daw and Sabkhet el Mouh plains. In most of the Upper Cretaceous, Paleogene and Pleistocene–Quaternary aquifers, the groundwater is brackish.

The Paleogene carbonate rocks act in the Palmyrean fold zone only locally as fresh water aquifers of limited extent, as outcrops are generally restricted to small areas along anticlinal flanks situated above main groundwater levels. The Paleogene aquifers below Pleistocene–Quaternary sediments in Ad Daw and Sabkhet el Mouh plains contain brackish to saline groundwater.

The general pattern of groundwater salinity in the southern Syrian steppe and Hamad can be characterized as follows:

Groundwater salinity is:

- Low to moderate (<1,000 mg/l TDS) in local fresh water occurrences in the Lower Eocene chert aquifer in the Wadi el Miyah, Wadi Sawab and Wadi Ruweishid areas
- Moderate to high (750–1,500 mg/l TDS) in the Eocene limestone – chert aquifer of the steppe area bordering the southern Palmyrean mountains
- Generally high (1,500–2,500 mg/l TDS) in the Paleogene chert aquifer in most of the central Hamad Plateau
- High to very high (1,500 to >3,500 mg/l) in several parts of the Hamad plateau (Sabaa Biar–Juwef–Rishe areas)

Main sources of the high groundwater salinity, which prevails in wide areas of the Hamad, are assumed to be evaporative enrichment during the recharge process and solution of rock material during very long retention periods of 24,000 and 36,000 years in carbonate and chert aquifers. The occurrence of groundwater with relatively low salinity appears related to recent groundwater recharge. Fresh water occurrences in Paleogene aquifers covered by the Jebel el Arab basalts are considered in Chap. 4.

An increase of groundwater salinity together with a change from oxic to anoxic conditions occurs, on the northern Arabian platform, in particular where groundwater from an overlying fresh water aquifer leaks through bituminous marly chalky rocks of the Maastrichtian aquitard into the underlying limestone–dolomite aquifer. These conditions are common in the eastern part of the northern Arabian platform, but is also observed in aquifers of the Judean highlands. "An increase in salinity and change from oxic to anoxic conditions are observed in the Upper subaquifer of the Judea group…at the western foothills of the Judea Mountains" "The deterioration in the water quality is explained as a result of seepage of more saline, organic-rich water from above… The latter is derived from the bituminous chalky rocks of the Mount Scopus Group, which confine the aquifer". "The incoming organic matter consumes the dissolved oxygen and allows bacterial sulfate reduction. The latter accounts for the H_2S in the aquifer, as indicated by sulfur isotopic analyses" (Gavrieli et al. 2002: 483).

On the east Jordanian limestone plateau, fresh groundwater with salinities of generally less than 600 mg/l TDS occurs in the Rijam (B4) aquifer of the Jafr area. Locally, the water quality is influenced by irrigation return flow leading to an increase of water salinity to 2,000–3,000 mg/l TDS. Electrical conductivity values of groundwater from the Rijam aquifer in the Wadi Sirhan catchment range from approximately 1,200–3,000 μS/cm. Locally elevated salinity (EC 3,600 μS/cm) is attributed to the presence of evaporite minerals in the limestone aquifer. Groundwater salinity in the Shalala (B5) aquifer appears to be, in general, similar to salinity in the Rijam aquifer.

3.4.2 Hydrochemistry of Paleogene Chalk Aquifers

Paleogene chalks and chalky limestones constitute a major component of the upper aquifer system in wide parts of the eastern plateaus of the northern Arabian platform. Aquiferous chalks extend over the Aleppo plateau, the southern Syrian steppe and Hamad, the east Jordanian limestone plateau, and synclinal structures of the Palmyrean zone. In the western mountain and rift zone, Paleogene chalk aquifers are found around Irbid in northern Jordan, in the Jenin sub-basin on the West Bank and in the Bekaa and the southwestern coastal zone of Lebanon.

The hydrochemical conditions of the chalk aquifers are somewhat different from the conditions in typical limestone and dolomite aquifers.

Chalks are generally fine grained, pure low magnesium carbonates with abundant intercalations of clay minerals. The chalk matrix provides a large internal surface area for water–rock contact and reaction. In the outcrop areas, congruent solution reactions predominate and the low Mg/Ca ratios in the rock are reflected in the groundwater hydrochemistry. Downgradient, incongruent solution and precipitation of carbonate, ion exchange, and, in confined systems, oxidation–reduction reactions predominate. Significant solution of carbonate occurs at shallow depth. The predominance of Ca and HCO_3 ions in groundwater of the outcrop areas can be attributed to the solution of calcite. Calcite saturation is usually attained within the unsaturated zone.

Hydrochemical changes within the aquifer comprise, in particular:

- Increase of Cl concentrations from solution of relict marine waters within the chalk matrix
- Conversion of Ca–HCO_3 water to Na–HCO_3 type water through cation exchange

The clay minerals within the chalk matrix function as ion exchange medium. “Clay minerals may exchange their Na^+ ions for Ca^{2+} and Mg^{2+}, causing the impoverishment of these cations in water”. “The ion exchange and adsorption processes are summarized by:

$$(1/2)\mathrm{Ca} + \mathrm{Na} \rightarrow \mathrm{Na} + (1/2)\mathrm{Ca}_x$$

where x is the adsorbing clay mineral” (Rosenthal 1987).

Within the chalk aquifer, water moves very slowly downgradient through the matrix, but more rapidly in the fissure system with a continuous exchange of water and solutes between the fissures and the matrix. “Thus, although the bulk of the water in the matrix is moving only very slowly, it is involved in the transport process because it is moving by continuous exchange through diffusion” (Downing et al. 1993: 58).

The chalk aquifers of the upper aquifer system in the west of the Aleppo plateau – the area between Hama and Aleppo – contain prevailingly Ca–HCO_3 type water with low to moderate salinity. The natural hydrochemical groundwater composition appears related prevailingly to:

- Input of dissolved substances from rainfall and enrichment by evaporative processes during recharge
- Dissolution of soil and rock carbonate through interaction with biogenic soil CO_2 and water

Mg/Ca ratios in fresh groundwater of the Aleppo plateau vary from 0.2 to 0.6. These ratios are higher than in groundwaters of the nummulitic limestone aquifers, but on average lower than in the Upper Cretaceous limestone and dolomite aquifers of the western mountain and rift zone.

The chalk aquifers on the plateau areas comprise, in general, an oxidized milieu.

Chalk aquifers covered by Neogene basalts on the Aleppo plateau contain Ca–Na–HCO_3 type water.

In morphologic depressions, in particular in the plains of closed basins, dissolved substances are enriched by evaporation or dissolution of evaporite minerals in sabkhas and in Pliocene lacustrine sediments.

In the Irbid area in northern Jordan, Paleogene chalks mainly of the Rijam (B4) formation contain fresh groundwater of Ca–HCO_3 type. Total salinity ranges from around 540 to 700 mg/l TDS. The anion composition changes from prevailingly HCO_3 predominance in the west around Ramtha with relatively high HCO_3 concentrations to higher Cl and lower HCO_3 concentrations in the more arid areas around Mafraq further east. Spring water generally has low salinities of around 300 mg/l TDS. Elevated salinities of up to 1,000 mg/l TDS found at some locations may be attributed to contamination from the surface in the phreatic aquifer, which is indicated by NO_3 concentrations of up to 264 mg/l.

According to Lloyd (1965), the chalk waters of the Belqa group aquifer in northern Jordan fall into the bicarbonate group "but with lower Ca and higher Na content than the equivalent limestone waters, probably indicating that the chalk is a less pure carbonate rock than the limestones of the area". Lloyd states, that the chalk groundwaters with a mean total salinity of 532 mg/l TDS "show a close grouping with a few analyses indicating the effects of metasomatic change".

In the Jafr area of the east Jordanian limestone plateau, the Rijam chalk (B4) aquifer contains fresh groundwater with a salinity of less than 600 mg/l TDS. The ionic composition is characterized by approximately equal percentages of Ca and Na and of HCO_3 and Cl. The major ion composition of the Na–Ca–Cl–HCO_3 type water indicates calcite dissolution with a significant evaporative Na–Cl concentration from recharge in an arid climate. Locally higher salinities of up to 3,000 mg/l TDS of Na–Cl type water are attributed to irrigation return flow. East of the Jafr area in downstream groundwater flow direction toward Wadi Sirhan, groundwater in the Rijam aquifer is brackish with salinities of 1.5 g/l TDS.

Paleogene chalks provide a relatively extensive aquifer in the Jenin sub-basin on the West Bank within the western mountain and rift zone. Groundwater discharging from the Jenin (Avedat) chalk aquifer in springs shows similar hydrochemical characteristics as the chalk aquifers in the Aleppo plateau in northern Syria. Groundwater salinity is, on average, 390 mg/l TDS; the chalk groundwaters are Ca–HCO_3 type waters. The chalk waters can be distinguished in their hydrochemical composition from groundwater of the Upper Cretaceous limestone and dolomite aquifer by:

- Higher Cl, SO_4 and Na concentrations
- Lower Mg concentrations

The chemical characteristics of the Paleogene chalk aquifer and the differences from the hydrochemical composition from groundwaters of the Judea aquifer can be related to the aquifer lithology: "The Judea Group is a high-transmissivity karstic sequence composed of dolomite and limestone beds, thoroughly flushed by soluble salts and of argillaceous components. Conversely, the Avedat Group is characterized by much lower transmissivities and is composed of limestones alternating with chalk and clay horizons" (Rosenthal 1987). Cation exchange and adsorption processes are important in the chalk aquifer with its high clay content.

Groundwaters in Paleogene chalk aquifers of the Lebanon mountain zone appear similar in their hydrochemical characteristics to chalk groundwaters of the Aleppo plateau and the Jenin area on the West Bank (Figs. 3.5 and 3.6; Table 3.3).

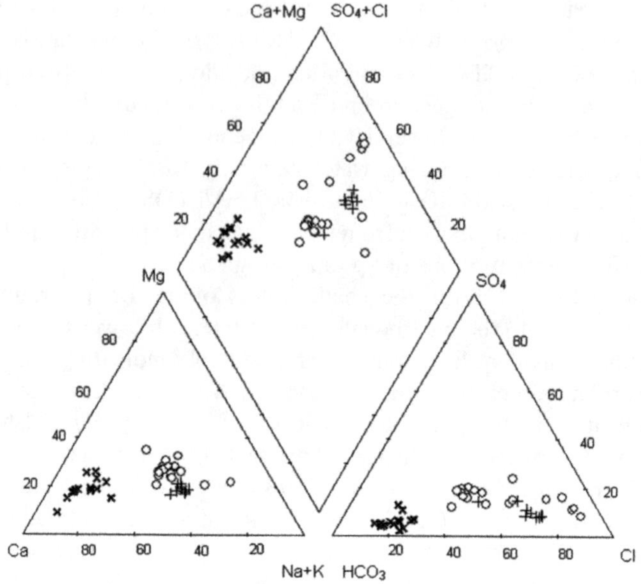

Fig. 3.5 Piper diagram: Groundwater samples from Paleogene chalk aquifers. × Hama-Aleppo area, northern Syria, samples of uncontaminated groundwater, data from Boeckh et al. (1970), ○ Rijam (B4) aquifer, Jafr area, southern Jordan, data from Hobler et al. (1991), + Rijam (B4) aquifer, Ramtha area, northern Jordan, data from Rimawi (1985)

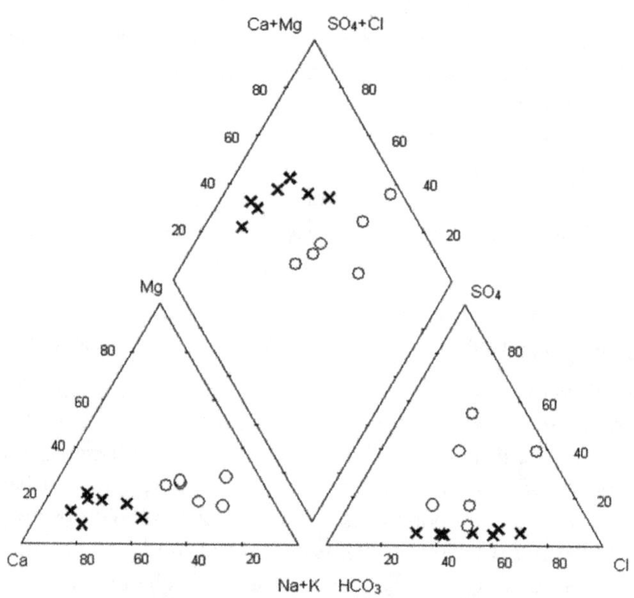

Fig. 3.6 Piper diagram: Groundwater samples from Paleogene chalk aquifers of the Aleppo plateau, northern Syria. Contaminated groundwater and groundwater in salt flat zones, data from Boeckh et al. (1970), × groundwater influenced by contamination from the surface, ○ groundwater in salt flat zones, salinity elevated through evaporation and dissolution of evaporites

Table 3.3 Hydrochemical parameters of chalk groundwaters in different areas of the northern Arabian platform

	TDS (mg/l)	HCO_3 (mg/l)	HCO_3 (meq%)	Mg/Ca
Aleppo plateau, fresh groundwater	400–700	200–250	70	0.15–0.4
Ramtha area	500–700	275–450	>60	0.6–0.9
Jafr	<600	100–280	40–60	0.4–1.0
Jenin	400	212	66	3
Lebanon	500	165–329	70	2

Data from Abumaizer (1996), Boeckh et al. (1970), Hobler et al. (1991), Rosenthal (1987), UNDP (1970)

The fresh water aquifers in Paleogene chalks on the northern Arabian platform are prevailingly phreatic aquifers with water levels at a few tens of metres below surface. These aquifers are, in many areas, endangered by contamination from the surface, in particular domestic waste and irrigation return flow.

3.4.3 *Hydrochemical Features of Cretaceous and Neogene - Quaternary Aquifers in Northern Syria*

3.4.3.1 Upper Cretaceous Aquifers of the Aleppo Plateau and the Jezire

Groundwater in the deeper Upper Cretaceous aquifer of the Aleppo area and the adjoining northern Syrian steppe is brackish with salinities of 3,000–4,000 mg/l TDS, high SO_4 concentration and elevated H_2S. The groundwater of the Upper Cretaceous carbonate aquifer shows a rather homogeneous hydrochemical composition over an extended area. SO_4, HCO_3 and Ca concentrations appear to be controlled mainly by a hydrochemical equilibrium, where concentrations of these three major constituents are approximately at a saturation level. The elevated H_2S contents reflect a low oxygen and strongly reducing environment in the confined aquifer. Variations of the hydrochemical composition are mainly related to varying Na and Cl concentrations (Fig. 3.7).

The sulfate source may be provided by evaporite layers in the Upper Cretaceous sedimentary sequence, the occurrence of which has been observed in Campanian carbonate rocks east of Hama and is indicated by extensive collapse structures in Jebel Shomariye in the northern Palmyrean mountains.

No significant hydrochemical impact of leakage from the upper aquifer into the Upper Cretaceous aquifer can be seen. Mixtures of groundwater from the brackish Upper Cretaceous aquifer and the overlying Paleogene chalk aquifer are found in some boreholes in the Aleppo–Hama area.

The brackish water tapped in the Upper Cretaceous aquifer on the Aleppo plateau is, at present, extracted from about 80 boreholes with depths between 400 and 775 m is used for irrigation as a substitute for irrigation from the depleting upper aquifer or in newly developed irrigation areas.

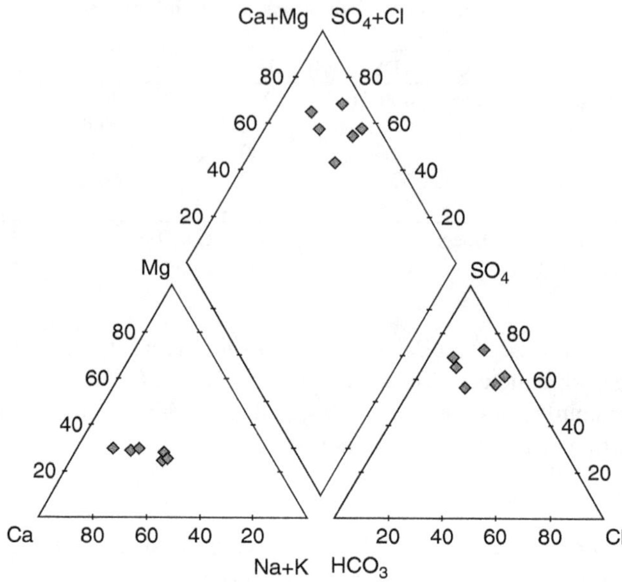

Fig. 3.7 Piper diagram: Groundwater samples from the deeper Upper Cretaceous aquifer of the Aleppo plateau. Data from Boeckh et al. (1970) and files of Ministry of Agriculture and Agrarian Reform, Jebel Al Hoss Agricultural Development Project, Aleppo

In the Khabour area, artesian wells discharge brackish, partly thermal groundwater from an anoxic aquifer with average H_2S concentrations of 7.5 mg/l. Methane is found in the water of some deep wells. The deeper brackish groundwaters are Ca–SO_4 type waters with salinities of 1,580–3,600 mg/l TDS. SO_4 concentrations are generally high with 700–2,000 mg/l, Cl concentrations vary from 350 to 700 mg/l, Mg/Ca ratios are around 0.6–0.7. The brackish groundwaters are similar in their hydrochemical composition to the groundwaters of the Upper Cretaceous aquifer of the Aleppo plateau. Water from some boreholes and springs appears to represent mixtures of fresh shallow groundwater and of the deeper brackish groundwater (Fig. 3.8).

3.4.3.2 Shallow Aquifers of the Aleppo Plateau

Miocene fractured and karstified limestones constitute, in wide areas west to northwest of Aleppo, an upper member of the shallow Paleogene–Neogene aquifer system. Groundwater in the Miocene carbonate rocks is generally Ca–HCO_3 type fresh water with a salinity of around 200–400 mg/l TDS.

Fresh groundwater in Pliocene lacustrine – fluvial deposits occurs in the surroundings of Sabkhet Jaboul. The groundwater is generally Ca–HCO_3 type water with a salinity of 300–500 mg/l TDS. Toward the centre of Sabkhet Jaboul, the groundwater becomes brackish.

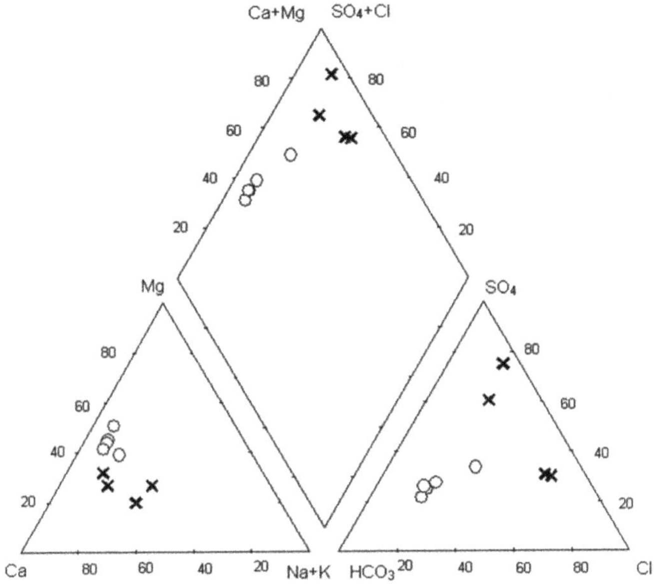

Fig. 3.8 Piper diagram: Groundwater samples from Paleogene aquifers in the Syrian Jezire. Data from Kattan (2002). ○ springs, × boreholes with admixture of deeper brackish groundwater

In the south of the Aleppo plateau (northern Syrian steppe), the upper aquifer system comprises, in addition to the main Paleogene aquifer member, aquiferous layers in Quaternary and locally in Maastrichtian sedimentary rocks. Groundwater in Quaternary aquiferous layers is generally brackish with salinities of 1,800–5,600 mg/l TDS. In the chalk–marl sequence of the Maastrichtian, aquiferous layers are found in cherts and silicified zones at shallow depth in the outcrop area of the Maastrichtian south of Khanaser and has been tapped in some boreholes in the south of the Aleppo plateau near the Palmyrean fold zone. The groundwater is generally brackish Na–Cl or Ca–SO_4 water with salinities of 1,300–5,600 mg/l TDS.

In the northern Syrian steppe around Resafe, shallow groundwater in Miocene gypsiferous deposits is brackish Ca–SO_4 to Na–Cl type water with salinities of 3,000–6,500 mg/l TDS.

3.4.3.3 Shallow Aquifers of the Euphrates–Mesopotamian Plains

In the Euphrates–Mesopotamian plain, fresh water and brackish water aquifers are generally restricted to the Paleogene–Neogene–Quaternary section of the sedimentary sequence. High volumes of discharge of fresh groundwater are sustained by subsurface inflow from the adjoining alpidic geologic units in the north.

Groundwater from the Paleogene Ras el Ain and Tell Abyad aquifers, issuing near the Syrian–Turkish border, is fresh Ca–(Mg)–HCO_3 type water with salinities of generally 400–500 mg/l. Groundwater issuing from the Ras el Ain aquifer around the Khabour river has salinities of 400–440 mg/l TDS; Mg/Ca ratios range from 0.86 to 1.2 (Fig. 3.8).

In most of the Jezire, brackish SO_4 type waters prevail with frequently high H_2S contents. In wide areas of the plains, evaporite layers in the Neogene Fars formation constitute a source of high sulfate concentrations in the groundwater.

3.4.4 *Hydrochemical Features of Aquifers in the Palmyrean Zone and the Southeast of the Northern Arabian Plate*

3.4.4.1 Palmyrean Fold Zone

Groundwater of the deeper Upper Cretaceous limestone and dolomite aquifer of the Palmyrean fold zone contains relatively high SO_4 concentrations at many places. The sulfate content may be attributed to dissolution of evaporite layers intercalated in the Mesozoic rock sequence, possibly in Campanian carbonate rocks or in older formations. The occurrence of evaporite deposits is indicated by large collapse structures on Jebel Shomariye in the northern Palmyrean mountains.

Brackish groundwater with elevated SO_4 concentration, salinity of 1,200–4,600 mg/l TDS and H_2S smell is found in the Upper Cretaceous in the wider surroundings of Palmyra and in the southern Palmyrean mountains at Basiri and Sawane. Near Palmyra, a brackish spring with 2,200 mg/l TDS and H_2S smell discharges from a cave into a small stream.

On the western surroundings of the Ad Daw plain, water with higher salinity and elevated temperature (Zamlet el Haber, 55°C) or steam (Abou Rabah) rise from the deep confined Upper Cretaceous aquifer.

A few perennial fresh water springs rise from the Upper Cretaceous limestone and dolomite aquifer at the southwestern border of the Ad Daw plain. The groundwater is of Ca–HCO_3 type with a salinity of 100–400 mg/l TDS.

In the Maqsam–Al Baida area, groundwater with a salinity of 900–1,100 mg/l TDS is extracted from wells in the Pliocene–Pleistocene sediments. This groundwater with relatively low salinity appears to leak from the Upper Cretaceous aquifer into the overlying basin deposits.

Groundwater salinity in the Pliocene–Pleistocene aquifer is, on average, considerably higher in a range of 1,000–14,000 mg/l TDS. The elevated salinity originates primarily from dissolution of evaporite layers in the lacustrine sediments. The wide-spread occurrence of evaporite minerals on the surface creates an increased salinity already in the runoff water in some streams, from which the shallow aquifer receives its main recharge.

In the eastern extension of the Palmyrean mountain zone northeast of Palmyra, groundwater occurrences are found in Upper Cretaceous to Paleogene carbonate formations at various locations (Arak, Hafne, Jebel Bishri area). The groundwaters are generally brackish with highly varying salinity and hydrochemical composition:

- Fresh water with a salinity of 800 mg/l TDS and H_2S smell issues in a spring from Eocene chalks and limestones at Arak.
- Brackish water with a salinity of 2,000–3,000 mg/l TDS of prevailingly Ca–SO_4 type has been tapped in Upper Cretaceous (mainly Campanian) limestones and cherts at Soukhne and at depth of 250–340 m below surface at Jebel Bishri.
- Water with 1,500–2,800 mg/l TDS in Pliocene deposits, leaking from the deeper Campanian aquifer, is extracted in shallow wells at Al Koum village.

3.4.4.2 Southern Syrian Steppe and Hamad

In the southern Syrian steppe and the Hamad, the separation between upper and deeper aquifer systems is discontinuous in many areas. In boreholes, water from the Paleogene aquifer and the Upper Cretaceous aquifer can frequently not be distinguished because of natural leakage or of mixing within uncased wells.

The groundwater is generally brackish, except for some fresh water lenses along larger wadi courses.

Fresh HCO_3 type water with no predominant cation occurs in the H4 area in northeastern Jordan. In general, the fresh to brackish groundwaters in the southern Syrian steppe show a slight tendency to Cl predominance, while the cations vary in a range of more or less equal percentages of Na, Ca and Mg. Mg/Ca ratios are mainly between 0.5 and 1.5.

Relatively high Na and Cl percentages occur in Paleogene and Upper Cretaceous aquifers of the Sabaa Biar – Juwef and Wadi al Miyah areas. Both areas constitute structural and paleogeographic depressions with considerable accumulation of marly deposits. Groundwater salinity is partly very high with TDS values up to 8,000 mg/l.

Deep groundwaters in the Upper Cretaceous aquifer are frequently Cl waters with relatively high SO_4 and HCO_3 concentrations and H_2S concentrations indicating anoxic conditions. In several wells in structural highs, SO_4 type waters have been encountered.

Main sources of the high salinity in wide areas of the Hamad region are assumed to be a solution of rock material during very long retention periods in carbonate and chert aquifers and enrichment by evaporation in areas with shallow groundwater occurrence (Figs. 3.9 and 3.10).

The hydrochemical composition of brackish groundwaters in the southern Syrian steppe and the Hamad shows high variations of major ion concentrations and no particular differentiation of specific hydrochemical groups. The diffuse aspect of the hydrochemistry of these groundwaters are probably related to a groundwater regime with slow groundwater movement and long retention periods, in which local conditions of groundwater recharge and flow and of the hydrochemical environment are dominant rather than wide ranging phenomena of groundwater flow and hydrochemical development.

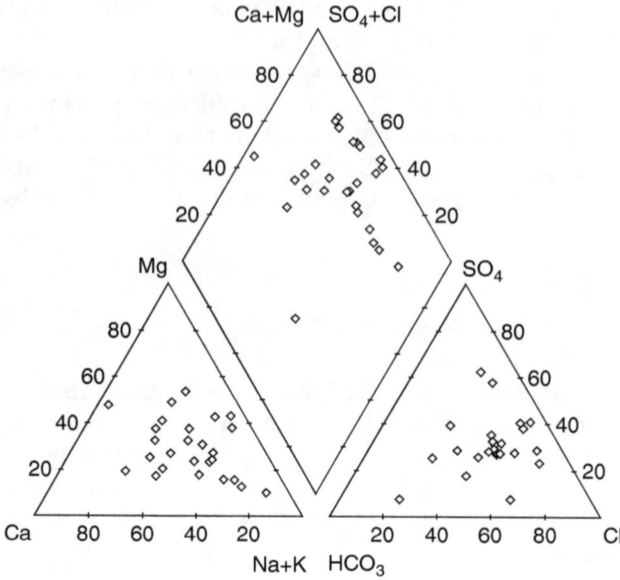

Fig. 3.9 Piper diagram: Groundwater samples from the southern steppe and Hamad in Syria. Data from ACSAD (1983)

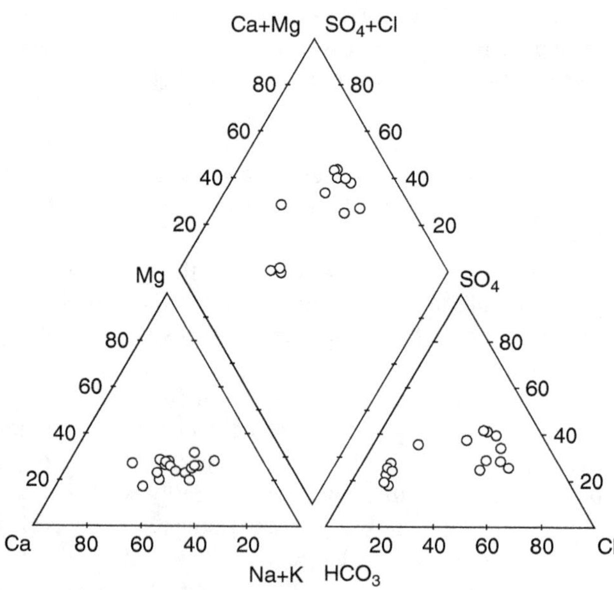

Fig. 3.10 Piper diagram: Groundwater samples from the Hamad in northeastern Jordan. Data from ACSAD (1983)

3.4.4.3 East Jordanian Limestone Plateau

Hydrochemical aspects of the Rijam chalk aquifer have been mentioned in Sect. 3.4.2.

Groundwater salinity in the Upper Cretaceous aquifer (A7/B2 aquifer) of the east Jordanian limestone plateau, increases from less than 600 mg/l TDS in the outcrop areas in the western highlands to 1,000–4,000 mg/l TDS in the Jafr area and to nearly 7,000 mg/l TDS in Wadi Sirhan. In general, the salinity of the groundwater increases along the main flow path toward east. The salinity increase is particularly remarkable in the sections, where the aquifer is confined under the Muwaqar (B3) marl aquitard. The water type changes from Ca–Mg–HCO_3 water in the western highlands to prevailingly Na–Cl water in the area east of Jafr.

The increase in NaCl is probably caused by dissolution in a poorly flushed aquifer and leakage through marls during long retention periods.

References. Abu-Ajamiieh (1967), Abumaizer (1996), ACSAD (1983), Boeckh et al. (1970), Downing et al. (1993), Droubi (1983), ESCWA (1999c), Faradzhev (1966), GITEC and HSI (1995), GTZ and NRA (1977), Hobler et al. (1991), Kattan (2002), Khouri and Agha (1979), Kozlov et al. (1966), Luijendijk and Bruggemann (2008), Medvedev (1966), Mikhailov (1966), Nativ et al. (1995), Oufland (1966a, b), Salameh and Khdier (1985), Ponikarov et al. (1967b), Protasevich and Maksimov (1966), Rosenthal (1987), Shatsky et al. (1969), UNDP (1970), United Nations (1982: 158 ff.), Wagner (1997), Wagner et al. (1982), Wolfart (1966).

3.5 Groundwater Age, Information from Isotope Data

3.5.1 ^{14}C and ^{3}H Data

The semi-arid plateau areas in the east of the northern Arabian platform receive very limited present-day recharge. The general recharge conditions are reflected in the statistical distribution of ^{3}H and ^{14}C values of random samples of groundwater from the sub-region:

From a sample set of 131 tritium analyses, ^{3}H is:

Below detection level in 114 samples (87%)
1.5 TU in 5 samples (4%)
3–7 TU in 6 samples (4.6%)
Significant with 16 TU to maximum 65 TU in 10 samples (7.6%)

In a set of 111 samples, ^{14}C values are:

Lower than 5 pmc, indicating ages of 24,000–36,000 years BP, in 48 samples (43%)
In a range 5.3–30 pmc, corresponding to ages of 10,000–24,000 years, in 45 samples (40.5%)

31–53 pmc, corresponding to ages of 5,000–10,000 years, in 14 samples (12%)
More than 60 pmc considered recent with retention periods of less than 4,000 years in 5 samples (4.5%)

More than 80% of the samples appear to represent Pleistocene recharge at periods of more than 10,000 years BP.

^{3}H contents in most samples from the deeper as well as from the shallow aquifer of the Aleppo plateau are below detection level. The groundwater in the deeper Upper Cretaceous aquifer may be assumed to be fossil; the shallow aquifer probably contains groundwater from Pleistocene and Holocene recharge.

^{14}C of spring water from the Paleogene Ras el Ain aquifer of the Syrian Jezire (one sample) is 34 pmc corresponding to an age of 9,000 years BP, ^{3}H values of samples from the aquifer are below detection level. ^{14}C values of deeper confined groundwater from the Khabour area in the Jezire range from 5.6 to 6.9 pmc, water ages are calculated, after correction for dilution factors and sulfate reduction, as 6,700–12,000 years BP.

The majority of ^{14}C water ages of samples from the southern Syrian steppe and the Hamad ranges from >20,000 years up to 40,000 years. The water of these samples is free of detectable tritium. Younger groundwater with ^{14}C ages of a few thousand years and ^{3}H values of 14–60 TU was found in wadis where recent indirect recharge takes place. These renewable resources, which generally extend as lenses with moderate salinity of less than 1,000 mg/l TDS along major wadis, have been explored e.g. in Wadi al Miyah and Wadi Murabaa in southern Syria and Wadi Muqat in northeastern Jordan.

For most groundwaters in the Palmyrean fold zone, retention periods of 10,000 to >20,000 years are indicated from ^{14}C data.

For the steppe area covering the Palmyrean fold zone, the southern Syrian steppe and the Hamad, the following general pattern of groundwater salinity in relation to groundwater ages is indicated:

- Old groundwater in most of the plain areas is brackish with a total salinity of 1,100–5,000 mg/l TDS and generally high Cl and SO_4 concentrations.
- Old groundwater with moderate salinity (500–2,000 mg/l TDS) occurs in fissured and karstified Upper Cretaceous (Cenomanian–Turonian) limestones and dolomites in the anticlinal structures of the Palmyrean Mountains, e.g. at Qaryatein, Sawane, Hafne and Arak. Along the southern margin of the Ad Daw basin, groundwater with relatively low salinity and high age (900–1,400 mg/l TDS, 20,000–30,000 years) has been explored in Pleistocene deposits. It is assumed that the fresh water occurrences, which have a limited potential and are quickly depleted during exploitation, originate from subsurface inflow from the southern Palmyrean mountains.
- The salinity of young groundwater varies according to surface and recharge conditions: Fresh water lenses with a salinity of 500–1,000 mg/l TDS are created by indirect recharge with limited evaporative enrichment of Cl and SO_4. Recent recharge with elevated salinity occurs in areas with Pliocene lacustrine sediments, e.g. the Ad Daw basin, where young groundwater with a salinity of up to

4,700 mg/l TDS is found. The relatively high groundwater salinity in these areas is caused mainly by dissolution of evaporites on the surface and within the lacustrine sediments.

In some areas, in particular in the Hamad of southern Syria and northeastern Jordan, ^{14}C values may be significantly lowered by oxidation of fossil organic matter in the aquifer or adjoining aquitards, accompanied by SO_4 reduction. The occurrence of such secondary processes is indicated by high HCO_3 concentrations, smell of H_2S and often relatively low (negative) $\delta^{13}C$ values.

In groundwater samples from the Rijam aquifer of the lower reaches of the east Jordanian limestone plateau, ^3H values are generally below detection level, ^{14}C values are low corresponding to water ages of 20,000 to >40,000 years BP.

3.5.2 *Stable Isotopes of Oxygen and Hydrogen*

3.5.2.1 Precipitation

$\delta^{18}O$ and δ^2H values of precipitation samples from the eastern part of the northern Arabian platform scatter rather closely to the MMWL with mean d values of +16.5 to +19.5‰ (stations Aleppo, Palmyra, Azraq). The general distribution of the data with a wide variation of $\delta^{18}O$ values between −3.65 and −7‰ does not differ significantly from the data distribution of precipitation samples in the western highland zone, indicating a dominant influence of Mediterranean meteoric conditions (Fig. 3.11).

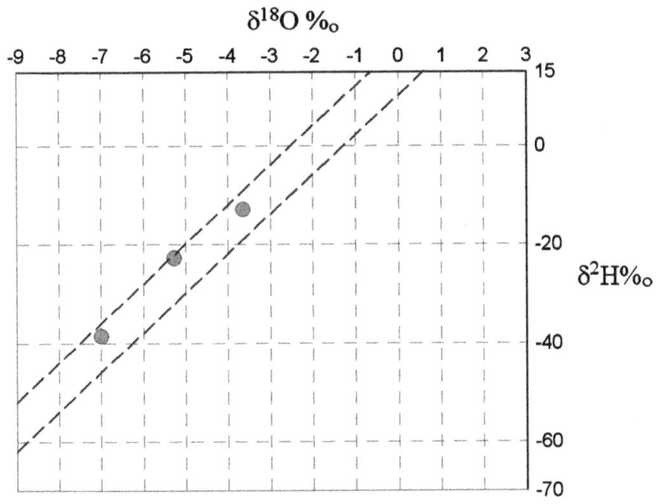

Fig. 3.11 $\delta^{18}O/\delta^2H$ diagram: Rain water samples from stations in the east of Syria and Jordan; stations Aleppo, Palmyra, Azraq. Data from Almomani (1996), Kattan (1996c)

3.5.2.2 Aleppo Plateau

In the Aleppo plateau and adjoining areas, the stable isotope composition shows:

- Significant differences between samples the from the deeper Upper Cretaceous aquifer and samples from the overlying shallow Paleogene aquifer
- Similarities between data of the deeper aquifer in the Aleppo plateau and the Upper Cretaceous aquifer of the Palmyrean mountain zone

$\delta^{18}O$ values of samples from the deeper aquifer of the Aleppo plateau are in a range of −7.5 to −8.5‰, scattering around a MWL with d around +13‰. The samples from the shallow aquifer have $\delta^{18}O$ values of −3.8 to −5.5‰ and d values of +9 to +2‰, showing a significant evaporative trend. Mixtures of the two water types may occur in uncased boreholes and in zones of groundwater leakage between the two aquifers.

The range of $\delta^{18}O$ and $\delta^{2}H$ values of groundwaters from the deeper aquifer of the Aleppo plateau corresponds to data ranges of groundwater samples in the Qalamoun area on the eastern slope of the Antilebanon mountains and from the Palmyrean mountain zone (Qaryatein, Sawane). It may be assumed, that the groundwater in the Upper Cretaceous aquifer on the Aleppo plateau as well as in the Palmyrean zone originates from recharge under a semi-arid climate corresponding to the present conditions on the eastern slope of the Antilebanon mountains. This climate was more humid than the present local climate, but apparently more continental than the present sub-humid climate in the mountains and highlands near the Mediterranean Sea.

The isotope data of groundwater in the upper aquifer system of the Aleppo plateau indicate a groundwater regime with low recharge under the present semi-arid to arid climate with a significant evaporative impact on the infiltrating water and relatively long retention periods (Fig. 3.12).

From the results of isotope hydrologic investigations, it can be concluded that:

- The deep aquifer in the Aleppo area is not recharged significantly through leakage of renewable groundwater in the shallow aquifer.
- The groundwater stored in the deep aquifer originates from distant recharge possibly in the northern Palmyrean mountains.
- The deep groundwater is fossil with an age of more than 10,000 years.

Isotope data from the upper aquifer (Paleogene to Neogene) indicate:

- Groundwater recharge during the Holocene with rising evaporative isotope enrichment with increasing aridity (d values below +10%).
- Limited components of actual recharge in the extracted groundwater (detectable tritium only in one of six samples).

The groundwater movement in the Upper Cretaceous carbonate aquifer of the Aleppo–Hama area is mainly directed toward the Orontes Valley (Al Ghab–Ar Rouj rift graben). The $\delta^{18}O$ values of around −6‰ of groundwater discharging in large springs of the Upper Cretaceous aquifer at the eastern border of Al Ghab

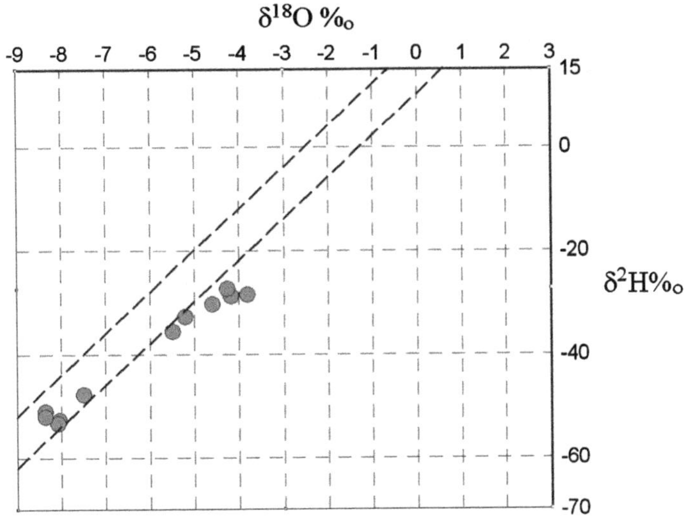

Fig. 3.12 $\delta^{18}O/\delta^{2}H$ diagram: Groundwater samples from the Aleppo plateau. Data from Wagner (1997), Wagner and Geyh (1999)

indicate, that the main proportion of the spring discharge originates from groundwater circulation sustained by present recharge in nearby aquifer outcrop areas and that the quantity of groundwater from the confined eastern parts of the catchment (^{18}O around −8‰) is comparatively very low.

3.5.2.3 Jezire

$\delta^{18}O$ values of fresh groundwater from the Ras el Ain aquifer on the northern rim of the Syrian Jezire are −6.4 to −6.7‰, d values vary between +14.4 and +15.9‰. $\delta^{18}O$ values of deeper confined groundwater in the Khabour area are significantly more negative, ranging from −7.2 to −8.2‰, corresponding d values lie between +15.5 and +20.3‰. As on the Aleppo plateau, the stable isotope values indicate recent groundwater recharge of the shallow fresh water and a Pleistocene origin for the deeper groundwater. Influences of evaporative isotope enrichment in the shallow groundwater appear less pronounced than on the Aleppo plateau.

The deeper confined groundwaters with $\delta^{18}O/\delta^{2}H$ ratios near the MMWL have corrected ^{14}C ages of more than 10,000 years and seem to be formed by direct infiltration of atmospheric precipitation with no or low evaporation. The groundwaters in the unconfined aquifer seem to originate from a mixture of rain water, snow melt and evaporated recycled water from irrigation in recharge zones between 700 and 950 m asl (Fig. 3.13).

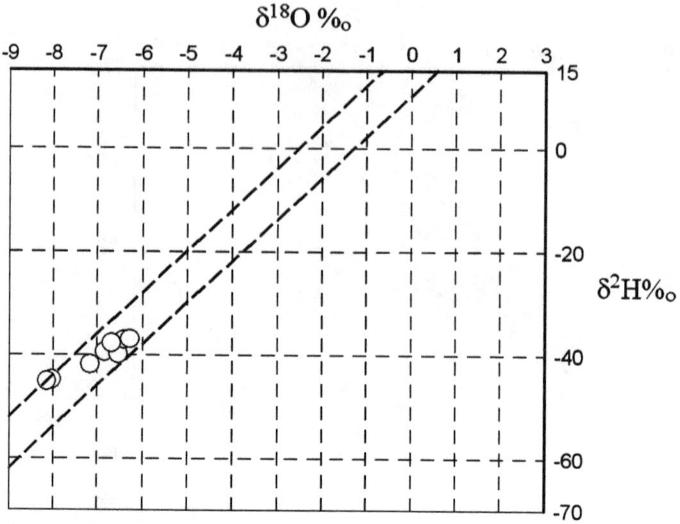

Fig. 3.13 $\delta^{18}O/\delta^{2}H$ diagram: Groundwater samples from the Syrian Jezire. Data from Kattan (2002)

3.5.2.4 Palmyrean Fold Belt

$\delta^{18}O/\delta^{2}H$ values of most groundwater samples from the Palmyrean fold belt scatter around a MWL with d +15‰ in a range of $\delta^{18}O$ between −6.9 and −8.9‰. The group of groundwaters with relatively uniform $\delta^{18}O/\delta^{2}H$ data comprises samples from the area between the Damascus plain and Qaryatein, from the southern Palmyrean mountains, the Ad Daw plain until Palmyra, and the eastern prolongations of the Palmyrean chains at Jebel Bishri. ^{14}C data of the group of samples vary from 1.1 to 26.9 pmc, indicating a wide range of retention periods from 10,000 to 34,000 years BP. ^{3}H values are, in general, below detection level.

The range of $\delta^{18}O$ and $\delta^{2}H$ values of these samples from the Palmyrean fold belt coincides approximately with average values of the Qalamoun area on the eastern flank of the Antilebanon mountains.

Exceptions from the general trend of $\delta^{18}O/\delta^{2}H$ data are found in a few samples, in which recent recharge is indicated from ^{3}H and ^{14}C data ($\delta^{18}O$ −5.88 to −6.03‰, ^{3}H 23–42 TU in 1980, ^{14}C 28.9–50.9 pmc) and an occurrence of fossil fresh water at Arak northeast of Palmyra ($\delta^{18}O$ −6.46 to −6.96‰, ^{14}C 1.5–4.8 pmc).

The variation of $\delta^{18}O$ values in the Palmyrean fold zone may reflect recharge at different altitude levels over a range of around 1,000 m (Fig. 3.14).

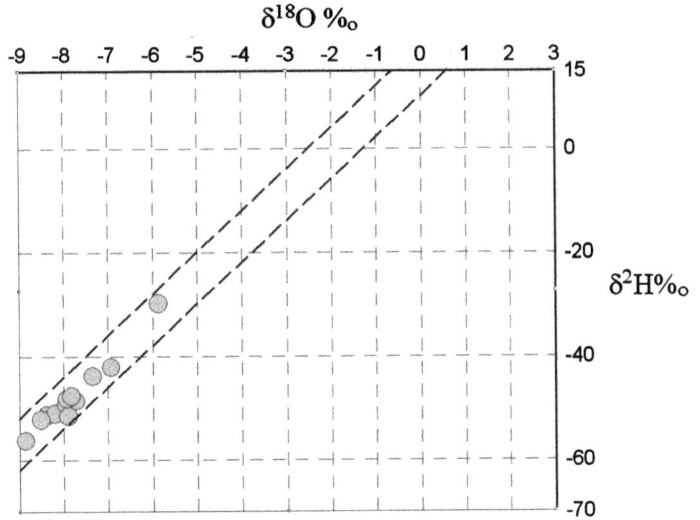

Fig. 3.14 $\delta^{18}O/\delta^2H$ diagram: Groundwater samples from the Palmyrean fold zone, Syria. Data from investigations of ACSAD–BGR (Wagner and Geyh 1999)

3.5.2.5 Badiye and Hamad

The tritium and stable isotope data from different aquifers in the Badiye and Hamad suggest a grouping into two main water types (Fig. 3.15).

- Fossil groundwater related to Pleistocene recharge with:

  ^3H: below detection level
  δ^2H: −37 to −60‰
  $\delta^{18}O$: −6 to −9‰
  D = +6 to +18‰

- Groundwater with evident impact of recent recharge with:

  ^3H: 5.3 to 74 TU
  δ^2H: −22 to −33‰
  $\delta^{18}O$: −4.7 to −6‰
  d = +8.8 to +17.3‰

The majority of samples apparently represents groundwater from Pleistocene recharge.

Apart from the mentioned two main water types, the aquifers, which comprise in some areas saturated sections of 100–200 m thickness, may contain groundwater of more complex origin (Figs. 3.16–3.17):

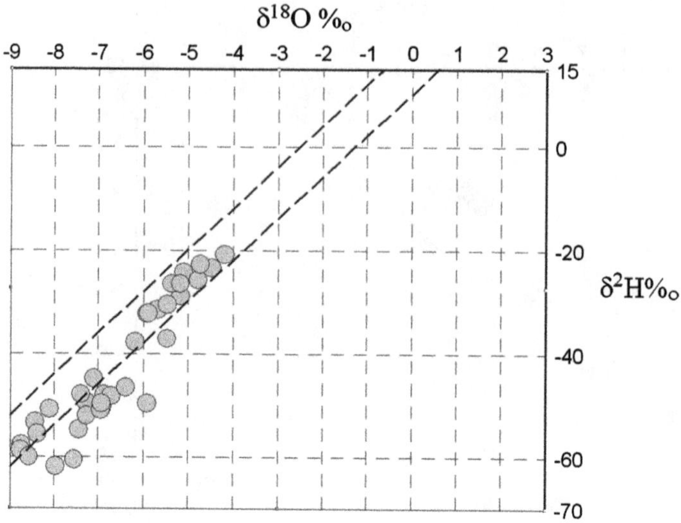

Fig. 3.15 $\delta^{18}O/\delta^{2}H$ diagram: Groundwater samples from the southern Syrian steppe and Hamad. Data from investigations of ACSAD–BGR (Wagner and Geyh 1999)

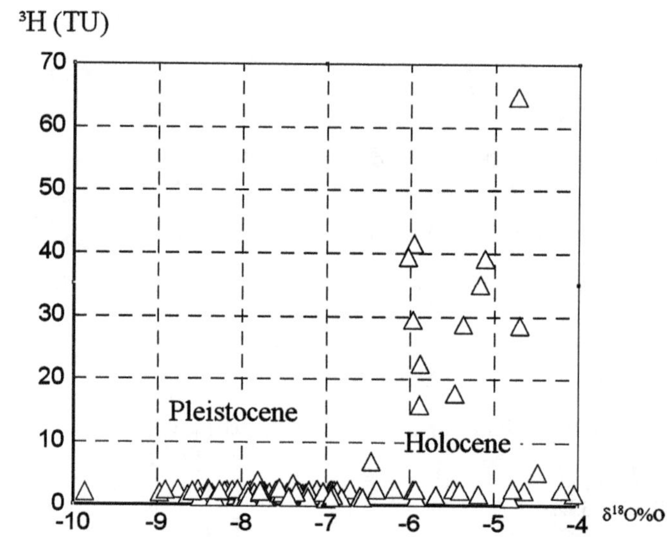

Fig. 3.16 $\delta^{18}O$ and ^{3}H values of groundwater samples from the southern Syrian steppe and Hamad. After Wagner and Geyh (1999)

- Groundwater from Holocene recharge but without or very little recharge during the past decades (^{3}H below detection level, $\delta^{2}H > -33‰$).
- Mixtures of Pleistocene and Holocene recharge with ^{3}H values between 18 and 35 TU and ^{14}C values between 22 and 36 pmc.

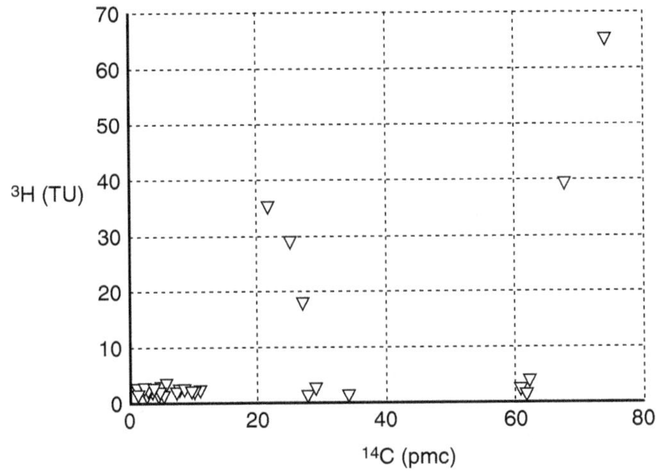

Fig. 3.17 $\delta^{18}O$ and ^{14}C values of groundwater samples from the southern Syrian steppe and Hamad. After Wagner and Geyh (1999)

Table 3.4 Ranges of $\delta^{18}O$ and d values in different areas of the eastern part of the northern Arabian platform

	^{18}O‰ range		d‰ range	
Aleppo plateau				
Upper aquifer	−3.8 to −5.5		+2 to +9	
Deeper aquifer (Upper Cretaceous)		−7.5 to −8.5		around +13
Jezire				
Paleogene aquifer	−6.4 to −6.7		+14.4 to +15.9	
Upper Cretaceous aquifer		−7.2 to −8.2		+15.5 to +20.3
Palmyrean fold belt				
General range		−6.9 to −8.9		around +15
Impact of recent recharge	−5.9 to −6.0		+17.3 to +15.7	
Southern Syrian steppe and Hamad				
General range		−6.9 to −9		+6 to +18
Impact of recent recharge	−4.7 to −6		+8.8 to +17.3	

3.5.2.6 East Jordanian Limestone Plateau

On the east Jordanian limestone plateau and on the northern edge of Wadi Sirhan, $\delta^{18}O$ values of the Paleogene–Neogene aquifer are in a range of generally −6.4 to −5.2‰ with d values between −1.6 and +15‰. With few exceptions, ^3H is below detection level, groundwater ages are generally high.

3.5.3 Pleistocene and Holocene Groundwater

The impact of paleo-recharge is particularly evident in the aquifers of the eastern part of the northern Arabian platform with its at present mainly arid climate conditions. The Pleistocene groundwater has more negative ^{18}O and ^{2}H values than the Holocene groundwater, but the range of deuterium excess values of Pleistocene and Holocene water is similar. We may conclude, that Pleistocene recharge took place in a cooler climate but precipitation during the Pleistocene had principally the same source as Holocene precipitation: storms originating over the eastern Mediterranean Sea (Table 3.4).

The Holocene groundwater, restricted mainly to the ranges of extensive wadi systems, comprises partly water recharged during the past few decades, but can also be several hundred or a few thousand years old. Samples collected from existing boreholes often represent water from various recharge events stored in the aquifer.

References. GITEC and HSI (1995), Kattan (1996b, 2002), Wagner (1997), Wagner and Geyh (1999).

Chapter 4
North Arabian Volcanic Province: Jebel el Arab–Golan–Al Harra

4.1 Geographic and Geologic Set-Up

4.1.1 Landscape and Climate

4.1.1.1 The Jebel el Arab Basalt Area

About 100 km southeast of Damascus and 100 km northeast of Aman, the volcanic mountain massif of Jebel el Arab rises above the plateaus of southern Syria and northern Jordan up to an elevation of 1,800 m asl. Jebel el Arab or Jebel Druze constitutes the core of a volcanic province, which extends from southwestern Syria across Jordan into the Al Harra basalt area of Saudi Arabia. The basalts of that North Arabian Volcanic Province originate from the regional volcanic activity which accompanied the tectonic events of the development of the Red Sea rift graben during the Tertiary to Quaternary. Further south on the Arabian Shield, volcanic rocks of comparable age extend over considerable areas in a general southeast–northwest direction along the eastern shoulder of the Red Sea graben. The basalt outcrops of the Jebel el Arab volcanic province continue the southeast–northwest trend of the Red Sea volcanics after a shift of about 200 km toward northeast. The Jebel el Arab basalt fields extend along the eastern flank of the tectonic Wadi Sirhan–Azraq depression and reach, in the northwest, the Damascus basin, the Golan heights and the rift graben at Lake Tiberias (Fig. 4.1).

The landscape of the Jebel el Arab basalt field shows two different faces: Rain-fed and irrigated agriculture on fertile soils with grain fields, vineyards and fruit trees in the west, and desert and steppe areas with bare basalt rocks and very limited vegetation in the east and south. A vast oasis adjoins the basalt field in the southwest on the Azraq plain (Qaa el Azraq).

Centres of civilization flourished in the western part of the basalt field since, at least, Nabatean times in the second century BC, and towns like Bosra, Suweida and Shahba were important provincial centres of the Roman Empire. The area owes its high level of agricultural and urban development to the relatively favourable climate conditions and to the occurrence of shallow groundwater and springs in aquiferous basalt layers.

W. Wagner, *Groundwater in the Arab Middle East*,
DOI 10.1007/978-3-642-19351-4_4, © Springer-Verlag Berlin Heidelberg 2011

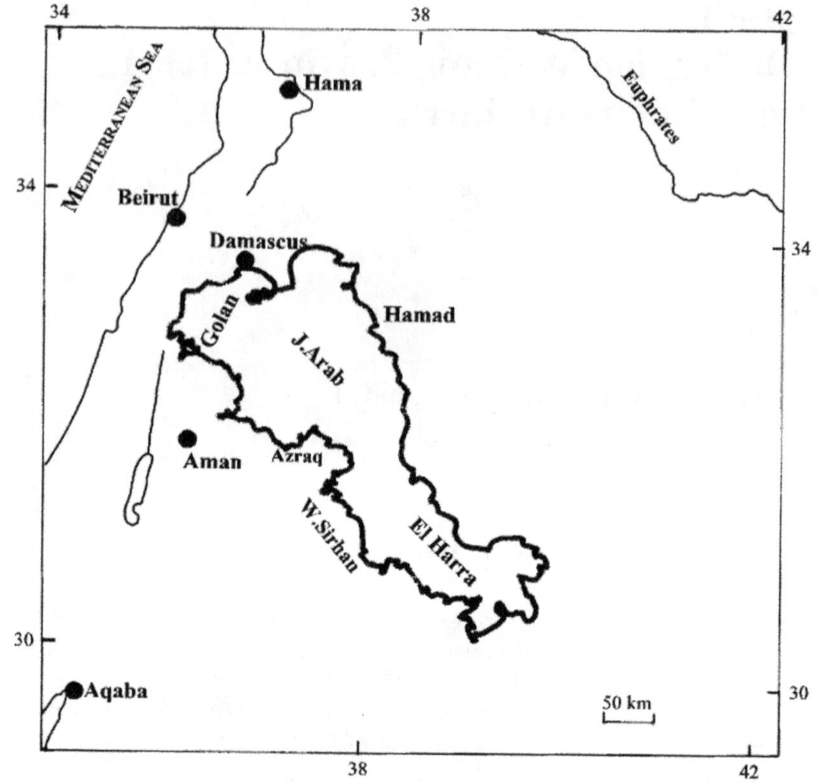

Fig. 4.1 North Arabian volcanic province ◯ basalt field

Moderately high precipitation rates originate from Mediterranean storms, which penetrate frequently until Jebel el Arab from the Lebanon–Antilebanon mountains and through the morphologic depression of Lake Tiberias–Yarmouk between Mount Hermon and the Ajloun highlands. The eastern and southeastern parts of the basalt field, situated at the lee-side of Jebel el Arab, have an arid climate and have been traditionally nomad lands.

The past few decades saw substantial agricultural developments and expansion of villages and towns, which were supported to a significant part by the construction and operation of boreholes tapping the basalt aquifer. Pipelines were built for the transport of water from springs and boreholes to urban centers within and outside the basalt area, e.g., for supplementing the city water supply of Aman and of Suweida.

A considerable number of investigations has been carried out in the basalt area of Jebel el Arab since about 1960. The author had the opportunity to participate in a study reviewing the existing information for a summary assessment of the groundwater resources of the basalt aquifer system jointly with colleagues from Jordan, Syria, Germany and the United Nations.

4.1.1.2 Morphology

The morphology of the basalt field extending over southwestern Syria and northern Jordan is dominated by the Jebel al Arab mountain massif, which is surrounded by plains and plateau like landscapes: the Hauran plain in the west, the northeastern desert of Jordan in the south, and the Hamad steppe and desert area in the east. About 2,500 km^2 of the Jebel el Arab mountain massif are situated above 1,000 m asl. Highest peaks of the basalt field are Tell Ghanie on Jebel el Arab (1,800 m), Tulul al Ashaqif in northeastern Jordan (900 m) and the top of the Golan heights with 1,199 m asl. The surrounding plains and plateaus descend with generally gentle slopes to 500 m asl in the west and south and 700 m in the east. Toward north, the basalts dip under sediments of the Damascus plain. Closed basins with seasonal lakes are located in morphological depressions at the fringes of the basalt field: lake Hijane (<600 m asl) in the Damascus plain, Manqaa ar Rahba near Zelaf (<600 m), Qaa el Azraq (<500 m), Hadhawdha in Wadi Sirhan (572 m, Fig. 3.4). In the west, the Yarmouk valley cuts into the basalt field to around 450 m asl. Prominent volcanic peaks are scattered over the Jebel el Arab mountain massif and the plateaus south and north of the massif.

The Hauran plateau grades in the west into the Golan heights, a basalt plateau which descends from an altitude of about 900 m asl on its northeastern edge to 250 m asl at the Yarmouk valley in the southwest. In the west, the Golan heights are delimited by faults with steep morphologic scarps on the boundary of the rift depression of Lake Tiberias and Houle. Along the escarpment the topographic elevation drops from 700 m asl at the edge of the Golan heights to 212 m below sea level at the shore of Lake Tiberias.

The boundary of the basalt field touches, in the west, the northwestern mountain and highland zone at Mount Hermon and the highlands of Ajloun, and, in the southwest, the east Jordanian limestone plateau. In the east, the basalts are adjoined by the sedimentary desert plateaus of the Hamad.

The surface drainage system of the Jebel el Arab basalt field comprises a network of wadis and mudpans. The wadi courses run, in general, radially from the Jebel el Arab mountain massif and Tulul al Ashaqif toward mudpans or seasonal lakes and, in the west, to tributaries of the Jordan river system. The basalt field extends over part of the Yarmouk and Zerqa river sub-basins in the west and over the catchments of several closed basins:

- The Damascus plain in the northwest
- The narrow Wadi Liwa between the Damascus and Yarmouk sub-basins
- The Azraq plain in the south
- The Hamad area in the east, where several mud pans act as local drainage centres

The eastern slopes of Jebel el Arab comprise the catchment of the extensive Wadi Sham drainage system, which ends in Manqaa ar Rahba near Zelaf.

The basalts of the Golan heights ("cover basalt") form a westward inclined plateau covering sedimentary rocks. In the southern Golan area, the basalt cover protects the soft sedimentary rocks from erosion and has led to the formation of a wide plateau. After the extrusion of the cover basalt, tectonic movements caused graben faults, lowering the erosion base by several hundred metres, incision of rivers and the formation of several gorges along the Yarmouk river and its tributaries Wadi Raqad and Wadi Hreer.

4.1.1.3 Climate

The climate of the Jebel el Arab basalt area is typical Mediterranean with a cold rainy season in winter and a dry and warm summer extending from May to October. Average annual precipitation ranges from 500 mm on the Jebel el Arab mountain massif and the Golan heights, and 250–300 mm in the Hauran plain to less than 100 mm in the south and east. Significant climate changes probably occurred about 10,000 years before present at the end of the pluvial climate phase, during which wetter climate conditions generally prevailed in the eastern Mediterranean region.

4.1.2 Geologic Structure

The geologic sequence of the Jebel el Arab basalt complex is composed of Neogene plateau basalts and Quaternary (Pleistocene–Recent) basaltic lava flows and shield volcanoes. Volcanic eruptions continued until historic times approximately 4,000 years ago.

The total thickness of Neogene to Quaternary basalts increases from less than 100 m on the fringes and in the southeastern and northeastern parts of the basalt field to more than 700 m on the foothills of Jebel el Arab.

4.1.2.1 Neogene Plateau Basalts

The Neogene plateau basalts are composed of single lava sheets of several metres thickness with intercalated soil or sedimentary layers. The lava sheets are crossed by basaltic dikes. A maximum thickness of about 1,500 m of Neogene basalt is assumed beneath the Jebel el Arab mountain massif (Krasnov et al. 1966). The unexposed feeder zones of Neogene basalt are probably located at the eastern shoulder of a rift basin structure, which may represent a northwestern prolongation of the Wadi Sirhan depression. The Neogene basalts are exposed in the southeast and east of the Jebel el Arab area and Al Harra; in the north and west of Jebel el Arab, the Neogene basalts are covered by the younger Quaternary basalts.

The Neogene basalts have prevailingly a Pliocene age. Middle Miocene effusive rocks, which are preserved in outliers in the vicinity of Damascus–Kiswe are represented mainly by basalt with some occasional tuff lenses with a visible thickness of 500 m at Kiswe. West of Zelaf, 30 m of basalt conglomerates and a 100 m thick basalt series of probable Miocene age underlie the Pliocene basalt. On the southern Golan heights, intercalations of basalt flows and intrusions of volcanic dikes are found in the Miocene Hordos formation of the Golan.

In the Lake Tiberias–Yarmouk area, the basalts overlie terrestrial deposits of lower Pliocene age, composed of conglomerates, sandstones, arenaceous marl and limestones (Ain Gev sands). The terrestrial deposits reach a thickness of 150–170 m near Lake Tiberias and thin out toward the Yarmouk river and Wadi Raqad. The thickness of the Pliocene basalts increases from a few tens of metres around Lake Tiberias to 200 m in Wadi Raqad.

A 30 to about 200 m thick sequence of basalt flows of late Pliocene age extends over wide areas of the Golan heights. The Pliocene basalts appear to originate mainly from extrusions in the Hauran–Jebel el Arab area, while secondary extrusions (plugs, small volcanoes, fissure dykes) are scattered over the Golan.

The thickness of the late Pliocene–early Pleistocene "cover basalt" increases from 30 m in the south of the Golan to 175 m toward north.

4.1.2.2 Quaternary Volcanics

The Quaternary volcanics comprise basaltic lava flows, shield basalts and basaltic cinder, scoria and tuff. Quaternary basaltic lava flows and Quaternary shield volcanoes extruded from point source feeders along NNW–SSE trending lineaments crosscutting the Neogene series. The total thickness of Quaternary lava flows varies from few metres to 150 m.

Quaternary lava flows are generally developed as narrow and relatively thin valley fillings in a pre-existing morphological relief. Quaternary shield basalts extend over larger areas and reach a maximum thickness of around 100 m. Surfaces of pahoehoe lava are frequently preserved on Quaternary lava sheets. Each of the individual shield basalts or lava flows has a thickness of about 15 m.

Recent basalts differ from the older ones by their exceptional freshness and form extensive plateaus in southwestern Syria, such as Diret et Touloul, As Safa, and the Hadar basalt sheet which extends in a wide band eastward from Mount Hermon. The Recent basalts overlie unconformably all the deposits of the Damascus depression, up to upper Quaternary limestones.

The Leja plateau in the northwest of Jebel el Arab is formed by Recent pahoehoe lavas departing from Tell Shihan volcano as a more than 45 km long and 6–12 km wide tongue. Near Shahba, block lavas of three recent volcanic cones cover an area of around 12 km^2 in a "disorderly chaotic piling of blocks of various forms and sizes" (Krasnov et al. 1966). The Recent volcanoes have well preserved volcanic cones with heights of 70–250 m.

In the area between Lake Tiberias, Yarmouk river and Wadi Raqad, middle–late Pleistocene volcanics comprise pahoehoe lavas and sheets of lava flows with a total thickness of a few metres to 25 m. On the eastern slope toward Lake Tiberias, the volcanics are overlain by detrital alluvial fans.

4.1.2.3 Subsurface of the Basalt Field

The subsurface of the Jebel el Arab mountain massif comprises probably:

- A sequence of Neogene shield basalt flows with a thickness of several hundred metres, which are covered partly by thin layers of Quaternary lava flows or basaltic tuff.
- The main feeder zone of Neogene basalts, which may constitute a thick sequence of basaltic rocks as vertical dikes or as sill-type intrusions into the sedimentary rocks.
- Layers or blocks of disturbed sedimentary rocks.

The basalt complex of the North Arabian Volcanic Province is underlain by Paleozoic to Neogene sedimentary rocks. The total thickness of sedimentary rocks above the Precambrian basement reaches in northern Jordan from 2.5 km to more than 8 km.

The sequence of Paleozoic to Jurassic rocks in northern Jordan and southern Syria is composed mainly of sandstones, siltstones, conglomerates and shales with intercalations of carbonate rocks in Syria. It can be assumed that a comparable sedimentary rock sequence underlies the basalt field at greater depth.

The Lower Cretaceous is represented by a sandstone facies in Jordan (Kurnub sandstone) and by detrital to carbonate deposits in Syria (Aptian–Albian).

Upper Cretaceous to Neogene sedimentary rocks are exposed in the surroundings of the basalt field. The Upper Cretaceous–Paleogene sedimentary sequence in northern Jordan and southern Syria can be subdivided into three major lithostratigraphic units:

- Upper Cretaceous limestones and dolomites, underlain in Jordan by an Upper Cretaceous chalk–marl formation
- Upper Cretaceous to Paleogene marls
- Paleogene chalks and marly limestones

The stratigraphic range of these units is not uniform over the area of northern Jordan and southern Syria. The deposition of carbonate rocks and of fine grained sediments, related to marine north–south transgressions, generally starts in the north at earlier stratigraphic stages than in the south.

Miocene to Recent deposits are intercalated, in some areas into the basaltic rock sequence. Neogene to Recent deposits with considerable thickness extend over the Damascus and Azraq plains at the northern and southwestern margins of the basalt field. Pliocene to Quaternary deposits, overlying the basalt in limited areas, comprise mud pan deposits, volcaniclastics and sediments in wadis and morphological depressions.

In most areas of the Jebel el Arab volcanic province, the basalts rest on Paleogene sedimentary rocks, which form the main outcrops in the areas adjoining the basalt field: the southern Syrian steppe and Hamad and the Jordanian limestone plateau. In the southwest of the Jebel el Arab basalt field, the Paleogene chalks and Upper Cretaceous–Paleogene marls are missing over a structural high, the eastern prolongation of the Ajloun dome, and the basalts lie directly above Upper Cretaceous limestones and dolomites.

References. Andrews (1992), BGR-WAJ (1994), ESCWA (1996), Krasnov et al. (1966), Michelson and Lipson-Benitah (1986), Safadi (1955), Schaeffer (1997), Wolfart (1966).

4.2 Aquifer Systems and General Groundwater Regime

4.2.1 Aquiferous Zones and Hydraulic Conditions

4.2.1.1 Major Aquiferous Sections

The sequence of Neogene to recent volcanic rocks and of Upper Cretaceous to Quaternary sedimentary rocks in the area of the Jebel el Arab basalt field comprises major aquifer sections in the following units:

- Various horizons with relatively high permeability within the basalt sequence
- Paleogene chalks and marly limestones: Rijam–Shalala (B4–B5) aquifer in Jordan, Eocene chalk and chert aquifer in Syria
- Upper Cretaceous limestones and dolomites: A7/B2 aquifer in Jordan, Cenomanian–Turonian limestone and dolomite aquifer in Syria

The Upper Cretaceous–Paleogene marls (Muwaqar formation or B3 in Jordan, Maastrichtian–Lower Eocene marls in Syria) constitute an extensive aquitard between the Upper Cretaceous and Paleogene aquifers.

Paleozoic to Lower Cretaceous formations underlying the basalt field at greater depth can be assumed to contain limited quantities of non-renewable brackish groundwater.

4.2.1.2 Productive Zones of the Aquifer System

Sections with relatively high productivity in the Jebel el Arab basalt sequence appear related to:

- The lower part of Quaternary to upper part of Pliocene volcanics in the Hauran plain (Ezraa–Deraa area)
- Parts of the Neogene and Quaternary basalts in the foothills of Jebel el Arab around Suweida, in Wadi Liwa and the Leja plateau northwest of Suweida
- The lower part of the basalt sequence, mainly Neogene basalts, which form a joint aquifer with Upper Cretaceous limestones and dolomites in the Halabat–Wadi Dhuleil areas and the northeastern desert in Jordan, and with Paleogene chalks in the Azraq area

Aquiferous horizons in the basalts with low to moderate productivity are found in wide parts of the basalt field. Of particular local importance are numerous, mainly perched, aquifers of limited extent in Neogene and Quaternary basalts of the Jebel el Arab mountain area.

The main basalt aquifer is probably related prevailingly to Neogene plateau basalts, infiltration conditions are, however, influenced in many areas by the overlying Quaternary lava flows.

The basalts and underlying Paleogene chalks and limestones (Rijam Formation) constitute, in general, one hydraulically connected aquifer system. Upper Cretaceous limestones and dolomites (Amman–Wadi Sir Formation) provide a deeper confined aquifer, but may be directly connected to the basalt aquifer in some areas, where the Paleogene has been eroded on the slope of the Ajloun anticline, such as Wadi Dhuleil.

In the eastern part of the Jebel el Arab basalt field, which extends into the Hamad steppe and desert area, Neogene plateau basalts appear to constitute a partly discontinuous aquifer in hydraulic connection with the underlying Paleogene chalk formation.

4.2.1.3 Hydraulic Conditions of Basalt Aquifers

Basalt aquifers are, in general, characterized by hydraulic anisotropy and discontinuous, heterogeneous aquifer properties. Large contrasts exist in hydraulic conductivity of different layers within the basalt aquifer systems. Relatively high permeabilities and preferential pathways are related to the boundary layers between individual basaltic flows and to joints and fractures resulting from cooling and tectonic stress. Lava flows form, after cooling, a solid mass with a coarse crust at the top. Blocky rock masses on the upper surface of lava flows, which are frequently associated with detrital stream deposits, produce a relatively high bulk porosity in most young basalts. Additional differentiation of permeability along boundaries of lava flows can be caused by buried soils and weathering at the top of basalt flows.

Porosity can be high in vesicular lava flows (10–50%), but in solid lava flows the effective porosity is generally less than 1%. Permeabilities of volcanic rocks range from very low values to 10^{-2} m/s. Young basalts have generally a higher permeability than older flows, where permeability is decreased by alterations related to weathering and deep burial and to the influx of cementing fluids (Matthess and Ubell 1983:236 f.).

In accordance with general concepts of groundwater occurrence and movement in basalts (Domenico and Schwartz 1990:70–71), the main components of groundwater flow in the basalt aquifer system in Jordan and Syria may be described schematically as follows:

- High horizontal flow along the tops of individual basalt flows
- Low vertical leakage through the interiors of basalt flows
- Leakage (vertical flow) along structural discontinuities: fractures and fissures caused by cooling and tectonic stress

Other general aspects to be considered in the study of the basalt aquifer system in Jordan and Syria is the occurrence and movement of groundwater in mountainous terrain, as altitude differences between main recharge and discharge areas reach up to 1,300 m. In mountainous terrain, the "water table may be considered a free surface whose depth and configuration depends on the interplay between infiltration and permeability distribution" (Domenico and Schwartz 1990:70). Where low-permeability materials alternate with higher-permeability units, downward percolating water in the unsaturated zone may become perched on low-permeability units (Domenico and Schwartz 1990:28–29).

4.2.1.4 Hydraulic Properties of the Jebel el Arab Basalt Aquifer

The Jebel el Arab basalt complex consists prevailingly of single sheet lava flows. The thickness of individual flows is reported to vary from 3 to 25 m, on average 5–7 m. The central parts of the basalt flows are generally constituted by well crystallized massive varieties, the upper and lower parts by coarse bubbled basaltoids. Up to 50% of individual flow sheets may be represented by porous varieties, but the pores are only to a limited extent interconnected. Permeable horizons occur prevailingly on the boundary zones between different basalt flows. Vertical hydraulic interconnections between these horizons are created by tectonic or cooling fractures.

In general, the permeability in the basalt aquifer system may decrease with depth. Water levels at more than 300 m below land surface in boreholes with significant well yields indicate, however, that productive aquiferous horizons extend to several hundred metres below surface.

Transmissivity values, obtained from single well tests, are reported to range from 1,000 to 10,000 m^2/d in the eastern part of the Hauran plain and the Leja plateau in Syria. In the area west of the road from deraa to Damascus the transmissivity decreases.

T values in the basalt aquifer in Jordan vary from 2 to 44,000 m^2/d.

Drilled wells have, in general, relatively high yields and specific capacities in the Azraq and Wadi Dhuleil areas and in the northern desert of Jordan east of Mafraq with well yields above 40 m^3/h and median Q/s values of 32–39 m^3/h/m. In these areas, the basalt and underlying Cretaceous limestones and dolomites or Paleogene chalks constitute a combined aquifer system. The relatively high well yields may be related to a rather high transmissivity of the aquiferous complex with a considerable thickness of carbonate rocks.

In the western part of Wadi Dhuleil and the northwestern part of Jebel el Arab, highly varying well yields are reported: <10 to >40 m^3/h. In the western Wadi Dhuleil area, the basalt constitutes only a thin top section of the aquifer complex formed mainly by Cretaceous limestones and dolomites. The median Q/s value is moderate with 11 m^3/h/m.

Well yields are relatively high at some locations around Suweida in the northwestern part of Jebel el Arab, which comprises a complex sequence of aquiferous sections within the basalt with one or two perched aquifer zones above the main basalt aquifer. In the area west of Jebel el Arab (Deraa–Ezra–Bosra), well yields and specific capacities are low to moderate, but relatively high specific capacities are indicated for a number of wells, in which zero drawdown is observed during operation.

Horizontal groundwater flow within the basalt aquifer system is generally restricted to horizons with significant permeability on the boundaries of individual lava sheets. These horizons may comprise about 10–20% of the total saturated thickness of the basalt aquifer. Assuming permeable layers over 20% of the saturated thickness of 100–300 m, T values of 1,000 m^2/d would correspond to permeabilities of 2×10^{-4} to 6×10^{-4} m/s. From the mean Q/s values ranging from 1 to 40 $m^3/h/m$ for different areas of the basalt aquifer system, transmissivities of around 30–1,300 m^2/d would be expected. In areas where the basalts form a hydraulically connected aquifer with underlying carbonate rocks, the relatively high well capacities may be related to a high transmissivity of the carbonate aquifers.

The following hydraulic conditions may be assumed:

- Permeability of the basalt aquifer: 20% of the basalt aquifer consists of layers that have an average horizontal permeability of 2×10^{-4} m/s, and permeability in 80% of the aquifer is less than 1×10^{-6} m/s. The resulting average horizontal permeability is 4×10^{-5} m/s.
- The hydraulic connections between the basalt/chalk aquifer and the underlying Upper Cretaceous limestone aquifer is, in general, characterized by a downward hydraulic gradient.
- The Upper Cretaceous–Paleocene marl aquitard between the basalt/chalk aquifer and the Upper Cretaceous aquifer is saturated with groundwater and there is an average difference of in the hydraulic potential of 10–20 m between the two aquifers. The thickness of the intercalated marls is 200–300 m with an average value of 250 m. The average vertical permeability of the marls is 1×10^{-9} m/s.

4.2.2 Recharge Conditions

The pattern of groundwater recharge in the Jebel el Arab basalt field depends mainly on the climatic conditions and on factors influencing the infiltration of rain and surface water, such as morphology, outcropping geological formations, soil cover. Relatively favourable recharge conditions are related to:

- Mean annual precipitation of >300 mm on Jebel el Arab, the western part of the Hauran plain and the Golan heights
- Outcrops of Neogene and Quaternary shield basalts with very limited soil cover, e.g. on the Jebel el Arab mountain massif and the Leja plateau, where the absence of a distinct surface drainage system indicates that surface runoff is of minor importance
- Particular sections of wadi systems, where runoff collected from extensive catchments can infiltrate

Significant groundwater recharge from present-day precipitation is evident, in particular, in shallow groundwater of Jebel el Arab from:

- The immediate response of spring discharge to the seasonal precipitation
- The low salinity and hydrochemical composition of the groundwater, which reflects recently recharged water (Sect. 4.4.2.1)
- Tritium values of 8.4–21.3 TU
- A $\delta^{18}O/\delta^{2}H$ relationship characteristic for present precipitation originating from the eastern Mediterranean Sea (Sect. 4.5.2.2)

Recent recharge is indicated also at the foothills of Jebel el Arab near Suweida and in the Wadi Liwa–Leja area northwest of Suweida through the wide-spread occurrence of perched groundwater and tritium values of 3.6–16 TU in some wells. The hydrochemical and isotope composition of groundwater samples in the Hauran Plain west of Jebel el Arab reflects an impact of evaporation during the recharge process. A significant component of indirect recharge within the plain can therefore be assumed, but the quantity of present recharge may be relatively small in comparison to the stored groundwater volume.

The surface drainage system, which can be distinguished on satellite images, indicates the following infiltration characteristics on surfaces of different types of basalt:

- Miocene–Pliocene plateau basalt and Pleistocene shield basalts: poorly developed drainage pattern – relatively high direct recharge and low surface runoff
- Quaternary valley – filling lava flows and Quaternary weathered tuff terrains: differentiated drainage pattern – low direct recharge, rapid surface runoff frequently ending in mud pans which are covered by fine grained sediments with low permeability

Recharge volumes are considered to be low to insignificant, in several areas:

- The eastern and southeastern arid parts of the basalt field
- Outcrop areas of valley filling lava flows
- Plains on which a thick soil has developed, e.g. in parts of the Hauran plain

Recharge in these areas appears to be restricted mainly local indirect recharge in wadi systems.

Most of the analyzed samples from boreholes in northern Jordan – the Mafraq area, the northeastern desert of Jordan and the basalt area north of Azraq – have an isotopic signature indicating recharge on the slopes of Jebel el Arab and groundwater movement over long distances with retention periods between 6,000 and 16,000 years (Sect. 4.5.2.2).

Fossil groundwater with very high groundwater ages of more than 20,000 years and stable isotope composition characteristic for recharge during previous pluvial periods is found in boreholes in the arid northeastern parts of the basalt region. Recent recharge occurs, however, in some wadi sections in the east of Jebel el Arab, e.g. in Wadi ash Sham in the Zelaf area.

Direct and indirect recharge amounts, in general, to 6% of precipitation and, in areas with less than 90 mm of mean annual rainfall, to 3% of precipitation.

References. Al-Kharabsheh (1995), Bajbouj (1982), Domenico and Schwartz (1990), ESCWA (1996), Gibbs (1993), Krasnov et al. (1966), Schaeffer (1997), Schelkes (1997), Selkhozpromexport (1974), Wolfart (1966), files of WAJ (transmissivity values).

4.3 Hydrogeologic Sub-Basins and Groundwater Flow Systems

4.3.1 Hydrogeologic Sub-Areas

The Jebel el Arab basalt aquifer complex comprises:

- Sub-regional aquifer systems with extensive groundwater flow regimes
- Various local flow systems at shallow to intermediate depth

Hydrogeologic sub-basins with extensive groundwater flow systems are the Yarmouk sub-basin and the Azraq sub-basin in the west and southwest of the basalt field.

Wadi Dhuleil constitutes a tributary hydrogeologic area of the mainly sedimentary aquifer system of Wadi Zerqa, situated on the western margin of the basalt field between the Yarmouk and Azraq sub-basins.

The eastern part of the basalt field covers an extensive area of basalt aquifers situated above and upstream of the Mesozoic–Paleogene sedimentary aquifer system of the Hamad. The extent of continuous basalt aquifers in that area is rather limited.

The Jebel el Arab mountain massif constitutes a main water divide area between four hydrologic sub-basins – Yarmouk, Dhuleil-Zerqa, Azraq and Hamad (Fig. 4.2) – corresponding approximately to major hydrogeologic sub-areas.

4.3.2 Jebel el Arab Mountain Massif

On the high mountain plateau of Jebel el Arab, situated above 1,300 m asl, outcrops of basalt flows of Pliocene age extend over about 400 km^2. These Pliocene basalts are composed of single lava sheets of several metres thickness with intercalated soil or sedimentary layers. In parts of the high plateau, the Pliocene basalts are overlain by volcanic tuff in cinder or scoriaceous cones. Quaternary lava flows extend, in particular, over the slopes of the mountain massif.

Numerous springs discharge from the Pliocene and Quaternary basalts at various levels between 1,000 and 1,600 m asl, mainly in the western and northern parts of the high plateau and on the western mountain slope. In the eastern part of the high plateau, shallow groundwater has been explored at depths of 20–50 m below surface with water levels at around 1,300–1,400 m asl.

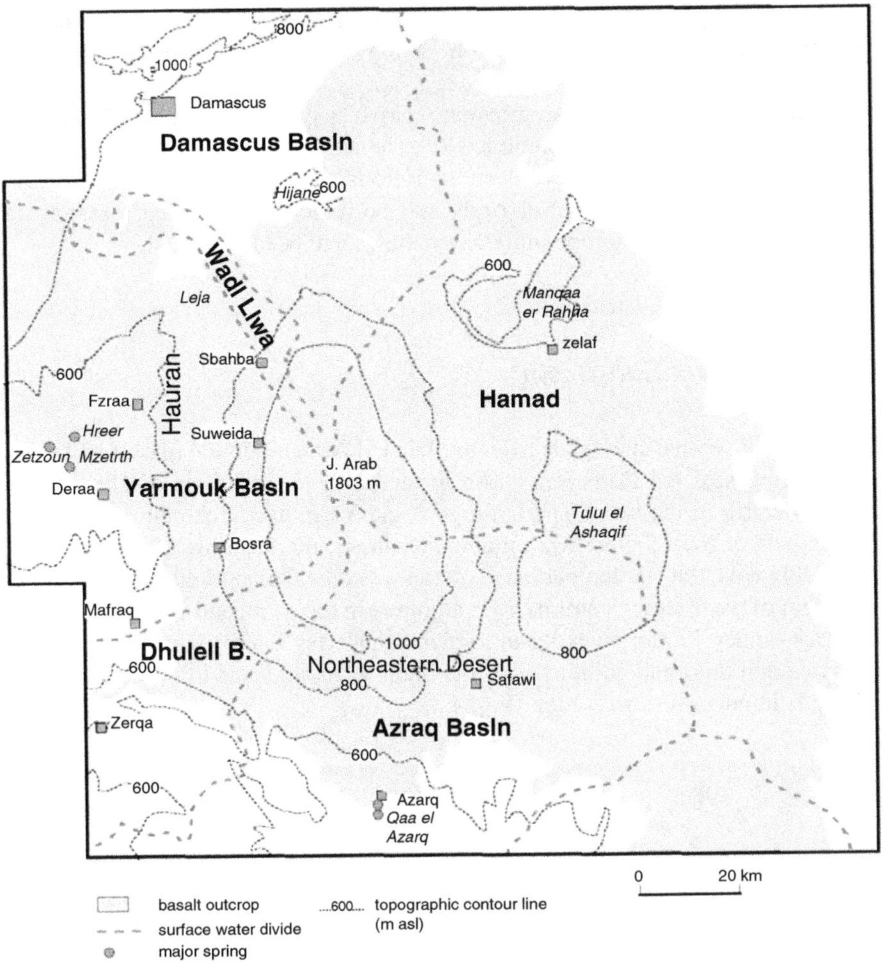

Fig. 4.2 Jebel el Arab basalt field, location map, main hydrologic basins

The springs issue at the boundary between individual lava flows. About 100 springs are observed in the Jebel el Arab area, most of which have a discharge of <1 l/s and dry up during summer. The mean discharge of larger springs is in the order of 12–20 l/s with high seasonal fluctuations. The mean total discharge of springs in the western part of Jebel el Arab and the adjoining mountain slopes in the Suweida area has been estimated at 3.6×10^6 m^3/a (Burdon 1954). The same order of magnitude of spring discharge can be assumed from reported summaries on spring discharge: 40 springs with mean discharge of 1.5 l/s and 70 springs with mean discharge of 0.5 l/s = 95 l/s or 3×10^6 m^3/a. Mean annual precipitation of 350 mm over the combined catchment areas of the springs, which extends over around 500 km^2, produces a volume of precipitation of 175×10^6 m^3/a. As spring discharge of 3×10^6 to 4×10^6 m^3/a corresponds to only about 2% of the mean annual volume of precipitation over the western Jebel el Arab mountain area, significant leakage of groundwater from shallow aquifers into deeper aquiferous horizons may be assumed.

The core of the Jebel el Arab mountain massif is probably occupied by basaltic dikes and sills of the main volcanic feeder zone and by highly disturbed blocks of sedimentary rocks. No extensive deep aquiferous zones are therefore expected in the deeper subsurface of Jebel el Arab, and no indications of deep groundwater circulation, e.g. through temperature anomalies, have been observed.

4.3.3 Yarmouk Sub-Basin

The Yarmouk sub-basin extends from the Jebel el Arab mountain massif in the east and the fringes of the Damascus basin in the north to the spring discharge zone around Mzeirib in the west. The hydrogeologic Yarmouk catchment includes the western part of the Jebel el Arab mountain massif, the foothills of Jebel el Arab around Suweida, the Hauran plain and, in the west, the Golan highlands. The basalt catchment of the Yarmouk sub-basin is adjoined in the south and west by the lower Yarmouk–upper Jordan river basin, which comprises mainly Upper Cretaceous limestone and dolomite aquifers with the main recharge areas in the Ajloun–Irbid highlands in northwestern Jordan (Fig. 4.3).

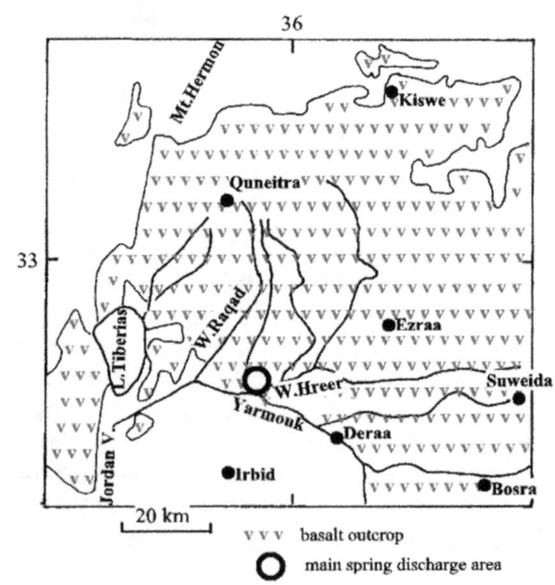

Fig. 4.3 Location map of Yarmouk catchment

4.3.3.1 Foothills of Jebel el Arab and Hauran Plain

The foothills of Jebel el Arab around Suweida are situated at altitudes between 800 and 1,000 m asl and are covered prevailingly by Lower Quaternary basalts which rest above Neogene basalts. The Hauran plain extends between the foothills of Jebel el Arab in the east and the Golan highlands in the west. Topographic elevations descend from around 800 m asl at the foothills to 500 m in the area near Deraa. West of Deraa, wadis of the Yarmouk river system are incised to elevations of less than 400 m asl. Basalt outcrops in the southern part of the Hauran plain are prevailingly represented by Lower Quaternary lava flows, which extend from volcanic eruption cones as valley fillings of a pre-existing morphologic relief. Details of the geological structure are, to a large extent, hidden under a thick soil cover. The plain is dissected by a well developed wadi system indicating a relatively important role of surface runoff.

The Pliocene–Quaternary basalts of the Deraa–Ezraa area form the principal productive aquifer system in the Syrian part of the basalt field. The water table of the main aquifer descends from around 560 m asl in the Jebel el Arab foothills near Suweida, more than 400 m below land surface, and 520 m asl in the north and east of the Hauran plain to 450 m in the southwest, where groundwater discharges in several large perennial springs. The main aquifer extends probably over the Neogene basalts and the lower section of Quaternary basalts. The thickness of the saturated main basalt aquifer decreases from 200 m in the east to around 50 m in the southwest.

4.3.3.2 Northern–Northeastern Part of the Yarmouk Sub-Basin

The northern and northeastern parts of the Yarmouk sub-basin appear to extend over surface water divides into Wadi Liwa and the Damascus plain. The northeast of the hydrogeologic Yarmouk Basin, including the Leja plateau and Wadi Liwa northwest of Jebel el Arab, comprises extensive outcrops of Quaternary shield basalts. On the Leja Plateau, which extends north of Suweida over around 300 km^2, a particularly well preserved 8–20 m thick shield basalt flow of Middle Quaternary age overlies the Lower Quaternary basalt.

Perched groundwater at some tens to some hundred metres above the main groundwater level occurs in many parts in the northeast of the Yarmouk sub-basin:

- Shallow groundwater which discharges in small springs south and northwest of Suweida
- Perched aquifers at intermediate depth in the area southwest and northwest of Suweida with water levels between 50 and 165 m below surface
- Aquifers at shallow to intermediate depth with water levels between 20 and 100 m below surface in the Leja plateau and Wadi Liwa

These local aquifers are sustained by present-day precipitation and extend in pervious layers of plateau basalts or shield basalts. Relatively favourable infiltration conditions and very limited runoff on the surface of these basalt types are indicated by a poorly developed drainage system. In particular the Leja Plateau shows no signs of significant surface runoff. Seasonal surface runoff occurs, however, in Wadi Liwa which forms a closed basin, in which surface runoff accumulates during the rainy season in a small lake near Braq. Groundwater from the local aquifers discharges, to some part, in small springs but major volumes of perched groundwater leak into deeper aquiferous layers reaching eventually the main basalt aquifer.

The main basalt aquifer in the Leja plateau and Wadi Liwa is situated between 150 and 370 m below ground surface (500–523 m asl) and groundwater flow is directed towards southwest into the hydrologic Yarmouk Basin.

In the Damascus plain adjoining the Yarmouk basin in the north, the basalts dip under aquiferous sands and gravels, which constitute a hydrogeologic system with large volumes of groundwater circulation. In pre-development state, most of the groundwater in the Neogene–Quaternary sedimentary and volcanic aquifers discharged in the area of seasonal lakes in the eastern part of the plain, Lake Hijane and Lake Ateibe. At present, the groundwater regime is dominated by artificial abstraction. To a minor part, groundwater flow in deeper aquiferous layers appears to be directed from the Damascus plain into the Yarmouk basin.

4.3.3.3 Golan Heights

The Yarmouk catchment extends in the north into the basalt area of the Golan heights. The volcanics of the Golan heights comprise mainly Neogene basalt flows. Quaternary basalts have a limited thickness of not more than 25 m and a low groundwater potential. A few springs emerge in the recharge areas of the Golan heights and numerous springs with low discharge of 0.5–1 l/s emerge in the Lake Tiberias area along the outcrop contacts between the basalt and underlying sedimentary rocks.

Groundwater flow in the aquiferous Neogene basalt is directed mainly toward south to the Mzeirib–Wadi Hreer spring discharge area and toward southeast, merging into the general groundwater flow of the Hauran plateau to the spring discharge area.

The Pliocene basalts of the Golan area comprise local aquifers at various levels. Groundwater in Lower Quaternary basalt aquifers flows toward Lake Tiberias and the Yarmouk spring discharge area.

4.3.3.4 Mzeirib–Wadi Hreer Discharge Area

Perennial springs in the Wadi Hreer–Mzeirib area rise, in general, on the boundary between Quaternary lava flows and underlying Paleogene sediments, which are represented in the Yarmouk area prevailingly by marls and marly chalks with low permeability. The discharge area comprises three major spring groups, Mzeirib and Chalalate Hreer with mean discharge of 1,400 l/s each and Zeizoun with a mean discharge of 800 l/s, and several smaller springs with mean discharge between 100 and 330 l/s. The discharge points are situated at elevations of 380–440 m asl. Total mean annual discharge of the springs is reported as 170 to 177 $\times$ 10^6 m^3.

4.3.3.5 Groundwater Balance

Groundwater balance calculations for the basalt aquifer system of the Yarmouk sub-basin lead to the following conclusions:

Subsurface flow from the eastern Yarmouk sub-basin – Jebel el Arab, Suweida foothills, Leja plateau, Wadi Liwa, Hauran plain – contributes about 60 $\times$ 10^6 m^3/a to the total spring discharge of 170 $\times$ 10^6 m^3/a in the Wadi Hreer–Mzeirib area. The calculated quantity of 60 $\times$ 10^6 m^3/a considers subsurface inflow from the Damascus basin and leakage losses of around 2.5 $\times$ 10^6 m^3/a into the deeper Upper Cretaceous aquifer.

A comparison of travel times calculated from hydraulic data with estimates of groundwater retention periods shows that the spring discharge is related to groundwater flow in the basalt aquifer as well as in the underlying Paleogene chalk aquifer. The combined saturated thickness of the basalt and chalk aquifers is 350 m, mean horizontal permeability 2.3 $\times$ 10^{-5} m/s, and travel times from different parts of the catchment to the discharge area are in the order of 3,200–9,200 years.

That groundwater flow with a mean retention period of 5,000 years is probably, to a large extent, sustained by present day recharge. Recharge is assumed, in the calculation, to correspond to 6% of the mean annual precipitation, which varies from around 250 to 500 mm in different parts of the catchment.

The calculated quantity of 60 $\times$ 10^6 m^3/a considers:

- A subsurface inflow into the Hauran plain of 35 $\times$ 10^6 m^3/a from west to northwest (Jebel el Arab, Suweida foothills and Leja plateau) and of 8.8 $\times$ 10^6 m^3/a from the north (Damascus plain and Wadi Liwa)
- Recharge of 11 $\times$ 10^6 m^3/a on the Hauran plain, or 6% of annual rainfall of 250 mm over an area of 750 km^2
- Leakage losses into the underlying Upper Cretaceous aquifer of 1.5 $\times$ 10^6 m^3/a

The remaining quantity of 110 $\times$ 10^6 m^3/a out of the total of 170 $\times$ 10^6 m^3/a of spring discharge is attributed to subsurface flow from the western part of the catchment: the Golan heights and Mount Hermon foothills.

4.3.4 Azraq Sub-Basin

4.3.4.1 Aquifer System

The Azraq sub-basin extends from the Jebel el Arab mountain massif in the north and the limestone plateau of eastern Jordan into the Azraq plain (Qaa el Azraq) in the centre of the basin. The northern part of the Azraq basin is occupied by outcropping basalts, outcrops in the southern part of the basin comprise mainly carbonate sediments of Upper Cretaceous to Paleogene age. The basalt field of the northern Azraq sub-basin descends from >1,000 m asl in the Jebel el Arab area in the north and around 900 m on the Tulul al Ashaqif in the east to 500 m asl in the Azraq plain.

Basalt outcrops in the Azraq catchment are Quaternary lava flows and Neogene and Quaternary shield basalts. Numerous volcanic cones are aligned in NNW–SSE trending chains in the area between the Azraq plain and Jebel el Arab. The surface drainage system in the basalt field of the Azraq basin shows major generally north–south oriented wadi systems, which extend prevailingly within the outcrops of Lower Quaternary lava flows. The wadi drainage system in the arid eastern part of the basin is directed to mudpans aligned in a morphologic depression zone between the slopes of Jebel el Arab and Tulul al Ashaqif.

Groundwater movement in the main basalt aquifer of the Azraq basin follows a general north–south direction towards the Azraq spring discharge area. Depth to groundwater exceeds 400 m in wide parts of the northeastern desert of Jordan, which is situated between Jebel el Arab and the Azraq plain. The main basalt aquifer is probably related mainly to Neogene plateau basalts, infiltration conditions are, however, influenced in many areas by the overlying Quaternary lava flows.

The basalts and underlying Paleogene chalks and limestones (Rijam formation) constitute, in general, one hydraulically connected aquifer system. Upper Cretaceous limestones and dolomites (Amman–Wadi Sir formation) provide a deeper confined aquifer, but may be directly connected to the basalt aquifer in some areas, where the Paleogene has been eroded.

4.3.4.2 Groundwater Balance

In the prevailingly semi-arid catchment of Azraq, hydrogeologic conditions are generally similar to those in the Yarmouk sub-basin, but recharge rates are lower in the dryer climate and are estimated at 2–18 mm/a on average. Spring discharge and evapotranspiration in the Qaa el Azraq was 16×10^6 m^3/a in the pre-development state and appears to have been sustained, to a substantial part, by present-day recharge, which occurs in particular in the mountainous northern parts of the catchment. An assumed infiltration of 6% of mean precipitation of 150–250 mm/a over 1,260 km^2 in the semi-arid northern parts of the catchment corresponds to an annual recharge of 13.5×10^6 m^3/a.

The spring discharge decreased significantly when operation of a well field tapping the basalt and Paleogene chalk aquifers started after 1980, and around 1991 the springs dried up completely.

In the vast arid eastern part of the catchment covering 4,500 km^2, recharge is estimated at 10.8×10^6 m^3/a from a recharge rate of 2.4 mm/a, of which around 4×10^6 m^3/a are assumed to be lost through leakage into the deeper aquifer.

Travel times of groundwater flow from recharge areas to the Azraq discharge area are calculated from hydraulic data as 5,700 to 16,000 years. These calculated travel times correspond, in general, to water ages derived from ^{14}C measurements. It can, however, not be excluded, that the groundwater of the basalt aquifer system reaching the Azraq area contains components of fossil groundwater, which is not related to present recharge and hydraulic conditions.

4.3.5 Wadi Dhuleil

The Wadi Dhuleil catchment covers a small section of the basalt field in the upstream part of the drainage system of Wadi Zerqa which is a tributary of the Jordan River. The topographic elevation of the catchment descends from >1,300 m at the Jebel el Arab mountain massif to 600 m asl in Wadi Dhuleil. The catchment is prevailingly covered by Lower Quaternary basalt flows. In the groundwater exploitation area at Wadi Dhuleil, the basalts are about 60 m thick.

The Upper Cretaceous to Paleogene marls and chalks are missing in the Dhuleil area and the aquiferous basalts are in direct contact with the Upper Cretaceous limestone–chert aquifer. The groundwater regime is highly disturbed by water extraction for irrigation and by irrigation return flow. Recharge over the catchment, which covers about 1,320 km^2, can be estimated at around 13×10^6 m^3/a. Present groundwater extraction exceeds the estimated natural recharge by about 100%. The deficit may be partly compensated by recharge from irrigation return flow, which leads, however, to serious water quality problems.

4.3.6 Hamad

In the eastern part of the Jebel el Arab basalt field, which extends into the Hamad steppe and desert area, Neogene plateau basalts appear to constitute a continuous aquifer in hydraulic connection with the underlying Paleogene chalk formation. Toward the eastern fringes of the basalt field, the basalts become unsaturated and the water level is situated in deeper carbonate formations of Mesozoic to Paleogene age. Groundwater movement is generally directed to the Euphrates–Gulf basin in the east. North of Jebel el Arab, some subsurface flow appears to occur from the Damascus basin into the basalt aquifer of the Hamad area.

References. Al-Kharabsheh (1995), Bajbouj (1982), Burdon (1954), ESCWA (1996), Krasnov et al. (1966: 37–38), Schelkes (1997), Selkhozpromexport (1974, 1986), Wolfart (1966).

4.4 Groundwater Salinity and Hydrochemistry

4.4.1 Salinity Distribution and Hydrochemical Environment

4.4.1.1 Groundwater Salinity

Salinity of the basalt groundwater is generally low to intermediate with 200–400 mg/l TDS in the Jebel el Arab area and in the basalt aquifer system west and south of Jebel el Arab. Higher groundwater salinities occur in the Hauran plain – up to 800 mg/l TDS – and in the joint volcanic–sedimentary aquifer on the southwestern fringes of the basalt field, where concentrations of total dissolved solids of up to several thousand mg/l are observed at some locations. Elevated salinities of more than 1,000 mg/l TDS appear to be caused mainly by irrigation return flow and by evaporative salt enrichment, e.g. in the Azraq plain. In the arid eastern parts of the basalt field, groundwater salinities generally range from 1,000 to 1,500 mg/l TDS, increasing to 6,000 mg/l towards southeast (Fig. 4.4).

4.4.1.2 Hydrochemical Conditions

The hydrochemical conditions of groundwater in the Jebel el Arab basalt aquifer system are, in general, characterized by the low chemical reactivity of the basaltic silicate rocks and by solution processes of carbonate and silicate minerals. Most of the groundwaters from the basalt aquifer system are Na–HCO_3 or Na–Cl type waters. Ca–HCO_3 type waters occur as spring water on Jebel el Arab.

The contribution of atmospheric input to the groundwater salinity can approximately be estimated from chemical analyses of rainwater from stations in Jordan and Syria. Shallow groundwater with low salinity (85–225 mg/l TDS) from Jebel el Arab shows an increase of Cl concentration in comparison to rainfall at Suweida and Ezraa by factors of 2–7 and of SO_4 concentrations by factors of 0–3. HCO_3 concentrations in groundwater are significantly higher than in rainwater. The hydrochemical composition of the low salinity groundwater is similar to the composition of flood waters from the Azraq area.

The increase of anion concentrations in the groundwater in comparison to rain water can be attributed to:

- Enrichment of substances dissolved in rainwater during infiltration
- Pedogenic inputs: reactions in the soil zone
- Lithogenic inputs: reactions with rock material within the aquifer or the overlying unsaturated zone

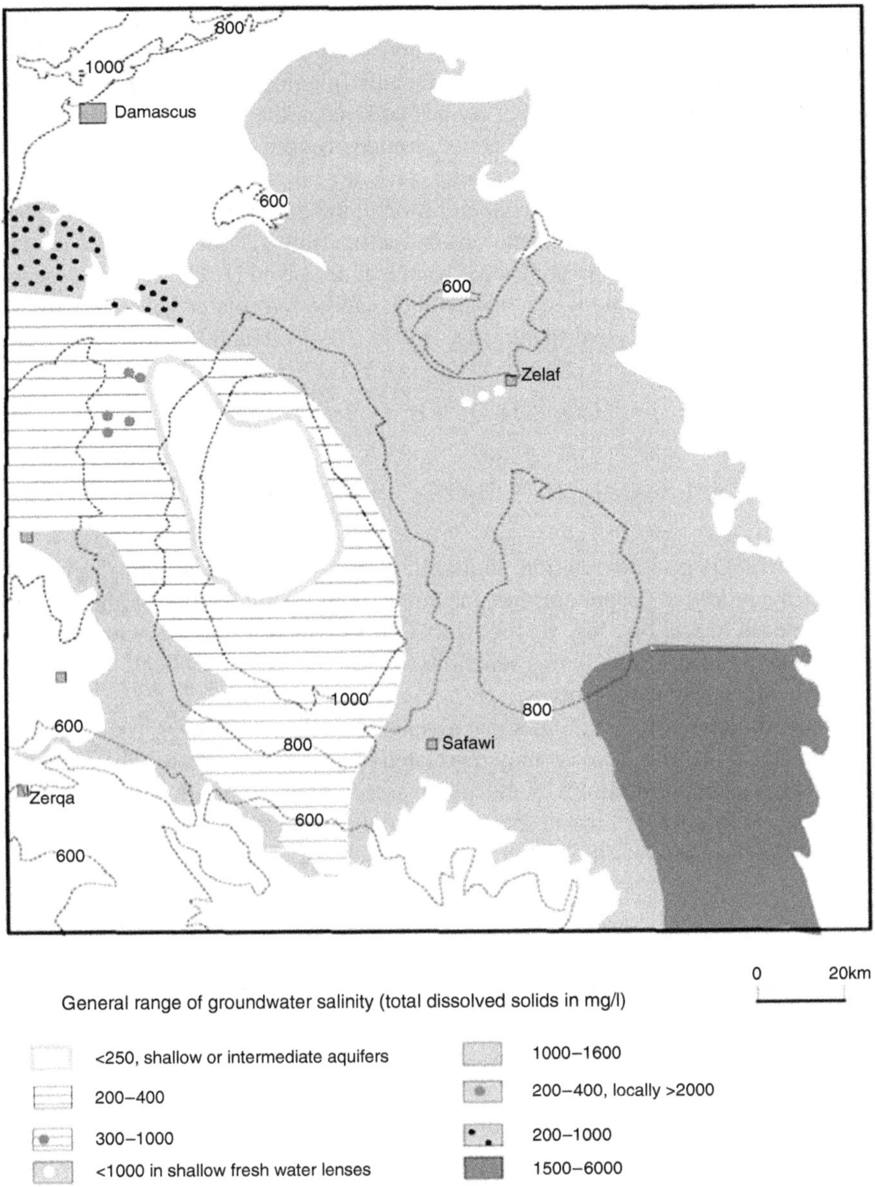

Fig. 4.4 Groundwater salinity distribution in Jebel el Arab basalt field from ESCWA (1996)

Enrichment through evaporation on the surface increases in particular Cl concentrations of the infiltrating water. Reactions of biogenic CO_2 produced in the soil zone with water and carbonate and silicate minerals leads to an increase in concentrations of HCO_3 and equivalent cations.

Solution of lithogenic components in groundwater of the basalts is predominantly controlled by reactions of water and dissolved CO_2 with rock material. Major mineral components of the basalts are silicates (plagioclase, potassium-feldspar, clinopyroxene, amphibole, biotite, olivine) and magnetite. Minor components are quartz, calcite, apatite, limonite/goethite, sulfides (pyrite, chalcopyrite) and some silicates, such as zeolite, epidote, chlorite, sericite.

A major portion of the HCO_3 concentration in the groundwater may be supplied by dissolution of calcite of volcanic origin and possibly also of carbonates of dust deposition on the surface. An additional source of dissolved HCO_3 may be provided from the process of silicate weathering, which can be formulated schematically on the example of plagioclase dissolution as follows (Appelo and Postma 1993: 204):

$$CO_2 + H_2O \rightarrow H^+ + HCO_3^-$$

$$CaAl_2Si_2O_8 + 2H^+ + H_2O \rightarrow Al_2Si_2O_5(OH)_4 + Ca^{2+}$$

Ca and HCO_3 go into solution while aluminum silicates (Kaolinite) remain as weathering products. Depending on the mineralogic composition of the basaltic rocks, release of Ca, Na, Mg, K, Si, Al and Fe as ions or molecular complexes is included in the silicate weathering reactions to a varying degree. Silicate weathering is, in general, a very slow process.

Concentrations of Cl and SO_4 show a general moderate increase in direction of groundwater flow. The increase may be related to a higher enrichment of dissolved substances during infiltration in the plain areas, where rainfall and velocity of surface runoff are lower than in the mountain area. Elevated groundwater salinity is generally correlated with elevated concentrations of Na and Cl.

4.4.2 Hydrochemical Features in Different Hydrogeologic Areas

4.4.2.1 Jebel el Arab Mountain Area

Water samples from springs and shallow wells in the Jebel el Arab mountain area are characterized by low ion concentrations and groundwater salinity between 85 and 225 mg/l TDS and are prevailingly Ca–HCO_3 type waters (Fig. 4.5). The samples represent perched groundwater in pervious zones of Pliocene basalt sheets or of Quaternary lava flows on the mountain slopes. Tritium contents of 8–21 TU indicate that the perched aquifers are replenished by recent recharge.

In general, the hydrochemical composition of shallow groundwater in the Jebel el Arab mountain area appears dominated by atmospheric inputs and evaporative enrichment of atmospheric inputs by a factor of around 5 and by reactions between soil CO_2, water and carbonate or silicate minerals.

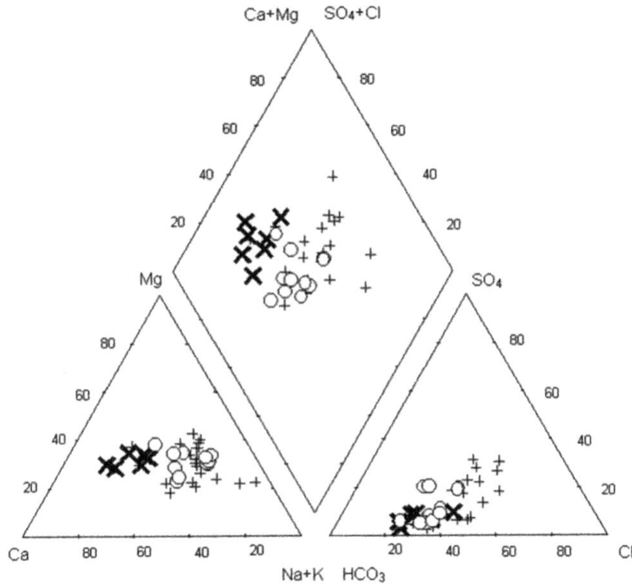

Fig. 4.5 Piper diagram: Groundwater samples from the basalt aquifer in the Jebel el Arab–Hauran–Mzeirib area, × shallow groundwater in the Jebel el Arab mountain area, + groundwater in the Hauran plain, ○ water of springs in the Mzeirib–Wadi Hreer area, data from Bajbouj (1982), ESCWA (1996)

4.4.2.2 Golan Heights

The water of springs from the Neogene and Quaternary basalt aquifers of the Golan heights is HCO_3 water with an average salinity of 475 mg/l TDS. Percentages of major cations are approximately equal with a slight predominance of Na or Ca, Na and Ca concentrations being higher than in precipitation water. Spring water from the Neogene basalt is generally Na–HCO_3 water with a Mg/Ca ratio around 1.0. The water of springs issuing from the Quaternary basalts is Ca–HCO_3 water with a Mg/Ca ratio of 0.7.

Calcium–sodium plagioclase, a major mineralogic component of the basalts, releases upon dissolution Na and Ca. Relatively high concentration of Mg in basalt waters is probably caused by the dissolution of olivine.

4.4.2.3 Hauran Plain and Mzeirib–Wadi Hreer Discharge Area

Samples from boreholes and springs from the Hauran plain and the groundwater discharge area around Wadi Hreer–Mzeirib can, in general, be characterized as Na–HCO_3 type waters with a salinity range of 250–800 mg/l TDS (Fig. 4.5). The composition of water samples shows a general zoning with:

- Relatively low Cl and SO_4 concentrations and total salinity (around 350 mg/l TDS) in boreholes in topographic higher parts of the plain (above 600 m asl) and in water from the large springs of Wadi Hreer–Mzeirib
- Moderate salinity of 500 mg/l TDS and somewhat higher Cl and SO_4 concentrations in boreholes at the topographically lower parts of the plain around Ezraa–Deraa

The rather low ion concentrations in the spring waters of Wadi Hreer–Mzeirib indicate that regional groundwater flow related to recharge in distant upstream areas of the catchment provides a major component of the spring discharge.

Comparatively high values of EC, HCO_3, Cl and NO_3 in groundwater at some locations of the Hauran plain are probably caused by anthropogenic sources of dissolved ion contents accompanied by an increase of HCO_3 and Ca contents through oxidation of organic carbon.

4.4.2.4 Mafraq–Azraq Area

On the southern and southwestern slope of Jebel el Arab, comprising the area east of Mafraq, the northeastern desert of Jordan and the Azraq area upstream of the Azraq plain, the basalt aquifer is connected with the underlying Paleogene chalk aquifer and, in the west where the Paleogene is missing, with the Upper Cretaceous limestone–chert aquifer. The basalt aquifer system here contains predominantly fresh groundwater of Na–HCO_3 to Na–Cl type with salinities of 250–700 mg/l TDS (Fig. 4.6). The fresh water is similar in its hydrochemical composition to water samples from the Ezraa area and is probably recharged prevailingly at higher parts of the catchment area. The hydrochemical composition of the groundwater appears dominated by atmospheric inputs enriched through evaporation and by dissolution of carbonate and silicate minerals. The composition of the fresh groundwater in the basalt as well as in the underlying carbonate rocks is clearly related to recharge in the basalt outcrop. No characteristics of typical carbonate rock water can be seen. A slight contribution of Na and Cl contents from the carbonate aquifer cannot be excluded.

Salinity of water extracted from the AWSA well field, situated directly upstream of the Azraq plain, is generally 300–460 mg/l TDS, with some values reaching up to 1,000 mg/l TDS. Fluctuations of the salinity and composition may be related to vertical or horizontal differentiation of the natural water quality, e.g. the occurrence of moderately higher contents of dissolved solids in water of the Paleogene chalk or in the basalt aquifer in areas surrounding the well field.

Samples from the springs Al Aura and Mustadhema, which are located at the rim of the basalt outcrop and have now dried up, show the same general hydrochemical characteristics as the water from AWSA boreholes (Fig. 4.7). Groundwater in the Azraq plain, which includes basalt water mixed with water infiltrated in the plain, has considerably higher salinity. Recent recharge and near surface groundwater in the plain are affected by evaporative enrichment of dissolved substances. The sabkha area of the Azraq plain contains saline groundwater.

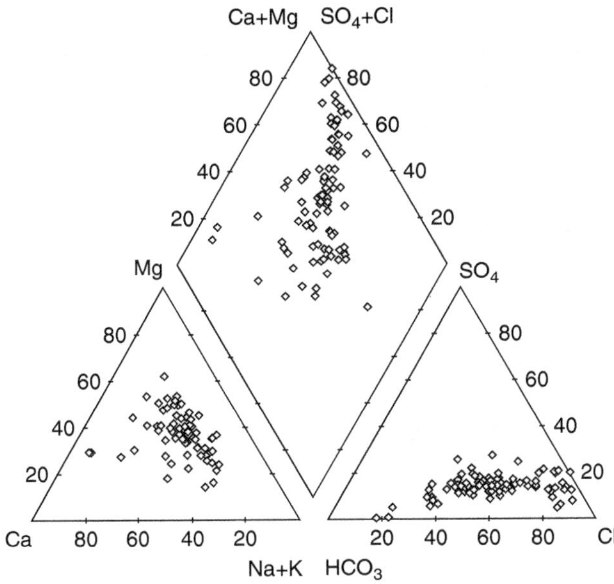

Fig. 4.6 Piper diagram: Groundwater samples from the basalt aquifer in the Mafraq area and northeastern desert of Jordan data from Rimawi (1985), Almomani (1996), ESCWA (1996)

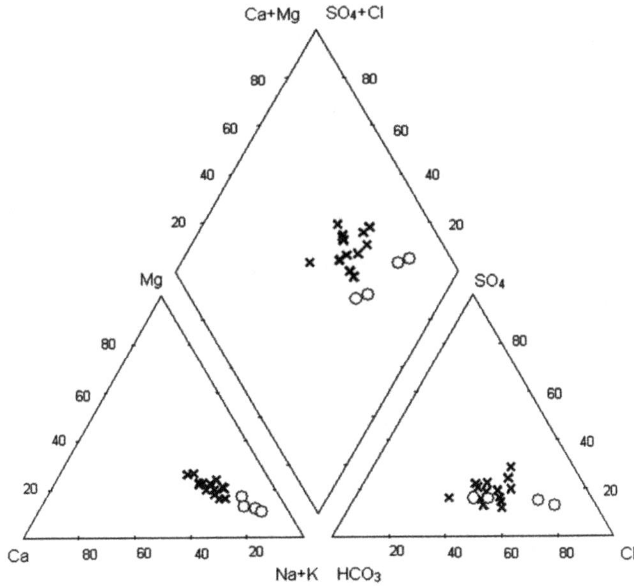

Fig. 4.7 Piper diagram: Groundwater samples from the basalt aquifer in the Azraq area, × AWSA well field, ○ springs in Azraq plain, data from Rimawi (1985), Rimawi and Udluft (1985), Al-Momani (1991), Almomani (1996)

A hydrochemical classification of groundwater in shallow aquifers of the Azraq depression has been presented by Rimawi and Udluft (1985). Four major hydrochemical groups have been distinguished:

- Typical basalt aquifer water with a mean salinity of 370 mg/l TDS
- Groundwater in carbonate and alluvial aquifers surrounding the basalt and in most parts of the Azraq plain with higher contents of dissolved solids than the basalt water and mean salinities of 800–1,200 mg/l TDS
- Hyper-saline Na–Cl water with Cl concentrations of 130 g/l in the central part of the Azraq depression, where pools of surface water evaporate

In the area east of Mafraq numerous boreholes, most of which tap aquiferous sections of the Pleistocene basalt and the underlying Upper Cretaceous carbonate aquifer (A7/B2 aquifer), extract groundwater mainly for irrigation. Salinity of samples collected from 84 boreholes between 1982 and 1992 ranges from 340 to 4,000 mg/l TDS. 70% of the evaluated water analyses are fresh water samples. The brackish groundwater, represented by 30% of the samples, is characterized by high Cl concentration and elevated Ca, Mg and Na contents. The relative cation contents are not significantly different from the fresh water of the basalt aquifer. HCO_3 concentration in a number of samples is low (<100 mg/l). NO_3 contents reach up to 75 mg/l in some samples; increased NO_3 contents appear to be correlated with high Cl and SO_4 concentrations, indicating that the main source of elevated ion concentrations is irrigation return flow (Fig. 4.8).

4.4.2.5 Wadi Dhuleil

In the Dhuleil area, about 80 production wells are in operation for irrigation and domestic supply. The wells tap mainly the Upper Cretaceous carbonate aquifer (A7/B2 aquifer), which is covered in part of the area by Pleistocene basalt. Depth to groundwater is generally between 50 and 100 m, the thickness of the basalt is <100 m.

Groundwater salinity of samples from the area with basalt cover (18 samples collected between 1985 and 1992) ranges from 400 to 6,000 mg/l TDS. Mean concentrations of all major ions, except for HCO_3, are higher than in other areas of the western part of the basalt aquifer system (Fig. 4.8). HCO_3 concentrations are generally low (around 100 mg/l). The occurrence of fresh water appears to be restricted, at present, to the more eastern – upstream – parts of the area. Historical water analysis data show that groundwater salinity and concentrations of major ions increased from low levels, before groundwater development started in 1961, to brackish water quality in many wells after 1970. Increasing NO_3 contents of the groundwater are generally correlated with high Cl and SO_4 concentration. The deteriorating quality of the irrigation water together with a diminution in the volume of the applied irrigation water led to high soil salinization. The high groundwater salinity and groundwater quality deterioration is apparently caused mainly by the rapid recycling of irrigation water: Part of the water applied for

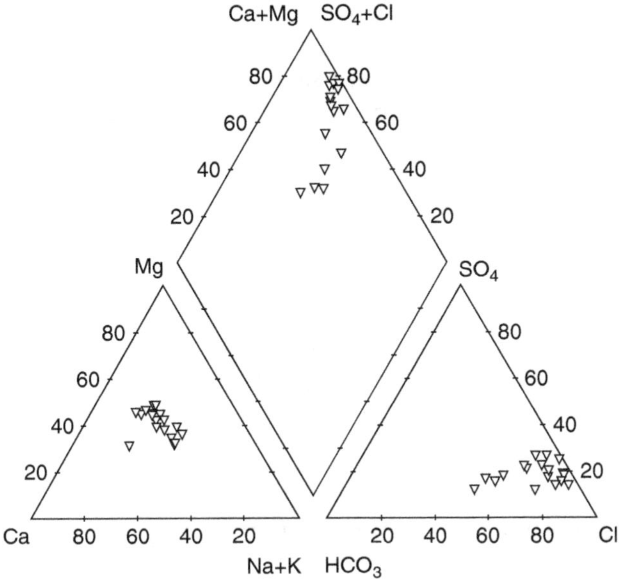

Fig. 4.8 Piper diagram: Groundwater samples from the basalt aquifer in the Dhuleil area, northern Jordan data from Rimawi (1985), ESCWA (1996)

irrigation seeps directly into the subsurface without longer retention in the soil zone. The reinfiltrating water shows no impact of reactions with soil CO_2 and has no detectable tritium. The hydrochemical composition of the reinfiltrating water is related to the dissolution of soluble salts from the saline soils and possibly to precipitation of less soluble carbonate constituents.

4.4.2.6 Hamad

The Hamad area and the eastern part of the Azraq sub-basin catchment extending northeast and east of Jebel el Arab have an arid climate with mean annual precipitation of 150 mm in the northwest decreasing to <70 mm southeast. The basalt thickness varies from a few tens of metres near the northern and eastern boundaries of the basalt outcrop to >400 m in the Tulul al Ashaqif area. In considerable parts of the northern and eastern extent of the basalt field, water levels of the main groundwater flow system are situated within the Paleogene chalk aquifer below the base of a relatively thin basalt cover.

Groundwater salinity in the basalt covered part of the Hamad area is generally >1,000 mg/l TDS, increasing to several 10,000 mg/l towards southeast. The groundwaters vary in their hydrochemical composition in wide ranges and include brackish water with Cl concentrations of up to 600 mg/l and SO_4 concentrations up to 950 mg/l. Fresh water to slightly brackish HCO_3 or Cl type water occurs along extensive wadi systems on the eastern and southern margin of the basalt field.

In general, the elevated groundwater salinity in the Hamad area and the eastern part of the Azraq basin can be attributed to the prevailing arid climate conditions with very low recharge rates, high evaporative enrichment of the limited quantities of infiltrating water, and low rates of groundwater circulation and flushing of aquifers. Fresh water occurrences appear to be restricted mainly to lenses along major wadi courses.

References. Abu-Sharar and Rimawi (1993), Al-Kharabsheh (1995), Almomani (1994), ESCWA (1996), Kattan (1996c), Nitsch (1990), Rimawi (1985), Rimawi and Udluft (1985), Rosenthal (1987), Salameh and Rimawi (1987a, b), Schelkes (1997), Selkhozpromexport (1974), Wolfart (1966).

4.5 Hydrogeologic Information from Isotope Data

Isotope hydrologic information on the groundwater from the basalt area is available from a number of investigations in the past decades (see references of this sub-chapter).

4.5.1 *^{3}H and ^{14}C Data*

Tritium values of 5–21 TU indicate recent groundwater recharge in the Jebel el Arab mountain area, on the western Golan heights, and at some locations in the Hauran plateau. From ^{14}C ages, contemporary recharge can be assumed for the Suweida area and the eastern parts of the Hauran plateau. The mean ^{14}C ages of the groundwater increase to a few thousand years towards central and southern parts of the Hauran plateau around Ezraa–Deraa. ^{14}C ages of groundwater in the Mzeirib–Wadi Hreer spring discharge zone are around 4,000–6,000 years (Fig. 4.9). Considering mean distances of groundwater flow of 40–60 km from the recharge areas, groundwater flow velocities may be estimated to be in the order of 10 m/a.

^{14}C data show that most of the groundwater of the Jordanian part of the basalt field are between 10,000 and 28,000 years old. Retention periods of groundwater discharging in the springs at Azraq are calculated from ^{14}C data at 8,000–15,000 years. For a 60 km long flow path, these ages correspond to a mean groundwater flow velocity between 4 and 7.5 m/a.

The ^{14}C ages of groundwater samples in the Hamad area are exceptionally high (19,000–28,000 years); the samples may represent fossil groundwater mainly from the underlying Paleogene chalk aquifer.

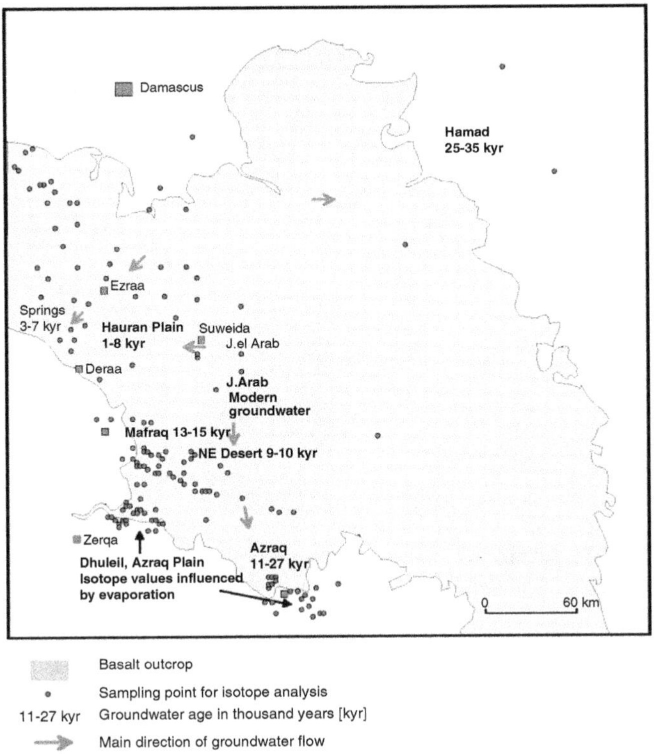

Fig. 4.9 ^{14}C water ages in the Jebel el Arab basalt field after ESCWA (1996)

4.5.2 *Stable Isotopes of Oxygen and Hydrogen*

4.5.2.1 Precipitation

Mean $\delta^{18}O$ and δ^2H values of precipitation water samples from the Jebel el Arab basalt area scatter close to the MMWL with d values around +20‰. $\delta^{18}O$ varies in a range of −5.27 to −7.41‰ (stations Azraq, Ezraa, Suweida, Kuneitra) (Fig. 4.10).

4.5.2.2 Groundwater

The isotopic composition of groundwater in the northwestern part of the basalt field shows a rather homogeneous development from the recharge area in Jebel el Arab over the Hauran plain to the Yarmouk spring discharge area.

The most negative $\delta^{18}O$ values between −6.5 and −8.8‰ were analyzed for samples of shallow groundwater in the Jebel el Arab mountain area. The deuterium excess of most of these samples exceeds +20‰ and corresponds to that of the present-day rainwater of Mediterranean origin.

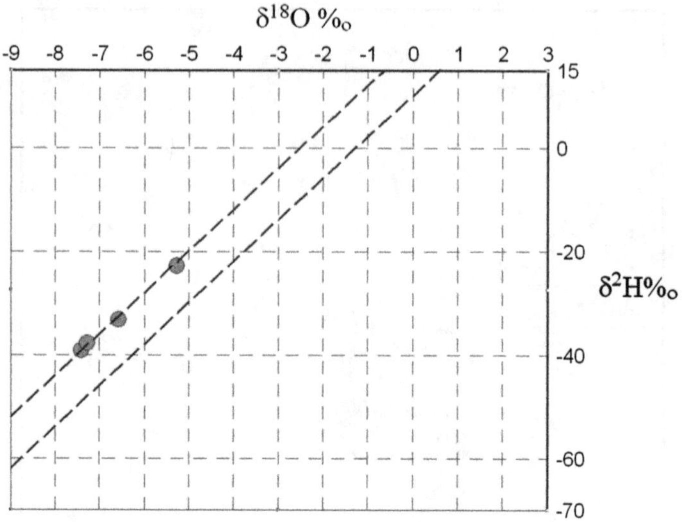

Fig. 4.10 $\delta^{18}O/\delta^{2}H$ diagram: Rain water samples from the Jebel el Arab basalt area stations Suweida, Ezraa, Azraq, data from Almomani (1996), Kattan (1996c)

The groundwater in the Jebel el Arab mountain area is similar in its isotopic composition to the groundwater of the higher ranges of Ansariye mountains and of the highlands of Jordan at Nuaime and may thus be interpreted to represent water recharged from Mediterranean rainfall at altitudes of, on average, 1,000 m asl.

In the Hauran plain and the foothills of Mount Hermon, $\delta^{18}O$ values range from −4.5 to −6‰, d values from +10 to +20‰ with a trend of increasing enrichment of heavy isotopes in direction of the groundwater flow toward south–southwest (Fig. 4.11). The trend to less negative $\delta^{18}O$ values from the Jebel el Arab mountain area and the Mount Hermon foothills toward the Hauran plain may be attributed to an altitude effect, but the increase of $\delta^{18}O$ values together with a general decrease of d values indicates a predominant impact of evaporative enrichment of heavier isotopes on the plain.

$\delta^{18}O$ values of water from springs in the Mzeirib–Wadi Hreer area vary around −5.5 to −6‰ with d values of +17‰. According to the isotopic composition, the water discharging in the springs may represent a mixture of groundwater from the Hauran plain and of groundwater recharged at higher altitudes on Jebel el Arab and on the foothills of Mount Hermon.

In the basalt field southwest and south of Jebel el Arab – areas east of Mafraq, northeastern desert of Jordan, Azraq area – $\delta^{18}O$ values range from −5.5 to −6.5‰ (Fig. 4.12). More negative values (−6.5 to −7‰) are found in some boreholes. d values show a wide scatter between +15‰ and >+20‰, values of >+18‰ were observed in particular in groundwater from the boreholes east of Mafraq, the northeastern desert of Jordan and the well field north of Azraq (AWSA well field). Water of the fresh water spring Al Aura, which has now dried up, had $\delta^{18}O$ values

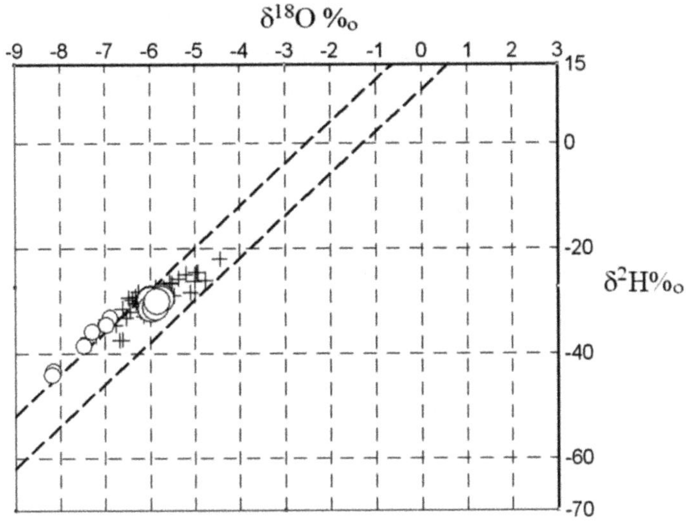

Fig. 4.11 $\delta^{18}O/\delta^2H$ diagram: Groundwater samples from the Jebel el Arab–Hauran–Mzeirib area data from ESCWA (1996), Kattan (1996b) ∘ Jebel el Arab springs, + Hauran, ○ Mzeirib springs

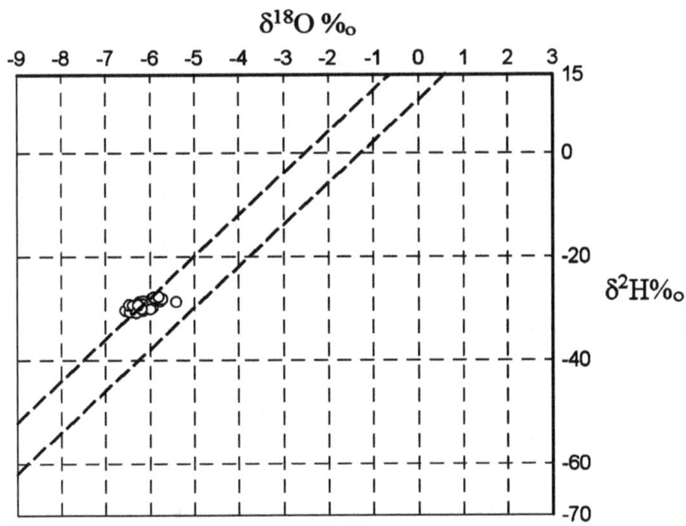

Fig. 4.12 $\delta^{18}O/\delta^2H$ diagram: Groundwater samples from the Mafraq–Azraq areas data from ESCWA (1996)

around −6‰ and d values around +18‰. These values are similar to the values of water of the Mzeirib springs in the Yarmouk catchment. The water discharging at Al Aura spring probably originates from recharge in various parts of the catchment area between Jebel el Arab and the northeastern desert.

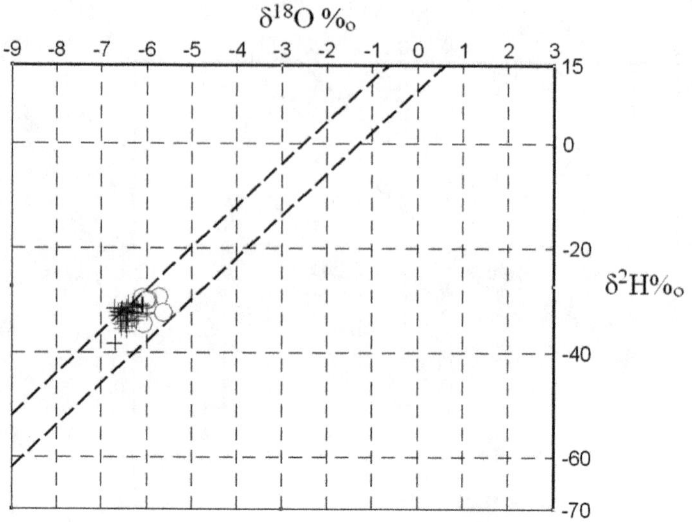

Fig. 4.13 $\delta^{18}O/\delta^{2}H$ diagram: Samples from springs and AWSA well field, + AWSA well field, o spring, data from Almomani (1996)

$\delta^{18}O$ values reported for the brackish springs, which issued previously in the Azraq plain (springs Souda and Kaisiye) varied in rather wide ranges with $\delta^{18}O$ between −9.4 and −19.4‰ and d values between −5.7 and −6.12‰, indicating an evaporative impact (Fig. 4.13).

In most samples of the groundwater from the basalt aquifer in the Damascus plain and the Hamad area north and northeast of Jebel el Arab, the stable isotope composition deviates significantly from that of the Jebel el Arab–Hauran areas toward more negative $\delta^{18}O$ values with d values around the GMWL.

4.5.3 *Impact of Irrigation Return Flow*

In several areas of the basalt aquifer system in Jordan, isotope signatures do not fit into the general picture. These deviations are, in particular, attributed to an admixture of irrigation return flow into the groundwater, which is indicated also by hydrochemical data. The agriculture area of Dhuleil appears to be particularly affected. The ^{14}C ages of 5,800–9,300 years are lower than those further north along the road between Mafraq and Safawi. The total salinity is between 400 and 6,000 mg/l TDS, HCO_3 concentrations are low, high nitrate concentrations are observed. The $\delta^{18}O/\delta^{2}H$ values reflect an influence of high evaporation rates; tritium is not found.

The admixture of irrigation return flow has a significant impact on isotope values: During flood irrigation, water evaporates and $\delta^{18}O/\delta^{2}H$ values are enriched. The evaporative moisture flux into the atmosphere prevents addition of tritium

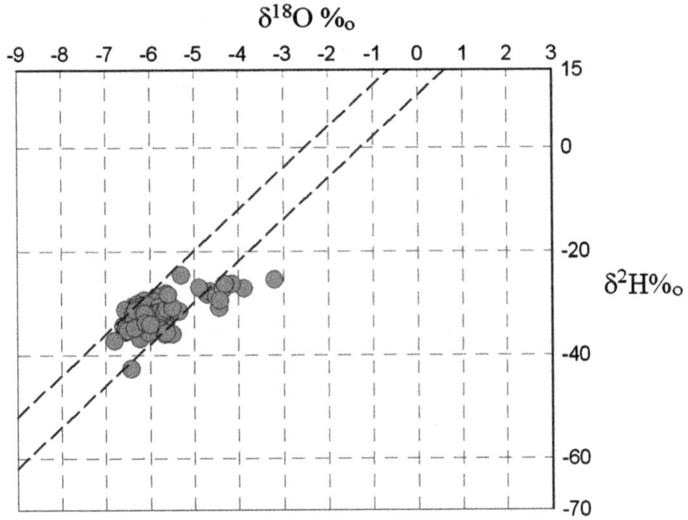

Fig. 4.14 $\delta^{18}O/\delta^{2}H$ diagram: Groundwater samples from Wadi Dhuleil and the Azraq plain data from Almomani (1996), ESCWA (1996)

from the atmospheric moisture, and ^{3}H remains below detection limit. During the seepage of water through the topsoil, carbon compounds are lost by the consumption of the plants while modern soil CO_2 with a high ^{14}C value from the root zone is transported into the aquifer. Accordingly, ^{14}C values in recycled groundwater of irrigation areas can be higher than values of unaffected groundwater.

In the Azraq plain (Qaa el Azraq), the isotopic composition of the groundwater is influenced by irrigation return-flow, seepage of isotopically enriched water from ponds and intensive groundwater extraction (Fig. 4.14).

4.5.4 Isotope Data and Groundwater Regime

In summary, the isotope data from the basalt aquifer system reflect the following main features:

- Significant recent recharge occurs in the Jebel el Arab Mountain area and its western foothills and on the foothills of Mount Hermon.
- Groundwater movement over large distances combined with local recent recharge shows evaporative isotope enrichment in the Hauran Plateau.
- The springs of the Yarmouk area discharge groundwater from different catchment areas with major recharge zones on the mountain zones of Jebel el Arab and the Mount Hermon foothills.

- Groundwater moves towards the Jordanian part of the basalt field over long distances from various parts of the slopes of Jebel el Arab.
- Dominant influences of irrigation return flow are evident in extensive irrigation areas.

In spring catchments on the foot of Mount Hermon, which reach into Mesozoic sedimentary aquifers adjoining the areas, an altitude gradient of $-0.23‰\ \delta^{18}O/100\ m$ is indicated. On flow paths from Jebel el Arab to the major spring discharge areas, altitude effects are not evident.

References. Almomani (1991, 1993, 1994, 1996), Arsalan (1976), ESCWA (1996), Kattan (1996b, c), Rimawi (1985), Rimawi and Udluft (1985), Verhagen et al. (1991: Chap. 3), Wagner and Geyh (1999), Zuppi (1986).

Chapter 5
Interior Shelf

5.1 Geologic and Geographic Set-Up

5.1.1 General Geologic Structure

After peneplanation of the Arabian Shield in the late Precambrian, detrital sediments were deposited on the outer slopes of the uplifted shield during the Paleozoic. Sandstone sedimentation continued until the Lower Cretaceous with various intervening periods of marine or lagoonal carbonate deposition. Widespread marine transgression since the Upper Cretaceous covered large parts of the Arabian Shelf with carbonate formations, leaving an outcrop belt of Paleozoic–Lower Cretaceous rocks in a structurally uplifted zone surrounding the Arabian Shield. That belt, in which Upper Cretaceous–Tertiary marine sediments were not deposited or have been eroded, is defined here schematically as "Interior Shelf" (Sect. 1.1.1.3).

The outcrop belt of Paleozoic to Lower Cretaceous prevailingly detrital rocks of the Interior Shelf surrounds the Arabian Shield in a semi-circle from the Tabuk–Disi area in the north over the Widyan basin margin and the Tuwayq mountains in the northeast and east into the Wajid basin in the southeast. Cambrian to Permian formations of the Interior Shelf overlie discordantly the crystalline rocks of the basement. The Interior Shelf is sub-divided into four major structural elements, which are, from north to south, the Tabuk–Disi segment, northern and southern Tuwayq segments, and Wajid sandstone segment (Figs. 5.1 and 5.2).

The Tabuk–Disi and Wajid segments, in the north and south of the Interior Shelf, comprise vast outcrops of Paleozoic sandstones and shales. On the northern and southern Tuwayq segments of the shelf, Paleozoic and Mesozoic formations are exposed consecutively in relatively narrow but very long outcrop strips.

The northeastern margin and the southern tip of the Interior Shelf are covered by the sand seas of the Great Nefud and Rub al Khali deserts.

The Paleozoic–Mesozoic formations of the Interior Shelf dip generally in gentle slopes toward east–northeast plunging under younger formations of the Interior Shelf and under the Upper Cretaceous–Paleogene–Neogene cover of the Arabian Platform. Uplift and tilting created the homoclinal structure of the Interior Shelf and caused erosion of deposits of the shelf during various phases between the late

W. Wagner, *Groundwater in the Arab Middle East*,
DOI 10.1007/978-3-642-19351-4_5, © Springer-Verlag Berlin Heidelberg 2011

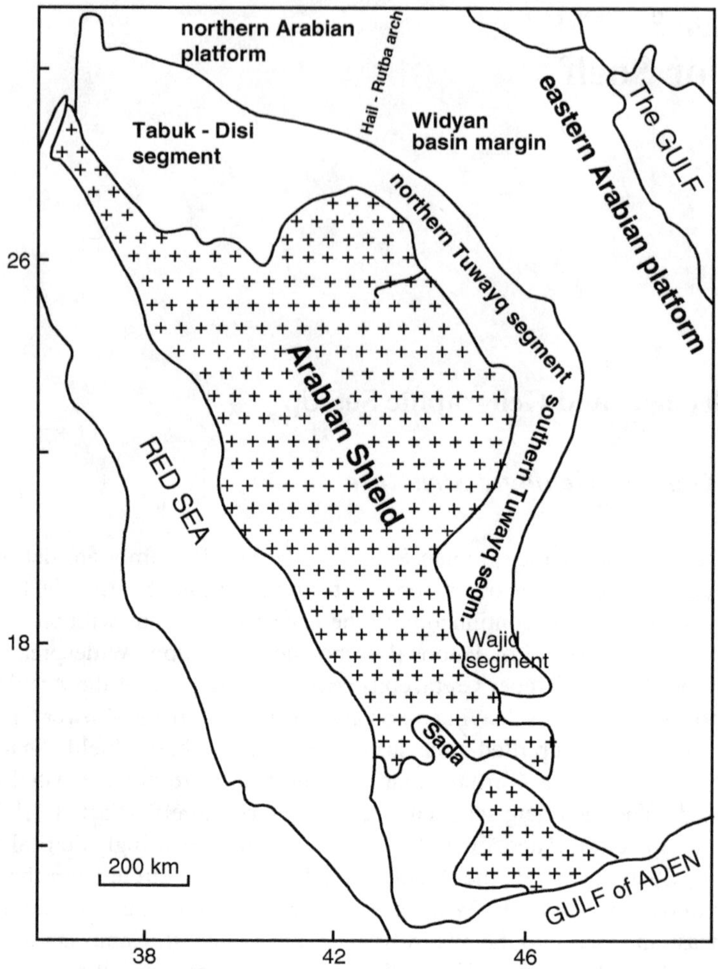

Fig. 5.1 Interior Shelf, main geologic units. After Edgell (1997), ESCWA (1999b)

Carboniferous and the Neogene. The most important uplift movements occurred during the early Oligocene and continued during the Miocene.

East of the clearly defined belt of the Interior Shelf, Paleozoic or Cretaceous sandstone formations are exposed in zones of structural uplift in erosional windows within the cover of the Cretaceous–Paleogene carbonate sequence of the Arabian Platform: the Hail–Rutba arch in the north and the Hadramaut arch in the south of the platform area. Sandstones of the Devonian Jauf formation and the Cretaceous Sakaka formation are exposed in the Sakaka and Jauf areas along the uplift zone of the Hail–Rutba arch, which adjoins here the margin of the Widyan segment of the eastern Arabian platform. On the Hadramaut arch, the Cretaceous Mukalla sandstones appear at or near the surface in the Wadi Hadramaut canyon. The sandstones are covered by

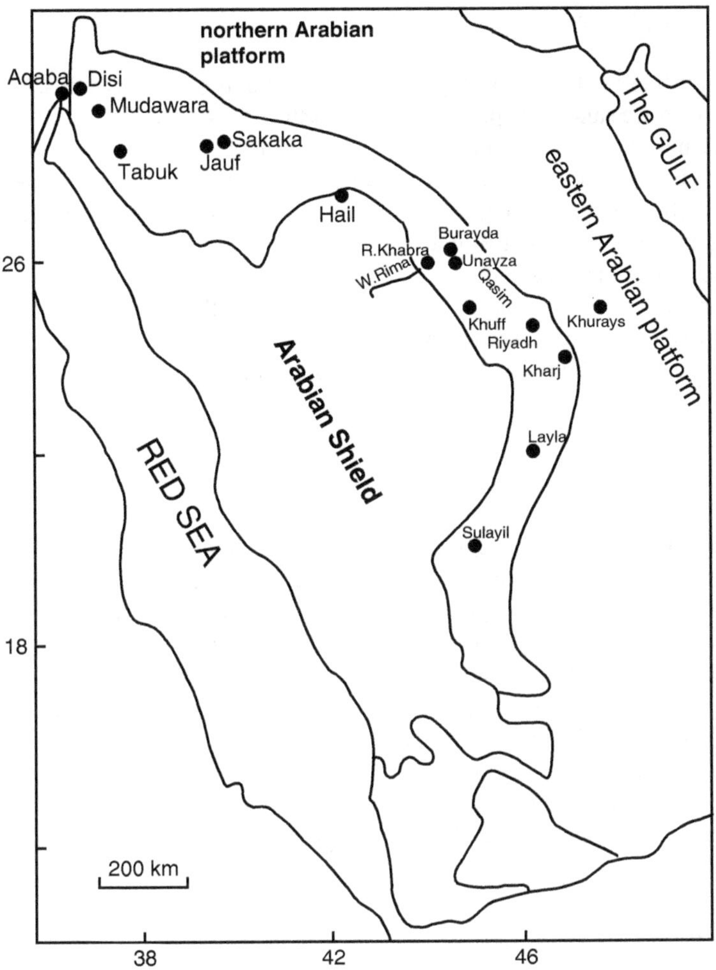

Fig. 5.2 Interior Shelf, location map

Quaternary sediments in wide parts of the canyon and by deposits of the Paleogene Hadramaut group in the plateau areas north and south of the canyon.

5.1.2 *Climate and Landscape*

5.1.2.1 Climate

The climate of the Interior Shelf is arid to hyper-arid. Mean annual rainfall ranges from 150 mm to less than 25 mm (Sect. 1.4.3.1). The catchment areas of some major wadi systems extend far into areas with mean precipitation of up to 250 mm/a within the Arabian Shield.

5.1.2.2 Morphology

The Interior Shelf extends, to a high percentage, over desert landscapes. Main types of landscapes are sandstone plateaus in the north and south of the Interior Shelf and elongated mountain belts with a series of prominent escarpments in the Tuwayq segments. Elongated sand dune belts extend between the mountain areas of the Interior Shelf. In the north and south, the mountain ridges disappear under the sand seas of the Great Nefud and Rub al Khali deserts.

The desertic plateaus and mountain ridges of the Interior Shelf are interrupted, in several areas, by zones of oases, such as Tabuk, Disi, the Qasim area in the north, Al Kharj in central Saudi Arabia, Al Aflaj in the south. Most of the oasis areas are, at present, irrigated by groundwater extraction from boreholes.

5.1.2.3 Tabuk–Disi and Wajid Segments

The Tabuk–Disi and Wajid segments in the north and south of the Interior Shelf, respectively, comprise extensive sandstone plateaus, which have been eroded to plains with local hills and ridges. The outcrops of Paleozoic sandstones form vast featureless plains, but also, in some areas like Wadi Rum in southern Jordan, bizarre landscapes of hills and inselbergs with steep erosional escarpments and gorges. The land surface of the sandstone plateaus in the Tabuk–Disi segment rises to peaks of 1,754 m asl at Jebel Rum northeast of Aqaba and 2,403 m at Jabal al Lawz northwest of Tabuk.

Surface drainage of ephemeral streamflow in wadis in the Tabuk–Disi segment is directed toward major closed basins: Mudawara, Tabuk, Wadi Sirhan.

5.1.2.4 Tuwayq Segments

The morphology of the northern and southern Tuwayq segments of the Interior Shelf is characterized by mountain ridges bordered by escarpments, which extend in a generally north–south directed up to 200 km wide belt between the Arabian Shield in the west and the Ad Dahna sand fields and the plateaus of the Arabian Platform in the east.

The escarpments correspond to the erosional edges of Permian–Mesozoic sedimentary beds, which dip gently toward east, and comprise, in a total length of 1,400 km, a series of five prominent and various secondary morphological steps: The most prominent morphologic step, the Tuwayq escarpment west of Riyadh, is formed by Upper Jurassic limestones and extends along 800 km at an average elevation of 840 m asl, 240 m above the adjoining plains in the west. Further east, the 250 km long Aruma escarpment, capped by Upper Cretaceous limestones, marks the boundary between the Interior Shelf and the Arabian platform. The summit of the Aruma escarpment is situated at 540 m asl, 120 m above the plain.

Several lower and shorter escarpments run through the western part of the Tuwayq mountain zone roughly parallel to the main Tuwayq escarpment. In general, resistant limestones form the summits of escarpments, while sandstones extend over plains between the mountain ridges. Sinkholes and solution cavities are found at the base of scarps, where anhydrite and gypsum have been dissolved. Elongated depressions between the mountain ridges comprise sand dune areas of limited extent – nefuds – but also belts of small oases along wadi courses following the morphologic depressions.

Three major wadi systems, originating in the crystalline areas of the Arabian Shield, dissect the Tuwayq mountains in a general west–east direction: Wadi Rima (or Ruma), Wadi Nisa – Wadi Sahba in the fracture zone of the central Arabian arch, and Wadi ad Dawasir. Development of the wadi systems started mainly in the Pleistocene. Along part of the breakthrough of the wadis through the Tuwayq mountains, deep gorges have developed.

5.1.3 Stratigraphic Sequence

The Interior Shelf comprises a sequence of Cambrian to Cretaceous sedimentary formations with a total thickness of up to 4,000 m. The deposits of the Cambrian to Carboniferous of the Interior Shelf are prevailingly sandstones with shale intercalations. After a regional unconformity in the late Carboniferous – early Permian, the nature of the sediments changed to limestone, dolomite and evaporites. The Triassic and Jurassic sediments comprise alternating series of shales, limestones and sandstones, and, at the late Jurassic, evaporitic deposits; during the early Cretaceous, clastic sedimentation prevailed on the Interior Shelf; Lower Cretaceous sandstones and shales overlie unconformably Paleozoic to Jurassic formations.

During the Upper Cretaceous, shelf carbonates were deposited over wide areas of the eastern Arabian platform. The upper limit of extent of the Upper Cretaceous shelf carbonates is here defined as the boundary between Interior Shelf and Arabian Platform.

The Cambrian to Upper Cretaceous sandstone complexes extend over the Tabuk and Wajid segments in the north and south of the shelf and form outcrops along some stretches of the Tuwayq mountains. Permian, Triassic, Jurassic and Lower Cretaceous carbonate–shale formation form major chains of the Tuwayq mountains.

5.1.3.1 Paleozoic Sandstone Formations

The stratigraphic succession of sedimentary rocks of the Interior Shelf starts with a Cambro–Ordovician sandstone complex, which was deposited mainly under fluviatile conditions. Shallow marine sediments form the top section of the Ordovician sandstones. The Cambro–Ordovician sandstones, denominated Disi formation or Rum group in Jordan, Saq formation in northern Saudi Arabia and Wajid sandstone in southern Saudi Arabia and Yemen, extend over the Tabuk–Disi segment in the north of the Interior Shelf and the Wajid segment in the south. The complex is composed mainly of medium to coarse grained sandstones with intercalations of thin beds of shale. The formation overlies the Precambrian crystalline rocks, the base of the sandstone complex is generally formed by a conglomerate. The thickness of the sandstone complex ranges from 200 m in the west to approximately 600 m in the Sada area in Yemen and to more than 2,000 m under the sedimentary cover of the eastern Arabian platform.

The Saq–Disi sandstone complex is followed, in northern Saudi Arabia and Jordan, by an Ordovician–Devonian sequence of alternating shales and sandstones: the Tabuq formation in Saudi Arabia and Khreim formation in Jordan, which reach a thickness of up to 1,700 m.

The Cambrian–Devonian Saq and Tabuk formations extend in outcrops at the western rim of the northern Tuwayq mountains, but are truncated on an erosional unconformity in central Saudi Arabia, where the Permian Khuff formation directly overlies the Precambrian basement (Khuff unconformity).

5.1.3.2 Devonian–Lower Cretaceous of the Tabuk–Disi Segment and the Hail–Rutba Arch

In the Jauf area on the margin of the Widyan segment of the eastern Arabian platform ("Widyan basin margin" Edgell 1997: 7), the Silurian–Lower Devonian Tawil member of the Tabuk formation is overlain by the Devonian Jauf formation, which is composed of limestones, dolomites, shales and sandstones with a total thickness of 285 m.

Lower Cretaceous sandstones with intercalations of siltstones and shale overlie the Paleozoic formations in the Sakaka and Jauf areas in northern Saudi Arabia and in southern Jordan (Sakaka formation in Saudi Arabia, Kurnub sandstone in Jordan). The thickness of the Sakaka and Kurnub formations ranges from 90 to around 200 m. In the surroundings of the Jawf area, the Upper Cretaceous Aruma formation and the Paleogene Hibr formation overlie, over erosional surfaces, the Sakaka formation or directly the Jawf formation. Outcrops of the Jawf and Sakaka formations on the Widyan basin margin are situated in a morphologic depression, which is adjoined, in the north, by a series of steep escarpments and, in the south, by the Great Nefud sand desert.

A more than 150 m thick complex of clastic rocks of Permo–Carboniferous age crops out in the centre of the morphologic Gaara depression in southwestern Iraq: the Gaara formation, composed of clay and quartz sandstone with 16 m saliferous marl and shale on the base. The Gaara formation is overlain in the Rutba–Gaara area by:

- The Upper Triassic Mulussa formation, 160 m dolomite, marly dolomite, sandstone, limestone
- A 180 m thick Jurassic sequence of dolomite, limestone, sandstone, argillite
- The Lower Cretaceous–Cenomanian Rutba sandstone
- Upper Cenomanian limestones with sand intercalations (Msad formation)
- Maastrichtian limestones and dolomites with marl intercalations

5.1.3.3 Permian–Lower Cretaceous of the Tuwayq Segments

On the northern and southern Tuwayq segments, Permian, Triassic, Jurassic and Cretaceous formations are exposed in relatively narrow but very long strips. In the sedimentary sequence of Permian–Mesozoic rocks of the Tuwayq mountains, marine shelf carbonates alternate with fluviatile, coastal and lagoonal deposits (Table 5.1). Limestones and dolomites of the Permian Khuff formation rest directly on the Precambrian crystalline rocks. The Permian and Triassic clastic rocks consists of nearly 1,000 m of alternating non-marine and marine clastic units with some thick calcareous sections. The sequence begins, at the base, with the Khuff formation: 250–600 m of limestones and dolomites, alternating with gypsiferous shale. The 280–650 m thick Lower Triassic Sudair formation is made up mainly of continental shale beds intercalated with siltstone, limestone and gypsum and, in the north, lenses

Table 5.1 Sedimentary sequence of the Tuwayq mountain zone. After Edgell (1997), MAW (1984), Powers et al. (1966)

Stratigraphic age	Formation	Prevailing lithology	Average thickness (m)
Lower Cretaceous	Wasia	Sandstone, shale	18–100
	Biyadh	Sandstone	425
	Buwaib	Calcarenitic limestone	18
	Yamama		50
	Sulay	Limestone	170
Jurassic	Hith	Anhydrite	90
	Arab	Argillaceous limestone, anhydrite	124
	Jubaila	Limestone, calcarenite	118
	Hanifa		113
	Tuwayq Mountains		200
	Dhruma	Sandstone with limestone interbeds	375
	Marat	Limestone, shale, sandstone	100
Triassic	Minjur	Sandstone with shale and mudstone	315
	Jilh	Limestone, shale, gypsum	100–400
	Sudair	Shale	115
Permian	Khuff	Limestones and dolomites	250–600
Cambrian–Ordovician	Saq/Wasia	Sandstone	Generally 400–500, eroded over central Arabian arch

of limestone. The middle–upper Triassic is represented by sandstones with shale and limestone intercalations (Minjur formation, 400 m). Toward south, limestones become dominant in the middle Triassic (Jilh formation). During the late Triassic, continental conditions prevailed over the Interior Shelf. In the central part of the Arabian Peninsula, upper Triassic and lower Jurassic sediments are absent.

After a wide-spread sedimentation gap from the late Triassic to early Jurassic, the sedimentary succession of the lower Jurassic begins with shales and shallow marine limestones (Marrat formation).

Sediments of Bathonian–Bajocian age, consisting of shallow water limestones and shale of some 500 m thickness, crop out along the Tuwayq mountains. Toward south, the sediments change into continental sandstones.

The middle–upper Jurassic sediments comprise mainly carbonates: limestones deposited on the continental shelf with high percentage of clastic materials, alternations of neritic limestone and dolomite with anhydrite horizons (Dhruma, Tuwayq, Hanifa, Jubaila, Arab formations). The late Jurassic is represented prevailingly by anhydrites with minor intercalations of limestones and dolomites (Hith formation). The sedimentary rocks of the late Jurassic extend in the Tuwayq mountains over a strip of about 1,000 km length.

Between the late Jurassic and early Cretaceous, most of the western part of the Arabian Peninsula was elevated; extensive erosion followed the orogenic movement, which started during the late Jurassic.

The Lower Cretaceous comprises detrital limestones, which are overlain by continental clastics of the Biyadh formation: sandstones with shale and conglomerate, and sandstones with carbonate intercalations. The clastic deposits comprising sandstones, conglomerates, quartzite, shale and siltstone indicate prevailing fluviatile and terrestrial sedimentation. The carbonates, mainly dolomites and oolitic limestones, appear related to shallow marine and coastal conditions (Table 5.1).

5.1.3.4 Jurassic–Cretaceous of the Hadramaut Arch

On the Hadramaut arch in southeastern Yemen, the Cretaceous Mukalla sandstones have been uplifted to near surface zones and are exposed in some deeply eroded wadis. The Mukalla formation includes a thick complex of fine to medium grained sandstones, which is overlain by shale, limestone, sandstone layers and underlain by a calcarenite, sandstone, limestone sequence. The formation may be correlated to the Cretaceous Tawila sandstones of the shield area in northwestern Yemen. The thickness of the Mukalla formation averages 300–400 m and probably reaches 1,000 m in the Shabwa area. The formation rests, in the west, on up to 3,000 m thick marly and calcareous Jurassic deposits. Further east, the formation overlies probably directly the Precambrian basement.

5.1.3.5 Upper Cretaceous–Quaternary

After an erosional phase, which continued until the middle part of the Cretaceous, a marine transgression spread over wide parts of the Arabian Peninsula during the late Campanian and Maastrichtian. The transgression left an extensive cover of carbonates on the eastern Arabian platform (Aruma formation) east of the Interior Shelf. On the Hail–Rutba arch northeast of the Interior Shelf, the cover of Upper Cretaceous marine deposits of the Widyan basin margin encloses outcrops of Paleozoic–Lower Cretaceous sediments in erosional/structural windows.

Accumulations of Pleistocene to Recent terrestrial deposits are found on the Interior Shelf in wadis and morphologic depressions. The wadis, which dissect the mountainous landscapes of the Interior Shelf, contain gravel and silt deposits with lenses of sabkha sediments.

References. Al-Sayari and Zötl (1978), Alsharhan et al. (2001), Beydoun (1991), Edgell (1997), Faroog and Rogers (1980), Helal (1965).

5.2 Main Aquifers

The Interior Shelf comprises very extensive sandstone aquifers with considerable thickness, the most important of which are the Cambro–Ordovician Saq–Disi and Wajid aquifers. Detrital deposits also form productive aquifers in the:

- Ordovician–Devonian Tabuk and Jauf formations
- Triassic Jilh and Minjur formations
- Lower Cretaceous Kurnub, Sakaka, Biyadh and Wasia formations
- Cretaceous Mukalla formation

In the Permian to early Cretaceous succession of shales, carbonates, sandstones and evaporites of the Tuwayq segments, aquicludes and aquitards alternate with aquiferous sections with minor or local importance. An extensive carbonate aquifer is found in the Permian Khuff formation. Pleistocene–Quaternary detrital deposits constitute shallow aquifers along major wadi courses and in morphologic depressions.

5.2.1 Cambro–Ordovician Sandstone Aquifers

5.2.1.1 Saq–Disi Aquifer

The Saq–Disi aquifer is composed mainly of medium to coarse grained sandstones with thin shale intercalations. The base of the aquifer is formed by the crystalline basement, on top the aquifer is followed by the Ordovician–Devonian Tabuk–Khreim formation, in the western highlands of Jordan directly by the Lower Cretaceous Kurnub formation. The thickness of the Saq–Disi aquifer increases from several hundred metres in the outcrop areas to more than 2,000 m toward east and northeast. In the southern Tuwayq segment, the Saq aquifer disappears in the Khuff area, where the formation is truncated by the Permian erosional unconformity.

The outcrop of the Saq–Disi formation extends over around 70,000 km^2 in an about 1,300 km long belt between southern Jordan and the Khuff area in central Saudi Arabia. The total extent of the Saq–Disi aquifer in the outcrop areas and in the subsurface is estimated at around 350,000 km^2. In the subsurface, the aquifer reaches far into the northwestern mountain and rift zone, the east Jordanian limestone plateau and the eastern Arabian platform.

The Saq–Disi aquifer is unconfined at and near the outcrop areas in the western part of the Tabuk–Disi segment (around Tabuk and on the eastern flank of the highlands of Jordan) and in the Tuwayq mountains. Depth to groundwater in the unconfined ranges of the aquifer is around 80–150 m.

In some areas near the boundary with the Arabian Shield, the Saq–Disi sandstones are unsaturated or contain local discontinuous groundwater lenses. Around Wadi Rum in southern Jordan, local groundwater occurrences in sandstone hills discharge above the impervious granite basement in numerous small seasonal springs and a few perennial springs.

In wide parts of the Tabuk–Disi segment, the Saq–Disi aquifer is confined under shale aquitards of the Tabuk–Khreim formation: the Hanadir shale in Saudi Arabia and Hiswa shale in Jordan. The confined Saq–Disi aquifer extends from the Saq–Disi segment of the Interior Shelf toward north and northeast under the northern Arabian platform. The top of the aquifer is situated at 400 m below land surface at Mudawara and descends to more than 1,000 m in the Jordanian limestone plateau and more than 2,000 m below land surface in Wadi Sirhan.

The confined Saq–Disi aquifer is under artesian conditions in some areas with flowing wells, e.g., in the Tabuk area and in Jordan south of the Karawi dike.

In the western highlands of Jordan east of the Dead Sea graben, the Khreim aquitard is missing and the Disi sandstones form a joint aquifer system with the overlying Kurnub sandstone.

5.2.1.2 Wajid Sandstone

In the south of the Interior Shelf, the Wajid sandstone extends over around 200,000 km^2 in southwestern Saudi Arabia and northern Yemen. It can be assumed, that the Cambro–Ordovician sandstone formations in the Tabuk–Disi and Wajid segments were deposited in a continuous sedimentary basin and were separated by uplift and erosion on the eastern margin of the Arabian Shield during the Carboniferous–Permian.

The Wajid formation is composed mainly of poorly cemented sandstones with a thickness of 200–900 m. In the Sada basin in Yemen, the Wajid sandstones occur in a downfaulted graben structure under a 30–60 m thick cover of Quaternary deposits.

These unconsolidated deposits are largely unsaturated, but they are effective in tapping surface water flows from the surrounding mountains and contribute thus to the recharge of the Wajid sandstones. The Wajid sandstones generally have low hydraulic conductivities, but their thickness under the Sada plain reaches 300–600 m. Continuous moderate transmissivities and large permissible drawdowns make the sandstone aquifer easily exploitable, but present abstraction rates exceed the average recharge by one order of magnitude.

5.2.1.3 Tabuk–Khreim Formation

The Ordovician–Devonian Tabuk–Khreim formation consists of alternating sequences of shales and sandstones. The Tabuk formation in Saudi Arabia comprises marine shales and continental to littoral sandstones in a total thickness of up to 1,700 m. The formation includes three main aquiferous sections separated by shale aquicludes or aquitards (Table 5.2).

The upper sandstone members are present mainly in the eastern part of the Tabuk segment. The sediments of the Khreim group in Jordan generally act as aquitard. Locally, however, groundwater can be exploited from aquiferous sandy layers, e.g., in the Mudawara area.

The Tabuk–Khreim formation extends over around 80,000 km^2 in Saudi Arabia and southern Jordan. The thickness of the Tabuk formation decreases in the Tuwayq mountain zone, where it is truncated by the Permian discontinuity south of Qasim. The equivalent Khreim formation in Jordan disappears toward west on the eastern slope of the Jordanian highlands. Apart from the outcrop areas in the west, the Tabuk sandstone aquifers are generally confined under shale members of the Tabuk formation.

5.2.1.4 Jauf Formation

In the Jauf area in northern Saudi Arabia, the Tabuk formation is overlain by the Devonian Jauf formation, an 85 m thick sequence of sandstones, limestones, dolomites and shales. The sandstone members of the formation provide an aquifer, the extent of which is mainly restricted to the western slopes of the paleogeographic

Table 5.2 Hydrostratigraphic scheme of Paleozoic–Lower Cretaceous formations in the Tabuk–Disi segment of the Interior Shelf. After Hobler et al. (1991), MAW (1984)

	Jordan			Saudi Arabia		
Stratigraphic age	Formation	Prevailing lithology		Formation	Prevailing lithology	
Lower Cretaceous	Kurnub	Sandstone, shale	Aquifer	Sakaka	Sandstone	Aquifer
Ordovician–Silurian	Khreim	Siltstone, shale, sandstone	Aquitard	Tabuk	Sandstone and shales	Alternating aquifers and aquitards
Cambro–Ordovician	Disi (Ram)	Sandstone	**Major aquifer**	Saq	Sandstone	**Major aquifer**

Hail arch. The main outcrops of the Jauf sandstones occur in the tectonic–morphologic Jauf depression, which is adjoined by steep 200 m high escarpments. In the depression, the Jauf sandstones form a shallow aquifer with water levels of less than 25 m below surface. Toward east, the Jauf formation dips to great depth below Carboniferous to Mesozoic formations; toward south, the sandstone outcrops disappear under the sand seas of the Great Nefud desert.

The Paleozoic Gaara sandstone is an important aquifer in the Rutba–Gaara area. The prevailingly calcareous Mulussa formation (Upper Triassic) forms an aquifer of local importance in the Rutba area.

5.2.2 *Permian–Jurassic Aquifers of the Tuwayq Segments*

5.2.2.1 Khuff Formation

The Upper Permian Khuff formation, consisting of limestones and dolomites with anhydrite intercalations, extends in the subsurface from the southern Rub al Khali over the Tuwayq segments into southern Iraq. The carbonate units of the formation act as generally brackish water aquifers with low productivity. Anhydrite aquicludes separate the formation into several aquiferous units (Table 5.3).

5.2.2.2 Triassic–Jurassic Formations

In part of the Tuwayq mountains, sandstones of the Triassic–Jurassic Jilh, Minjur, Marat and Dhruma formations form an aquifer complex with the Minjur sandstones

Table 5.3 Hydrostratigraphic scheme of Paleozoic–Cretaceous formations in the Tabuk–Sakaka area. After Alsharhan et al. (2001), MAW (1984)

Stratigraphic age	Formation	Prevailing lithology	Thickness (m)	
Cretaceous	Sakaka sandstone	Sandstone	42	Aquifer in Sakaka area
Devonian	Jauf formation	Limestone, shale, sandstone	300	Aquifer in Al Jauf area, local aquitard
	Tawil sandstone	Sandstone	190	Upper Tabuk aquifer
Silurian	Sharawa sandstone	Sandstone	380–765	
	Qusaiba shale	Shale	50–200	Aquitard
	middle Tabuk sandstone	Sandstone	83–242	Middle Tabuk aquifer
Ordovician	Raan shale	Shale	14–100	Aquitard
	lower Tabuk sandstone	Sandstone	54–390	Lower Tabuk aquifer
	Hanadir shale	Shale	54–100	Aquitard
Cambrian	Saq	Sandstone		Deep brackish water aquifer

as main productive aquifer unit. The extent of the Minjur sandstones is limited mainly to central Saudi Arabia. In the Riyadh area, the Minjur formation consists of a 400 m thick sequence of coarse grained quartz sandstones with shale and sandstone intercalations. The thickness of the formation decreases toward north and south. The Minjur aquifer complex includes, in some areas, sandstone layers of underlying or overlying formations, which are developed, in different parts of the Tuwayq mountains, in carbonate–shale or in detrital facies.

East of Riyadh and in the Qasim area, sandstones of the Triassic Jilh formation form the lower member of the Triassic–Jurassic aquifer complex. The general base of the aquifer complex is formed by the Triassic Sudair shale aquiclude underlying the Jilh formation. Sandstone layers at the base of the Lower Jurassic Marat formation constitute an upper member of the Minjur aquifer. The top of the Minjur aquifer complex is generally formed by shales, siltstones and aphanitic limestones of the Marat formation.

In the Qasim area, the Jilh aquifer unit consists of 180–250 m thick fine to medium grained sandstones.

The Upper Jurassic sequence of limestones, argillaceous limestones and anhydrites (Tuwayq mountain limestone, Hanifa, Jubaila and Arab formations, Hith anhydrite) contain, in the Tuwayq segments, a succession of aquiferous carbonate sections with intercalated aquitards or aquicludes. The productivity of the aquiferous carbonate is generally low, the groundwater is prevailingly brackish to saline. Limited amounts of fresh to brackish water are extracted from wells in Wadi Hanifa.

5.2.2.3 Lower Cretaceous Carbonate Formations

The Lower Cretaceous carbonate formations in central Saudi Arabia (Sulaiy–Yamama–Buwaib formations) provide generally no aquifers with economic importance, with a few local exceptions:

- Calcarenites of the Sulaiy formation form a local aquifer with brackish groundwater (1,000–3,000 mg/l TDS) in Wadi Sulaiy, 20 km southeast of Riyadh.
- In the area of the Layla lakes (Uyun al Aflaj), dissolution of anhydrites of the Hith formation has caused collapse structures in the overlying Lower Cretaceous carbonates. Groundwater discharge from the karstic carbonate formations is collected in an area with a number of small lakes, extending previously over 4 km^2.

5.2.3 Cretaceous Sandstone Aquifers

The lower Cretaceous of the Interior Shelf comprises sandstone aquifers of considerable thickness:

- Kurnub aquifer in Jordan
- Sakaka aquifer on the Widyan basin margin
- Biyadh–Wasia aquifer in the Tuwayq segments
- Mukalla sandstone on the Hadramaut arch

5.2.3.1 Kurnub Sandstone

The Kurnub formation is composed mainly of fine to medium grained sandstones with siltstone and clay intercalations. The sandstones are exposed in the southern desert of Jordan on an escarpment, which may be considered the eastern boundary of the Interior Shelf. Toward north and east, the Kurnub sandstones dip under the Upper Cretaceous–Paleogene cover of the northern Arabian platform and extend, in the subsurface, far into the platform area of the shelf. In the platform area, outcrops of the Kurnub formation occur near Amman in eroded anticlinal and domal structures and in deeply incised wadis on the Dead Sea–Jordan escarpment.

The thickness of the Kurnub sandstones generally ranges from 100 to 300 m. The top of the formation descends from the outcrop areas to depths of a few hundred metres below land surface in the platform areas and to more than 1,000 m below land surface in the Azraq area.

The Kurnub sandstones are generally unsaturated in the outcrop areas. In most of its extent, the Kurnub aquifer is confined under aquicludes of overlying younger formations, in particular shales of the Upper Cretaceous Ajloun formation.

The saturated thickness of the confined Kurnub is generally in the order of 150–200 m. The thickness of the formation decreases toward east and southeast and the sandstones disappear on the southern edge of the Sirhan depression.

The Rutba sandstone, together with sandstone intercalations in the overlying Msad formation, have been tapped west and northwest of Rutba at depths of 230–290 m. The groundwater appears to be generally unconfined to semi-confined. Specific well yields range between 0.4 and 9 m^3/h/m.

5.2.3.2 Sakaka Sandstone

In the area between the Turayf and Tuwayq segments of the Interior Shelf, the homoclinal structure of the shelf is modified by the uplift of the Hail–Rutba arch, which separates the northern Arabian platform from the eastern platform. Outcropping Paleozoic–Cretaceous sandstones form aquifers on the uplift structure in the Rutba–Gaara and Jawf areas.

The Cretaceous Sakaka sandstone formation constitutes an aquifer in the Jawf and Sakaka areas in northwestern Saudi Arabia. The thickness of the Sakaka aquifer, which overlies the Devonian Jauf formation, is around 200 m.

5.2.3.3 Biyadh and Wasia Formations

Sandstones of the Lower Cretaceous Biyadh and Wasia formations constitute an extensive aquifer complex in the Tuwayq mountains and, in the subsurface, over wide parts of the eastern Arabian platform. The Biyadh formation consists of quartz sandstones with shale and conglomeratic layers. The Wasia formation comprises mainly sandstones with shale intercalations and some limestone layers. The thickness of the Wasia–Biyadh aquifer reaches up to 600 m. In eastern and northern Saudi Arabia, the Biyadh and Wasia sandstones form one hydraulically connected aquifer system. Further downdip on the eastern Arabian platform, hydraulic connection between the two formations is restricted by shale aquitards in the top sections of the Biyadh formation.

References. Al-Sagabi (1978), Alsharhan et al. (2001), Bassam (1998), Bazuhair (1989), Edgell (1997), Hobler et al. (1991), Lloyd and Pim (1990), Margane et al. (2002), MAW (1984), Sharaf and Husein (1996), Sowayan and Allayla (1989), Sunna (1995), WAJ-ODA (1994).

5.3 Groundwater Regimes

5.3.1 Hydraulic Parameters

Permeabilities of the sandstone aquifers of the Interior Shelf are generally moderate. The high potential of some of the sandstone aquifers, such as the Saq–Disi aquifer and the Biyadh–Wasia aquifer, is related to the great saturated thickness and their huge extent. Hydraulic conductivities are rather moderate, reported values range from 3×10^{-6} to 10^{-4} m/s, but the aquifer thicknesses of 400 to >3,000 m in the Saq–Disi aquifer and of 200–900 m of the Wajid aquifer support a high productivity. Transmissivities of the Cambro–Silurian sandstone aquifers are reported as 80–3,000 m^2/day for the Saq aquifer, 300–1,000 m^2/day, with an average of 700 m^2/day, for the Disi aquifer, and 50–1,800 m^2/day for the Wajid aquifer.

In the Paleozoic Tabuk sandstone aquifer complex, hydraulic conductivities range from 2×10^{-6} to 5×10^{-5} m/s, the total thickness of aquiferous sections reaches 400–900 m. Transmissivities for each of the three aquifer members of the Tabuk formation are in the order of 18–2,600 m^2/day. Similar transmissivity values are found in the Lower Cretaceous Wasia aquifer; a mean transmissivity of 664 m^2/day is estimated for the Biyadh sandstone aquifer. Transmissivities of the Lower Cretaceous Kurnub sandstone aquifer are rather low with 20–80 m^2/day. Mean hydraulic conductivity of the Kurnub sandstones is in the order of 10^{-5} m/s, of the overlying Khreim aquitard in ranges of 10^{-7}–10^{-9} m/s.

Transmissivities in carbonate formations of the Interior Shelf are generally low. High transmissivities in the order of 2,800 m^2/day are found in the Upper Jurassic Arab formation in areas, where solution openings and collapse structures in evaporite layers create an elevated hydraulic conductivity.

The productivity of the Jawf aquifer is generally low to moderate. The Sakaka sandstones, overlying the Jawf aquifer in parts of the Al Jawf area and of the platform area further east, constitute an aquifer with relatively high productivity and storativity.

5.3.2 Groundwater Recharge

Present-day groundwater recharge rates in the arid Interior Shelf, with mean annual precipitation of 100 mm, are very low. For the extensive outcrop area of the Saq aquifer, a recharge volume of 310×10^6 m^3/a has been estimated. The Biyadh–Wasia aquifer in central Saudi Arabia may be replenished by 480×10^6 m^3/a from infiltration of runoff from the Tuwayq mountains and subsurface inflow in wadis (data of 1967). For the central Tuwayq area, a recharge volume of 7–8 $\times 10^6$ m^3/a has been estimated, corresponding to an infiltration of 5% of the precipitation over a recharge area of 1,500–1,800 km^2.

For the Riyadh and Al Kharj areas in the central Tuwayq zone, recharge rates were calculated from meteorologic data. Mean annual rainfall is between 40 and 60 mm with main rainfall events from January to April. Recharge occurs mainly during January–February with minor recharge contributions during November–December. Recharge in March–April is insignificant because of high evaporation rates. The calculated recharge rates are 12.7 mm/a for the Saq aquifer at Riyadh (data for 1964–1986) and 3.8 mm for the Wasia–Biyadh aquifer at Al Kharj (data for 1970–1983), or 22% and 9% of mean annual rainfall, respectively. Other estimates of recharge for the same area range from 3 to 6.5 mm/a with a weighted average of calculated recharge of 4.0 mm/a for the central Tuwayq area and Al Kharj, or around 10% of annual precipitation.

A dynamic model of the annual water balance for local climatic and soil parameters was applied for estimates of recharge rates for the Minjur and Wasia sandstones aquifers in the Riyadh area in central Saudi Arabia. The estimates indicate a probable median annual recharge of 15 mm for the Minjur outcrop area (11.5% of the mean annual precipitation), and of 3 mm for the Wasia outcrop area (3.8% of the mean annual precipitation). The Minjur aquifer is exposed in an area west – upstream – of the Tuwayq mountains with mean annual rainfall of 130 mm, the Wasia outcrop area, east of the Tuwayq mountains, has a mean annual rainfall of 80 mm. The calculated recharge volumes are 31.4×10^6 m^3/a for the outcrop area of 2,092 km^2 of the Minjur aquifer and 2.4×10^6 m^3/a for the outcrop area of 807 km^2 of the Wasia aquifer.

Table 5.4 Estimated recharge rates on the Interior Shelf in central Saudi Arabia

Area	Aquifer	Estimated recharge rate (mm/a)	References
Central Tuwayq and Kharj	Wasia–Biyadh aquifer	4.0	Subyani and Sen (1991)
Riyadh area		12.7	
Wadi as Sahba	Wasia–Biyadh aquifer	3–5	Sogreah (1968)
Central Saudi Arabia	Wasia–Biyadh aquifer	5.2 6.5	SMMP (1975) BRGM (1976)
Central Saudi Arabia	Minjur aquifer Wasia	15 3	Caro and Eagleson (1981)
Khurais area	Sand dunes	20.0	Dincer et al. (1974)

Recharge rates in the sandstone and sand dune areas in north central Saudi Arabia may be around 1–1.5 mm/a; recharge rates in the southern parts of the Interior Shelf are probably even lower. For sand dune areas with a mean rainfall of 75 mm/a, recharge rates of 1.5% of rainfall during average and wet years have been calculated or approximately 1,500 m^3/km^2/a.

Reported estimates of recharge rates are summarized in Table 5.4.

5.3.3 Groundwater Flow Systems and Flow Volumes

In the sandstone aquifers of the Interior Shelf, groundwater moves generally from the outcrop areas in direction of the homoclinal dip of the strata toward east–northeast.

The Cambrian–Triassic aquifers of the Tuwayq mountain zone have been classified in Saudi Arabia as rejecting aquifer systems: Low hydraulic conductivity in the overlying aquicludes prevents significant leakage into the aquifer systems of the eastern Arabian platform downstream of the Tuwayq zone. Groundwater from the rejecting aquifers discharges in springs and into shallow aquifers within the Tuwayq zone. In the platform area, the rejecting aquifers constitute deep aquifers with prevailingly stagnant brackish or saline groundwater.

The Wasia–Biyadh aquifer is considered a depleting aquifer system: Groundwater moves in a generally eastward direction from the Tuwayq zone into the eastern Arabian platform, where the Wasia–Biyadh formations constitute a deep aquifer. Groundwater leakage from this deep sandstone aquifer into the overlying aquifer system of the platform occurs mainly in zones of tectonic fracturing or dislocations.

In the Tabuk–Disi segment of the Interior Shelf, groundwater flow in the Saq–Disi aquifer is, in general, directed from the outcrop areas toward north–northeast into the adjoining limestone plateau of the northern Arabian platform. The Lower Cretaceous Kurnub aquifer contains a distinct groundwater flow system in the east Jordanian limestone plateau, separated from the underlying

Disi aquifer by the Khreim aquiclude. In the western highlands of Jordan, the Disi and Kurnub aquifers become hydraulically connected with the disappearance of the Khreim formation and constitute a single deeper sandstone aquifer complex. Groundwater from the Disi–Kurnub aquifer discharges in springs and leakages along the Dead Sea graben and in wadis on the western escarpment of the highlands of Jordan.

In the Jafr area, the northward groundwater flow in the deeper sandstone aquifer turns westward to the Dead Sea graben.

The groundwater flow system in the Saq–Disi aquifer extends in the Tabuk–Disi segment of the Interior Shelf and the adjoining areas of the northern Arabian platform over 74,000 km^2 from the Tabuk area in the south to the northern tip of the Dead sea.

Locally, the groundwater movement is modified in the Paleozoic–Lower Cretaceous sandstone aquifer by structural features: A structural high in the area between Disi and Quweira in southern Jordan separates the Wadi Yutm groundwater basin with southeastward flow toward the Red Sea coast from the main Saq–Disi groundwater basin with general northward groundwater flow. The northwest–southeast trending Kharawi dyke in southern Jordan acts as a hydraulic barrier causing local deviations of groundwater flow. Artesian groundwater flow is found in an area around Mudawara south of the dyke.

The following estimates of groundwater flow through aquifers of the Interior Shelf, sustained by present-day recharge, have been made:

Saq aquifer, total flow in Saudi Arabia	290×10^6 m^3/a
Disi aquifer, inflow from outcrops on the Interior Shelf into limestone area in Jordan	50×10^6 m^3/a
Wajid aquifer	500×10^6 m^3/a
Lower, middle and upper Tabuk aquifers	140×10^6 m^3/a
Jauf aquifer	15×10^6 m^3/a
Khuff aquifer	80×10^6 m^3/a
Jilh aquifer	60×10^6 m^3/a
Minjur aquifer in northern Saudi Arabia	100×10^6 m^3/a
Wasia–Biyadh aquifer	480×10^6 m^3/a
Mukalla aquifer	500×10^6 m^3/a

The estimates indicate rather high figures of groundwater flow volumes and of present-day recharge. Considering the vast extent of the various aquifer complexes, rates of groundwater flow and recent recharge appear, however, small corresponding to only a very minor part of the large volumes of the prevailingly brackish groundwater stored in the aquifers. Any larger groundwater exploitation therefore leads to mining of the fresh water resources.

Groundwater velocities in the sandstone aquifer complex have been estimated to be around 2 m/a.

The following estimates of stored volumes of groundwater in different aquifers on the Interior Shelf and their extension into the adjoining eastern Arabian platform have been made:

Saq	280×10^9 m^3	
Wajid	205×10^9 m^3	
Tabuk (three members)	205×10^9 m^3	
Jauf	100×10^9 m^3	
Khuff	30×10^9 m^3	Brackish groundwater
Jilh	115×10^9 m^3	Brackish groundwater
Minjur	460×10^9 m^3	In the Sudair–Riyadh–Aflaj area, fresh to slightly brackish groundwater
Dhruma	180×10^9 m^3	Brackish groundwater

References. Bazuhair (1989), Caro and Eagleson (1981), Edgell (1997), El Naser and Gedeon (1996), FAO (1979), Hobler et al. (1991), Lloyd and Pim (1990), Sharaf and Husein (1996), Subyani and Sen (1991), WAJ-ODA (1994), Wolfart (1961).

5.4 Groundwater Salinity and Hydrochemistry

5.4.1 Sandstone Aquifers

On the Interior Arabian Shelf, two major types of sandstone aquifers with different lithologic–hydrochemical environments can be distinguished:

- The Cambro–Silurian sandstone sequence of prevailing continental origin composed mainly of quartz sandstones of the typical Nubian sandstone facies.
- Paleozoic–Mesozoic sandstones of epicontinental to shallow marine origin comprising interlayering sandstones, limestones and shales.

More or less pure continental sandstones constitute the Saq–Disi aquifer in northern to central Saudi Arabia and southern Jordan and the Wajid aquifer in southern Saudi Arabia and Yemen.

Paleozoic–Mesozoic sandstone aquifers deposited under coastal marine to fluviatile conditions include the Tabuk, Jauf, Kurnub, Biyadh and Minjur aquifers.

5.4.1.1 Saq–Disi Aquifer

The Saq–Disi aquifer consists generally of quartz sandstones with siliceous cements and small amounts of kaolinite. The hydrochemical composition of groundwater in the Disi aquifer in part of its outcrop areas is typical for an aquifer matrix with low hydrochemical reactivity. Near the outcrop area in Jordan, groundwater salinity is generally low with 200–300 mg/l TDS in the unconfined sector, increasing slightly to 300–400 mg/l in the confined sector. The low salinity groundwater is Ca–HCO_3 type water with mean anion concentrations of around 120 mg/l HCO_3, 40 mg/l Cl, 30 mg/l SO_4.

Similar hydrochemical conditions prevail in the Saq aquifer within and near the outcrop area in Saudi Arabia around Tabuk, where salinity is generally less than 400 mg/l and frequently less than 200 mg/l TDS.

The Ca–HCO_3 type groundwater with low salinity is found, in particular, in wide parts of the Tabuk–Disi segment between Tabuk, Hail and southern Jordan. On the Widyan basin margin and the northern Tuwayq segment, Na–Cl type groundwater prevails with moderate to elevated salinities (500–1,000 mg/l TDS, up to 1,500 mg/l TDS in some areas).

Most of the low salinity Ca–HCO_3 type groundwater contained in the aquifer appears to originate from Pleistocene recharge, retention periods being estimated at 10,000–30,000 years (Sect. 5.5.1). Present-day recharge derived from wadi runoff from the basement is characterized by Ca–Na–HCO_3 or Na–Cl type groundwater.

Generally low HCO_3 concentrations between 100 and 200 mg/l indicate a very limited impact of carbonate dissolution. Silicate hydrolysis may be a major source of HCO_3 concentrations.

Brackish Na–Cl type groundwater is found in the Saq–Disi aquifer on the Arabian Platform downstream of the aquifer outcrop and around wadi courses, where Holocene recharge from runoff water is accompanied by elevated salinity.

In undisturbed conditions, the hydrochemical pattern in the aquifer near the outcrop area appears dominated by:

- Fossil Ca–HCO_3 or Na–HCO_3 type groundwater
- Mixing of Na–Cl type groundwater with low to moderate salinity with the fossil groundwater

On the Arabian Platform, where the Saq–Disi sandstones are overlain by a thick sequence of younger deposits, the groundwater salinity increases in direction of groundwater flow and the water is generally Na–Cl type water. In southern Jordan, the Disi aquifer is overlain by marly deposits of the Ordovician–Silurian Khreim group, which contain in aquiferous layers Na–Cl water with salinities of 900–9,000 mg/l TDS.

A gradual salinity increase is observed over the around 250 km long distance between the Disi outcrop area and the discharge zone in the Dead Sea valley: In the unconfined and confined parts of the aquifer south – upstream – of the Kharawi dyke, groundwater salinity ranges from 200 to 300 mg/l TDS. North of the dyke, the groundwater salinity in the Disi aquifer increases from 250 to 400 mg/l TDS in direction of groundwater flow – toward north and then west – to around 700 mg/l in the Maan area, 1,350 mg/l in Wadi Mujib and 1,000–3,000 mg/l TDS in the Dead Sea area.

The Ca–HCO_3 type groundwater changes to Na–Cl type water at a salinity of around 700 mg/l TDS. The change in salinity and of hydrochemical composition is attributed mainly to leakage of brackish water from overlying aquifers or aquitards.

In northern Saudi Arabia, Na–Cl type water has been tapped in the Saq aquifer through deep wells far away from the outcrop of the formation. The main source of

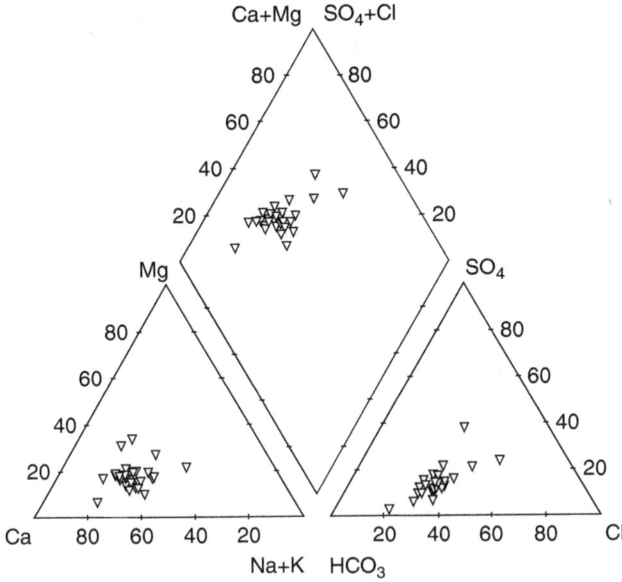

Fig. 5.3 Piper diagram: Groundwater samples from the Disi aquifer, southern Jordan. Data from Hobler et al. (1991)

the salinity increase is assumed to be downward leakage from overlying brackish water aquifers, especially in the Qasim and Sirr areas.

Elevated salinities are found at the outlets of some large wadis entering the sandstone outcrops from the area of the Arabian Shield. Groundwater in wells tapping the Saq aquifer at the downstream ends of Wadi Rima and Wadi Rishe in the Unayza–Burayda area is generally brackish Na–Cl type water with salinities ranging from 1,000 to 8,700 mg/l TDS.

According to Lloyd and Pim (1990) "lower Cl concentrations occur in some parts of the aquifer further down the hydraulic gradient, than under the outcrop area. This is attributed to the diminution of recharge with time. The lower concentrations represent significant past recharge ~10–30,000 years B.P. and the higher concentrations relate to more recent recharge derived from runoff from the basement under predominantly arid conditions; and also salination from localised groundwater evaporation".

Elevated salinity in the Saq aquifer along Wadi Rima appears to be caused primarily by evaporative salt concentration in a sporadic wadi flow regime. Nitrate concentrations are high with an average of around 200 mg/l.

The drainage area of Wadi Rima extends over around 71,000 km^2 and is the largest stream basin in Saudi Arabia, reaching far into the Arabian Shield. Wadi Rima crosses the outcrop of the Saq sandstones along a stretch of about 40 km. At present, wadi runoff reaches the Saq outcrop area during a few days generally once or twice a year. During the pluvial periods of the Pleistocene, the aquifer was probably filled with groundwater of low salinity. With the significant reduction of the volume of wadi flow in the arid Holocene climate, mainly fine grained sediments are deposited along the wadi channel and the flood plain. Surface outflow from the wadi is restricted by sand dune barriers, and water flooding the wadi is, to a considerable extent, evaporated leaving a cover of soluble salts, in particular NaCl and gypsum. NaCl is flushed through following floods into the groundwater, while sulfate salts are mainly accumulated in the wadi sediments.

In the Riyadh el Khabra area, a plain at the confluence of tributary wadis with Wadi Rima, groundwater in the Saq aquifer is brackish Na–Cl type water with TDS values between 1,600 and 8,700 mg/l. The occurrence of brackish groundwater is attributed mainly to irrigation return flow, as:

- Groundwater salinity is relatively low in newly constructed wells, but increases in the following years.
- Salinity in the upper layers of the aquifer tends to be higher than in deeper sections of the aquifer and a decrease of water salinity is generally observed from the beginning of pumping to later hours of operation.

NO_3 concentrations reach up to 370 mg/l.

5.4.1.2 Lower Cretaceous Sandstone Aquifers

Fresh groundwater with salinities of 600–1,000 mg/l TDS occurs in the Sakaka sandstones in the Al Jauf–Tabuk area. In the morphologic Jauf depression, fresh Ca–HCO_3 type groundwater is probably related to recent recharge in the ramifying wadi system. Na–Cl type water with salinities of up to 3,000 mg/l TDS, which has been tapped in the Permian–Lower Cretaceous sandstone aquifer complex of the Jauf depression, may represent fossil groundwater.

Groundwater in the Kurnub sandstone aquifer in southern Jordan is prevailingly brackish. Groundwater salinity increases from 1,400 mg/l TDS on the northern margin of the Interior Shelf to nearly 3,000 mg/l in the Jafr area on the east Jordanian limestone plateau.

The groundwater appears to be generally Na–Cl type water. In the subsurface, fresh groundwater occurs in the Kurnub sandstone aquifer on the southwestern slopes of the western highlands of Jordan with EC values of 530 μS/cm in Wadi Hasa; EC values increase to 600–800 μS/cm further north on the highlands and, on the Jordan valley escarpment, to around 3,000 μS/cm in hot springs and up to 7,000 μS/cm in boreholes.

Groundwater in Cretaceous sandstone aquifers of the Tuwayq segments is prevailingly brackish (Fig. 5.4). Water in the Minjur aquifer has salinities of 1,400 mg/l TDS near Riyadh and 1,600 mg/l at Sudair. The salinity increases toward south and reaches 4,000 mg/l TDS at Al Aflaj.

Fresh to slightly brackish groundwater occurs in the Biyadh–Wasia aquifer in Wadi Nisa (500–900 mg/l TDS), Khurais (550–1,500 mg/l TDS) and Al Kharj (>1,000 mg/l TDS). The salinity of the aquifer increases with depth and toward the

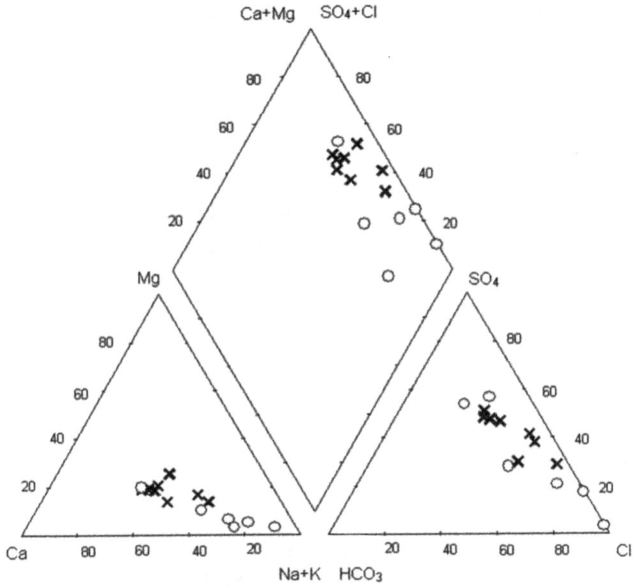

Fig. 5.4 Piper diagram: Groundwater samples from the Minjur and Wasia aquifers, Saudi Arabia. ○ Minjur aquifer, × Biyadh–Wasia. Data from Otkun (1972) aquifer

confined sections of the aquifer under the eastern Arabian platform; in many offshore areas of the Gulf, the formation contains brines.

5.4.2 *Carbonate Aquifers*

Groundwater in the Permian–Cretaceous carbonate aquifers of the Interior Shelf is generally brackish.

In the Upper Permian Khuff formation, composed of limestones and dolomites with some anhydrite, groundwater salinity ranges from 3,800 to 6,000 mg/l TDS.

The Middle Triassic Jilh aquifer contains water with 3,800–5,000 mg/l TDS; the elevated salinity is caused mainly by dissolution of $CaSO_4$ from anhydrite and gypsum layers.

In calcarenites of the Lower Cretaceous Sulaiy aquifer, groundwater with a salinity of 1,000–3,000 mg/l TDS is found 20 km southeast of Riyadh.

References. Al-Sagabi (1978), Al-Sayari and Zötl (1978), Bassam (1998), Edgell (1997), El Naser and Gedeon (1996), Handy and Alomani (1984), Hobler et al. (1991), Jado and Zötl (1984: 187 ff), Lloyd and Pim (1990), Salameh and Rimawi (1984, 1987b), Sharaf and Husein (1996), Sowayan and Allayla (1989), van der Gun and Ahmed (1995), WAJ-ODA (1994).

5.5 Recent and Pleistocene Groundwater: Information from Isotope Data

5.5.1 Groundwater Age

Groundwater in the Paleozoic–Cretaceous sandstone aquifers of the Interior Shelf is prevailingly fossil. The following ranges of groundwater ages have been calculated from ^{14}C data:

- Saq aquifer generally 22,000–28,000 years
- Wajid aquifer in Saudi Arabia generally >30,000 years
- Minjur aquifer on outcrop 15,500 years, 100 km downstream further east 34,800 years
- Wasia–Biyadh aquifer:
  - Near the outcrop at Kharj 8,000 years
  - At Khurais 16,000 years
  - 70 km east of the outcrop 20,760 years
  - At Abqaiq 250 km east of the outcrop 22,500 years
- Sakaka aquifer in the Tabuk–Jauf area 21,000–>30,000 years

^{3}H is below detection level in most samples from the Paleozoic–Mesozoic sandstone aquifers.

In the outcrop area of the Disi sandstone aquifer, tritium values of >4 TU were analysed in groundwater from a few wells in southern Jordan, "but other outcrop wells at similar groundwater elevations do not have tritiated water. This is seen as demonstrating that modern recharge occurs but only with a variable distribution in time and location and is consistent with the hypothesis that sporadic recharge to the system is through infiltration loss from run-off during flash floods" (Llloyd and Pim 1990, data mainly of 1970s).

In samples taken from the Disi aquifer of southern Jordan in 1987 to 1993, tritium contents were below detection level.

Most groundwaters in the outcrop areas of the Disi aquifer are "modern" in ^{14}C terms with values of 38.1–41.9 pmc, but also old waters are present under the sandstone outcrops with values of 18.1–22.8 pmc. ^{14}C ages of 6,000–20,000 years are found in the confined area close to the outcrop.

Relatively low Cl concentrations (<100 mg/l) in some parts of the confined aquifer are assumed to represent recharge before about 10,000–30,000 years, while higher Cl^{-} concentrations occur in groundwater related to more recent recharge under predominantly arid conditions.

In the area of Riyadh el Khabra, where Wadi ar Rima crosses the Interior Shelf on the northern rim of the Tuwayq mountain zone, ^{3}H values of groundwater in wells tapping the Saq sandstone aquifer are low, ranging from less than detection level to 2.5 TU. In water from a deep well tapping the Minjur sandstone aquifer, ^{3}H was below detection level (data of 1974).

In the central Tuwayq mountain zone, significant ^{3}H values are generally restricted to shallow groundwater in wadi sediments. In Wadi Hanifa, which crosses the Interior Shelf in the Riyadh area, tritium values of 54–74 TU were found in groundwater of the 60–80 m thick wadi sediments (data from 1973 to 1974). In the shallow aquifers of the Wadi Hanifa catchment "all waters having tritium values of more than 10 TU were collected from upper and central parts of valley provided with both a nearby and a remote recharge area, whereas waters with less than 4 TU leave parts of broad valleys or else come out of Al Kharj basin far away from the recharge area" (Al-Sayari and Zötl 1978: 226).

In Wadi ad Dawasir, groundwater from deep artesian wells and from shallow wells reaching the Wajid sandstone aquifer has low ^{3}H values (less than detection level to 2.8 TU). ^{3}H values of samples from shallow wells in the wadi sediments range from 1.5 to 45 TU (data of 1975).

5.5.2 Stable Isotopes of Oxygen and Hydrogen

The clusters of $\delta^{18}O$ and $\delta^{2}H$ values of groundwater from the Interior Shelf show several specific scatters in different areas, apparently related to Holocene or pluvial Pleistocene climate conditions and various sub-regional meteoric regimes. $\delta^{18}O$ values of Pleistocene groundwaters are generally in a range of −5.5 to −8‰; groundwaters originating from Holocene recharge have significantly less negative $\delta^{18}O$ values in a range of −1.25 to −3.3‰.

$\delta^{18}O$ values of groundwater in the Disi sandstone aquifer of southern Jordan are in a range of generally −6.5 to −7.2‰ with d values scattering around +12‰ (Fig. 5.5). "The groundwaters do not correlate directly with the Mediterranean rainfall type relationship" (Lloyd 1981). Recent recharge of Mediterranean origin is indicated in some samples from the Disi sandstone outcrop area in Jordan by d values of around +20‰.

The general pattern of $\delta^{18}O/\delta^{2}H$ values in the Disi aquifer appears comparable to the distribution in aquifers of the semi-arid to arid plateau areas of eastern Jordan and southeastern Syria, indicating an origin from precipitation in a generally dry Pleistocene–Holocene climatic environment.

Wadi ad Dawasir crosses the Interior Shelf on an about 150 km long west–east stretch in the southern Jebel Tuwayq area over outcrops of the Wajid sandstones, Jurassic formations and Quaternary clastic sediments. Groundwater with low salinity from artesian boreholes tapping the Paleozoic–Mesozoic aquifers have $\delta^{18}O$ values between −6.06 and −8.02‰, d values scatter around +6‰ (Fig. 5.6). The most negative $\delta^{18}O$ values are found in groundwaters of Jurassic aquifers downstream of the Wajid outcrop area. In brackish groundwater of shallow wells in the Fara area, the following ranges of $\delta^{18}O$ values were found:

- Wells tapping the Wajid sandstone aquifer below a cover of Quaternary sediments: $\delta^{18}O$ −4.39 to −5.68‰ (d on average of +3.8‰)

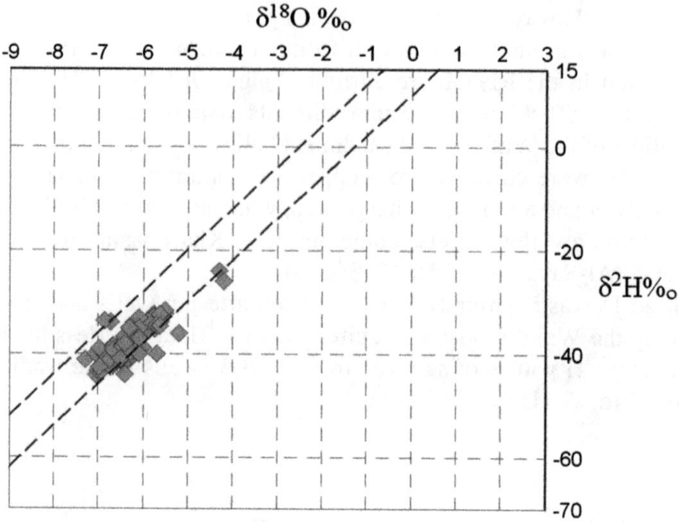

Fig. 5.5 $\delta^{18}O/\delta^{2}H$ diagram: Groundwater samples from the Disi aquifer in southern Jordan. Data from El Naser and Gedeon (1996)

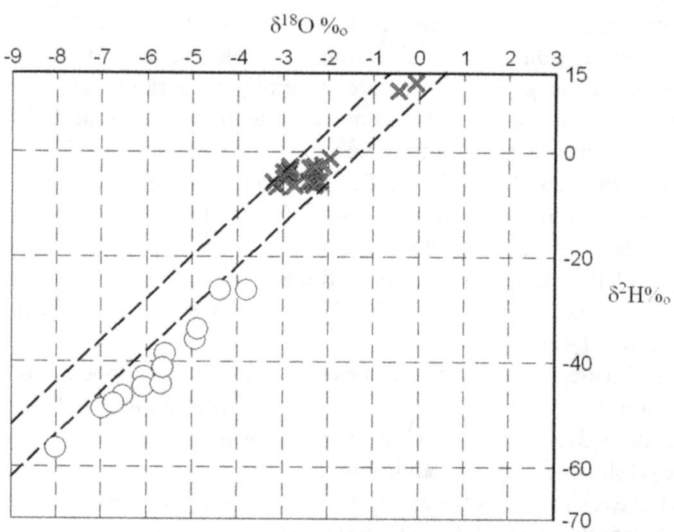

Fig. 5.6 $\delta^{18}O/\delta^{2}H$ diagram: Groundwater samples from the Tuwayq mountains. × Groundwater in wadi sediments, ○ fossil groundwater in sandstone aquifers. Data from Al-Sayari and Zötl (1978)

- Groundwater from Quaternary sediments: $\delta^{18}O$ −3.61 to −4.89‰ (d +2.6 to +6.3‰)
- One sample with high ^{3}H (45 TU): $\delta^{18}O$ −2.0‰ (d +10.3‰)

"…the enrichment of isotopes as well as the mineralization in hand-dug wells may be explained by an admixture of recent shallow groundwater to old sandstone water"..."the mineralization of wells in Wadi Ad Dawasir most likely occurs on account of mixing of recent, more or less salty and shallow groundwater with qualitatively superior old waters originating in the Wajid Sandstone" (Al-Sayari and Zötl 1978: 250, 252). In detail, variations of the relationship between isotopic composition and salinity of the groundwater appears to be complicated by different mixing proportions and salinization processes.

Brackish groundwater extracted from the Minjur aquifer in a deep well in Wadi Hanifa (salinity 1,200 mg/l TDS) has a $\delta^{18}O$ value of −5.46‰ and a d value of +4.8‰. These values, deviating significantly from values of groundwater in shallow wells in wadi aquifers of Wadi Hanifa, apparently represents Pleistocene recharge.

Groundwater from wadi aquifers in Wadi al Hawta and Wadi Birk in the central Jebel Tuwayq area have d values of +18 to +19.9‰ and $\delta^{18}O$ values between −2.8 and −3.3‰, which may be representative for the general range of Holocene groundwaters of the Arabian Peninsula. From the stable isotope data, generally moderate groundwater salinity (450–670 mg/l TDS) and significant ^{3}H values (up to 30 TU), recent recharge without considerable evaporation impact may be assumed. "Both δD and $\delta^{18}O$ increase together with the tritium content, whereas the water temperature falls with the rising content of tritium and of stable isotopes. It may thus be concluded that warmer waters that are free from tritium and contain less D and $\delta^{18}O$ become mixed with cool affluents enriched with tritium and stable isotopes. Presumably, these tritium-containing affluents have a relatively short infiltration time, which points to a nearby recharge area. They probably flow in from the rather steep and short tributaries in the valley flanks of the breaching. Their recharge area of the older, tritium-free component may be the area west of the breaching" (Al-Sayari and Zötl 1978: 224).

$\delta^{18}O/\delta^{2}H$ values from the Sakaka sandstone aquifer in the Sakaka–Jauf area in northwestern Saudi Arabia deviate significantly from the pattern in the Disi aquifer: d values are negative varying mainly between −5.1 and −13‰, $\delta^{18}O$ values range from −0.6 to −1.3‰. The samples represent fossil groundwater with calculated ages of 21,000 to >36,000 years.

Samples of brackish fossil groundwater (12,000 to >35,000 years) in Paleogene–Quaternary aquifers in southern Wadi Sirhan have $\delta^{18}O$ values of −1.13 to −2.68‰ and d values of −4.4 to −11.1‰. These values possibly result from mixing of groundwaters with different age.

$\delta^{18}O$ values of groundwater from the Saq sandstone aquifer in the Riyadh al Khabra plain on Wadi Rima range from −1.25 to −2.37‰, d values from +1.5 to +8.06‰. These values are probably affected by evaporation effects of irrigation return flow (Al-Sayari and Zötl 1978: 188, 192).

Groundwaters in the Quaternary aquifers of the Wadi Hanifa drainage system in the central Jebel Tuwayq area (Wadi Hanifa, Wadi Nisa, Wadi al Luhy, Al Kharj plain) have $\delta^{18}O$ values between −2.0 and −3.3‰ and d values between +12 and +16.6‰. The group of samples with rather uniform stable isotope composition represents fresh to brackish groundwaters with salinities between 440 and 3,000 mg/l TDS and highly different tritium values in different wadi sections.

The groundwater may be assumed to originate from Holocene to Recent recharge under a Mediterranean type meteorologic regime with, however, less negative $\delta^{18}O$ values than groundwaters on the northern Arabian platform. An impact of evaporative isotopic enrichment of infiltrating wadi runoff and of irrigation return flow may have modified the stable isotope values.

References. Al-Sayari and Zötl (1978), El Naser and Gedeon (1996), Lloyd (1980a, 1981), Lloyd and Pim (1990), Salameh and Rimawi (1984, 1987b), Wagner and Geyh (1999).

Chapter 6
Eastern Arabian Platform

6.1 Geologic and Geographic Set-Up

6.1.1 General Geologic Structure

The eastern Arabian platform occupies the southeastern part of the Arabian Shelf. In the north and south, the platform grades into deep sedimentary basins: In the north to northeast, the platform is adjoined by the Mesopotamian–Euphrates basin on the margin with the alpidic Taurus–Zagros mountains and the Gulf basin. In the south, the Rub al Khali constitutes a large intra-platform basin. The eastern Arabian platform adjoins, in the west, the Interior Homocline and is bounded in the northwest by the Rutba uplift structure, which separates the eastern and northern segments of the Arabian Platform. The southern margin of the eastern platform is formed by the anticlinal structure of the Hadramaut arch; in the southeast, the platform borders the allochthonous units of the Oman mountains and extends into the central plateau of Al Wusta area in Oman. Structurally, the eastern Arabian platform includes, along its western and southern margins, the lower part of the "Interior Homocline" with gentle slopes of the strata toward the Euphrates–Gulf–Rub al Khali basin centres.

Continuing the homoclinal structure of the Interior Shelf, the strata of the eastern Arabian platform dip from the western boundary of the platform gently toward east–northeast. The general homoclinal slope of the sedimentary succession grades in the north–northeast into the Euphrates–Mesopotamian sedimentary basin and, in the east, into a sub-horizontal undulating structure. Zones of gentle anticlinal uplifts cross the platform area between the Tuwayq segments of the Interior Shelf and the Gulf coast in approximately south–north direction.

Acording to the general geologic structure as well as to main morphologic–hydrographic features, the eastern Arabian platform can be sub-divided into three main geologic units (Fig. 6.1):

- Segments dipping into deep Neogene–Quaternary sedimentary basins in the north and south: the Mesopotamian–Euphrates basin and the Rub al Khali basin, respectively

W. Wagner, *Groundwater in the Arab Middle East*,
DOI 10.1007/978-3-642-19351-4_6, © Springer-Verlag Berlin Heidelberg 2011

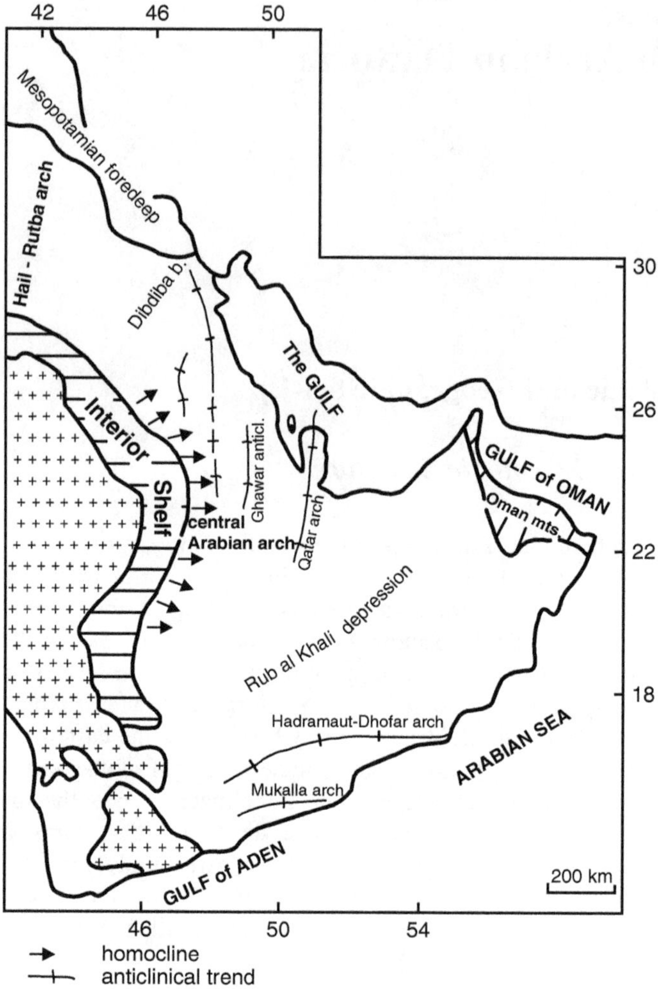

Fig. 6.1 Main geologic–structural units of the eastern Arabian platform. After Alsharhan et al. (2001), Edgell (1997)

- The relatively horizontal platform segment between the Tuwayq zone of the Interior Shelf and the Gulf coast in eastern Saudi Arabia, or middle Gulf segment

The upper margins of the different segments of the eastern platform are occupied by, in comparison to the Euphrates–Gulf–Rub al Khali depressions, relatively uplifted zones. In the northwest, structurally higher zones west of Salman grade toward east into the Euphrates and Dibdiba depressions. Approximately north–south oriented structurally positive trends follow toward east (Muqala, Wafra–Burgan, Umm Gudair, En Nala trends, Ghawar anticline, Damam and Bahrain domes, Qatar arch, Edgell 1997).

The southern part of the eastern platform is occupied by the Rub al Khali basin, which plunges toward northeast into the Gulf. The northern boundary of the Rub al Khali basin is formed by the central Arabian arch, a gentle anticlinal structure. From the western, southern and southeastern margins of the Rub al Khali, Paleogene formations dip under the Neogene–Quaternary cover of the central parts of the basin. The eastern margin of the Rub al Khali basin is covered by sand and gravel accumulations on the foot of the Oman mountains.

The area covered by the Rub al Khali desert corresponds to a large geologic basin structure within the platform, which developed mainly during the Tertiary. The basin forms a broad northeast dipping syncline, which is bordered in the northeast to southeast by structural uplifts: central Arabian arch, Interior Shelf, Hadramaut arch and Oman mountains.

The Rub al Khali basin merges in the north at the Dubai coast into the marine Gulf basin. During the Paleocene–Eocene, a deep depression (Ras el Khaima basin, Alsharhan et al. 2001) covered the northeastern edge of the Rub al Khali basin, and a vast marine transgression expanded over the southern part of the platform including the presently uplifted fringes of the Rub el Khali basin.

6.1.2 Climate, Morphology and Hydrography

6.1.2.1 Climate

The climate of the eastern Arabian platform is arid to hyper-arid. Mean annual precipitation ranges from 150 to less than 25 mm (Sect. 1.4.3.1). Rainfall is mainly related to sporadic events of low atmospheric pressure, approaching the Arabian Peninsula from northwest.

In interior Oman, infrequent high intensity rainfall events are associated with low pressure systems moving across Oman or tropical storm/cyclone events tracking westward across the Arabian Sea. Rainfall from tropical cyclones occurs, on average, once in 2 years.

6.1.2.2 Morphology

The morphology of the eastern Arabian platform is dominated by desert plateaus, sand and gravel plains, sand dunes and sand seas. Fluvial and coastal plains mark the margins of the platform along the Euphrates river and the Gulf coast. From the western and southern margins, limestone plateaus descend in gentle slopes toward the Euphrates–Gulf depression and the Rub al Khali sand desert. The Al Widyan limestone plateau in the northwest is dissected by various wadis, which run in a general northeast direction toward the Euphrates plain.

Main types of landscapes of the eastern Arabian platform are (Figs. 6.2, 6.3 and 6.4):

Limestone plateaus on the western and southern slopes of the Euphrates–Gulf–Rub al Khali basin, comprising the Widyan plateau in the northwest, the Suman plateau in the west and, in the south, the slopes of the Hadramaut and Dhofar highlands with the plateaus of the northern Jols in Yemen and of Nejd and Al Wusta in Oman

The Gulf coastal plains and the Euphrates plain which extends in the southeast into the Dibdiba and Kuwait plains

Sand dune areas and sand seas with huge dimensions of the Rub al Khali and the Great Nefud deserts, and elongated strips of eolian sands of Ad Dahna and Al Jafura

Alluvial sand and gravel plains adjoining, in particular, the western slope of the Oman mountains

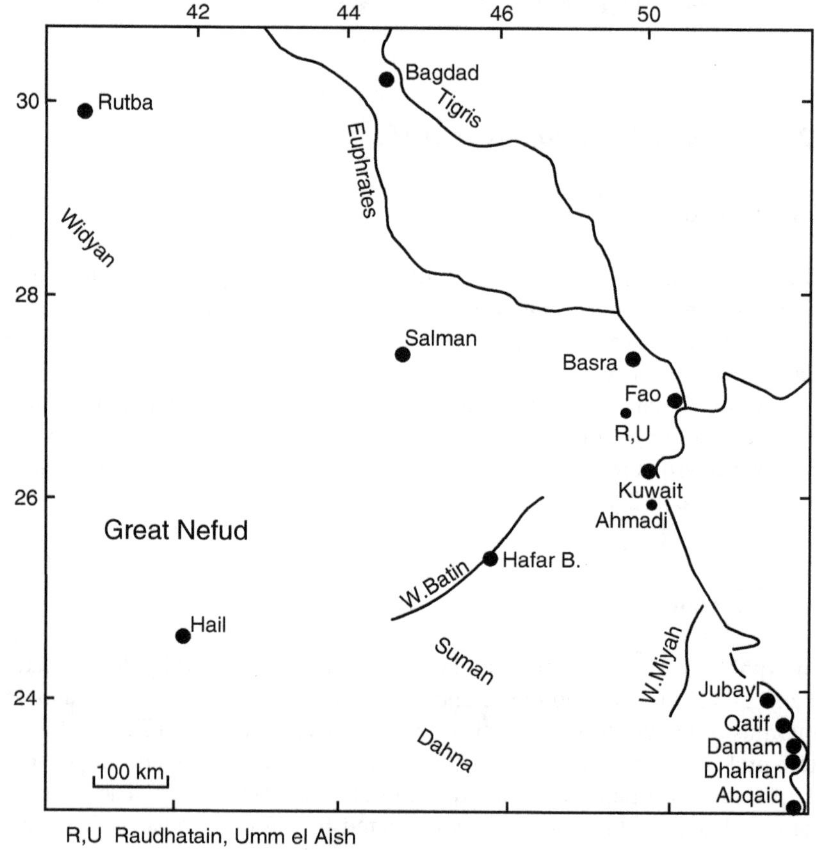

Fig. 6.2 Eastern Arabian platform, location map, northern part

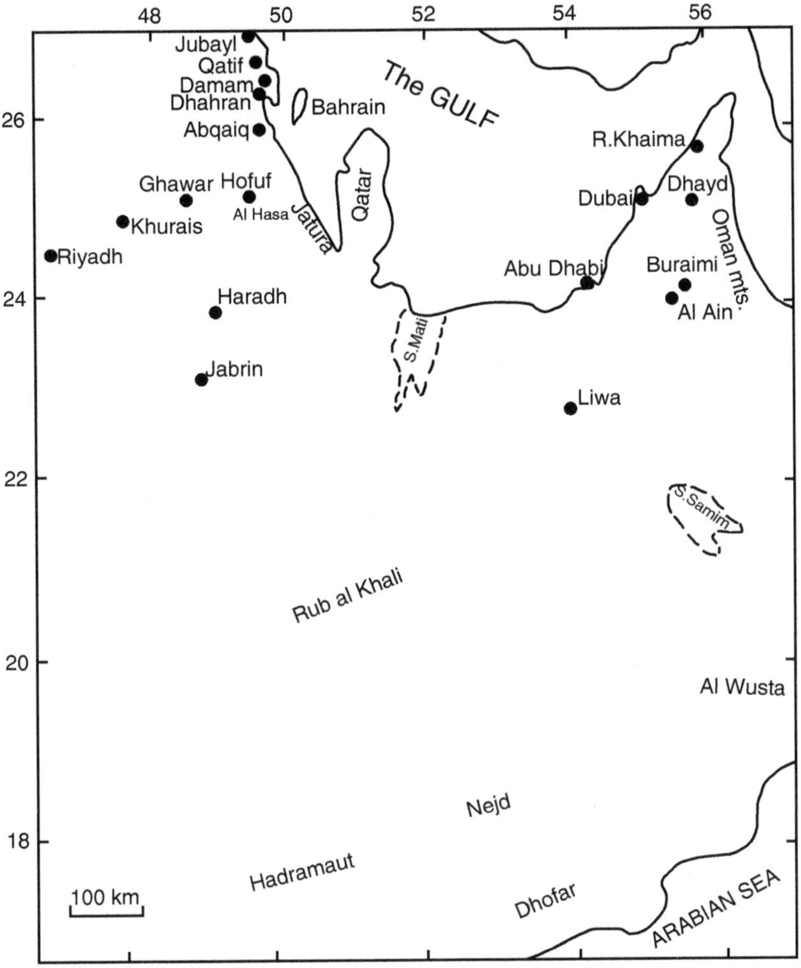

Fig. 6.3 Eastern Arabian platform, location map, southern part

6.1.2.3 Euphrates–Shatt el Arab Sub-Basin

As a hydrologic–morphologic unit, the Euphrates–Shatt el Arab sub-basin covers the northern part of the hydrologic basin of the Gulf (Arabian side). The sub-basin comprises the Widyan plateau at the northeastern margin of the Great Nefud and parts of the southern desert of Iraq and extends over the Dibdiba plain into Kuwait and northeastern Saudi Arabia. In general, the sub-basin comprises plateau – like landscapes with a general gentle slope from the Widyan area at around 500 m asl toward the Euphrates valley and the Gulf coast in the north and northeast.

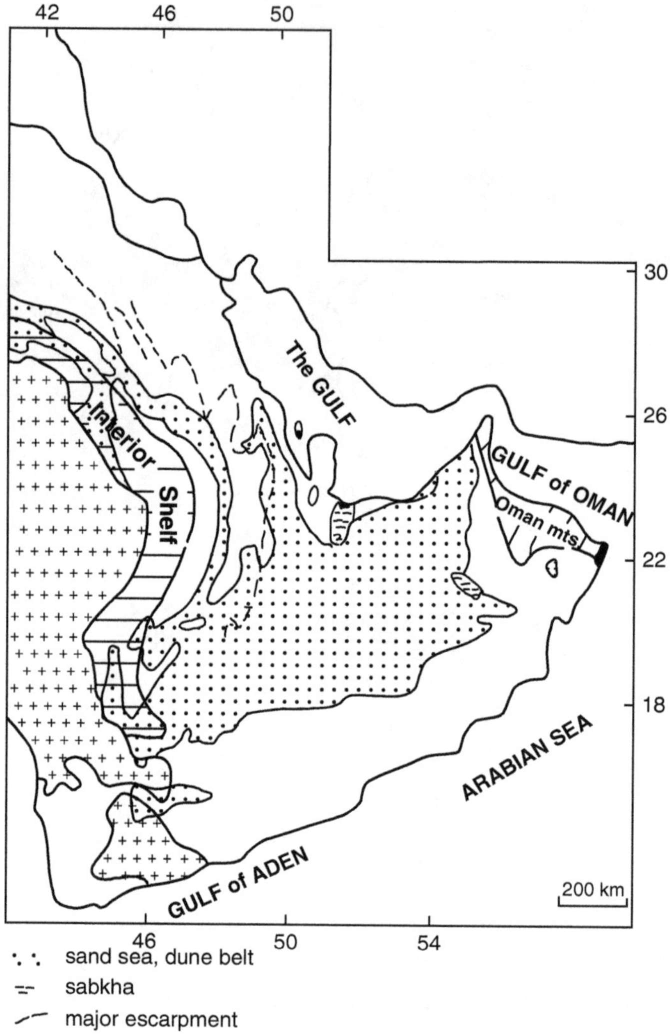

Fig. 6.4 Geomorphologic features on the eastern Arabian plate. After Alsharhan et al. (2001), ESCWA (1999b), Wohlfahrt (1980)

The Widyan segment of the eastern Arabian platform constitutes an approximately 600 km wide plateau, which is covered by Upper Cretaceous–Paleogene carbonate rocks and dips gently toward northeast to the Euphrates plain. The plateau is crossed by various northeast directed wadi systems, which disperse in the Euphrates plain.

The plains in northern Kuwait are covered by fluvial and estuarine deposits.

The Euphrates river runs in its about 500 km long lower course along the northeastern boundary of the eastern Arabian platform. Stream flow in the lower stretch of the Euphrates river is predominantly controlled by dams in Turkey, Syria and Iraq. Inflow from ephemeral runoff in tributary wadis on the right bank of the river is very limited along its lower course, but significant volumes of groundwater discharge in springs within the lower Euphrates valley.

The Shatt el Arab arises from the confluence of Euphrates and Tigris rivers at El Qurna in southern Iraq, from where it flows as a 600–2,000 m wide river over a distance of 110 km into the Gulf. The annual average stream flow in Shatt el Arab is reported as 21×10^9 m^3 at Basra and 35×10^9 m^3 at Al Fao on the Gulf coast.

6.1.2.4 Middle Segment of the Eastern Arabian Platform

The middle segment of the eastern Arabian platform, between the Euphrates–Shatt el Arab sub-basin in the north and the sub-basin of the Rub al Khali sand desert in the south, comprises a 500 km long and around 250 km wide area of prevailingly plateau landscapes. This middle part of the Euphrates–Gulf–Rub al Khali basin extends from the Ad Dahna sand dune belt on the boundary of the Interior Shelf in the west to the Gulf coast in the east.

Ad Dahna forms an up to 100 km wide and around 1,300 km long strip of sand seas and sand dunes between the Great Nefud and Rub al Khali deserts, bordered in the west by the Aruma escarpment of the Tuwayq mountain zone and, in the east, by the 80–250 km wide approximately north–south trending Suman plateau. Elevations of the Suman plateau range from 400 to 250 m asl, its eastern edge forms a prominent escarpment indented by ancient streams with the base of the cliff at 150–200 m asl.The Suman plateau is crossed by two major wadi courses: the Wadi Rima–Wadi Batin system in the north and Wadi Sahba in its southern part. The plateau disappears, at its southern edge, under the sand seas of the Rub al Khali and grades to the northwest and southwest into sand and gravel plains of Wadi Batin–Dibdiba and Hofuf.

Some zones of oases, like Wadi al Miyah and Jabrin, are interspersed within the prevailingly desertic landscapes of the midle segment of the eastern Arabian platform.

The coastal area between Shatt el Arab and Jubayl comprises low rolling plains covered with a thin mantle of sand. A belt of drifting sands and dunes merges in the south into the sand area of Al Jafura. In the coastal area south of Jubayl, sabkhas and sand areas alternate with outcrops of Eocene to Neogene carbonate rocks.

Some positive topographic features, such as the Damam dome and the Abqaiq and Dukhan anticlines, rise a few tens of metres above the flat desert landscapes.

6.1.2.5 Rub al Khali

The Rub al Khali is the largest continuos sand desert in the world with an area of around 640,000 km^2.

The western part of the Rub al Khali is covered mainly by sheets of fine sand and is situated at an elevation of about 600 m asl. To the east, the desert drops to 180 m asl, high sand dunes alternate with sand sheets and salt flats.

The predominantly eolian landscape of the eastern Rub al Khali is superimposed on a flat relict marine coastal plain. The elevation of the plain is about 70–100 m asl with dunes rising up to 100 m above the plain. Active and fossil sabkhas are scattered, in the eastern Rub al Khali, throughout the dune fields. Extensive sabkha areas at low topographic elevations on the eastern margins of the Rub al Khali form a common base level for surface and subsurface drainage from the Oman mountains in the east and also for groundwater movement from the Rub al Khali. The most conspicuous of these morphologic depressions is the in north–south direction 100 km long Umm as Samim sabkha area on the border between Saudi Arabia and Oman, which covers 2,500–3,000 km^2 at an average elevation 60 m asl.

The southern margins of the eastern Arabian platform are occupied by limestone plateaus of the northern slopes of the highlands of Hadramaut and Dhofar (Nejd) and the Al Wusta central plateau of Oman. The plateaus are crossed by a number of wadi systems, which run out in the sand seas of the Rub al Khali.

Major gravel plains extend along the western foot of the Oman mountains (western gravel plain in Oman and United Arab Emirates) and on the entrance of Wadi Sahba into the coastal plain in the north of the Rub al Khali.

In the southeast, the Rub al Khali grades into the central plateau of Al Wusta, situated between the Nejd area and the Oman mountains. The core of Al Wusta consists of a broad low plateau at elevations between 120 and 240 m. Several elongated up to 30 km long depressions are found on the plateau, which may represent structural features and the remnants of ancient drainage courses. To the southeast, the plateau is dissected by a number of coastward trending, normally dry wadis, which are deeply incised, with cliffed walls up to 50 m high.

The southeast of the plateau, where the wadis drain toward the Arabian Sea, is situated outside the hydrologic Rub al Khali sub-basin, but constitutes part of the Arabian Platform.

6.1.2.6 Hydrography

Most wadis, which reach the eastern Arabian platform from the Arabian Shield and the Interior Shelf, disappear on the margins of the sand sea and plateau areas on the upper reaches of the platform. Only the Wadi Rima–Wadi Batin drainage system constitutes an over a long distance continuous wadi course from the Hejaz highlands into the Euphrates plain.

Several wadi systems dissect the Widyan segment in the northwest of the platform and disperse on the margins of the Euphrates plain. On the southern margin of the Rub al Khali, a number of larger wadi systems cross the Hadramaut and Nejd plateaus until they disappear on the limits of the sand seas. Major wadis originating on the Arabian Shield end, after crossing the Interior Shelf, on the upper margin of the eastern Arabian platform on barriers of sand dunes of Ad Dahna, Rub al Khali and Ramlat as Sabatayn. On the eastern margin of the Rub al Khali, wadi courses of the Oman mountains end in gravel plains of the mountain foreland.

Gravel fans of paleo-river systems extend over wide areas of the eastern Arabian platform. Large gravel fans spread out, in particular, from the at present generally dry courses of Wadi Rima–Wadi Batin, Wadi Sahba and Wadi ad Dawasir (Fig. 6.5).

Gravel accumulations of Wadi Rima–Wadi Batin cover the Dibdiba plain over a surface of about 60,000 km^2 in parts of southeastern Iraq, northeastern Saudi Arabia

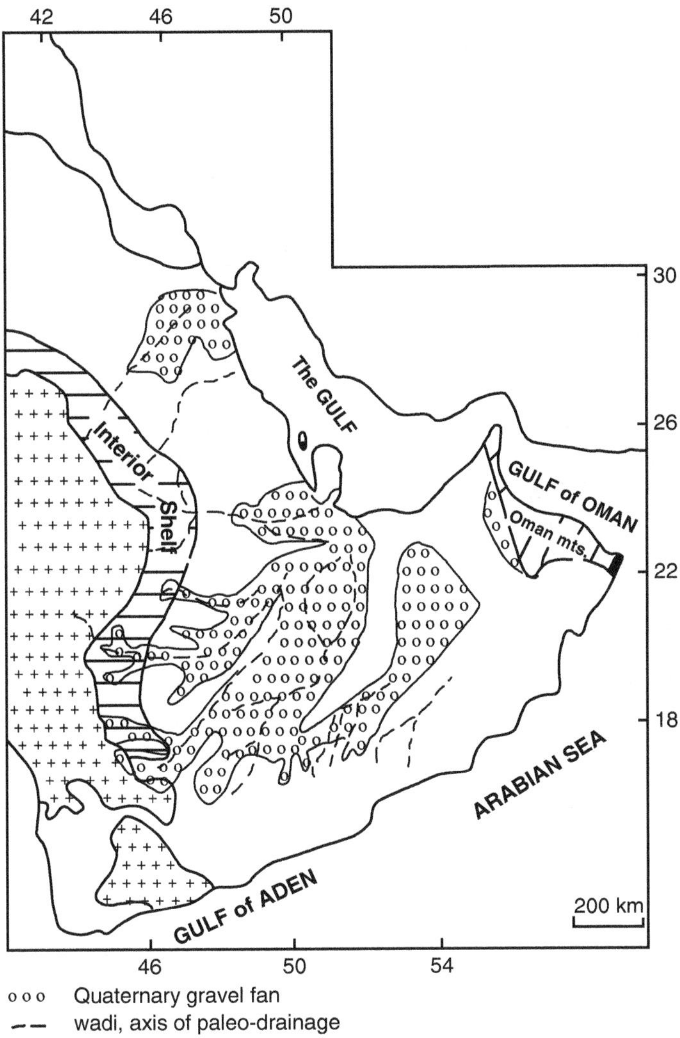

Fig. 6.5 Eastern Arabian platform: gravel fans, paleo-drainage system. After Al-Sayari and Zötl (1978), ESCWA (1999b)

and Kuwait. The Dibdiba fan is cut through by the northeastward directed Wadi al Batin, the flat continuation of Wadi Rima.

The gravel fan of Wadi Sahba reaches from the Kharj plain in the Tuwayq mountain zone in eastward direction until the Gulf coast north and south of the Qatar peninsula. Wide parts of the fan are covered by sand dunes of Al Jafura.

Gravels of Wadi ad Dawasir extend from the Tuwayq mountains along the northern border of the Rub al Khali over more than 1,000 km to the Gulf coastal area between Qatar and Sabkhet al Matti. The alluvial fan of Wadi Dawasir is covered by desert pavement, eolian sand and sand dunes.The fluviatile sediments of the Wadi Dawasir fan were deposited mainly by drainage systems descending from the southern Asir highlands in the southwest and from the Hadramaut mountains in the south. The large alluvial fan of Wadi ad Dawasir is covered by desert pavement, eolian sand and dunes.

Wadi ad Dawasir continues from the Interior Shelf for about 70 km toward the east into the accumulation plains as a 10 km wide flood channel. At present, the wadi channel carries only episodic local surface runoff.

The vast gravel accumulations of the eastern Arabian platform are attributed to humid periods during the upper Pliocene to lower Pleistocene, when stream runoff had sufficient transport energy to deposit coarse erosional material, derived from the highlands of the Arabian Shield, on the plains of the platform. Deposition of the wadi channels fillings occurred mainly during the early Quaternary. The thickness of the channel deposits reaches 70 or more metres. Over wide areas, however, the paleo-river gravels constitute only a thin surficial cover.

Under the present arid climate conditions, morphologic remnants of the paleo-wadis act as drainage systems over rather short distances after sporadic rainfall toward local morphologic depressions. In many areas, the previous wadi systems of the eastern Arabian platform have been separated by large sand dune belts into independent local drainage areas.

6.1.2.7 The Arabian Gulf

The Arabian Gulf or Persian Gulf is a shallow epicontinental sea with a maximum depth of 110 m. Salt flats (sabkhas) are common along the shoreline, Sabkhat Matti being the largest sabkha with an area of 6,000 km^2 and an extent of 120 km in north–south direction.

The shoreline of the Gulf is dominated by supratidal sabkhas, sand bars and carbonate sands. Bordering that zone to the south is a 30–120 km wide zone of active dunes, often resting directly over a gravel surface.

6.1.3 Sedimentary Cover of the Eastern Arabian Platform

The thickness of sedimentary formations on the eastern Arabian platform reaches 3 km on its western boundaries and around 20 km in the Gulf area. Information on the stratigraphic and sedimentologic features of that thick sequence of sedimentary rocks, the major part of which corresponds to Paleozoic–Mesozoic formations, can be found in Murris (1984) and Alsharhan and Nairn (1997).

The upper section of the sedimentary cover, which is of main interest with a view to the groundwater regime, comprises primarily marine Cretaceous to Neogene formations and terrestrial Neogene–Quaternary deposits. Marine transgressions over the eastern Arabian platform during the Upper Cretaceous–Paleogene werde accompanied by the deposition of marine carbonates, which rest, in wide areas, on a Mesozoic sandstone complex (Biyadh and Wasia formations) (Table 6.1).

6.1.3.1 Cretaceous

The succession of carbonate deposits starts with the Upper Cretaceous Aruma formation, which is composed mainly of limestones with chalky and dolomitic layers and, at the top, shale and dolomite intercalations. The thickness of the formation varies, in general, from 60 to 140 m, but increases to 600 m toward north. Along the Tuwayq mountain zone, the western limit of the Aruma outcrop forms a prominent escarpment, corresponding to the schematically defined boundary betwen the Interior Shelf and the eastern Arabian platform. The Tayarat formation in Kuwait and southern Iraq corresponds in its stratigraphic age and lithology largely to the Aruma formation. Extensive outcrops of the Aruma formation cover the Widyan area on the northwestern margin of the eastern Arabian platform. In the east of the Rub al Khali segment along the boundary with the Oman mountains, the Upper Cretaceous deposits consist of shale and carbonate formations (Fiqa, Juweiza, Simsima formations).

6.1.3.2 Paleogene

The Paleogene of the eastern Arabian platform, which is summarily denominated Hasa group, comprises a sequence of prevailingly carbonate rocks, which is generally sub-divided into three formations: Umm er Radhuma, Rus and Damam formations.

The Paleocene *Umm er Radhuma formation* is composed mainly of massive limestones and dolomitic limestones, which are karstified in part of the outcrop areas. The Umm er Radhuma outcrops surround the eastern Arabian platform on its fringes from southern Iraq across central Saudi Arabia to the northern slopes of the Hadramaut and Nejd highlands in the south in a more than 2,000 km long semicircle, which is, however, interrupted by the cover of the Ad Dahna and Rub al Khali sand deposits. In central Saudi Arabia, the Umm er Radhuma outcrops form the easternmost of the west facing escarpments of the Tuwayq mountain belt.

The Umm er Radhuma formation is covered by younger Paleogene to Neogene formations on most of the eastern Arabian platform, but the formation appears on the surface in limited outcrops on the Gulf coast at Qatar and Bahrain.

Table 6.1 Hydrostratigraphic scheme of the eastern Arabian platform after Alsharhan et al. (2001), Edgell (1997), ESCWA (1999b), Zubari (1997)

Stratigraphic age			Euphrates sub-basin		Middle Gulf segment	Rub al Khali sub-basin	
			West	East			Southern margin
Quaternary						Gravel plain aquifer: **main aquifer** on the eastern margin of the sub-basin, fresh to brackish water	
Tertiary	Neogene			Dibdiba: aquifer, fresh to brackish water Kuwait: brackish water aquifer	Hofuf: generally unsaturated Dam: aquifer with low productivity Hadrukh: *aquifer*		
	Paleogene	*Damam aquifer*	**Major aquifer**	Deeper brackish water aquifer	**Major aquifer in Gulf coastal area** Mainly brackish water	Generally deep brackish water aquifer	**Aquifer** Fresh to brackish water
		Rus formation	*Major aquitard*				Aquitard, locally aquiferous
		Umm er Radhuma aquifer	**Aquifer**, fresh to brackish water	Deep brackish water aquifer	**Major aquifer**	Deep brackish water aquifer	Generally deep brackish water aquifer
Upper Cretaceous			Aruma (Tayarat): aquitard/brackish water aquifer Wasia sandstone: deep brackish water aquifer				

The thickness of the Umm er Radhuma formation increases from 240 m in Wadi Batin toward east and south to more than 600 m near the Gulf coast in northern Kuwait and east of Hofuf. Over anticlinal structures at Ghawar, Bahrain, Qatar, the thickness of the formation is reduced to around 300 m.

The Lower Eocene *Rus formation* is characterized by marls and chalky limestones with layers of chert, gypsum and anhydrite. The Lower–Middle Eocene *Damam formation* consists of an alternation of chalky limestone, dolomite and shale.

Outcrops of the Rus and Damam formation are found in extensive exposures on the northern and southern margins of the eastern Arabian platform: in southwestern Iraq and on the northern slopes of the Dhofar and Hadramaut highlands, in a narrow strip in the west of the platform, and in limited areas of the Gulf coast area: a quarry at Ahmadi in Kuwait, structural domes at Damam, Bahrain and Qatar, a number of small patches between the sand dunes of the southeastern Rub al Khali and in local uplifts at the margin of the Oman mountains, e.g. Jebel Hafit.

The thickness of the Damam formation increases from 75 m in Wadi al Miyah area toward the Gulf coast and generally ranges from 90 to 300 m in Iraq, Saudi Arabia and the Gulf coast area. In parts of the Rub al Khali, the thickness increases to more than 400 m, and in the west of the United Arab Emirates the formation grades into a 500 m thick sequence of shales.

In anticlinal structures, the thickness of the formation is reduced and, on the crests of the Damam and Bahrain anticlines, the formation has been eroded.

In Kuwait, the Rus formation consists of a 20–200 m thick sequence of anhydrite and limestones, the 60–200 m thick Damam formation comprises a basal unit of nummulitic limestone with shales, which is overlain above a karstified unconformity by chalky, marly and dolomitic limestones and cherty limestones at the top of the formation.

In the Gulf coast area, the Damam formation is sub-divided into five members (from top to bottom):

Alat member:	Chalky and dolomitic limestones with chert bands in the top part, dolomitic marl at the base, 25–70 m
Khobar member:	Limestones and dolomites, dolomitic marl at the base (“orange marl”), 25–60 m
Alveolina limestone:	Thin layer of limestone interbedded with marl and shale
Saila and Midra shale:	5–15 m (In Bahrain “shark tooth shale”)

The *western and southern slopes of the paleogeographic Rub al Khali basin* comprise an up to 200 km wide plateau area with outcrops of Paleogene formations. Paleocene, lower and middle Eocene, as well as upper Tertiary sedimentary rocks thicken toward the centre of the basin.

In the mountains and plateaus of Dhofar, Nejd and northern Hadramaut in southern Oman and eastern Yemen, the Paleocene to Eocene sequence of carbonate rocks is denominated Hadramaut group and comprises the following main lithological units (from top to bottom):

Southern Oman	Eastern Yemen
Andur or Qara formation	Habshiya formation
Limestone with chert, more argillaceous (equivalent of Damam formation)	Shale and marl with interbedded chalk and limestone (equivalent of Damam formation)
Rus formation	Rus formation
Chalky limestone, marl, evaporite beds	Massive gypsum with chert bands and marly horizons
	Jeza formation
	Calcareous shale, thin beds of limestone, in places chalky or marly
Umm er Radhuma formation	
Limestones and dolomitic limestones with chalky horizons and marl an shale beds	

The Damam, Rus and Jeza formation are exposed in wide areas of the plateaus of the Dhofar and Hadramaut highlands, while outcrops of the Umm er Radhuma formation are restriced to wadi beds.

6.1.3.3 Neogene to Quaternary

The Paleogene is overlain, in wide parts of the eastern Arabian platform, by Neogene to Quaternary sediments. A regression of the sea from the Arabian Shelf during the middle Eocene was followed by an uplift of the platform during the upper Eocene–middle Miocene.

From the middle Miocene to Pleistocene, terrestrial to shallow marine sediments were deposited on the eastern Arabian shelf.

The middle Miocene deposits comprise limestones, anhydrite, salt, mudstone and marl, and are overlain by sandy limestones and fluvial sediments, mainly sandstones and conglomerates.The Pleistocene sediments are made up prevailingly from fluvial sandstones, gravels and sandy mudstone.

The Neogene cover of the eastern Arabian platform is sub-divided into the following stratigraphic–lithologic units (from top to bottom):

Pleistocene		Gravels
Pliocene	Kharj formation	Conglomerates, sandstones, limestones (limited to an area south of Riyadh)
Miocene	Hofuf formation	Calcareous duricrust, conglomerate, argillaceous and sandy limestone, total 95 m at Hofuf
	Hadrukh formation	Marly sandstone, sandy marl, sand, clay and sandy limestone, 20–120 m
	Dam formation	Marl, clay with interbedded sandstone and limestone, 90 m at Damam dome reef limestone, calcarenitic limestone

The Al Wusta central plateau in interior Oman is covered, in wide parts, by limestones, marls and evaporites of the Oligocene–Miocene Fars formation, overlying carbonates of the Paleogene Hadramaut group.

During the late Pliocene to early Pleistocene, the regression of the sea combined with a humid climate phase caused the erosion of large wadis and the filling of wadi channels with detrital sediments. Extensive fluvial deltas developed during relatively wet phases in the upper Pliocene and lower Pleistocene; major gravel accumulations are found in the areas of Wadi Batin in the north of the eastern Arabian platform and of Wadi as Sahba and Wadi ad Dawasir. Wadi channel deposits reach thicknesses of more than 70 m. The top of the gravel deposits is overlain by mainly fine clastic sediments with gravel layers. Accumulation of sand dunes started at the end of the Pliocene to early Pleistocene and continued throughout the Quaternary. In the east of Rub al Khali, basal sand dunes were covered by the sea during the late Pliocene. Numerous extensive sabkhas in the eastern Rub al Khali are probably remains from the marine phase.

On the eastern margin of the Rub al Khali, a vast gravel plain extends as a narrow long strip between the Oman mountains and the sand dune fields. The gravel plain adjoins the western foothills of the Oman mountains from Ras al Khaima in the north to the Al Jaw plain near Al Ain and to central Oman. The plain is covered by an around 60 m thick sequence of sand and gravel with thin interbeds of silt and clay. Deposits of gravel extend to distances up to 70 km from the mountain margins.

References. Ahmed and Kraft (1972), Al-Awadi et al. (1998), Al-Hajari and Kendall (1992), Al Mashadani (1984), Al-Sayari and Zötl (1978), Alsharhan (1989), Alsharhan and Nairn (1997), Alsharhan et al. (2001), Bakiewicz et al. (1982), Bramkamp et al. (1963), Cherif and El Deeb (1984), Clark et al. (1987), Edgell (1997), ESCWA (1999b), Hughes Clarke (1988), Mukhopadhyay (2003), Murris (1984), Rizk et al. (1997), Robertson (1992), Sayed and Al-Ruwaih (1995), Wohlfahrt (1980).

6.2 Multi-Aquifer System of the Euphrates–Gulf–Rub al Khali Basin

6.2.1 General Hydrogeologic System

The eastern Arabian platform comprises the hydrologic Euphrates–Gulf–Rub al Khali basin, a huge, partly artesian basin, in which Cretaceous sandstones, Cretaceous–Paleogene carbonate formations and Neogene deposits form a complex multi-aquifer system, which, despite its large horizontal dimension, shows a rather uniform general set-up. The system of Cretaceous to Neogene–Quaternary aquifers with circulating groundwater has been sub-divided into three sections (FAO 1979):

- Upper aquifer: Neogene sediments and Paleogene Damam formation
- Middle aquifer: Paleogene Umm er Radhuma formation and permeable parts of the overlying Rus and underlying Aruma formations
- Lower aquifer: Wasia–Biyadh formations (“Cretaceous sandstone aquifers”)

Productive aquifers are found, in particular, in the sequence of Paleogene carbonate rocks, which are exposed in an outcrop belt extending for around 1,500 km from southern Iraq to the eastern province of Saudi Arabia and into the Hadramaut–Dhofar highlands on the southern boundary of the Rub al Khali desert. The partly karstified carbonate aquifers reach below Neogene–Quaternary sediments toward east and north to the Euphrates valley and the Gulf coast.

East of the main outcrop belt, the Paleogene formations are again exposed or uplifted to near surface in several anticlinal structures: the Ghawar anticline, Damam dome, Bahrain dome and Qatar uplift. In some of these outcrop areas, lenses of shallow locally recharged groundwater are found in the uplifted Paleogene formations, e.g., in Bahrain and Qatar.

In the regional aquifer system, groundwater moves generally from the outcrop areas of the Umm er Radhuma and Damam formations on the upper fringes of the hydrogeologic basin to discharge zones at topographically low elevations. Discharge zones in springs and sabkhas are located, in particular, in:

- The Euphrates Valley, Shatt el Arab and the Gulf coast in the northern part of the basin
- Oases, sabkhas and the sea shore in the coastal area of the Gulf
- Sabkha areas within the Rub al Khali and on the gulf coast of Abu Dhabi north of the Rub al Khali

According to the location of major discharge zones, the Euphrates–Gulf basin can be divided into three sub-basins corresponding to the main structural geologic units or segments. These sub-basins are delimited vaguely along groundwater stream lines rather than by sharp water divides.

We may distinguish:

- The Euphrates–Shatt el Arab sub-basin with groundwater discharge in the Euphrates valley and groundwater flow into the northern Gulf coastal area
- The middle Gulf sub-basin with groundwater flow to the spring discharge zone in the Hasa oasis and on the Gulf coast and to an extensive sabkha belt parallel to the Gulf coast
- The Rub al Khali sub-basin

The Paleogene aquifer system constitutes one of the most important groundwater reservoirs of the Arabian Peninsula. The stored groundwater is, however, prevailingly fossil, present-day recharge is very limited according to the arid climate conditions. Lenses of fresh to slightly brackish water are sustained by recent recharge under karstic surfaces in the Gulf area, e.g., in Qatar.

6.2.2 Hydro-Stratigraphic Sequence

As mentioned above, the multi-aquifer system of the eastern Arabian platform comprises three main aquiferous sections (Table 6.1):

- A lower sandstone aquifer unit
- The Umm er Radhuma aquifer as a middle aquifer section
- An upper aquifer section with the Damam formation as the main aquiferous unit

The sequence of Paleogene rocks includes the two most important aquifers of the eastern Arabian platform: the Umm er Radhuma aquifer and the Damam aquifer. The two aquifers are separated by the Rus formation which consists of evaporites, marls and limestones.

The Umm er Radhuma formation is composed prevailingly of limestones and dolomites, which are partly karstified, and is, in general, an aquifer with considerable productivity. Shale intercalations can form aquitards within the aquifer. The Rus formation comprises anhydrite, limestone and shale and forms an aquitard with many discontinuities, which are related to erosion over anticlinal structures, aquiferous sections in limestones, and dissolution of evaporites.

The Damam aquifer is divided in the Gulf area around Dhahran, Bahrain and Qatar into two aquiferous units: a lower unit: Khobar aquifer, and an upper unit: Alat or Abruq aquifer; the aquiferous units are separated by marly sediments of the Alat formation. The Rus formation, which generally acts as aquitard between the Umm er Radhuma and Damam aquifers, forms a joint aquifer with the Umm er Radhuma formation in some areas, where it is represented by fissured carbonates or where dissolution of evaporite layers has created secondary permeabilities.

The Umm er Radhuma aquifer extends into the upper part of the Cretaceous Aruma limestones. Shales of the lower part of the Aruma formation generally act as aquitard, separating the Umm er Radhuma–Aruma aquifer from Cretaceous sandstone aquifers.

In the north of the eastern Arabian platform, the Upper Cretaceous Tayarat carbonate formation constitutes a separate aquifer below the Gharra beds at the base of the Umm er Radhuma formation, an aquitard which is composed mainly of marly limestone, dolomite and anhydrite.

The deeper sandstone aquifer consists of the Lower Cretaceous Biyadh and Wasia sandstone formations, which dip from the outcrop belt on the Interior Shelf under the sedimentary cover of the eastern Arabian platform.

6.2.3 Cretaceous–Tertiary Aquifers

6.2.3.1 Cretaceous Aquifers

The Paleogene is underlain on the eastern Arabian platform prevailingly by sandstones of Paleozoic to Cretaceous age (Wasia, Biyadh sandstones, sandy facies of Aruma formation, Mukalla/Tawila sandstones in Yemen). The thickness of these sandstone formations, which have a relatively high permeability, is several hundred metres. Groundwater salinity is moderate in some areas on the boundary between Interior Shelf and eastern Arabian platform but increases to high values in the centers of the hydrogeologic sub-basin.

The Upper Cretaceous Aruma and Tayarat formations, composed mainly of limestones, dolomites and chalk, constitute an aquifer of secondary importance in the northern parts of the eastern Arabian platform. The 400–600 m thick Aruma formation acts, in many areas, as an aquitard below the Umm er Radhuma aquifer together with the shaley upper sections of the Wasia formation.

6.2.3.2 Umm er Radhuma Aquifer

The Umm er Radhuma aquifer, the most productive section in the multi-aquifer system within the middle Gulf sub-basin, comprises the Umm Er Radhuma formation and, in some areas, permeable parts of the Aruma and Rus formations.

The Umm er Radhuma is mainly a limestone sequence with dolomitic, shale and marl intercalations. The thickness of the Umm er Radhuma formation increases from about 240 m in the outcrop area at Wadi Batin and 490 m in Wadi al Miyah to 700 m at the Gulf coast. Over structural highs, such as the Ghawar anticline and the Damam dome, the formation thickness thins to around 300 m. The top of the Umm er Radhuma formation has a general dip from >400 m asl in the outcrop area in the west to around 220 m below sea level in most of northeastern Saudi Arabia.

The Umm er Radhuma formation is exposed in karstified outcrops on the uplifted margins of the sub-basin.

In wide parts of the outcrop areas, the formation is unsaturated or contains discontinuous groundwater occurrences. In southwestern Iraq, the Umm er Radhuma comprises mainly discontinuous perched aquifers or shallow aquifers with low productivity on the Waqsa elevation. In the west (areas west of and around Salman and on the western margin of the Wadi Batin depression), the Umm er Radhuma formation provides a generally confined aquifer with moderate productivity below the Rus aquitard. The top of the aquifer is situated in these areas at around 50–100 m below land surface.

In the plateau areas of the middle Gulf basin in northeastern Saudi Arabia, the Umm er Radhuma formation constitutes a confined aquifer with moderate to high productivity.

In deep structural depressions, the Umm er Radhuma aquifer is situated at depths of several hundred metres below land surface and contains brackish to saline groundwater. In the centre of Dibdiba depression and under the Kuwait plain, the top of the aquifer is located at depths of more than 600 m below land surface.

On the northern slopes of the Hadreamaut highlands in eastern Yemen, the Umm er Radhuma formation is represented by limestones and dolomites with joints and fissures, but with limited karst development. The 170 to more than 200 m thick formation provides an aquifer with moderate to poor productivity. Further south along the Hadramaut arch, the formation is generally unsaturated. In several areas, perched groundwater occurrences rest on unpermeable layers within the Umm er Radhuma formation. In the Nejd plateau in southern Oman, the Umm er Radhuma formation comprises two main aquiferous zones in highly solution fissured sections:

- A confined aquifer with, in some areas, high productivity in the upper Umm er Radhuma formation
- The principal aquifer in the top section of the lower Umm er Radhuma formation, a generally confined karstic aquifer

Aquiferous fracture zones with relatively low productivity are found in the lower Umm er Radhuma formation below the karstified top section.

In eastern Yemen, a major productive zone of up to 60 m thickness appears to be situated at the top of Umm er Radhuma/base of Jeza formation. In the western part of the Hadramaut plateau, the Umm er Radhuma formation is underlain, in general, by Upper Cretaceous sandstones (Wasiya, Mukalla/Tawila sandstones). Towards north and east, the Upper Cretaceous appears to comprise a more clayey lithology: limestones, marls, shales and clastics of the Mahra group.

The Umm er Radhuma aquifer extends under much of the Rub al Khali basin. In the Rub al Khali of Saudi Arabia, the aquifer is 150–400 m thick and is artesian in some areas. In the south of the United Arab Emirates, the Umm er Radhuma aquifer is 300–550 m thick. In the Liwa area in the northern Rub al Khali, the top of the Umm er Radhuma is situated at 985 m below surface and the aquifer contains highly saline water.

In most of the Gulf coastal area, the Umm er Radhuma aquifer is confined, but in the core of anticlinal structures, it forms an unconfined aquifer connected with aquiferous sections of the Rus formation.

6.2.3.3 Rus Formation

The Rus formation, a complex of anhydrite, limestone and shale, acts mainly as an upper aquiclude or aquitard for the Umm er Radhuma aquifer. The formation is aquiferous, where it is developed in carbonate facies and comprises thick sequences of limestone beds and, in particular, where evaporites of the formation have been removed by dissolution, such as in Qatar and Bahrain.

Aquiferous sections are found in the Rus formation:

- In Qatar and Bahrain in limestones and at places, where dissolution of evaporites has created secondary permeabilities.
- In fissured limestones at the top of the formation in the Nejd plateau in southern Oman.

In most of the coastal area of Saudi Arabia, the Rus formation along with the Midra and Saila shales and the Alveolina limestone members of the Damam Formation, form a confining bed that separates the Umm er Radhuma aquifer from the Khobar aquifer. This confining bed consists mainly of chalky limestones, anhydrites, marls and shales with a thickness range from 10 to 110 m.

In eastern Yemen, the Rus formation comprises generally a 50–250 m thick evaporitic sequence of bedded to massive gypsum and anhydrite with intercalations of thin beds of chalky porous limestone and is underlain by the Jeza formation.

At some locations, karstified sections of the Rus formation contain perched groundwater above the underlying Jeza formation.

In the northern part of the Hadramaut plateau, the top of the Rus formation comprises an unconfined aquifer composed of 100 m thick limestones at 50–140 m below land surface.

The Jeza formation consists of shales or calcareous mudstones with intercalations of calcareous or gypsiferous bands. The permeability of the 40–130 m thick formation is generally very low. The formation is exposed over large areas of the Jol plateau where it restricts recharge to underlying units and causes rapid runoff following rainfall.

In the western part of Hadramaut, aquiferous layers in the Jeza formation are connected with the underlying Umm er Radhuma aquifer.

6.2.3.4 Damam Aquifer

The Damam aquifer provides the most important aquifer of the Euphrates–Shatt el Arab sub-basin. Outcrops extending over considerable areas west of Salman probably constitute the main recharge areas of the aquifer. Recharge occurs prevailingly through infiltration of local runoff on karstic surfaces and in morphologic depressions and wadi beds. The thickness of the saturated aquifer increases from a few metres in the morphologically higher parts of the outcrop area to 30–80 m in the more northern parts of the Salman zone, where the depth to groundwater of the main Damam aquifer reaches 60–120 m below surface. In the Euphrates depression, the Wadi Batin plain and under the Kuwait plain, the Damam aquifer is confined below an aquitard composed of Neogene marls.

In the central parts of the Dibdiba depression, the Damam formation constitutes a deep aquifer with its top descending from around 100 m asl at the border between Saudi Arabia, Iraq and Kuwait to 300 m below sea level in the north.

In Kuwait the Damam aquifer provides a productive reservoir of brackish groundwater. The thickness of the Damam formation increases from about 120 m in the southwest to about 300 m in the northeast of Kuwait. The top of the formation dips from 140 m asl (75 m below ground surface) to 150 m below below sea level under Kuwait Bay and more than 300 m below sea level in northeastern Kuwait. The general dip is interrupted by a dome structure in eastern Kuwait, where the top of the Damam limestone is exposed at the surface at Ahmadi.

The main Damam aquifer in Kuwait consists of chalky and karstified limestones of the upper and middle units of the formation.

The Damam aquifer in Kuwait is generally confined under an aquitard, which is composed of a cherty chalky limestone unit of the upper Damam together with the basal clay and calcrete layers of the overlying Kuwait group. The thickness of these aquitard layers ranges between a few meters to about 30 m, but hydraulic connections exist between the Damam aquifer and the overlying Kuwait aquifer. The lower shaley and anhydritic part of the Damam formation together with the anhydritic Rus formation forms a barrier to the upward flow of water from the underlying saline Umm er Radhuma aquifer. The eastern boundary of the brackish Damam aquifer coincides with a saline water front running approximately parallel to the Gulf coast.

In some parts of the Gulf coast area in Saudi Arabia, the Damam formation provides an important aquifer with relatively high well yields. The productivity of the aquifer is attributed, in particular, to considerable upward leakage from Umm er Radhuma aquifer.

In the Gulf coast area, the Damam aquifer is formed by two limestone complexes of the Damam formation, the Khobar and Alat members. The lower part of the Alat member consists of dolomitic marl with thin limestone interbeds and is up to 35 m thick. These marly beds, also known as the "orange marl", act as a leaky aquitard between the Khobar and Alat aquifer sections. In wide areas, the Damam aquifer is confined below thick argillaceous Neogene sediments. The base of the Damam aquifer is formed by the low permeability shales of the Damam formation, together with the Rus anhydrite.

The thickness of the Damam formation increases from around 70 m in the Wadi al Miyah area toward the Gulf coast, but in anticlinal structures the thickness of the formation is reduced. Variations of the formation thickness are shown in Table 6.2.

On the southern margin of the Rub al Khali sub-basin in northern Hadramaut, the Habshiye (Damam) formation is composed of paper shales, gypsiferous marls, sandstones, chalky limestones and siltstones in a thickness of 40 m to >350 m. The formation is generally unsaturated and contains locally perched aquifers, from which limited quantities of groundwater are extracted through shallow wells in some places.

The Damam aquifer probably extends under wide parts of the Rub al Khali in Saudi Arabia with a maximum thickness of about 90 m. Water levels of the Khobar member are found at depths of 35–60 m below surface. The Alat member constitutes an aquifer with low productivity in the northern and eastern parts of the Rub al Khali in Saudi Arabia at depths of about 12–20 m below surface. In some parts of the western Rub al Khali, the Damam formation has been eroded.

The Damam aquifer continues northward into the United Arab Emirates and possibly extends in the subsurface over most of Abu Dhabi Emirate. An increase in

Table 6.2 Thickness and position of the Damam aquifer after Al-Sayari and Zötl (1978), Zubari (1997)

Area	Thickness (m)			Top of aquifer depth bls (m)
	Alat	Khobar	Orange marl	
Wadi al Miyah	25	45		80–250
Gulf coastal area	30–50	50–65	6	100
Gulf coast near Damam	100	60	15	
Al Hasa	20	25–35		230
Damam and Bahrain anticlines				
Crest	Missing	Missing	9	Near surface
Slopes	20	30–50		
Northern–central Qatar		25		Near surface
Southwestern Qatar	25			25

thickness and a major change of the lithology to marls occurs east of a line Al Ain–Abu Dhabi. No productive aquifer has been found in these marly Paleogene sediments (Padbeh formation) in the northern Emirates. Near the western border of the northern Oman mountains, alluvial deposits overlie directly the Upper Cretaceous Juweiza Formation and the Paleogene appears to be missing.

At Jebel Hafit, which rises as an approximately north–south trending anticline from the gravel plain south of Al Ain, fractured Paleogene limestones contain thermal brackish groundwater. The groundwater has been tapped in a well field at the foot of the 1,240 m high Jebel Hafit.

Much of Al Wusta in central Oman is underlain by the Tertiary Hadramaut and Fars groups, which consist predominantly of marine limestones, marls and thin interbedded evaporites, and terrestrial sequences in the uppermost Fars group. Beneath the central plateau, Tertiary limestones and marls form an unconfined aquifer with the water table generally occurring 100–150 below ground surface.

6.2.3.5 Neogene Aquifers

Neogene deposits extent over wide areas of the eastern Arabian platform. The marine to terrestrial Neogene sediments contain alternating layers of marly sandstone, sandy marl, calcarenite and limestone. Some intercalations with higher permeability are important as local aquifers or as components of the sub-regional aquifer system.

On the Suman plateau, the Neogene includes clastic deposits with evaporite intercalations and karstified carbonates. The Neogene deposits are here generally unsaturated, but act as collectors of recent recharge to the hydraulically connected underlying Paleogene aquifer complex.

The Neogene sedimentary sequence overlying the Damam aquifer in eastern Saudi Arabia and the Gulf area comprises two significant aquifers: the detrital Hadrukh aquifer and karstified limestones of the Dam formation. The Hadrukh sandstones have their highest permeabilities in the area southwest of Hofuf. An erosional gap over the Ghawar anticline allows groundwater leakage from the Umm er Radhuma aquifer into the Hadrukh aquifer. The sediments of the Hofuf and Kharj formations are generally situated in the unsaturated zone.

At Al Hasa near Hofuf in eastern Saudi Arabia, the Dam Formation comprises fissured and karstified limestones, from which a significant number of springs supplies water to the 200 km^2 wide Al Hasa oasis area. Major quantities of the spring discharge at the oasis appear to ascend from the underlying Umm er Radhuma aquifer.

In Kuwait, the Damam aquifer is covered by Miocene–Pliocene sandstones and conglomerates of the Kuwait group, which is sub-divided into three formations: Ghar, Lower Fars and Dibdiba formations with a total thickness of a few metres in the west, about 100 m in the north and 420 m in northeast Kuwait. The middle and lower formations of the group constitute an extensive water table aquifer (Kuwait aquifer).

In the Raudhatain and Umm el Aish areas in northern Kuwait, the Dibdiba formation contains shallow fresh water lenses in morphologic depressions.

In the Al Wusta central plateau in interior Oman, main aquiferous sections are found in the limestone, marl, evaporite sequence of the Fars formation and in formations of the underlying Paleogene Hadramaut group.

6.2.4 Gravel Plain Aquifers

The northern Oman mountains are adjoined, in the west, by a gravel plain, which extends as a 10–40 km wide strip from Ras el Khaima in the north to Al Ain (Al Jaw plain) in the south. The surface of the gravel plain descends in a gentle slope of around 1‰ from east to west.

The gravel plain connects the sand and gravel fans of major wadis, which enter the foreland on the western margin of the Oman mountains. In the west, the gravel plain is delimited by the cover of sand seas of the Rub al Khali. On some stretches, the gravel plain is interrupted by protruding sand dunes or by hills with outcropping Tertiary formations, e.g., in Jebel Hafit.

The formation of the alluvial fan aquifer began in the middle Miocene with the deposition of erosional material from the uplifting Oman mountains continues today. The gravel fan aquifer is composed mainly of partly cemented ophiolite clastics. The unconsolidated cover of the gravel plain consists of an up to 100 m thick sequence of alluvial sand and gravel with thin interbeds of silt and clay, with a general gradual change from relatively coarse grained sediments in the wadi fans to more fine grained deposits toward west and in the areas between the main fans. The alluvial deposits extend under the sand dune fields to distances of about 70 km from the western boundary of the Oman mountains.

The western gravel plain (western Bayada) in the Dhayd area comprises a productive Quaternary aquifer composed mainly of alluvial deposits with a thickness of 20 m to some tens of metres. The aquifer is intensively used for irrigation and has largely been depleted around Dhayd.

The aquifer is underlain, in most areas, by the Miocene Fars formation, composed mainly of continental deposits: siltstones, marl, claystones, mudstones and evaporites. The evaporites are generally lenses of gypsum of a few metres in thickness. The thickness of the Fars formation varies from 0 near the Gulf coast, where it has been eroded, to nearly 200 m near the Oman border.

In the Dhayd area, the gravel fan aquifers is underlain by Upper Cretaceous sediments (Juweiza formation) comprising a sequence of carbonates and shales (flysch deposits), the thickness of which reaches several hundred metres (>800 m in one borehole). The Juweiza formation constitutes an aquifer with low to moderate productivity, which is in hydraulic contact with the overlying Quaternary aquifer.

In the Al Ain–Buraimi area, several wadis or alluvial plains extend along the eastern margin of the western gravel plain on the entrance of wadis from the ophiolite mountains (Wadi Safwan, Al Jaw plain). Groundwater and surface water from the mountain catchments drain in "gaps" through the mountain front. In the north and west of Al Ain, present-day wadis are located between northeast–southwest trending sand dunes. Buried paleo-drainage networks contain saturated alluvial fill and may constitute major fresh water tongues.

The alluvium of Wadi Safwan forms an elongated trough west of the mountain front with a saturated thickness of up to 40 m.

The Al Jaw plain extends southeast of Al Ain over around 400 km^2 between the western margin of the Oman mountains and the eastern border of Jebel Hafit. The plain is covered by Quaternary–Recent sandy gravels with argillaceous and calcareous cements. The Quaternary sediments are underlain, in most of the Al Jaw plain, by the Oligocene–Miocene Fars formation, which consists of clay, marl, limestone, gypsum and halite. In the north of the plain, the Quaternary deposits rest on the Paleogene Damam formation. Alluvial conglomerates occur along the hill border of the plain; the west and south of the plain is partly covered by low sand dunes.

Aquiferous horizons are found in the Quaternary–Recent sediments and within the Fars formation. Surface and groundwater replenishing the aquifer enter the plain through two main channels.

On the Dakhiliye or interior foreland of the Oman mountains, alluvial plains cover a synclinal structure between the southern slope of Jabal Akhdar and the Adam mountains, in which Mesozoic rocks re-emerge from the plain in an anticlinal uplift 80 km south of the main Oman mountain massif.

Up to 80 m thick Quaternary alluvial deposits fill north–south trending wadi systems within Jebel Akhdar and extend over large alluvial fans on the plains adjoining the mountains in the south.

The alluvial deposits are composed mainly of coarse to fine grained gravels of varying lithological composition, sand silt and clay. Evaporitic salts and gypsum occur at the surface and in the alluvial sequence.

The Quaternary deposits overlie or adjoin rocks of the Hawasina nappe, mainly turbiditic limestones and radiolarites, carbonates of the Hajar supergroup and Semail ophiolites.

The alluvial sediments constitute an unconfined aquifer in the northern piedmont area (Wadi Halfayn). On the alluvial plains further downstream, the aquifer is semi-confined to confined under up to 10 m thick silt deposits, which act as local confining layers.

The Quaternary aquifer is recharged from surface flow, which is ephemeral and occurs only after occasional heavy rainfall events.

References. Abderrahman et al. (1995), Ahmed and Kraft (1972), Al Ansari et al. (1995), Al Awadi and Mukhopadhyay (1995), Al-Awadi et al. (1998), Allam (1995), Al Mashadani (1995), Al-Murad and Zubari (1992), Al-Murad (1994), Al-Mutawa (1993), Al-Ruwaih (1996b), Alsharhan et al. (2001), Amer et al. (1990), Bakiewicz et al. (1982), BRGM (1977), Burdon and Al-Sharhan (1986), Clark et al. (1987), Dabbagh et al. (1988), Edgell (1997), ESCWA (1999b), FAO (1979), GEOCON (1995), IWACO (1986), Macumber (1995), Martin and Krapp

(1977), Matter et al. (2006), MAW (1984), Mukhopadhyay (1999), Mukhopadhyay et al. (1994b), Otkun (1973, 1974, 1975), Pike (1985), Qutob (1978), Read (1995), Rizk et al. (1997), Robertson (1992), Sayed and Al-Ruwaih (1995), Zubari (1997).

6.3 Groundwater Regimes

6.3.1 Hydraulic Parameters

6.3.1.1 Umm er Radhuma Aquifer

The Umm er Radhuma formation provides the most productive aquifer in the middle Gulf segment of the eastern Arabian platform with transmissivity values between 5,000 and 11,000 m^2/d. The distribution of hydraulic conductivity values within the aquifer appears to be controlled mainly by lithologic variations and the development of secondary permeability in fractures, fissures and solution openings. The limestones of the Umm er Radhuma formation are generally fine grained, the occurrence of higher transmissivities depends on the degree of dolomitization and the development of secondary openings.

Moderate transmissivities of 360–6,000 m^2/d are found in the Umm er Radhuma aquifer in wide parts of eastern Saudi Arabia, relatively high T values are reported from Wadi al Miyah (10,000–36,000 m^2/d), Damam and Qatif areas (14,000–55,000 m^2/d).

Relatively low permeabilities prevail in the northern parts of the eastern platform, where the Umm er Radhuma formation is developed in a more argillaceous, anhydritic facies.

Transmissivities are particularly high, where karstification affected the upper units of the aquifer, in particular in the coastal area of the middle Gulf segment and along the crests of anticlinal uplifts. High transmissivities on the crest of the about 100 km wide Ghawar anticline of 52,700 m^2/d decrease to values between 475 and 950 m^2/d on the eastern flank of the anticline and of 59 m^2/d outside the anticlinal structure.

Around Salman in southern Iraq, transmissivities range from 10 to 1,700 m^2/d.

Mean specific capacity values of boreholes tapping the Umm er Radhuma aquifer (C-aquifer) in the Nejd in southern Oman are reported as 17 m^3/h/m, transmissivities are in a range of 500–4,000 m^2/d.

The relatively high transmissivities of the Umm er Radhuma aquifer, found in several areas, are related, in particular, to the considerable aquifer thickness, which is generally in a range of several hundred metres.

6.3.1.2 Umm er Radhuma–Rus Aquifer in Bahrain and Qatar

Transmissivity values of the Rus–Umm er Radhuma aquifer in Bahrain range from 1,210 to 40,000 m^2/d. Testing of wells completed in the residual Rus formation in central Bahrain, where the anhydrite is removed by solution, showed a wide range of T values between 110 and 13,630 m^2/d, related to the high local variation in degree of solution of the anhydrite.

In Qatar, transmissivity varies between 2 and 4,500 m^2/d for the Rus aquifer and between 232 and 8,433 m^2/d for the Umm er Radhuma aquifer.

6.3.1.3 Damam Aquifer

Although the Damam aquifer is thin compared to the Umm er Radhuma aquifer, it has been tapped by thousands of wells in Saudi Arabia, Bahrain, Kuwait, and Qatar because of the shallow depth to the water and reasonably high permeabilities. The aquifer has, however, limited storage capacity.

Transmissivities of the Damam aquifer in the middle Gulf segment and the Euphrates–Shatt el Arab segment range from <10 to 45,000 m^2/d. In Kuwait, the transmissivity of the Damam aquifer is controlled by zones of karstification, dolomitization, faulting and fracturing and varies from 13 to 4,640 m^2/d, with a mean of 481 m^2/d (data from 194 wells). Generally, the transmissivity data in Kuwait scatter around low values of about 100 m^2/d with higher transmissivities at major anticlinal and fault structures.

The degree of secondary porosity related to karstification and fissuring of the Damam formation gradually decreases in Kuwait from southwest to northeast, comformable to the topographic slope and groundwater flow direction. Accordingly, a general decrease of transmissivity and of productivity of well fields is observed in direction of groundwater flow from southwest to northeast. Significant local variations of transmissivity are observed on domal structures, which may be related to an increase of the intensity of fracturing on the anticlines.

Calibrated T values for the Alat and Khobar members of the Damam aquifer in the Gulf coastal area range between 600 and around 45,000 m^2/d (Alat 610–34,700 m^2/d, Khobar 660–45,500 m^2/d). Values of hydraulic conductivity of the aquifer from tests in eastern Saudi Arabia and Bahrain are reported as 10^{-5} to 10^{-4} m/s for the Alat aquifer and 10^{-7} to 10^{-2} m/s for the Khobar aquifer. Generally, permeabilty is relatively low in the Alat aquifer and is higher in the karstified and fissured Khobar aquifer, particularly on structural highs.

In a similar trend as in the Umm er Radhuma aquifer, T values increase toward the crest of the Ghawar anticline.

Transmissivity values of the Damam aquifer in different areas of the middle Gulf segment are reported as:

Al Hasa area (Hofuf)	14.7–3,024 m^2/d
Alat member on the Damam dome	27–200 m^2/d
Alat member in Bahrain average	360 m^2/d
Khobar member in Bahrain up to	41,000 m^2/d
Alat (Abarug) member in Qatar	<10–1000 m^2/d, mean 355 m^2/d

6.3.1.4 Paleogene–Neogene Aquifers in Southern Iraq

The data of well yields and specific capacities in southern Iraq show moderate to high values for the Damam and Umm er Radhuma aquifers. Transmissivity values reported for the Damam and Umm er Radhuma aquifers around Salman zone range generally from 3–1,500 m^2/d and 10–1,700 m^2/d, respectively. From boreholes tapping the Umm er Radhuma aquifer together with the underlying Tayarat aquifer, transmissivity values of 220–570 m^2/d are reported.

6.3.1.5 Neogene Aquifers in Saudi Arabia and Kuwait

In the Neogene sedimentary sequence of the middle Gulf sub-basin, fresh groundwater occurs in the detrital sediments of the lower part of the Hadrukh formation and in karstified limestones of the Dam formation. Transmissivities of the Hadrukh aquifer around Hofuf vary from 60 m^2/d to 3,500 m^2/d, and transmissivities of the overlying Dam aquifer are around 1,000 m^2/d on the Ghawar anticline. In other parts of the middle Gulf sub-basin, transmissivities of the Neogene aquifer appear to be lower.

The saturated thickness of the Kuwait group aquifer increases from a few metres in the southwest of Kuwait to 400 m in the northeast; the average transmissivity increases accordingly from 10 $m^2/$ to 1,500 m^2/d.

6.3.1.6 Permeabilities of Confining Layers

The lower boundary of the Umm er Radhuma aquifer corresponds over most of the eastern Arabian platform to a shale and marl sequence within the Aruma formation, which forms an aquitard. Vertical permeabilities of 33 core samples of the Aruma shale ranged between 10^{-10} and 6×10^{-9} m/s.

Vertical permeabilities of the anhydrite, limestone, shale sequence of the Rus aquitard overlying the Umm er Radhuma aquifer in Saudi Arabia are in the order of 10^{-9} to 10^{-8} m/s.

In Bahrain and most of the coastal area of Saudi Arabia, the Rus formation along with the Midra and Saila shales and the Alveolina limestone members of the Damam formation form a confining bed that separates the Umm er Radhuma aquifer from the Khobar aquifer. A vertical permeability of the confining marls between the Damam aquifer and the Umm er Radhuma aquifer is reported as 4×10^{-9} m/s for one core test of the shales in Bahrain. During pump tests of the Umm er Radhuma aquifer in Bahrain, no impact on the piezometric head in the overlying Khobar aquifer was observed.

The lower unit of the Alat member, the Alat marl or orange marl, forms a discontinuous confining bed in some areas, that separates the Khobar and Alat aquifers. Vertical leakage values of the Alat marl range from 10^{-10} to 10^{-9} m/s.

The Neogene complex, overlying the Damam aquifer in most of the Gulf coast area, functions as an aquifer for lateral flow and as an aquitard for vertical flow. Its aquitard characteristics are determined by the clay and shale beds: shales of the Hadrukh formation and marls of the Dam formation. The vertical hydraulic conductivity of the shales, which are not distributed continuously, ranges between 10^{-12} to 7×10^{-9} m/s.

6.3.2 *Groundwater Recharge*

Groundwater recharge rates on the arid eastern Arabian platform are, in general, very low and, in some areas, practically negligible. Even low recharge rates can, however, produce significant volumes of groundwater recharge on the very extensive outcrop areas of some of the aquiferous formations. Estimates of mean recharge volumes indicate e.g., 1×10^9 m^3/a for the Umm er Radhuma aquifer in eastern Saudi Arabia and 200×10^6 m^3/a for the Neogene aquifers of eastern Saudi Arabia.

6.3.2.1 Southern Iraq

In southern Iraq, mean annual rainfall ranges from 94 to 140 mm. Maximum precipitation during 24 h can exceed 40 mm. Recharge on the outcrops of Paleogene carbonate formations may occur prevailingly through runoff infiltration in local depressions and wadi beds.

From water level fluctuations in the Wadi Batin plain, the following recharge estimates for the Neogene Dibdiba aquifer, which overlies the Paleogene Damam aquifer, have been made:

- Upper Dibdiba aquifer, sand and gravel:
  Recharge in wet year 9.72 mm
  Recharge in dry year 4.05 mm
  Recharge in percentage of annual rainfall 3.8–14.3%

- Lower Dibdiba aquifer, sandstone with sand and claystone:
  Recharge in wet year 7.6–12 mm
  Recharge in dry year 2.5–7 mm
  Recharge in percentage of annual rainfall 4.85–8%

Average recharge on the Umm er Radhuma and Damam aquifers in the Salman zone in southwestern Iraq are estimated from water level observations at 6–7.5 mm per year, according to the following data:

Mean water level fluctuation	0.41 m/a
Mean storage coefficient	0.015–0.02

Recharge has, however, to be expected to vary in wide ranges locally and in different parts of the area according to the infiltration conditions.

6.3.2.2 Kuwait

Under the climatic and morphologic conditions of Kuwait, significant runoff occurs only in some wadi systems in the northern part of the country. Mean annual rainfall, related mainly to storm events between November and April, is about 110 mm for the years 1962–1986 with a variabilty from 31 mm (1964) to 242 mm (1976). Potential evapotranspiration averages 2.3 m/a. Amount, duration and distribution of rainfall are generally not sufficient to produce significant runoff or direct infiltration. In the north of Kuwait, some recharge occurs from runoff in extended wadi systems which converge into the morphologic depressions of Raudhatain and Umm al Aish, creating isolated fresh water lenses in the Kuwait group aquifer floating on brackish to saline water. Average annual recharge in the Raudhatain–Umm al Aish area is estimated at 850,000 m^3. The recharge has no remarkable impact on the water quality in the deeper Damam aquifer.

6.3.2.3 Northeastern Saudi Arabia

Recharge to the Umm er Radhuma aquifer of northeastern Saudi Arabia has been calculated for an area, where the aquifer is covered mainly by sand dunes of Ad Dahna and gravel plains. The model calculations consider rainfall, infiltration rate, evaporation, and soil moisture deficiency, and is applied to four basic environments with different recharge conditions:

- Ponding in areas northwest of Hafr al Batin with low infiltration rates and high evaporation rates of ponded runoff
- Ponding adjacent to sand dunes
- Infiltration through sand dunes
- Infiltration in gravel plains

Computed annual recharge volumes for data of 27 years (1952–1978) show a wide variation in different years, ranging from $8.3 \times 10^6\ m^3$ to $4 \times 10^9\ m^3$ over the total investigated area of 216,000 km^2. This reflects, to some extent, directly the variation of rainfall, but a certain minimum rainfall is required before any recharge can take place.

Mean annual recharge calculated for the different recharge conditions are:

- Ponding adjacent to Ad Dahna sand dunes: $620 \times 10^6\ m^3$ over an area of 34,900 km^2 or 17 mm
- Direct infiltration through Ad Dahna sand dunes: $203 \times 10^6\ m^3$ over an area of 53,300 km^2 or 4 mm

- Ponding on the gravel plain northwest of Hafr el Batin: 195×10^6 m^3 over an area of 75,600 km^2 or 2.6 mm
- Direct infiltration east of Ad Dahna: 29.7×10^6 m^3 over an area of 57,000 km^2 or 0.05 mm

Calculation of direct infiltration through the gravel plain east of Ad Dahna indicate significant recharge in only three out of 27 years. Minimum rainfall for recharge in that area is around 60 mm.

Bakiewicz et al. (1982) report results of recharge estimates for different aquifers of northeastern Saudi Arabia for the same period (1952–1978): Umm er Radhuma, Damam and Neogene aquifers.

Recharge occurs as:

- Direct infiltration on the outcrop of the aquifer
- Direct infiltration of rainfall on pervious superificial deposits overlying the sub-outcrop of the aquifer, reaching the aquifer through percolation in the unsaturated zone
- Infiltration of run-off to stream beds or of run-off ponded over the outcrop of the aquifer

Recharge was estimated from rainfall amounts, intensity and duration, evaporation and evapotranspiration, infiltration intakes, field capacity of the soils and run-off measurements.

The estimates of recharge volumes give the, by far, highest figures for the Umm er Radhuma aquifer with a mean of 1.05×10^9 m^3/a over the 27 years. Estimates of mean annual recharge are 2×10^6 m^3 for the Damam aquifer and 20×10^6 m^3 for the Neogene aquifer. The recharge estimates for the Umm er Radhuma aquifer in individual years vary between 8×10^6 m^3 and 4.08×10^9 m^3. For the Damam aquifer, significant recharge of $>8 \times 10^6$ m^3 was calculated for only 9 out of 27 years, with a maximum of 149×10^6 m^3 in 1 year. For the Neogene aquifer, zero recharge was calculated for 14 of the 27 years, the maximum estimate for 1 year was 1.3×10^9 m^3.

Additionally, groundwater replenishment may occur from transfer of water by vertical flow from either shallower or deeper aquifers. The Umm er Radhuma aquifer is, in particular, replenished through vertical leakage from the lower aquifer (Wasia/Biyadh) and through diffuse leakage through the Aruma formation. Replenishment by vertical transfer from adjacent aquifers is expected to be rather constant due to the buffer effect of the huge groundwater storage in the major aquifers.

Studies of groundwater recharge on the karstified outcrop of the Umm er Radhuma formation were made at the As Sulb plateau in eastern Saudi Arabia. The Umm er Radhuma formation is partly covered by sand dunes, but the karstified limestones are directly exposed over wide ranges. Numerous sinkholes, collapse structures, karst shafts and corrosionally extended joints favour infiltration conditions. During episodic rainfall, surface runoff discharges over short distances into karst openings. Detailed studies in test areas proved a groundwater recharge of 47% of the average precipitation or a yearly average of 44 mm. Groundwater recharge from surface runoff into caves takes place from rainfall events with a precipitation rate exceeding 4.5 mm per half an hour. On average, four rainfall events with such an intensity may occur. Although the generalization of the observations for larger areas may be doubted, the investigations indicate that a considerable amount of recharge can be expected from runoff related to high-intensity and short duration rain storms on open karst surfaces.

6.3.2.4 Southern Oman

Modern recharge in the outcrop area of the Lower Umm er Radhuma aquifer in the Nejd in southern Oman is estimated to be in the order of 16.25 mm/a.

6.3.2.5 Gravel Plain Aquifer

Groundwater replenishment on the upper margin of the gravel plain aquifer of the United Arab Emirates was estimated as 0.31 m^3/s over a 150 km long stretch in the Dhayd–Al Ain area, including subsurface inflow from fractured ophiolites and inflow in wadis through gaps in the mountain foothills. Recharge from rainfall on the gravel plain was estimated at 0.57 m^3/s. Groundwater discharge from the gravel plain aquifer includes vapour loss by evaporation with an estimated rate of 0.57 m^3/s and discharge to the Gulf of around 0.35 m^3/s.

Recharge from pecipitation on the lower reaches of the gravel plain is thought to be negligible. During infrequent events of flood flow, recharge from the unsaturated zone may occur from transmission losses in wadi channels. Upward leakage from the underlying Fars formation into the gravel aquifer is a significant source of groundwater salinity in the gravel plain.

From throughflow calculations along a 65 km wide front of the gravel plain in the United Arab Emirates, recharge from wadi flow infiltration was estimated to correspond to 14% of rainfall in the catchment within the adjoining ophiolite mountains, or 4 m^3/s.

6.3.3 Groundwater Flow Systems and Flow Volumes

The Euphrates–Gulf–Rub al Khali basin can be divided, as stated previously, into three major hydrogeologic sub-basins:

- The Euphrates–Shatt el Arab sub-basin
- The midle Gulf sub-basin
- The Rub al Khali sub-basin

The upper boundaries of the hydrogeologic sub-basins are defined by the limit of saturation of Upper Cretaceous–Paleogene aquifers in the uplifted western and southern margins of the eastern Arabian platform. The sub-basins, defined in that

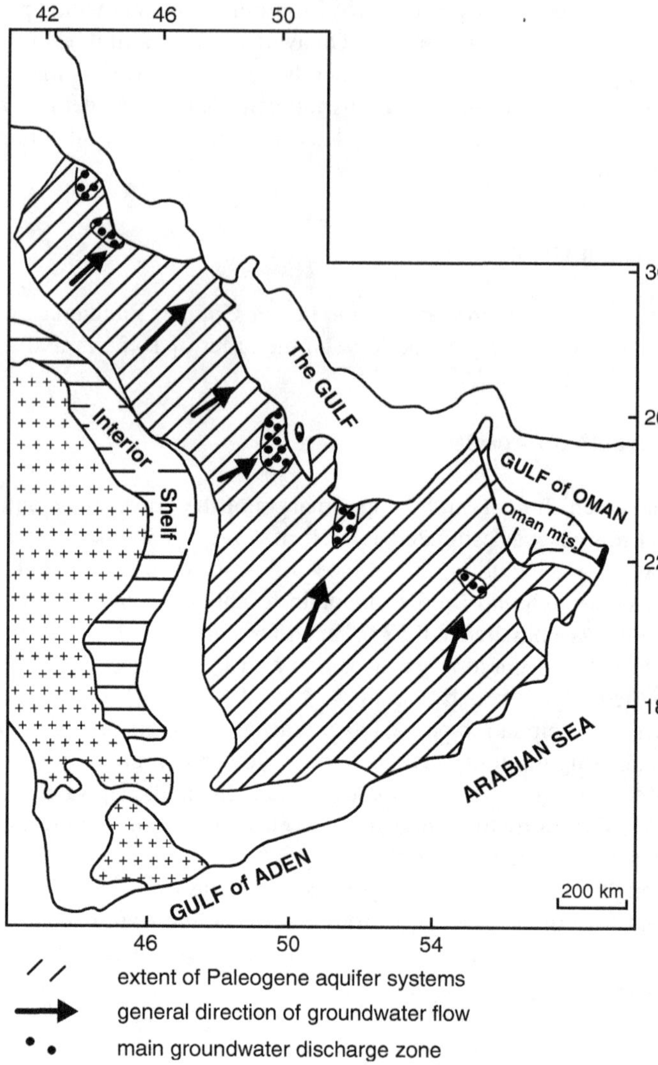

Fig. 6.6 Eastern Arabian platform: groundwater flow systems. After ESCWA (1999b)

sense, receive some subsurface inflow from aquifers of the Interior Shelf, in particular the Wasia–Biyadh sandstone aquifer (or "Cretaceous sandstone aquifer") and from leakage of perched aquifers, which occur locally on the basin margins.

6.3.3.1 Euphrates–Shatt el Arab Sub-Basin

Groundwater flow in the Umm er Radhuma, Damam and overlying Neogene aquifers of the Euphrates–Shatt el Arab sub-basin is directed, in general, from southwest to northeast. In the Euphrates valley, groundwater discharge occurs in large springs, into lakes and into natural depressions. In the northeast of the sub-basin, groundwater movement appears to be directed from the plateau area eastward into the topographically lower plain areas: Wadi Batin–Dibdiba depression and the Kuwait plain.

Main groundwater discharge zones are located along the Euphrates right bank and is attached to the Euphrates fault zone.

Probably, some of the groundwater discharge in the Euphrates valley is contributed through subsurface inflow from Pliocene–Quaternary aquifers of the Mesopotamian lowlands of the northern Arabian platform.

Groundwater discharge volumes from the Umm er Radhuma and Damam aquifers in the northern part of the Euphrates–Gulf sub-basin are in the order of 1.56–1.8 m^3/s or around 50 million m^3/a, comprising:

• Pumping from about 50 boreholes in the Salman zone (Table 6.3)	100–115 l/s
• Discharge from large springs in the Euphrates valley	964–1,120 l/s
• Discharge from small springs on the plateau area	100 l/s
• Evaporation from the groundwater fed Sawa lake	400–475 l/s

A major amount of the total groundwater flow discharges in springs along the Euphrates valley.

The total discharge of main springs in 1979 to 1981 was 1,120 l/s or 35.33×10^6 m^3/a (Tables 6.3 and 6.4, Fig. 6.7).

Parallel to the Gulf coast in Kuwait, a belt of saline water forms the eastern boundary of the Damam brackish water aquifer. The overlying Kuwait group aquifer is in direct contact with the sea along the coastal areas of Kuwait. Along a static saline water wedge, upward flow of groundwater from the Damam aquifer through the Kuwait group aquifer occurs in this area, with an ultimate groundwater discharge into the Arabian Gulf. The coast of northeastern Kuwait thus constitutes the lower end of the sub-regional groundwater flow system.

Groundwater discharge in Kuwait includes, at present, substantial extraction of brackish water. Groundwater is abstracted mainly from the Damam aquifer (Abdaliya, Sulaibiya, Wafra, and Shegaya B, C, and D well fields) and, to a minor degree, from the Kuwait group aquifer. Until 1989, the abstraction rate

Table 6.3 Well yields in Paleogene–Neogene aquifers in southern Iraq (after Al Mashadani 1995, values from a total of 104 wells)

Area	Aquifer	Average well yield (m^3/h)	Specific well capacity, average ($m^3/h/m$)
Hwara–Maneya	Umm er Radhuma	162	High
Wagsa	Umm er Radhuma	3.1	0.06
Salman	Damam	267	High
	Umm er Radhuma	33.8	32.5
Wadi Batin	Neogene	236	1.25
	Damam	142	1.46
	Umm er Radhuma	8.9	0.6

Table 6.4 Discharge of main springs in the Euphrates valley 1979/1980 (after Mashadani 1995, locations on Fig. 6.7)

Spring	Mean discharge (l/s)	Annual discharge (10^6 m^3)	Annual fluctuation Q_{max}/Q_{min}
Ain Ruhba	103	3.3	1.5
Ain Assaf	537	169	1.5
Ain Said Saed	29	0.9	101
Ain Rehaima	566	1.8	24
Ain al Ruhba	119	3.7	1.5
Ain al Azeyah	292	0.9	2.7
Ain Ghayem	143	4.5	1.6
Ain Jassam	103	3.3	1.2
Total	1,120	353	

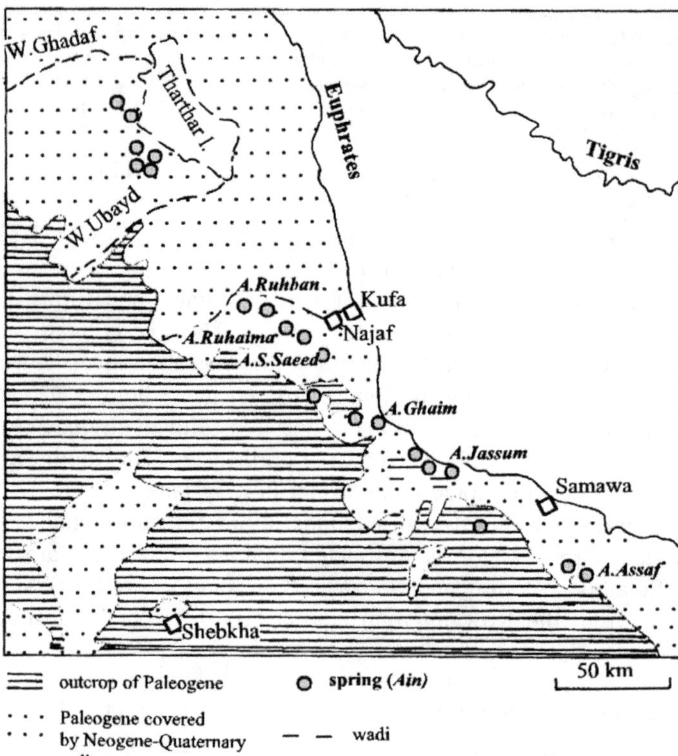

Fig. 6.7 Major springs in the Euphrates valley. After Al Mashadani (1995), ESCWA (1999b)

from the well fields reached 110×10^6 m^3. These rates do not include the abstraction through private wells, e.g., at Wafra. In 1990–1992 the production rates were reduced during the Second Gulf War, in 1993, the production rates were about 120×10^6 m^3.

6.3.3.2 Sub-Regional Groundwater Flow in the Middle Gulf Segment

Groundwater flow in the Umm er Radhuma and Damam aquifers follows in the middle Gulf sub-basin, in general, the topographic slope from the outcrop areas in the west to the coastline in the east. In the outcrop areas of the aquiferous formations on the western fringes of the sub-basin, the groundwater surface is situated at 200 m below land surface in a range of 240–400 m asl. The general groundwater flow pattern comprises a zone with relatively steep hydraulic gradient at and adjacent to the aquifer outcrop areas (around 4‰), a middle zone on the Suman plateau with gradients of <1‰, and an eastern zone with again steeper gradients of around 2‰ in the coastal plain.

Under the coastal plain at topographic elevations of a few metres above sea level, groundwater from the Umm er Radhuma aquifer leaks into the overlying Damam aquifer, from which water discharges into the coastal sabkha belt.

In detail, the groundwater regime is complicated by lateral and vertical variations of the hydraulic properties, by leakage between the different aquifers, in particular upward leakage in structural highs, the occurrence of stagnant water bodies with high salinity and of shallow groundwater lenses recharged from local rainfall.

Significant deviations from the general groundwater flow pattern occur:

- At the Ghawar anticline where groundwater probably leaks from the deeper Wasia–Biaydh aquifer into the Umm er Radhuma aquifer, causing a local flattening of the hydraulic gradient
- At groundwater discharge areas, in particular around Hofuf

In the Umm er Radhuma aquifer, the salt water front generally runs parallel to the Gulf coast and groundwater leaks into the overlying Damam aquifer; the Damam aquifer extends into the coastal area of Dhahran, Bahrain and Qatar.

Prior to the operation of extensive well fields, which started around 1940, natural discharge from the aquifer system was occurring mainly by evapotranspiration in shallow water table areas, evaporation in sabkhas and spring discharges. At present the main discharge component is abstraction by wells (Table 6.5).

The Alat and Khobar aquifers are mainly exploited in the towns along the coastal belt of the Gulf. Most of the water pumped from the Alat aquifer is used for irrigation and lifestock. The water of the Khobar aquifer is applied to more than 80% to irrigation. In the Greater Dhahran area, the Umm er Radhuma aquifer is exploited for industrial and domestic use and landscape irrigation.

Groundwater extraction from the Umm er Radhuma supports large irrigation schemes in northeastern Saudi Arabia, in particular the King Faisal irrigation project at Haradh.

The Neogene aquifer is used mainly for irrigation purposes in the Al Hasa area. Thirty-two major spring outlets occur in the Neogene aquifer, and thousands of private wells tap the aquifer within an area of 200 km^2. The Neogene aquifer receives significant water quantities through upward leakage from the Umm er

Table 6.5 Groundwater extraction from Tertiary aquifers in northeastern Saudi Arabia after Abderrahman et al. (1995), Zubari (1997)

Area	Aquifer	Groundwater abstraction $\times 10^6$ m^3/a	Year	Groundwater abstraction $\times 10^6$ m^3/a	Year
Coastal belt	Alat	493	1967	101	1990
	Khobar	122	1967	326	1990
Abqaiq	Khobar	33	1967	288	1990
Greater Dhahran	Umm er Radhuma	178	1967	373–685	1980–1990
Wadi el Miyah	Khobar			408	1994
	Alat			133	1994
Wadi el Miyah	Umm er Radhuma	2.8–3.6			
Haradh	Umm er Radhuma	90			
Al Hassa	Neogene	319	1977	608	1990
	Damam	505	1979		
	Umm er Radhuma			20	

Radhuma aquifer. Since 1969 wells in the Al Hasa area tap also the Damam and Umm er Radhuma aquifers (Table 6.5).

6.3.3.3 Rub al Khali Sub-Basin

The southern part of the eastern Arabian platform is occupied by the huge Rub al Khali sub-basin, which is characterized by extremely arid conditions.

The Rub al Khali constitues a hydrologic sub-basin with a paleo-river system directed to the Gulf coast of the United Arab Emirates between the northern Oman mountains and Sabkhet el Matti.

The boundaries of the hydrogeological basin of Paleogene aquifers of the Rub al Khali may be defined as follows:

- In the north: transition into the middle segment of the Gulf basin in the area of the central Arabian arch (Wadi as Sahba)
- Northwest: boundary of outcrops of Paleogene (Umm er Radhuma) and Cretaceous (Wasia and Biyadh sandstones) formations in the southern Tuwayq area
- Southwest: inferred western limit of extent of Paleogene formations below Neogene–Quaternary sediments
- South: water divide between Rub al Khali and Arabian Sea
- East: boundary between Arabian Shelf and Oman mountains
- North: Gulf coast of the United Arab Emirates

In the gravel plain aquifer on the northeastern margin of the Rub al Khali sub-basin, groundwater flows in the upper Tertiary to Quaternary sand and gravel aquifer from the western border of the Oman mountains in west to nothwest direction to sabkha discharge zones near the Gulf coast of the United Arab Emirates.

Most of central and southern Oman occupies a large groundwater basin, which stretches from the Oman mountains southward to the Dhofar mountains and, in the west, into the Rub al Khali desert. Sub-regional groundwater flow in mainly Tertiary aquifers passes from recharge zones within and adjacent to the Oman mountains and from the Dhofar mountains and the Nejd plateau in the south to groundwater discharge within the Rub al Khali. A major discharge zone of the aquifer system is considered to be the Umm as Samim sabkha.

Groundwater movement in the multi-layer aquifer system of Paleogene rocks in southern Oman is generally directed from the Jabal Qara area in the south towards north. On the Hadramaut highlands in the southwest of the Rub al Khali sub-basin, groundwater in the Tertiary aquifers moves toward north–northeast and discharges probably in inter-dune sabkhas in the Rub al Khali.

References. Al Ansari et al. (1995), Al Awadi and Mukhopadhyay (1995), Al-Awadi et al. (1995), Al-Harthy et al. (1995), Allam (1995), Al Mashadani (1995), Al-Murad and Zubari (1992), Al-Ruwaih (1996a, b), Al-Saafin et al. (1990), Alsharhan et al. (2001), Al Sulaimi et al. (1992), Bakiewicz et al. (1982), BRGM (1977), Burdon and Al-Sharhan (1986), Caro and Eagleson (1981), Edgell (1997), ESCWA (1999b), FAO (1979), Falkner (1994), GDC (1980), GEOCON (1995), Hötzl (1995), Hötzl et al. (1993), Imes and Wood (2007), Italconsult (1971), Martin and Krapp (1977), Matter et al. (2006), Mukhopadhyay (1999), Mukhopadhyay et al. (1994b), Otkun (1973, 1975), Parsons (1963), Pike (1985), Rasheedudin et al. (1989), Rizk et al. (1997), Sayed and Al-Ruwaih (1995), Zubari (1997).

6.4 Groundwater Salinity and Hydrochemistry

6.4.1 Groundwater Salinity in Multi-Aquifer Systems of the Euphrates–Gulf–Rub al Khali Basin

The eastern Arabian platform constitutes a huge geologic basin, which coincides approximately with the hydrologic Euphrates–Gulf–Rub al Khali basin. The basin comprises extensive multi-aquifer systems composed mainly of Paleogene–Neogene carbonate formations and detrital Neogene–Quaternary sediments.

Groundwater in the main aquifers of the Euphrates–Gulf–Rub el Khali basin is predominantly fossil and brackish. Fresh water occurs in very limited areas in outcrop zones of the Paleogene Um er Radhuma and Damam aquifers in central to northern Saudi Arabia, southern Iraq, Hadramaut, Dhofar, and in gravel plain aquifers. Generally, groundwater salinity increases from the outcrop areas of the aquiferous formations in direction of groundwater flow reaching >3 g/l TDS in most of the topographically lower parts of the basin and >10 g/l TDS near the Gulf coast, under the Dibdiba plain in the northeast, and in wide areas of the Rub al Khali sub-basin (Fig. 6.8).

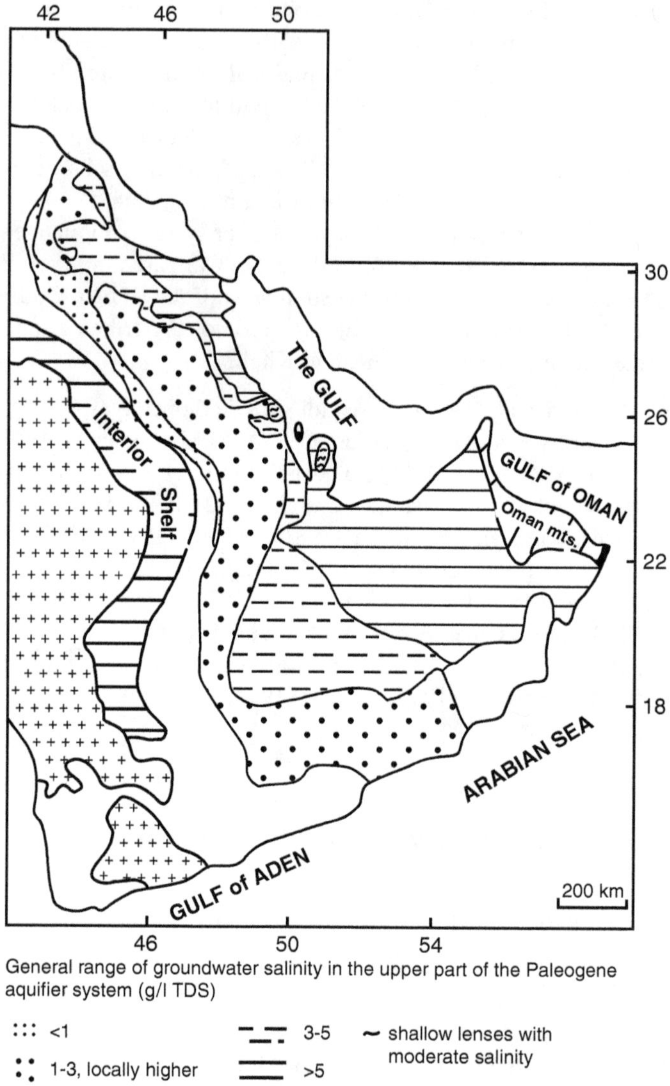

Fig. 6.8 Eastern Arabian platform: groundwater salinity distribution. After ESCWA (1999b)

An essential topographic–hydrologic feature, controlling the hydrochemical regime in particular in the middle Gulf groundwater sub-basin, is the sabkha discharge line, "which exists around the Upper Aquifer's 100 m piezometric contour and from whence the water character becomes a chloride water". "The sabkhahs and associated springs behave as an open basin discharge horizon where the groundwaters arrive at a sulphate endpoint". "... as water is discharged at the sabkhah line, the sulphate concentration becomes intolerably high ... and sulphate is deposited"(FAO 1979: 212, 214).

The general pattern of groundwater salinity distribution is modified, in particular, by the occurrence of brackish groundwater with moderate salinity (<3 g/l TDS) along the middle part of the Gulf coast in the Damam area ("tongue of fresh water" FAO 1979) and of lenses with relatively low groundwater salinity in Bahrain, Qatar, and in the Liwa area in the northern Rub al Khali.

The pattern of groundwater quality and hydrochemistry in the very extensive Euphrates–Gulf–Rub al Khali basin is reviewed in the following sections, applying a generalized zonation into hydrogeologic–hydrochemical environments:

- Fresh water occurrences on the upstream margin of the basin
- The Umm er Radhuma aquifer
- The Damam aquifer
- Neogene and Quaternary aquifers
- Multi-aquifer discharge zones

6.4.2 *Paleogene Aquifers*

6.4.2.1 Fresh Water Occurrences on the Upstream Margin of the Paleogene Aquifer System

Groundwater with salinity of less than 1,000 mg/l TDS is found in limited areas of the Umm er Radhuma and Damam outcrop belts on the upstream margin of the basin. Main water types with moderate salinity in the aquifer outcrop areas are:

- HCO_3 waters, which are relatively rare and always related to wadi flow and local infiltration of precipitation to aquifer outcrop and recharge areas
- SO_4 waters, which have dissolved wind blown gypsum, gypsum sand and local evaporite deposits
- Cl waters from adjacent aquifer discharge, in particular from the Wajid sandstone and the Wasia–Biyadh aquifer

Examples of Ca–HCO_3 type water in the Um er Radhuma aquifer of Wadi al Miyah in Saudi Arabia are mentioned by Otkun (1975). In southern Iraq, fresh water of Ca–HCO_3 type or Ca–Mg–HCO_3 type is found:

- In perched aquifer sections in the outcrop zone of the Damam aquifer
- In zones of runoff infiltration in morphological depressions or wadis, e.g., in Wadi Batin

SO_4 waters prevail in the recharge areas of the Paleogene–Neogene aquifers of the Euphrates and middle Gulf sub-basins. "Dissolution of Ca–SO_4 occurs in most of the recharge areas due to dissolving of windblown gypsum and gypsum sand". "These regions are in a highly evaporative environment where windblown gypsum and gypsum sands are widespread. Dissolution of gypsum during periods of recharge introduces sulphate rich waters into the groundwater regime. In a sense the gypsum is a cyclic salt" (FAO 1979: 210).

Groundwater salinity in the Umm er Radhuma aquifer is mainly moderate to slightly brackish in the outcrop area of Paleogene formations on the Nejd plateau (southern Oman) with electrical conductivity values of $<2,000\ \mu S/cm$. Fresh groundwater with a salinity of 500–1,000 mg/l TDS occurs in an aquiferous zone at the top Umm er Radhuma/base Jeza formation in the northern Hadramaut plateau (southern Yemen) with locally higher salinities of up to 2,500 mg/l TDS.

6.4.2.2 Umm er Radhuma Aquifer

Groundwater salinity in the Um er Radhuma aquifer generally increases from moderate values of around 1,000 mg/l TDS in the recharge areas in direction of groundwater flow to the discharge areas. The groundwater salinity reaches around 4,000 mg/l TDS in the coastal zone and rises to values $>10,000$ mg/l TDS in many areas near the coast. The increase in salinity is generally accompanied by a change of Ca–HCO_3 type water to Na–SO_4 water and finally Na–Cl water.

Deviations from the general pattern of increasing salinity in direction of groundwater flow are observed in several areas. Occurrences of anomalously low salinity are generally interpreted as showing preferential paths of groundwater flow. Areas of anomalously high salinity appear to be related to areas of particularly low hydraulic conductivity and very slow groundwater movement. In some parts of the Umm er Radhuma outcrop area e.g., east of Khurais, groundwater salinity reaches more than 3,000 mg/l.

Salinities of groundwater tapped in boreholes in the unconfined Umm er Radhuma aquifer in the outcrop area in the Suman plateau (northeastern Saudi Arabia) range generally from 1,200 to 1,800 mg/l TDS. The aquifer receives, in that area, present recharge through infiltration of surface flow into karst openings during ephemeral rain storms.

HCO_3 and Mg percentages of the investigated groundwaters on the Suman plateau are in rather narrow ranges around 10 meq% HCO_3 and 25 meq% Mg, while SO_4/Cl and Ca/Na ratios vary over relatively wide ranges; SO_4 concentrations reach up to 850 mg/l, Cl concentrations up to 424 mg/l, HCO_3 concentrations range from 93 to174 mg/l. No detectable tritium has been found in the samples, but an impact of actual infiltration from the surface is indicated by high seasonal variations of groundwater salinity in some boreholes and significant nitrate concentrations of up to 70 mg/l.

The water extracted from the boreholes appears to be mainly brackish groundwater stored in the aquifer with varying, but generally small admixtures of recent recharge with moderate salinity. The most diluted waters pumped from boreholes are Ca–SO_4 to Na–Ca–SO_4 type waters with a salinity of 1,000–1,220 mg/l TDS (Fig. 6.9).

The low HCO_3 concentrations reflect the low CO_2 production in an arid environment with poor soil development.

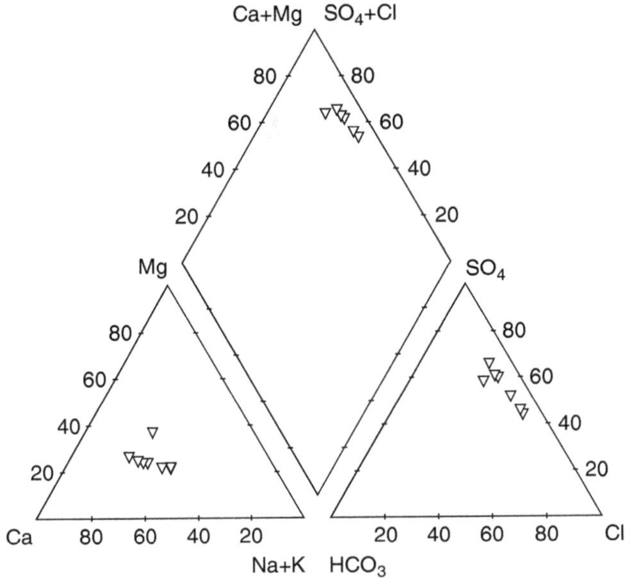

Fig. 6.9 Piper diagram: Groundwater samples from the Umm er Radhuma aquifer of the Suman plateau, Saudi Arabia. Data from Hötzl et al. (1993)

Generally, groundwater in the Umm er Radhuma aquifer is brackish to saline. This is attributed to various sources of groundwater salinity:

- Dissolution of wind blown gypsum and gypsum sand dispersed on the surface, carried in ephemeral runoff and accumulating in closed depressions
- Dissolution of evaporite layers in Paleogene formations, in particular the Rus formation, and in Neogene sediments
- Upward leakage of Na–Cl water from the underlying sandstone aquifer
- Evaporative enrichment of dissolved solids in sabkha discharge zones
- Ancient sea water intrusion, which may have reached up to the present sabkha discharge line
- Recent sea water intrusion along the Gulf coast

In several groundwater exploitation areas, the natural groundwater salinity is additionally elevated by human activities: contamination from the surface, irrigation return flow, initiation or enhancement of leakage of brackish water from adjoining aquifers.

The impact of these various sources of groundwater salinity leads to a differentiation of the hydrochemical pattern within the flow regime of the Umm er Radhuma aquifer, and the different sources of groundwater salinity create a wide range of hydrochemical water types. The increase of total salinity is, in general, accompanied by a change from HCO_3 and SO_4 type waters with moderate salinity in the outcrop areas to brackish Ca–SO_4, Na–SO_4 and Na–Cl type waters further downstream.

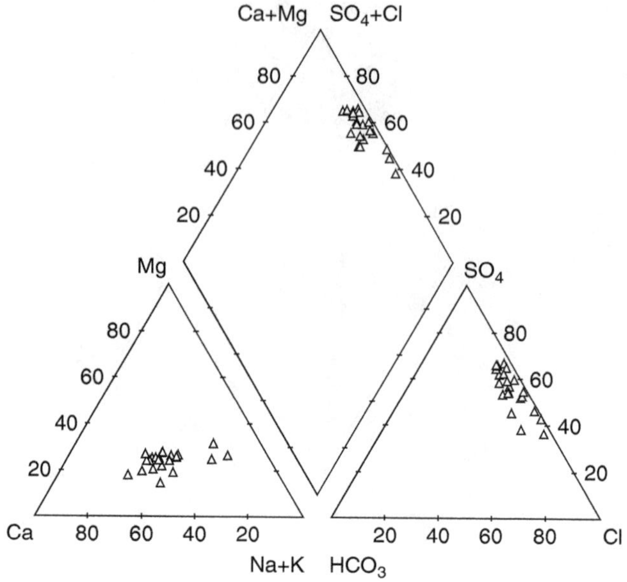

Fig. 6.10 Piper diagram: Groundwater samples from the Umm er Radhuma aquifer of eastern Saudi Arabia. Data from Sen and Al-Dakheel (1986)

Predominant hydrochemical water types in the Umm er Radhuma aquifer are Ca–SO_4 and Na–Cl waters. Percentages of HCO_3 and Mg ions are low, apart from low salinity groundwaters in very limited parts of the aquifer outcrop, as mentioned above. Trilinear Piper diagrams show a scatter of Umm er Radhuma groundwater samples in a wide range for the Ca, Na and SO_4, Cl percentages, while Mg and HCO_3 remain at relatively low percentages of generally <30 meq% and <20 meq%, respectively (Fig. 6.10).

Sulfate dominant waters are found in recharge areas of unconfined aquifer zones, in isolated perched water table aquifers, and in sabkha discharge zones. Chloride waters occur down gradient of the sabkha discharge line. The distribution of Cl and SO_4 waters in the downstream parts of the Gulf groundwater basin possibly reflects an ancient sea water intrusion phase and flushing by a sulfate rich water from the present groundwater flow system. Groundwaters with TDS concentration of less than 3,000 mg/l are bicarbonate or chloride waters, in the salinity range of 3,000–6,000 mg/l TDS sulfate waters predominate and in the range of 6,000–50,000 mg/l TDS and above chloride waters.

Groundwater salinities of $>5{,}000$ mg/l TDS at shallow to intermediate depth are generally found only directly near the coast, where, under probably stagnant conditions, salinity levels reach 10,000 and more mg/l TDS.

Salinities of groundwater extracted from the Umm er Radhuma aquifer at Haradh for the King Faisal irrigation project are relatively low with 909–1,320 mg/l TDS.

Relatively low salinities of <4,000 mg/l are observed in the coastal zone between Qatif and Dhahran. In the Greater Dhahran area (Dhahran and Khobar cities), groundwater salinity in the Umm er Radhuma aquifer ranges from 3,000 to 4,176 mg/l TDS, with an average of 3,545 mg/l. Na, Cl, and SO_4 are predominant ions. Increasing groundwater exploitation caused an increase of average salinity of pumped water by about 22% from 2,750 mg/l TDS in 1979 to 3,545 mg/l TDS in 1990.

6.4.2.3 Damam Aquifer

The pattern of groundwater salinity and hydrochemical composition in the Damam aquifer is, in principle, similar to conditions in the underlying Umm er Radhuma aquifer. The groundwater salinity increases, in general, from 1,000 mg/l TDS in the outcrop areas to 5,000 mg/l TDS in the coastal zone and to more than 10,000 mg/l TDS directly along the Gulf coast. The outcrop zone of the Damam formation in Saudi Arabia is narrow and groundwater replenishment may be strongly influenced by leakage from underlying or overlying aquiferous units. In the Euphrates sub-basin in the north of the eastern Arabian platform, the Damam formation is exposed over considerable areas, while the underlying Umm er Radhuma formation constitutes a deeper aquifer, in which groundwater circulation is very restricted in the downstream parts of the sub-basin: the Euphrates plain, Dibdiba depression and Kuwait plain.

In the outcrop areas of the Damam formation, groundwater with moderate salinity is found in various locations in the southern desert of Iraq. The fresh water of the outcrop areas develops downstream into brackish groundwater, and groundwater discharging in the springs along the Euphrates valley is prevailingly Ca–SO_4 type water with salinities between 2,000 and 5,000 mg/l TDS (Fig. 6.11). As in the groundwater of the Umm er Radhuma aquifer, percentages of HCO_3 and Mg are in narrow ranges of 10 meq% HCO_3 and around 25 meq% Mg, while Ca, Na, SO_4 and Cl are dominant ions with a wider variation of ion ratios.

Some springs discharge Na–Cl water with higher salinities of 12,000–19,000 mg/l TDS (Ain Joza, Ain Ridha, Sawa lake).

In northeastern Saudi Arabia near the Kuwait border, the Alat aquifer contains water with less than 3,000 mg/l TDS, while salinities in the deeper Khobar sub-aquifer are higher than 6,000 mg/l TDS.

Groundwater extracted from the Damam aquifer in well fields in Kuwait near the Iraqian border shows similar hydrochemical composition as the spring water in the Euphrates valley (Figs. 6.12 and 6.13.). In well field Shagaya D, where the top of the around 200 m thick Damam formation is situated at 400 m below surface under Miocene–Pleistocene deposits of the Kuwait group, brackish water with salinities of 2,650 to around 3,000 mg/l TDS (data of 1989) is SO_4 type water with approximately equal percentages of Ca and Na; HCO_5 percentages are low, Mg percentages in the range of 25 meq%. Variations in anion ratios correspond mainly to different

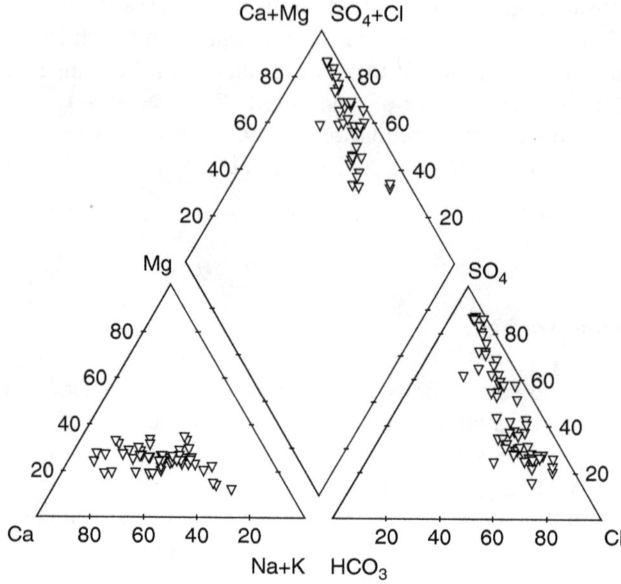

Fig. 6.11 Piper diagram: Water samples of springs in the Euphrates valley, Iraq. Data from Al Mashadani (1995)

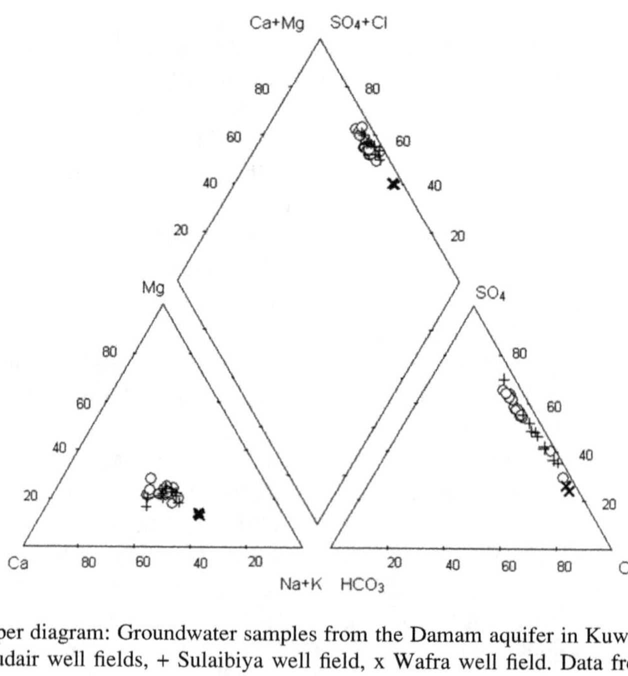

Fig. 6.12 Piper diagram: Groundwater samples from the Damam aquifer in Kuwait. O Shagaya and Umm Gudair well fields, + Sulaibiya well field, x Wafra well field. Data from Al-Ruwaih (1995)

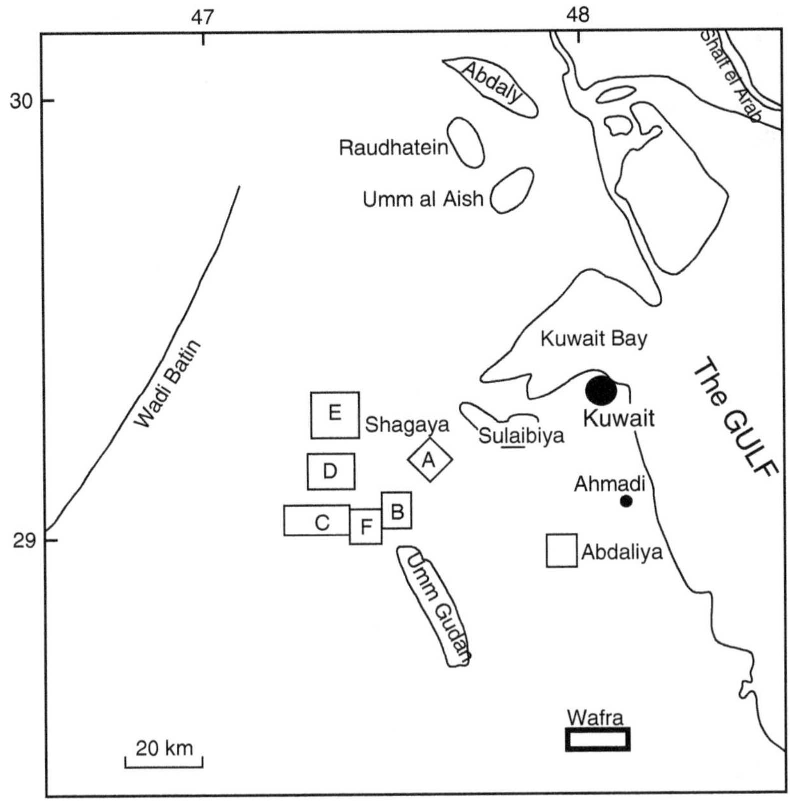

Fig. 6.13 Well fields in Kuwait. After Al-Ruwaih (1995), Sayed and Al-Ruwaih (1995), Alsharhan et al. (2001)

Cl concentrations, HCO_3 and SO_4 concentrations appear to be determined mainly by saturation levels.

Salinity ranges in other well fields tapping the Damam aquifer in Kuwait are, with a general salinity increase from west to east:

Shagaya C	2,630–2,900 mg/l TDS
Umm Ghudair	2,700–3,500 mg/l TDS
Shagaya B up to	5,400 mg/l TDS
Sulaybia	3,500–8,000 mg/l TDS

The extracted water in Al Shagaya and Umm Gudair areas in southwestern Kuwait is Na–SO_4 and Ca–SO_4 type water.

Toward the bay of Kuwait, salinity increases to 40,000 mg/l TDS and to >100 g/kg in northeastern Kuwait.

The hydrochemical conditions have been interpreted "as a flushing of the aquifer by infiltrating meteoric waters down as far as present sea level". "This is followed by a zone where present pressure heads are close to those of the Arabian Gulf sea-water so that sea-water may totally replace meteoric water in the aquifer". Further downstream in a zone with salinities of 10,000–40,000 mg/l TDS, pressure heads are close to the sea level and the aquifer may be strongly affected by sea water in a sector of the aquifer filled "probably with water introduced during the post-Oligocene marine–lagoonal transgression" (Burdon and Al Sharhan 1986:391).

On the eastern margin of the Rub al Khali sub-basin, brackish groundwater occurs in fractured and cavernous limestones of Paleocene–Eocene age within the local anticlinal structure of Jebel Hafit near Al Ain. The water with a salinity of 3,900–6,900 mg/l TDS appears to represent a brackish water lens recharged from local rainfall overlying saline water with >15,000 mg/l TDS. The brackish groundwater is Na–Cl type water with Cl concentrations of 1,900–3,000 mg/l and SO_4 concentrations between 450 and 780 mg/l.

6.4.3 *Neogene and Quaternary Aquifers*

6.4.3.1 Neogene Aquifers

In the western desert of Iraq, corresponding to the southwestern part of the Euphrates groundwater sub-basin, groundwater in Upper Fars sandstones and conglomerates of the Dibdiba formation has salinities of 500–2,500 mg/l TDS.

Groundwater in the Neogene Kuwait group aquifer in Kuwait is brackish to saline except for shallow fresh water lenses in the Raudhatain and Umm el Aish areas.

The following salinity ranges and water types are found in the Kuwait group aquifer in different areas of Kuwait:

In northern Kuwait:	Na–Cl water with 4,000–13,000 mg/l TDS
	Na–SO_4 and Ca–SO_4 water with 3,500–6,500 mg/l TDS at some locations
	Na–Cl water with 100 g/kg TDS in the lower part of the aquifer
In southern and central Kuwait:	Na–Cl water with 4,000–10,000 mg/l TDS
	Ca–SO_4 to Na–Cl water with 3,000–5,000 mg/l at some locations

Groundwater in the Neogene aquifers of eastern Saudi Arabia is generally brackish with salinities around 4,000 mg/l TDS. In the multi-aquifer discharge zone of Hofuf, groundwater salinity in the upper Neogene aquifer varies around 1,400–2,000 mg/l TDS.

6.4.3.2 Gravel Plain and Sand Dune Aquifers

The electrical conductivity of groundwater samples collected from gravel plain aquifers in the United Arab Emirates west of the Oman mountains varies between 252 μS/cm east of Al Ain and 173,000 μS/cm in a sabkha area along the Al Ain–Abu Dhabi road (data of 1996). Groundwater salinity increases from <1,000 mg/l in the east of the Al Ain area in the direction of groundwater flow toward west–northwest.

The development of groundwater in the shallow aquifer of the western gravel plain of the United Arab Emirates from fresh water at the mountain foot near Al Ain to saline water near the Gulf coast is caused mainly by infiltration of water, which is heavily enriched in dissolved solids through evaporation, and by upward leakage of brackish to saline water from the underlying Fars formation. Salt water intrusion as a result of heavy groundwater pumping is noticed in several areas between Dubai and Al Ain.

A salinity increase from the mountain foot toward the coastal plain is also obvious in water percolation in the unsaturated zone. Cl concentrations in the unsaturated zone near the Oman mountains are around 630 mg/l along wadi courses and 970 mg/l in areas between the wadis. 10 km downstream from the mountain foothills, Cl concentrations in the unsaturated zone of wadis reach 1,450 mg/l, at 36 km from the mountain front 3,750 mg/l, and near the coast 8,050 mg/l.

Near the Oman mountains, the aquifer is characterized by Mg–HCO_3 and Ca–HCO_3 water. The hydrochemical character changes in flow direction with increasing salinity to Ca–SO_4 and Mg–SO_4 water types and finally to Na–Cl water type.

In the alluvial plain south of Jebel Akhdar in Oman, groundwater evolves from low mineralized Ca–Mg–HCO_3 or Mg–Ca–HCO_3 type water (<600 mg/l TDS) to more mineralized Na–Mg–(Ca)–Cl–SO_4 type water (TDS maximum 1,940 mg/l). In the downstream part of the alluvial plain, groundwaters from different catchments gradually merge and a strong increase in the Mg, Ca, SO_4, Na and Cl concentrations is observed. In groundwater originating from limestones of Jebel Akhdar, Ca–Mg–HCO_3 water prevails with Mg/Ca ratios <1. Mg/Ca ratios of groundwater moving into the plain from ophiolite mountains are 1.17–3.28.

Brackish groundwater in the sand dune aquifer of the Liwa area with salinities in the order of 3,000–7,000 mg/l TDS is characterized by the Na–Cl type water.

6.4.4 Groundwater Discharge Zones

In spite of the wide variations of groundwater salinity and of major ion concentrations in the multi-aquifer system, groundwater samples show a rather homogeneous hydrochemical composition in specific areas of the eastern Arabian platform.

In particular in some major discharge zones of brackish groundwater, ion ratios are in relatively narrow ranges.

Main groundwater discharge areas are, in particular, the zone of springs on the right bank of the Euphrates river and the Hasa and Qatif oases in the middle Gulf sub-basin. The springs along the Euphrates river issue prevailingly groundwater from the Damam aquifer (Fig. 6.7).

In the Al Hasa oasis area, brackish water discharges from a multi-aquifer system including the Umm er Radhuma, Damam and Neogene aquifers. The discharge area is covered mainly by karstified marly limestones of the Pliocene Hofuf formation; groundwater issues into solution caves and karst tubes. Water of flowing springs and water extracted from shallow excavations and up to 100 m deep boreholes is used for irrigation.

Groundwater salinity of the discharging and artificially extracted groundwater at Al Hasa is in a range of 1,300–2,400 mg/l TDS. The groundwater is prevailingly Na–Cl type water; Ca/Na ratios and SO_4/Cl ratios as well as total salinity show local variations. Generally, groundwater salinity increases within the oasis area in direction of the hydraulic gradient from southwest to northeast; the salinity increase is accompanied by an increase of Na, Ca, Cl and SO_4 concentrations and a shift to higher Na and Cl percentages. The pattern of groundwater temperatures suggests, that groundwater with high temperatures of up to 40°C in the southwest are related to fast upward leakage from deeper parts of the aquifer complex, while the relatively cooler and more saline groundwaters in the northeast of the oasis area comprise larger components of groundwater moving through the Neogene aquifer.

Intensive groundwater abstraction for irrigation causes seasonal fluctuations of groundwater salinity and a continuing increase of average annual salinity levels in spring and well water. For the period 1976–1987, a rise of the verage salinity value of all spring waters of the oasis from 1,400 to 1,700 mg/l TDS has ben observed.

In Tertiary aquifers of the Qatif area near the Gulf coast, groundwater salinity is with 1,800–3,200 mg/l TDS higher than in the Hasa area. Major ion ratios of groundwaters from Tertiary aquifers in the Qatif area are similar to ratios in the Al Hasa area with a tendency to higher Cl percentages of groundwaters in the Qatif area (Figs. 6.14 and 6.15).

References. Abderrahman (1989), Abderrahman and Rasheeduddin (1994), Abderrahman et al. (1995), Al Awadi and Mukhopadhyay (1995), Al-Awadi et al. (1995), Al-Harthy et al. (1995), Allam (1995), Al Mashadani (1995), Al-Ruwaih (1984, 1993, 1995), Al-Sayari and Zötl (1978), Alsharhan et al. (2001), Bakiewicz et al. (1982), BRGM (1977), Burdon and Al-Sharhan (1986), Clark et al. (1987), Edgell (1997), ESCWA (1999b), FAO (1979), Gedeon et al. (1995a, b), GEOCON (1995), Handy and Alomani (1984), Hötzl (1995), Hötzl et al. (1993), Imes and Wood (2007), Khalifa (1997), Martin and Krapp (1977), Matter et al. (2006), Mukhopadhyay (2003), Naimi (1965), Otkun (1973, 1974, 1975), Rizk and El-Etr (1997), Rizk et al. (1997), Sen and Al-Dakhel (1986), Zubari (1997).

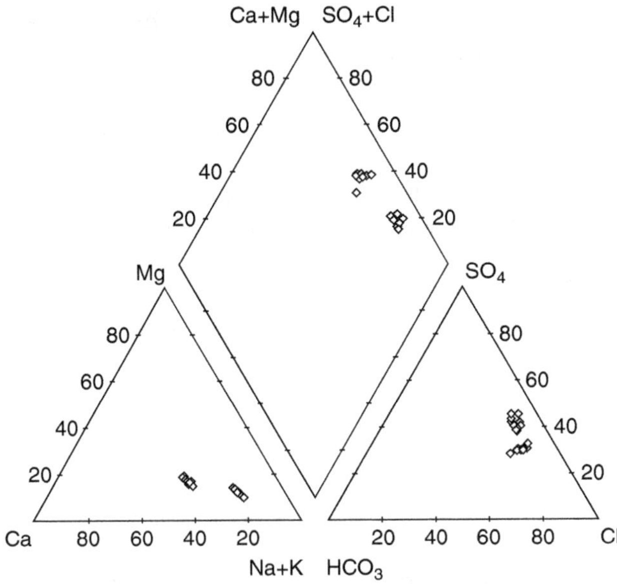

Fig. 6.14 Piper diagram: Water samples from the Al Hasa discharge area, eastern Saudi Arabia. Data from Otkun (1974)

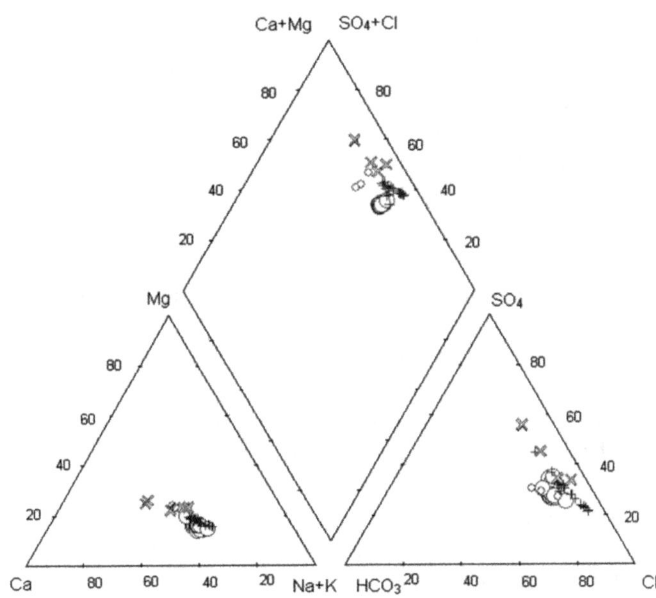

Fig. 6.15 Piper diagram: Groundwater samples from Tertiary aquifers in discharge and exploitation areas of eastern Saudi Arabia. O Al Hasa groundwater discharge area, + Qatif groundwater discharge area, o Haradh boreholes (Umm er Radhuma aquifer), **x** Wadi al Miyah boreholes. Data from Al-Sayari and Zötl (1978)

6.5 Shallow Groundwater Lenses Above the Brackish Main Aquifer System

Shallow groundwater lenses occur above the main groundwater system of the eastern Arabian platform in some areas, where morphologic conditions or open karst surfaces favour local recharge from the sporadic rainfall events. The sub-regional groundwater flow system with brackish to saline water is overlain, in these areas, by lenses of water with lower salinity, which are fed by recent rainfall. These lenses with fresh to brackish groundwater are found, in particular, in Paleogene aquifers in Bahrain and Qatar, and in Neogene deposits in northern Kuwait and interior Oman.

6.5.1 Bahrain

The main island of Bahrain is formed by an anticline, the core of which is occupied by an outcrop of the Rus formation. In the eastern and central parts of the island, extensive solution of anhydrite layers of the Rus formation has created a karstic aquifer, which is in hydraulic connection with the underlying Umm er Radhuma aquifer.

The Rus–Umm er Radhuma aquifer is under unconfined conditions at the eroded core of the Bahrain anticline. On the flanks of the anticline, the aquifer is confined by shales of the lower Damam formation and, in some areas, by shales and anhydrites of the upper Rus formation.

The upper 5–10 m of the Khobar member of the Damam formation are highly permeable and form the main productive aquifer in Bahrain. The aquifer is separated from the underlying less permeable Alat limestone aquifer by the "orange marl" aquitard.

The Damam aquifer is in direct contact with the sea level in the east of Bahrain and in the offshore areas north of Bahrain.

On the northwestern coast of Bahrain, the Damam aquifer receives inflow of groundwater, which moves from the Dhahran area over a distance of 24 km under the Bay of Bahrain. Piezometric levels are around 6 m asl at the coast of Dhahran and 3 m asl in the northwest of Bahrain. The water level of the shallow brackish water lens descends from 6 m asl in the anticlinal core of the Rus–Umm er Radhuma aquifer in the centre of Bahrain to around 3 m asl in marginal areas of the island, where groundwater leaks into the overlying Damam aquifer.

The inflowing groundwater in the Damam aquifer from Saudi Arabia has a salinity of 2,500–4,000 mg/l TDS. The salinity in the shallow brackish water lens was around 2,000–3,000 mg/l TDS in a pre-development stage.

Previously brackish water springs with around 2,900 mg/l TDS discharged in the northern part of Bahrain main island. The total discharge was estimated at 1.5 m^3/s.

The springs on the main island have nearly completely ceased to discharge, but submarine springs and springs on small islands still exist.

Inflow of brackish groundwater in the Damam aquifer from the coast in Saudi Arabia is estimated at 110×10^6 m^3/a. Groundwater abstraction (218×10^6 m^3/a) greatly exceeds the natural groundwater flow. Since 1989, brackish groundwater is extracted from the Umm er Radhuma aquifer for a desalination plant (24×10^6 m^3/a).

Most of the groundwater at shallow depth in Bahrain is Na–Cl water representing mixtures of the subsurface inflow from Dhahran with intruding sea water or upward leaking saline water.

In the southwest of Bahrain, saline water with >10,000 mg/l TDS occurs in the vicinity of an extensive sabkha.

At a few locations on Bahrain, water of Ca–SO_4–Cl type with a salinity of <2,000 mg/l TDS is found in the shallow brackish water lens. In general, the salinity of the shallow groundwater is in ranges of 8,000–10,000 mg/l TDS. The salinity increases with depth to about 15,000 mg/l at around 200 m below land surface and to more than100 g/kg in brines in the Aruma formation underlying the Um er Radhuma formation.

At the eastern coast of Bahrain island, the Damam aquifer affected by sea water intrusion.

6.5.2 *Qatar*

Qatar peninsula is formed by an anticlinal structure with outcrops of the Rus formation in its core in the northern part of the peninsula. The flanks of the anticline along the coastline and in the south of the peninsula are covered by the Damam formation.

Dissolution of evaporite layers in the Rus formation has created secondary permeability and has lead to surface collapse structures favouring recharge from local runoff, which, in spite of the arid climate with average rainfall of 75 mm/a, has produced a relatively extensive fresh water lens. In the outcrop areas of the Rus formation, the fresh water lens extends within calcareous and cavernous gypsiferous deposits of the Rus formation and underlying Umm er Radhuma limestones. The fresh water lens in the Umm er Radhuma–Rus aquifer is sustained by recharge from rainfall, which is collected in surface depressions and discharging into solution collapse structures.

To a minor extent, recharge from local rainfall also occurs on karst surfaces of the Damam formation in the southern part of the peninsula.

The aquiferous parts of the Rus formation are connected hydraulically with the Umm er Radhuma aquifer in the north central part of the peninsula, and, in some areas, with the overlying 10–30 m thick Khobar member of the Damam formation. In the southern areas of Qatar peninsula, the Rus and Umm er Radhuma formations are separated by basal shales of the Rus formation.

In most of Qatar peninsula, the Rus formation is overlain by limestones of the Damam formation. In the southwestern part of Qatar (Abu Samra area), the upper limestone member (Abarug member) of the Damam formation provides an aquifer, which is locally confined under a sequence of clays, marls, limestones, and shales of the Neogene Dam formation.

The Abarug sub-aquifer in the southwest of Qatar receives water through subsurface flow from the equivalent Alat aquifer in eastern Saudi Arabia. Groundwater flow in the Damam aquifer is directed from the border between Saudi Arabia and Qatar in the southwest to the coastal area in the northeast. In the Rus–Umm er Radhuma aquifer, groundwater flows from local recharge mounds in the north and south of the peninsula toward the coast. The hydraulic head in the recharge mounds is around 6–9 m asl.

At the western coast of Qatar, Sabkhat Dukhan constitutes an extensive area of groundwater discharge through evaporation.

The shallow fresh water lens in Qatar is underlain by brackish and salt water. The fresh water body within the central parts of northern Qatar extends down into the upper 20–50 m of the Umm er Radhuma formation. The saturated thickness of these lenses reaches about 100 m in central Qatar and gradually decreases towards the coastal areas and the south. The depth to groundwater varies, in general, from 30 to 90 m.

Within the recharge mound of the shallow groundwater lens in the Rus–Umm er Radhuma aquifer, groundwater salinity is generally <1,000 mg/l TDS. Parker and Pike (1976) estimated that the shallow groundwater lens with a salinity of <2,000 mg/l TDS extends over about 2,180 km^2 (about 20% of the land area of Qatar) and contains about 2.5×10^9 m^3 of freshwater.

Deeper groundwater with higher salinity leaks upward into the fresh water lens along major structural features. Near the coast, the freshwater lens within the Rus formation is bordered by an interface with the sea water.

The groundwater salinity in the outcrop area of Rus carbonate rocks ranges from 400 to about 4,000 mg/l. HCO_3 groundwaters prevail in the low salinity aquifer, indicating a hydrochemical predominance of carbonate dissolution. The occurrence of HCO_3 groundwater is restricted to the northern part of the peninsula, while sulfate and chloride waters dominate in other areas of Qatar. "Progressively down the hydraulic gradient in the Rus F. gypsiferous dissolution exceeds carbonate dissolution as the waters become saturated with respect to carbonate minerals and anhydrite becomes more prevalent in the matrix". "Sulphate waters dominate and high concentrations develop" (Lloyd et al. 1987).

In the centre of Qatar, a zone of high salinity water marks the transition between the carbonate and sulfate facies of the Rus formation.

The salinity of the Abarug aquifer ranges from 3,000 to 6,000 mg/l TDS with high sulfate concentrations. The salinity increases toward east and northeast in direction of groundwater flow up to 10,000 mg/l near the coast.

Elevated sulfate concentrations in the southern part of Qatar are derived from the dissolution of evaporites of the Rus formation and from windblown gypsum, gypsum sand and anhydrites, which may originate from an ancient sabkha system.

In general, the groundwater salinity increases with depth and toward the coast line with a change to Cl type waters.

At about 180 m below sea level (210 m below ground surface), the brackish water aquifer in Qatar is underlain by saline water with more than 15,000 mg/l TDS. The high salinity is attributed to dissolution of anhydrite, upward leakage of saline water from the Umm er Radhuma aquifer and seawater intrusion, intensified by heavy extraction.

6.5.3 Northern Kuwait

In northern Kuwait, fresh water lenses are found in shallow sand and gravel deposits under morphologic depressions in the Raudhatein and Umm al Aish areas.

The plain of northern Kuwait is covered by deposits of the Miocene–Pleistocene Dibdiba formation, which constitutes the upper section of the Kuwait group. The Dibdiba formation comprises an upper member of gravels and sand with gypsum layers and a lower member composed of medium to coarse sand and clayey sand with gypsum and carbonate cement. The Dibdiba formation is underlain by the Miocene Ghar and Fars formations.

The morphologic Raudhatein and Umm el Aish depressions constitute the centres of extensive, well developed drainage systems with a dendritic pattern of long, closely spaced and mainly toward northeast directed stream channels. The main channels of the system of shallow wadis terminate in the two local depressions.

The Raudhatein depression, extending over 4–5 km in east–west direction and 15 km in north–south direction, has a drainage area of about 670 km^2. The depression is situated at around 38 m asl. 12 wadis drain into the Raudhatein depression, the highest elevation of the catchment is 136 m asl.

The drainage system of the Umm al Aish depression covers 450 km^2, topographic elevations of the catchment area descend from 115 to 30 m asl.

Surface runoff from occasional rain storms create local fresh water lenses within the depressions. The shallow aquifer in the depressions is composed of unconsolidated sand and gravel sediments, which are cemented in a gypsiferous and carbonatic zone near the surface ("gash zone").

Groundwater salinity in the shallow lens at Umm al Aish varied from 500 to 1,500 mg/l TDS in 1966 and from 400 to 2,000 mg/l TDS in 1977. In 1983, salinities of 300–600 mg/l TDS were measured in eight wells in the centre of the depression.

The low salinity groundwater in the southwestern and central parts of the depression are mainly Ca–HCO_3 and Ca–Cl type waters. On the downstream margins in the east of the depression, the groundwater becomes brackish Na–Cl type water.

In the centre of the Raudhatein depression, salinity of shallow groundwater is between 300 and 1,500 mg/l TDS. Predominant ions are Na and SO_4 with wide variations of Ca/Na ratios and of anion ratios (Fig. 6.16).

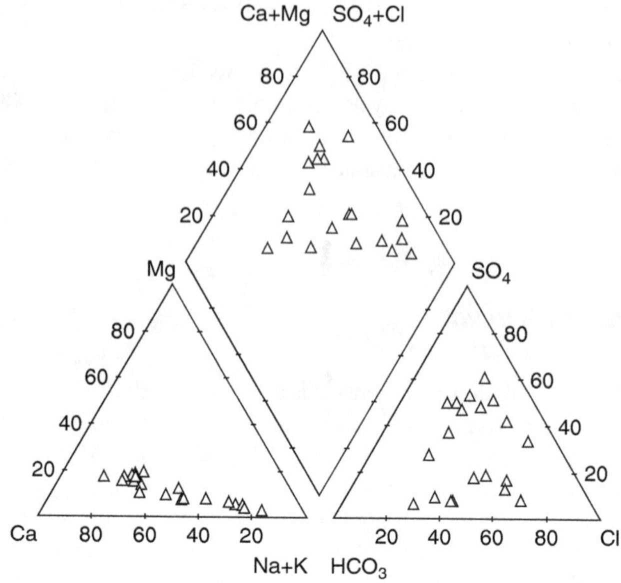

Fig. 6.16 Piper diagram: Groundwater samples from fresh water lenses in northern Kuwait. Data from Al-Ruwaih (1984, 1985)

6.5.4 Al Wusta Area in Oman

In the hyper-arid Al Wusta area in Oman, fresh water lenses occur above the main brackish to saline aquifer in areas where rapid recharge of stream runoff is favoured by the morphologic conditions.

"In such settings recharge is largely independent from mean annual rainfall and instead occurs in response to occasional intense rainfall events. The extreme aridity enhances rather than hinders recharge, in that there is very little vegetation or soil present to retard infiltration; it is further assisted by the parched landsurface. Water infiltrates rapidly, with minimal evaporation despite high daily temperatures, with minor impact on groundwater quality" (Macumber 1995).

The fresh to brackish water lenses are found in two types of morphologic depressions:

- Closed depressions on the flat plateaus of the central Al Wusta area
- Depressions along major eastward draining wadis

Local fresh to brackish water lenses on the plateau areas with salinities of 250–5,000 mg/l TDS are found mainly in sediments of the Fars formation. Salinity in the underlying extensive aquifer system in Tertiary formations is >16,000 mg/l TDS.

In the Maabar depression, a 10–20 m thick fresh water lens extends in Tertiary limestones with the water table at 92 m below surface.

The eastward flowing wadis run from broad shallow depressions on the plateau areas through 30–60 m deep and 200–600 m wide gorges to the coastal area. Within the gorges, the valleys are covered by an around 20 m thick alluvial gravel fill, which enhances rapid infiltration of sporadic flood flow. Shallow groundwater lenses extend from the plateau section with salinities of 600–1,800 mg/l TDS into the gorge section, where salinities are often <350 mg/l TDS.

The water table in the underlying extensive aquifer is situated at 100–140 m below surface, salinities range from 6,000 to 40,000 mg/l TDS. The fresh to brackish water lenses and the deeper saline water are contained in Tertiary carbonate aquifers of the Fars, Damam and Rus formations.

References. Al-Hajari (1990), Al-Mahmood (1992), Al-Ruwaih (1985), Alsharhan et al. (2001), Al-Youssef (1994), ESCWA (1999b), FAO (1979), Harhash and Yousif (1985), Heim (1928), Lloyd et al. (1987), Macumber (1995), Macumber et al. (1995), Metcalf and Eddy (1980), Robinson and Al Ruwaih (1985), Zubari (1997, 1999), Zubari et al. (1997).

6.6 Groundwater Age and Paleohydrology: Information from Isotope Data

6.6.1 Groundwater Age

The distribution of ^{3}H and ^{14}C values in groundwater samples from the eastern Arabian platform reflects the groundwater regime, which is characterized by the predominance of fossil groundwater stored in the main aquifers and very limted present-day recharge.

^{3}H values in rain water samples on the Suman plateau in northeastern Saudi Arabia were 8–9 TU in 1986–1989, mean ^{3}H in precipitation at Bahrain was 6 TU in 1995.

^{3}H in groundwater samples from the eastern Arabian platform is generally below detection level. ^{3}H values above 5 TU were found in a few samples from outcrop areas of Paleogene aquifers in Saudi Arabia and Oman and in shallow fresh water in the gravel plain on the eastern margin of the Rub al Khali sub-basin.

On the plateau areas of northeastern Saudi Arabia (middle Gulf segment of the eastern Arabian platform), significant tritium values are found at a few locations on or near the Umm ar Radhuma outcrop and, at shallow depth, in the Damam and Neogene formations.

Infiltration of local flood water into karst openings on the Umm er Radhuma outcrop of the Suman plateau appears to be generally not sufficient to create measurable ^{3}H levels in the groundwater. The volume of present-day recharge is probably low in comparison to the stored groundwater volume.

In aquifers of Kuwait, detectable tritium was found only in shallow fresh water lenses in the depressions of Umm al Aish and Raudhatein.

Out of 15 groundwater samples collected in Bahrain, tritium was above detection level (12 TU) in one sample, which was taken from the brackish water lens of the Rus–Umm er Radhuma outcrop.

In Qatar, significant ^{3}H values of up to 20 TU occur in the Rus outcrop zone in the north of the peninsula, indicating that the source of the fresh water lens in that area is modern rainfall. In the southern part of Qatar, ^{3}H is generally below detection level.

On the Nejd plateau in southern Oman, ^{3}H values in the uppermost aquifer (A aquifer, Damam and top of Rus formation) range from less than detection limit to 9 TU, indicating the occurrence of recent recharge in some areas. The aquifer is phreatic under outcrops in wadi courses, where it receives recharge during sporadic rainfall. In the deeper Umm er Radhuma aquifer, ^{3}H is below detection level.

In northeastern Saudi Arabia, ^{14}C ages of groundwater samples range generally from 20,000 to 26,000 years BP for the Umm er Radhuma aquifer and from 16,000 to 20,000 years BP for the Damam aquifer. The ^{14}C ages in the Paleogene aquifers increase generally from the outcrop areas in the west in direction of groundwater flow toward east and northeast. Deviations from the general pattern of ^{14}C data distribution are explained by leakage between different aquifers.

Out of 160 ^{14}C analyses of water samples of sedimentary aquifers in eastern Saudi Arabia, 47% have calculated ages ranging between 18,000 and 28,000 years. The data tend "to center around a calculated age of about 20,000 years" (Shampine et al. 1979).

^{14}C water ages of groundwater from the Paleogene–Neogene aquifer system in the Hasa and Qatif oasis areas in eastern Saudi Arabia are in a range of 22,000 to >34,500 years.

In the Damam and lower Kuwait group aquifers of Kuwait, ^{14}C values of 2.5–4.3 pmc correspond to groundwater ages between 17,000 and 27,000 years.

^{14}C data of groundwater in Bahrain indicate ages between 6,000 and >17,000 years for the Damam aquifer and of 7,000–14,000 years for the Rus–Umm er Radhuma aquifer.

In the Damam aquifer of Qatar, groundwater ages of 20,000–26,000 years were calculated from ^{14}C data for deep and shallow groundwater.

On the western gravel plain of the United Arab Emirates – west of the Oman mountains – tritium values of up to 12.4 TU were found in groundwater near the mountain foothills.

On the Nejd plateau in southern Oman, mixtures of old groundwater with components of modern groundwater are indicated in shallow aquifers by modern ^{14}C ages and low tritium values. The ^{14}C age of the groundwater in the upper Umm er Radhuma formation (B aquifer) ranges from 9,000 to 11,000 years in the southeast and to about 20,000 years further downstream. Estimates of the groundwater age for the "C aquifer" (top of lower Umm er Radhuma formation) range from 4,000 to 16,000 years in the south close to the Dhofar mountains and to 30,000 years in the northwest towards the more central parts of the Rub al Khali. In the "D aquifer" (lower Umm er Radhuma) ^{14}C ages of groundwater are in the order of 9,000–26,000 years. Hydraulic connections with overlying aquifers are assumed.

An apparent age–distance trend in the Paleogene aquifers from the Dhofar mountains toward the Rub al Khali basin implies that recharge originates in the Umm er Radhuma outcrops of the Dhofar mountains.

6.6.2 *Stable Isotopes of Oxygen and Hydrogen*

6.6.2.1 Precipitation

Data of stable isotopes of oxygen and hydrogen in rain water have been recorded at Bahrain meteorologic station since 1963. The weighted mean values (1963–1991) are −2.27‰ for δ^{18}O and −7.2‰ for δ^{2}H (Yurtsever 1992) with a deuterium excess of +11.0‰. If values of low intensity rainfall are disregarded, the isotope data of rainfall at Bahrain fit a local meteoric water line of

$$\delta^2\mathrm{H} = 8 \times \delta^{18}\mathrm{O} + 10.9.$$

6.6.2.2 Holocene Recharge

δ^{18}O/δ^{2}H values of samples from shallow groundwater lenses in the Gulf coastal area are similar to values of rain water inBahrain (δ^{18}O −2.5‰, d +10‰). Average δ^{18}O and d values in these lenses, sustained by Holocene recharge, are:

Northern Kuwait	δ^{18}O − 2.9‰ to −3.5‰	d +15‰
Bahrain	δ^{18}O − 2.5‰	d +15‰
Qatar	δ^{18}O − 2‰ to −3‰	d +10‰

The δ^{18}O/δ^{2}H values of occurrences of Holocene groundwater on the eastern Arabian platform are comparable to values of modern groundwater in aquifers of the Tuwayq segments of the Interior Shelf, where δ^{18}O values range mainly from −1.8‰ to −3.5‰ with d values between +8‰ and +19‰.

On the Suman plateau in northern Saudi Arabia, where recent recharge is mixing with older stored groundwater, δ^{18}O values vary between −4.38‰ and −3.22‰. d values of most samples scatter around +10‰ (Fig. 6.17). In areas where infiltration of rain water is favoured on open karst surfaces, average δ^{18}O values are around −4‰, while δ^{18}O values of around −3.5‰ prevail, where the Umm er Radhuma aquifer is covered by Neogene deposits and dune sands. d values of +3.3‰ are observed in an area upstream of the karst outcrop. The water samples from the karst outcrop areas with δ^{18}O values around −3.5‰ and d values around +10‰ appear to contain relatively high proportions of Holocene recharge.

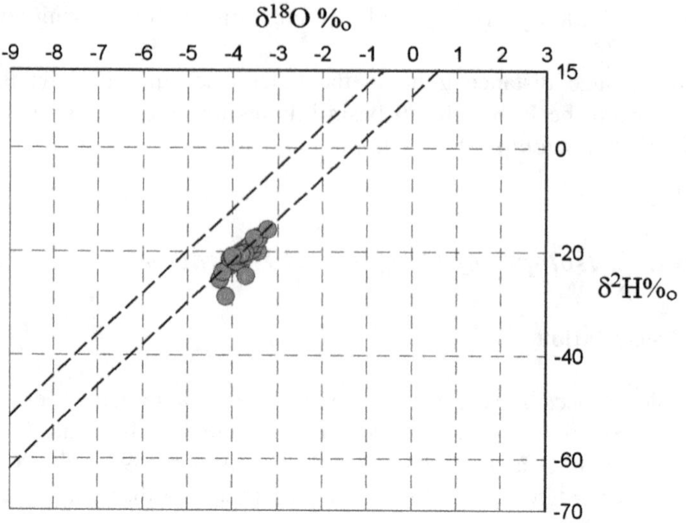

Fig. 6.17 $\delta^{18}O/\delta^{2}H$ diagram: Groundwater samples from the Umm er Radhuma aquifer of the Suman plateau, Saudi Arabia. Data from Hötzl et al. (1993)

6.6.2.3 Pleistocene Recharge in Eastern Saudi Arabia (Middle Gulf Sub-Basin)

Groundwaters of the eastern Arabian platform, which originate from Pleistocene recharge, have generally more negative $\delta^{18}O$ and $\delta^{2}H$ values than the Holocene groundwaters, and d values around 0‰.

According to a statistical evaluation of stable isotope data of main aquifers of the eastern Saudi Arabia (Shampine et al. 1979), mean $\delta^{18}O$ values of groundwater of the Umm er Radhuma, Khobar, Alat and Neogene aquifers are −4.6‰ to −5.4‰, mean $\delta^{2}H$ values −35.0‰ to −39.5‰, and the corresponding d values +1.8 to +5‰. In general, $\delta^{18}O$ values of around −2.5‰ occur in the outcrop areas of the aquiferous formations and become more negative downdip to about −6.5‰. "The higher values in the outcrop areas probably represent an average of the values in modern precipitation, and indicate modern recharge". "It would seem... that the large variation measured throughout a given aquifer are mostly due to paleoclimatic temperature changes" (Shampine et al. 1979: 455, 457).

In the Umm er Radhuma and Damam aquifers of the catchment area of the Hasa oasis, the delta values become more negative from the outcrop area toward the coast. The $\delta^{18}O$ and $\delta^{2}H$ values range from −4 to −7.5‰ and −31 to −50‰, respectively. The most negative values ($\delta^{18}O$ <−7‰, $\delta^{2}H$ <−40‰) possibly reflect upward leakage of water from the deep Wasia aquifer on the Ghawar anticline. The deep groundwater may have its origin in distant recharge areas on the Interior Shelf.

Mean $\delta^{18}O$ values are −4.5‰ in groundwater from the Umm er Radhuma aquifer of Haradh, in the Damam aquifer on the coast between Dhahran and

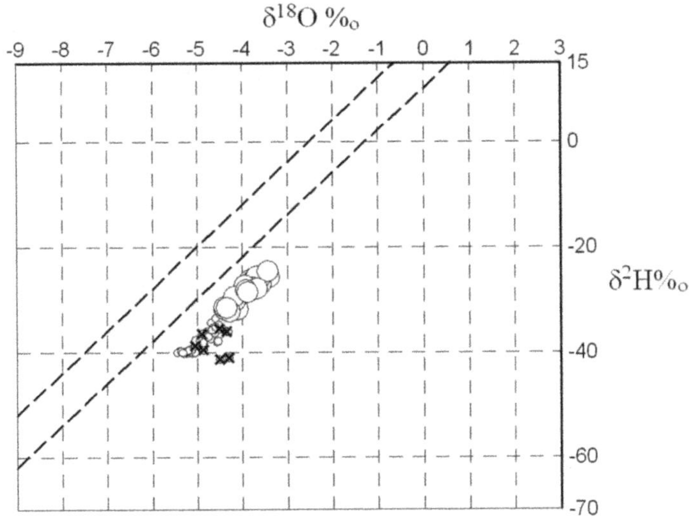

Fig. 6.18 $\delta^{18}O/\delta^2H$ diagram: Groundwater samples from Tertiary aquifers in eastern Saudi Arabia. O Qatif, o Hasa, x Haradh. Data from Al-Sayari and Zötl (1978), Otkun (1974, 1975)

Qatar, and in subsurface inflow into Bahrain and Qatar. On average less negative $\delta^{18}O$ values around −3.9‰ are found in the Damam aquifer of the coastal area at Qatif and in subsurface inflow into Kuwait. More negative $\delta^{18}O$ values prevail in the Hasa oasis area, which are attributed to discharge of deeper groundwater from distant recharge areas (Fig. 6.18).

Local variations of the delta values are observed within Al Hasa oasis in the Hofuf area, where about 160 brackish springs discharge from Neogene sediments within an area of around 250 km^2. The aquiferous Neogene sediments have a thickness of 180 m and overlie the Paleogene Damam and Umm er Radhuma aquifers. The distribution of groundwater temperatures and $\delta^{18}O$ values appears to reflect a complex groundwater flow system with upward leakage of hot groundwater. The deeper hot groundwater has more negative $\delta^{18}O$ values than groundwater from the shallow aquifer sections with lower temperature. A temperature gradient of the spring water from 45°C west of Hofuf to 30°C in the east follows the general direction of groundwater flow and the trend to less negative $\delta^{18}O$ values.

Mixtures of different proportions of deeper and shallow groundwater result in $\delta^{18}O$ ranges of −5.1‰ to −5.45‰ in the northern part of the Hofuf area, of −4.85‰ to −5.2‰ in the western part, and less negative $\delta^{18}O$ values of −4.6‰ to −4.8‰ east of Hofuf (Otkun 1974).

6.6.2.4 Kuwait

In Kuwait, $\delta^{18}O$ values of brackish groundwater vary from −3.1‰ to −4.5‰ in the Paleogene Damam aquifer and from −1.4‰ to −2.1‰ in the Neogene lower

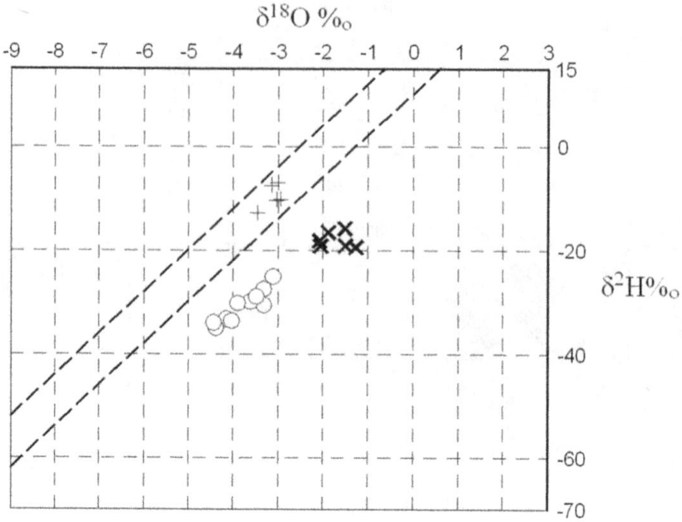

Fig. 6.19 $\delta^{18}O/\delta^{2}H$ diagram: Groundwater samples from Tertiary aquifers in Kuwait. O Damam aquifer (Shagaya and Sulaybiya well fields), x Kuwait aquifer (Wafra well field), + shallow groundwater, Dibdiba formation in northern Kuwait. Data from Yurtsever (1992)

Kuwait group aquifer (Fig. 6.19). In both aquifers, d values are around 0‰ and groundwater ages around 22,000 years.

Groundwater in the confined Damam aquifer is replenished through subsurface inflow from recharge areas in Saudi Arabia, while the overlying Kuwait group aquifer has been recharged within a more local radius.

$\delta^{18}O$ values of shallow groundwater lenses in northern Kuwait are −2.9‰ to −3.5‰, and d values of +15‰ are higher than values of the Pleistocene groundwaters of the Damam and lower Kuwait aquifers.

6.6.2.5 Bahrain and Qatar

Groundwaters of the Umm er Radhuma, Rus and Damam aquifers in Bahrain show two different stable isotope signatures: The bulk of water samples taken from the Damam aquifer and the confined Umm er Radhuma aquifer has $\delta^{18}O$ values between −3‰ and −5‰ with d values scattering around +3‰. A few samples from the Rus–Umm er Radhuma outcrop area have $\delta^{18}O$ values of −2.2‰ to −3.12‰ with d values at +10‰ (Fig. 6.20).

The more negative values coincide with the average value of −4.5‰ in the Damam aquifer in the coastal area of Dhahran in Saudi Arabia and can be attributed to submarine inflow of fossil groundwater. The less negative $\delta^{18}O$ values and d values of −10‰ in the Rus–Umm er Radhuma outcrop area resemble values of present-day precipitation in Bahrain and are assumed to represent recent recharge.

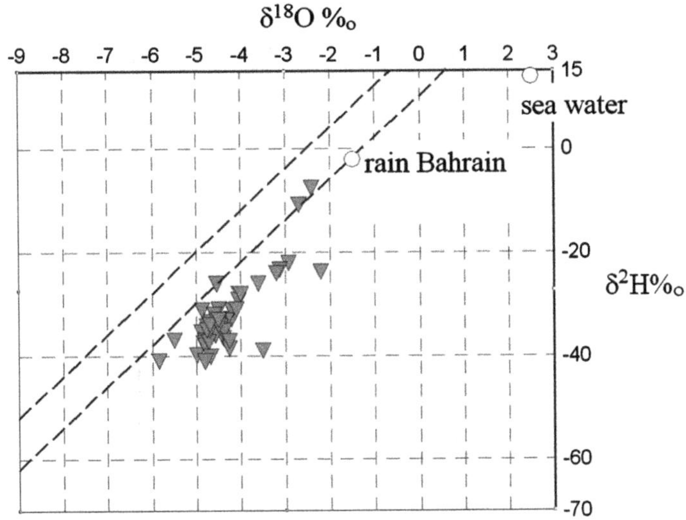

Fig. 6.20 $\delta^{18}O/\delta^2H$ diagram: Groundwater samples from Tertiary aquifers in Bahrain. Data from GDC (1980)

$\delta^{18}O$ values of −3‰ on the northwestern coast of Bahrain island indicate an impact of sea water intrusion.

$\delta^{18}O/\delta^2H$ values of groundwater in Qatar can be explained by a mixing of several components:

- Deep artesian groundwater with $\delta^{18}O$ values around −5.4‰ and groundwater from the Damam aquifer in the southern part of Qatar with $\delta^{18}O$ values from −3.8‰ to −4.8‰, both with d values around 0‰
- Groundwater in the northern part of Qatar with $\delta^{18}O$ values between −1.5‰ and −2.9‰ with d values around +10‰
- Groundwater in the coastal area affected by sea water with $\delta^{18}O$ values up to +2.5‰

The groundwater with less negative values in the northern part of the peninsula corresponds to the fresh water lens sustained by recent recharge. The more negative $\delta^{18}O$ values in the south and at greater depth represent the inflow of fossil brackish groundwater from Saudi Arabia.

In a $\delta^{18}O/\delta^2H$ diagram, data points plot close to mixing lines between fossil and recent groundwater and between groundwater and sea water (Fig. 6.20).

A relationship between the delta values of the stable isotopes and the salinity of the shallow groundwater is partly masked by evaporative isotope enrichment, in particular in water ponded in surface depressions as well as by dissolution of evaporites of the Rus formation. Salinization of recently recharged water indicated by the delta values and the presence of tritium by dissolution of evaporites is highly variable.

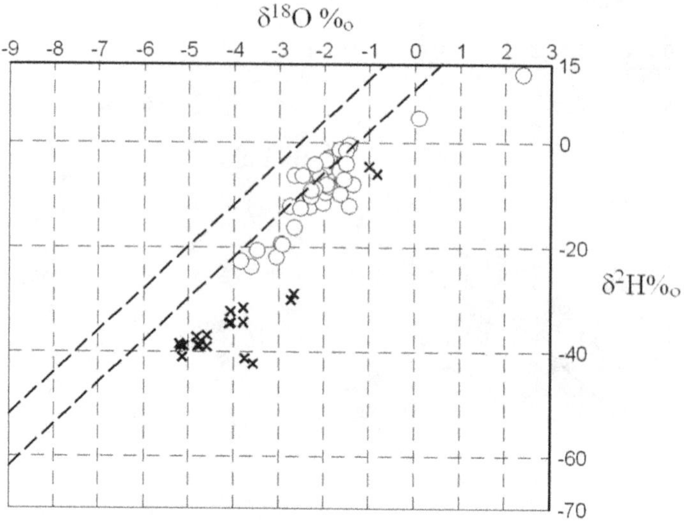

Fig. 6.21 $\delta^{18}O/\delta^2H$ diagram: Groundwater samples from Tertiary aquifers in Qatar. O wells in northern and central Qatar, x wells in southwestern Qatar and wells tapping deeper confined groundwater. Data from Yurtsever and Payne (1979)

6.6.2.6 Margins of the Rub al Khali

$\delta^{18}O/\delta^2H$ values of groundwater from the western gravel plain of the United Arab Emirates in the area around Dhaid–Mudam form a cluster around the GMWL (d = +10‰) with $\delta^{18}O$ between −1.3‰ and −2.5‰ with an obvious trend to evaporative enrichment to lower d values and $\delta^{18}O$ values, which are les negative than −1‰. The impact of isotopic enrichment is still more pronounced in groundwater samples from the gravel plain in the Al Ain area, where $\delta^{18}O$ values vary from −1.9‰ to −0.3‰ and d values from +10‰ to around 0‰ (Fig. 6.22).

Stable isotope values of groundwater in the sand dune area of Liwa form a cluster around $\delta^{18}O$ +2‰ and δ^2H between 0‰ and +6‰.

In the sand dune area of Ramlat ash Sharqiye in northern Oman, $\delta^{18}O$ values of +2‰ indicate a strong evaporative enrichment of direct recharge from modern precipitation.

The groundwaters on the gravel plain and in the sand dune areas appear to originate prevailingly from Holocene recharge. The $\delta^{18}O/\delta^2H$ values show a deviation from the values of present precipitation, which can be attributed to evaporative isotopic enrichment in the arid climate. The isotopic enrichment is particularly pronounced in the sand dune area, where the rain water is highly concentrated by evaporation before percolation into the aquifer.

Stable isotope values of groundwater in the alluvial aquifers of interior Oman show an enrichment (from −3‰ to −3.5‰ $\delta^{18}O$) through evaporation which characterizes runoff originating from localized summer orographic/convective

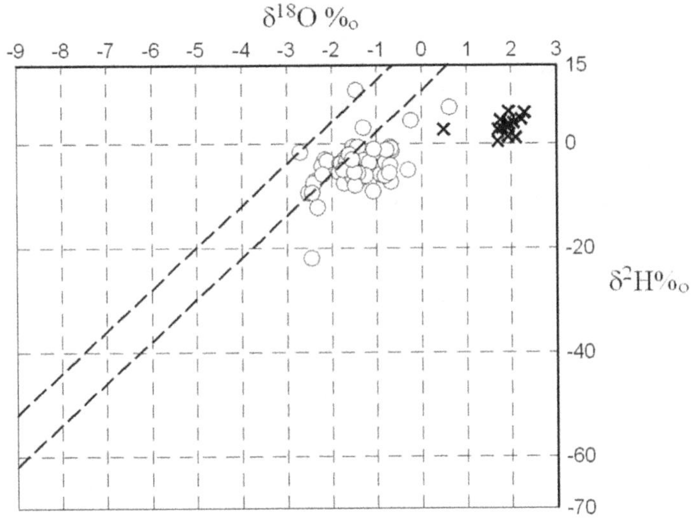

Fig. 6.22 $\delta^{18}O/\delta^{2}H$ diagram: Groundwater samples from aquifers in the western gravel plain and the sand dune area of the United Arab Emirates. O western gravel plain between Dhayd and Al Ain, x Liwa sand dune area. Data from Ministry of Agriculture and Fisheries, Dubai

storms. Despite their irregularity, such storms seem to be more effective in recharging the interior alluvial aquifers than the rains associated with winter low pressure systems.

The groundwater from the Nejd plateau in southern Oman is generally more enriched in $\delta^{18}O$ and $\delta^{2}H$ than groundwater from Al Wusta, indicating a greater input of less depleted rains, which probably are related to isotopically enriched cyclonoc precipitation with a possible contribution of monsoonal precipitation in the Dhofar mountains. The isotopic character of the fossil groundwater from the main Nejd aquifers is similar to the character of modern Nejd groundwaters. The $\delta^{18}O/\delta^{2}H$ of the groundwaters appear to be influenced mainly by the geographic distribution of rainfall types rather than by paleo-climatic variations.

The stable isotope data indicate that groundwater in the shallow A aquifer is replenished by cyclonic or depression type storm events tracking form northern or central Oman. The groundwater shows no signs of evaporation. Obviously, the lack of a significant soil cover and the outcrop of aquifer rocks in many wadis favor rapid infiltration of surface runoff. The $\delta^{18}O$ values for the A aquifer vary from −7.5‰ to −2‰. The $\delta^{18}O$ values for the deeper B and C aquifers vary from −6‰ to −4‰. It is assumed that these groundwaters were recharged during the Pleistocene mainly by convective storm events, which were more frequent than at present. No indications on recharge from monsoonal rainfall with less negative $\delta^{18}O$ values have been found.

6.6.2.7 Interpretation

In a tentative interpretation, the distribution of values of stable isotopes of oxygen and hydrogen in aquifers of the eastern Arabian platform may be summarized as follows:

The isotopic composition of Holocene recharge is influenced by air moisture reaching the eastern Arabian platform from the Mediterranean Sea and from the Indian Ocean–Arabian Sea. Present-day precipitation from eastern Mediterranean meteorologic events is characterized by d values of +22‰ and $\delta^{18}O$ values of −5.29‰ at sea level. Present-day precipitation on the Gulf coast at Bahrain has mean $\delta^{18}O$ values of −2.5‰ with a deuterium excess (d) of +10‰.

Average $\delta^{18}O$ and d values of aquifers with prevailing Holocene recharge are generally in a range of −2‰ to −3.5‰ $\delta^{18}O$ and of −10‰ to +15‰ for d. $\delta^{18}O$ values in the Umm er Radhuma aquifer are around −3‰ in the outcrop area on the western margin of the platform. These values are similar to values of groundwaters in wadi aquifers of the Tuwayq segments of the Interior Shelf, adjoining the eastern Arabian platform in the west, ($\delta^{18}O$ mainly −1.8 to −3.5‰, d +8 to +19‰), which are recharged from recent runoff infiltration. Significant deviations from the general range of Holocene groundwaters to less negative $\delta^{18}O$ values and lower d values are found in particular in the east of the Rub al Khali sub-basin, where stable isotope values are influenced by evaporative enrichment and geographic variations of the meteorologic regime.

Groundwaters originating from Pleistocene recharge, prevailing on the eastern Arabian platform, have generally more negative $\delta^{18}O$ values than the Holocene groundwaters (generally −4 to −7.5‰) and d values around 0‰.

In the topographically lower part of the platform adjoining and including the Gulf coastal area, a general zonation can be seen for $\delta^{18}O$ values of Pleistocene groundwaters, all characterized by low ^{14}C values and d values around 0‰: In the area between Haradh–Al Hasa and the coast between Dhahran and Qatar as well as in subsurface inflow to Bahrain and Qatar, $\delta^{18}O$ values vary around −4.5 to −4.9‰. Mean $\delta^{18}O$ values of groundwaters of the Damam aquifer in the area from Qatif to Kuwait are less negative with, on average, −3.8‰. Significantly less negative are average $\delta^{18}O$ values of Pleistocene groundwaters of the Neogene aquifer in southern Kuwait with −1.8‰.

The spatial variations of $\delta^{18}O$ values of the Pleistocene groundwaters are probably related to the groundwater flow and recharge conditions with more negative values in the sub-regional long-distance groundwater flow system and less negative values in areas, where recharge was more strongly influenced by moisture sources from the Gulf.

References. Akiti et al. (1992), Al-Sayari and Zötl (1978), Alsharhan et al. (2001), Clark et al. (1987, 1995), Hötzl (1995), Hötzl et al. (1993), Imes and Wood (2007), Jado and Zötl (1984), Jordan et al. (1980), Khalifa (1997), Kulaib (1991), Macumber et al. (1997), Matter et al. (2006), Otkun (1974), Robinson and Al Ruwaih (1985), Shampine et al. (1979), Wagner and Geyh (1999), Yurtsever (1992), Yurtsever and Payne (1979).

6.7 Gypsum, Sabkhas, Sand Dunes: Particular Hydrogeologic Features of the Shelf Area of the Arabian Peninsula

Since the beginning of the Holocene, an arid to hyper-arid environment dominates the eastern Arabian platform and the Interior Shelf. Apart from the limitation of groundwater recharge from the low rainfall, several physical features related to the arid climate have a significant impact on the hydrogeologic conditions: wide spread occurrence of evaporite minerals, development of salt flats in morphologic depressions, accumulation of wind blown sand in dunes and sand seas.

The low rates of groundwater recharge result in low hydraulic gradients and low rates of flushing of the aquifers. The limitation of flushing has implications on the hydrochemical groundwater regime in many areas, where evaporite formations within the sedimentary sequence constitute sources of groundwater salinization.

6.7.1 Influence of Evaporite Formations on the Hydrogeologic Conditions

Gypsum and anhydrite layers were deposited during various paleogeographic stages on the shelf area of the Arabian Peninsula. The deposits of Infracambrian salt basins, which extend in the subsurface over the eastern parts of the Arabian Shelf, are generally situated below the zone of active groundwater circulation.

Dessication of the carbonate platform of the Arabian Shelf from the Permian to early Jurassic lead to the formation of evaporite basins. Again during the Paleocene to early Eocene, wide parts of the eastern Arabian platform were covered by a shallow evaporitic shelf and, during the Miocene, evaporitic conditions occurred in coastal–terrestrial environments of the Euphrates–Gulf basin. Anhydrite and gypsum intercalations are found, in particular, in the Permian Khuff formation, the Lower Triassic, and the Upper Triassic Arab formation. The Upper Jurassic Hith formation consists mainly of massive anhydrite. The Lower Eocene Rus formation is composed prevailingly of anhydrite, gypsum and marl, except for paleogeographic structural highs, where the formation includes mainly dolomite and limestone sequences. Evaporitic deposits are found in Miocene formations in wide parts of the eastern Arabian platform, e.g., in the Middle Miocene Upper Fars formation.

The evaporitic layers in the sedimentary sequence of the shelf area of the Arabian Peninsula act, to a large extent, as aquicludes or aquitards between mainly carbonatic aquifers and constitute an important source of groundwater salinization in adjoining aquifers. The evaporitic Rus formation constitutes a particularly important sub-regional aquiclude/aquitard, which separates the middle and upper aquifer complexes of the multi-aquifer system of the eastern Arabian platform.

A certain impact on the groundwater quality may be attributed to the widespread occurrence of dispersed evaporite minerals on the eastern Arabian platform, mainly eolian or evaporation deposits. The predominance of sulfate waters in many areas may be explained, to some extent, by the dissolution of surficial salts during the infiltration process.

In a few areas, the aquiferous conditions are enhanced though solution and collapse structures in near surface evaporite formations. In the Layla lakes area (Uyun al Aflaj) in southern Saudi Arabia, dissolution of anhydrite layers of the Hith formation has caused collapse openings and brecciation of the overlying Sulayi, Yamama and Buwaib formations with an increase of secondary permeabilities of the affected formations. In Qatar, dissolution of anhydrites of the Rus formation has created solution openings and collapse depressions improving the recharge conditions and permeabilities of the outcropping Rus–Umm er Radhuma aquifer.

6.7.2 Hydrogeologic Features of Sabkhas

Salt flats, in Arabic sabkha or khabra, constitute a particular morphologic feature in semi-arid and arid zones, which are characterized by low volumes of surface runoff, limited erosional energy and development of internal drainage zones. Sabkhas are situated in morphologic depressions and have, in general, a nearly flat surface and a several metres thick cover of sabkha deposits: clay, silt and muddy sand, frequently with salt incrustations.

Sabkhas can be differentiated into inland, wadi and coastal sabkhas. Larger inland sabkhas generally occupy the morphologically lowest parts of tectonic basin or graben structures and constitute the centre of closed hydrologic basins. Inland sabkhas with extensive catchment areas are found in the Rub al Khali subbasin and on the northern Arabian platform on plateau areas between the hydrologic Mediterranean and Euphrates basins (Jaboul, Matah, Qa el Azraq, Wadi Sirhan, Jafr, the previous Hijane and Ateibe lakes on the Damascus plain). Wadi sabkhas constitute morphologically flat stretches along wadi courses or on the confluence of different side wadis and are covered by sabkha type sediments, alternating with fluvial wadi sediments. Wadi sabkhas are found on the Hamad plateau (Wadi Ruweishid), within the Jebel el Arab basalt field, and in larger wadi systems of the Interior Shelf.

Coastal sabkhas constitute flood plains, which extend over parts of many coastal plains and which may be inundated seasonally from the sea. Coastal sabkhas are common along the shoreline of the Gulf from Kuwait to the United Arab Emirates. Some sabkhas, which originated as coastal salt flats, are now found many kilometres inland at elevations as high as 150 m asl.

Many sabkhas act as groundwater discharge zones draining sub-regional or local aquifers. To a large extent, groundwater discharge in sabkhas occurs through evaporation; on several sabkhas of the Arabian Plate, seasonal or perennial springs are located on the rim of the salt flats. Springs and surface runoff create seasonal or ephemeral lakes in several sabkha zones, but many of these lakes have dried up because of intensive groundwater extraction in the catchment areas.

A considerable number of sabkhas on the Arabian Platform is situated at altitudes above the present groundwater levels and are possibly remnants of hydrologic environments with wetter climate conditions.

Sabkhas connected to the present active hydrologic cycle are characterized by the occurrence of saline shallow groundwater, salt crusts or encrustations on the surface and within the surficial deposits, and a general absence of vegetation.

Sabkhas in present or previous coastal zones act as major groundwater discharge zones for the multi-aquifer regime of the eastern Arabian platform. A sabkha discharge zone extends in a belt parallel to the Gulf coast from Kuwait to the southwestern end of the Gulf coast near Qatar and continues into the coastal strip of Abu Dhabi and Dubai.

The groundwater salinity generally changes from brackish levels upstream of the sabkha discharge zone to saline water under the sabkha flat and down the hydraulic gradient from the discharge zone. The high salinity is caused by evaporative salt enrichment and present or previous sea water intrusions.

Coastal sabkhas represent the essential discharge area for regional groundwater flow systems. The water is highly saline with Cl type water and generally long residence times.

Larger sabkhas in Qatar at Sauda Nathil and Jaww as Salama in the south of the perninsula occupy depressions of 18–22.5 km^2, which lie close to sea level and are even lower locally. The origin of these depressions seems to be related to the dissolution of fractured limestones.

On the Gulf coast of Abu Dhabi, sabkhas are situated on average about 2 m asl and have a flat surface with a gradient of about 0.1‰. The 1 m thick typical sabkha deposits are composed of algal and dolomitic crusts above a layer of anhydrite; halite crusts appear seasonally on the surface. The typical sabkha deposits are underlain by 10 m thick Quaternary eolian silica and carbonate sands and Tertiary formations composed mainly of marl and gypsum. A deeper confined aquifer is probably connected with the sub-regional groundwater flow system of Paleogene aquifers of the Rub al Khali sub-basin.

The topography of the sabkhas in the coastal plain of Abu Dhabi is controlled by the situation of the water table at less than 1 m below surface and by wind deflation.

The sand layer below the sabkhas contains a brine with >300 g/kg TDS. Salinity in the artesian Tertiary aquifer is 104–286 g/kg TDS.

The hydrologic budget of the costal sabkhas is dominated by evaporation and recharge from rainfall; lateral groundwater inflow and upwelling groundwater are minor components within the water budget. Evaporation from the water table constitutes a large outflow of water that is replaced by inflowing groundwater or sea water. Upward leakage of groundwater into the sabkhas has been estimated at 4–5 mm/a.

The largest sabkha in the Gulf coast area is Sabkhet el Matti on the boundary of Qatar and Abu Dhabi, which extends over around 6,000 km^2. Much of the sabkha area is covered by partly cemented dune sand and is undergoing slow deflation. The whole sabkha is salt encrusted.

Many sabkhas with generally elongated shape are scattered between the sand dunes of the eastern Rub al Khali.

Sabkhet Umm es Samim extends as a 100 km × 30 km wide salt flat over a large morphologic depression on the eastern rim of the Rub al Khali west of the foothills of the Oman mountains. The sabkha constitutes a discharge zone for groundwater inflow in Paleogene aquifers from a large catchment in the southwest and for different aquifers on the west slope of the Oman mountains in the east.

The sabkha, which runs parallel to the Oman mountain range, is situated at an average elevation of 60 m asl 200 km from the nearest coast. The sabkha "is a salt encrusted playa, which may have developed in a natural basin or deflation hollow, where the groundwater table is very close to, or reaches the surface. In this situation, efflorescence or capillarity evaporation causes crystallization of evaporites from groundwater" (Alsharhan et al. 2001: 12). Possibly, the sabkha occupies the area of a Pleistocene lake, which dried up during the Holocene.

A preliminary satellite interpretation reveals sand-dune covered evaporites north, west and south–west of the centre of the sabkha. Alluvial gravel plains east and northeast of the centre show a drainage pattern with alluvial channels, which indicate former surface runoff into the sabkha. Near the centre of the sabkha, a paleo-plateau covered by evaporites appears to be a residual feature of the bottom of an evaporated water body, which filled the sabkha area probably until the late Pleistocene or early Holocene.

Within the sabkha area, two types of sand accumulations can be seen:

- Older sand fans on the northwestern margin of the sabkha, which probably belong to the first cover of aerial sediments after desication of the surface water.
- Younger sand dunes, which migrated from southeast to northwest over the whole sabkha area and cover also part of the alluvial channels.

The sand cover shows, that no significant surface runoff reaches the sabkha under the present arid climate.

6.7.3 Recharge in Sand Dunes

Wide parts of the shelf area of the Arabian Peninsula are covered by eolian sand deposits. Sand dunes and sand seas of huge dimensions extend over the Rub al Khali and Great Nefud deserts and Ad Dahna, elongated sand fields are found in the Tuwayq segments of the Interior Shelf and the Jafura near the Gulf coast.

The sand seas and dunes of the Great Nefud cover about 57,000 km^2 in northern Saudi Arabia.

Ad Dahna is a belt of reddish sand, about 1,300 km long and 25 km wide, extending between the Great Nefud in the north and the Rub al Khali sand seas in the south.

Sand dunes cover about 74% of the United Arab Emirates. The surface of the sand covered area on the northern margin of the Rub al Khali rises in elevation from sea level at the Gulf coast to 250 m asl at Liwa in the south of the United Arab Emirates.

The sand dunes of the Rub al Khali increase in height from a few metres in the north to more than 200 m in the south. Some of these giant linear sand dunes are up to 260 km long; the dunes follow mainly a NNE trend at distances of 2–6 km.

In the eastern Rub al Khali, sand dunes alternate with sabkha deposits. Eolian deposits consist of medium to very fine-grained sand with silt and are composed of quartz, carbonates and heavy minerals. The sabkha deposits between the dunes are composed of thin sand and silt deposits and reach a thickness of 50–150 m.

Ramlat as Sabatayn, a sand desert with a maximum west–east extent of some 350 km, is situated southwest of the Rub al Khali between the Yemen highlands and the eastern plateau area of Yemen. Rainfall and vegetation are nearly absent, except along the margins of Ramlat as Sabatayn, where rivers bring water from adjacent mountain and upland zones.

Groundwater in the sand dune areas is generally brackish. In some areas of the Rub al Khali, a brine extends in shallow aquiferous lenses. Groundwater with moderate salinity is found only in some areas in confined aquifers below the sand deposits and as shallow groundwater on the fringes of the sand cover, where episodic surface runoff occurs from adjoining highlands.

The occurrence of actual recharge in the desertic sand areas of the eastern Arabian platform is indicated e.g., in the Liwa area by ^{3}H values of 5–10 TU. In irrigation areas at Liwa, return flow causes high NO_3 concentratiosn of 250–500 mg/l.

Estimates of recharge rates in sand dune areas, which have been made for several areas of the eastern Arabian platform, can be summarized as follows:

Umm er Radhuma sub-crop covered by sand dunes:	2–2.3 mm/a
Sand dunes in the Qasim area	30 mm/a on sand dunes 0 at swale sites betwen the dunes
Ad Dahna	4 mm/a
Ramlat ash Sharqiye, Oman	<10 mm/a

The estimates indicate, that some recharge can occur through deep percolation of rain water in the sand dune areas, depending on frequency and intensity of episodic precipitation. Data of salinity and of δ^{18}O and δ^2H, e.g.,in the Liwa area, show, that the water infiltrating in the sand dunes is strongly affected by evaporitic enrichment.

Recharge on sand dunes is, in most cases, direct, i.e. rain falling on the sand area infiltrates directly into the subsurface. The flow through the unsaturated zone can be considered generally as piston flow (Lerner et al. 1990: 39). The presence of moisture in the sand layers depends, apart from the climate, on the grain size distribution of the sand.

Mean grain size in sand dunes often varies between 0.15 and 0.4 mm. Sand dunes with coarser grains have a low field capacity and let rainwater infiltrate to a considerable depth.

References. Al-Sagaby and Moallim (1996), Alsharhan et al. (2001), Beydoun (1991), Caro and Eagleson (1981), Dincer et al. (1974), ESCWA (1999c: 43 ff.), Sanford and Wood (2001), Schaeffer (1997), Shampine et al. (1979), Al-Turbak et al. (1996), van der Gun and Ahmed (1995), Zubari (1997).

Chapter 7
Arabian Shield

7.1 Geology, Morphology and Climate

7.1.1 Geology

7.1.1.1 Precambrian Shield

The Arabian Shield, the uplifted basement of the Arabian Plate, occupies about 770,000 km^2 in the west and southwest of the plate. The 9–15 km thick basement sequence is composed of gneiss, metasediments and volcanics and has been intruded extensively by granitic plutons. The oldest rocks exposed on the shield are represented by the "basement gneiss", which occurs mainly in diapiric or intrusive structures. Sedimentary and volcanic rocks, metamorphosed to greenschist facies, comprise schists, marble, quartzite, conglomerate, greywacke, metadiorite, metagabbro, amphibolite. The Precambrian sequence is penetrated by mainly granitoid and granitic plutonic bodies and dikes and by stocks, pipes and batholits, ranging from gabbro to granitic lithology.

The shield was affected by two major orogenic cycles: the Hijaz orogeny about 600 million years ago and the Nejd orogeny in the late Precambrian–early Paleozoic. The Nejd orogeny created a southeast–northwest trending system of transcurrent faults, which extend as tectonic lineaments over several hundred kilometres length over the surface of the shield (Fig. 7.1).

In Yemen, the Precambrian basement comprises four distinct major rock divisions:

- Plutonic rocks:
  Mainly elongated syntectonic granitic or granodioritic intrusions
  Post-tectonic granitic intrusions with an approximately circular shape
  Metamorphic basic and ultramafic intrusive rocks, anorthosite in greenschist facies
- A low to medium grade metamorphic unit with schists, metavolcanics, meta-ultramafite and metasediments
- A medium to high grade metamorphic unit including orthogneiss, amphibolite, amphibolite gneiss
- Migmatites, migmatitic amphibole and biotite gneiss

W. Wagner, *Groundwater in the Arab Middle East*,
DOI 10.1007/978-3-642-19351-4_7, © Springer-Verlag Berlin Heidelberg 2011

By the end of the Precambrian, most of the Arabian Plate constituted a deeply eroded peneplain, and the shield has been a rather stable land mass since that time.

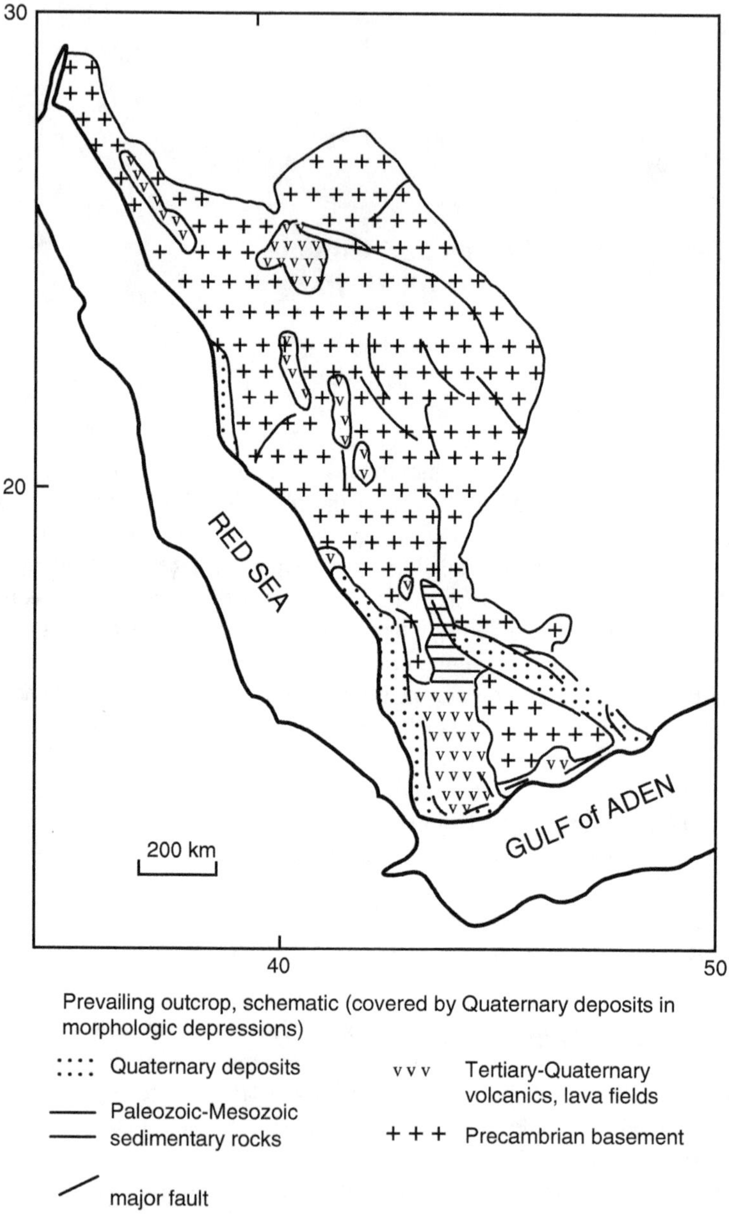

Fig. 7.1 Arabian Shield, outcrops of main geologic units. After Alsharhan et al. (2001), Edgell (1997), Kruck et al. (1996), van der Gun and Ahmed (1995), Wohlfahrt (1980)

7.1.1.2 Mesozoic–Tertiary Sedimentary Rocks

During the Jurassic, the Arabian Shield began to break up, at its southern end, into basin and uplift segments. Subsiding zones of weakness were inundated by relatively narrow sea invasions during the Mesozoic to early Paleogene, and marine and terrestrial sediments were deposited in the zones of marine transgressions. A particularly deep sedimentary basin, the Marib–Al Jawf basin in the north of Yemen, was filled with about 5,000 m thick Upper Jurassic to Cretaceous sediments.

The tectonic break-up of the southern part of the Arabian Shield resulted in the isolation of a southern block of outcrops of the Precambrian basement from the main crystalline shield area through a zone of sedimentary–volcanic basins, which comprises the Marib–Al Jawf basin and most of the western highlands of Yemen. The sedimentary basins of the southern Arabian Shield are delimited from the outcrops of the crystalline basement by major fault systems: the southwest–northeast trending Sada graben system, the WNW-ESE trending Al Jawf fault and the east–west trending Wadi Mawr fault (Kruck et al. 1996, Fig. 2).

In the northwest of Yemen, the Permian Akbra shales overlie discordantly basement rocks and the Paleozoic Wajid sandstone of the Interior Shelf. The up to 130 m thick Akbra formation is composed of predominantly glacial and glacio-lacustrine deposits: varved, laminated mudstones, siltstones, shales with intercalated sandstones and boulder beds.

The Jurassic–Paleogene of the Yemen highlands comprises the following sedimentary sequence:

- Lower–middle Jurassic Kohlan sandstone: An on average 60 m and up to 100 m thick sandstone formation with conglomerate and shale
- Upper Jurassic–Lower Cretaceous Amran limestone: Carbonate rocks and marls with sand an shale intercalations, and gypsum and coal seams at the top, with a thickness of 320 m in the outcrop area around Amran and of several thousand metres in the Al Jawf basin
- Upper Cretaceous Tawila sandstone: 150–320 m thick formation of sandstones with shale intercalations
- Paleogene Medj-Zir sandstone: 120–200 m thick sandstones deposited in a basin around Sanaa

7.1.1.3 Tertiary–Quaternary Volcanics

An uplift of the Afro–Arabian dome in the present East African–Red Sea rift zone was accompanied by intense volcanic activity, which extended over a belt ranging from western Saudi Arabia over the Gulf of Aden into Afar and Ethiopia. The Red Sea rift, separating the Arabian sector of the Nubo–Arabian Shield, was initiated in the Miocene and resulted in the disengagement of the present southern Arabian Peninsula from the main block of the African Shield.

Tertiary volcanic rocks of the southern Arabian Shield (Yemen volcanics) include basalts, rhyolites and tuffs with sedimentary intercalations and reach a thickness of 1,500–2,000 m. The volcanic activity continued into the Quaternary (Aden volcanics).

Various granitic bodies intruded the southern Arabian Shield in the Tertiary.

In Saudi Arabia, basaltic lava flows, which erupted during the late Tertiary to Quaternary along north–south fractures, cover part of the Precambrian basement rocks. The volcanic and intrusive formations, related to the Tertiary rift tectonics, have been differentiated into three principal rock units:

- The mostly volcanic and volcanogenic Sita unit and cogenetic dike rocks of the Damm unit
- Thick dike intrusions of the Gumaya unit, which postdates the Sita unit

The largest basalt field in Saudi Arabia is Harrat Rahat, which extends on the Arabian Shield between Taif and Madina over a length of 300 km in north–south direction parallel to the Red Sea coast. The volcanic rocks cover morphologic depressions which are filled with detritic continental sediments of probably Oligocene age.

7.1.1.4 Quaternary Sediments

During the Quaternary, wadis and morphologic depressions were filled with alluvial and terrace deposits. Sheet wash gravels extend over plain surfaces. Quaternary detrital deposits fill wadi courses and various morphologic basins in the highlands of Yemen.

7.1.2 Morphology

7.1.2.1 Mountains, High Plateaus and Escarpments

The Arabian Shield accompanies, along its western rim, the Red Sea coast between Aqaba and the Gulf of Aden for more than 2,000 km. Major topographic units of the shield are a 100–150 km wide belt of highlands to mountains in the west, parallel to the Red Sea coast, and a 500–600 km wide high plateau east of the coastal mountain chains. The mountain belt in the west comprises the Hijaz mountains in the north with moderately high elevations of up to 1,800 m asl and the high Asir mountains in the south with an average elevation of 1,800–2,400 m, reaching up to 3,133 at Jabal Soda near Abha in Saudi Arabia and to 3,620 m at Jabal Nabi Shuayb near Sanaa. In the north, the shield ends in the Midyan mountains (the land of the biblical Midianites), which form an around 100 km wide spur between Red Sea–Gulf of Aqaba and the Tabuk segment of the Interior Shelf.

The 40–140 km wide Hijaz mountain belt grades in the east into the central Arabian high plateau, which is situated at altitudes between 1,000 and 1,500 m asl with a general gentle eastward slope to the Tuwayq mountains of the Interior Shelf. The central Arabian high plateau forms, together with the northwestern part of the Tuwayq segment of the Interior Shelf, the highlands of central Saudi Arabia: Al Nejd. In the southeast, the highlands of the shield are adjoined by the Wajid basin of the Interior Shelf.

The morphology of the southern part of the Hijaz highlands is characterized by ridges and deep canyons; further north the topographic altitude and ruggedness of the morphology decrease.

The up to nearly 600 km wide Nejd high plateau constitutes, in wide areas, a mosaic of pediments and desert plains with protruding inselbergs and volcanic cones. Vast basalt fields, such as Harrat Rahat, are spread over the western part of the Nejd plateau.

The western rim of the coastal mountains of the shield drops in steep escarpments to the generally narrow coastal plain bordering the Red Sea.

The Asir mountains of southwestern Saudi Arabia extend toward south into the highlands of Yemen, a more than 200 km wide belt of mountains and intermountain basins with altitudes generally above 1,500 m asl and mountain peaks above 3,000 m The highlands are adjoined by the western escarpment, a mountain and hill zone with deeply incised wadis and steep slopes toward the Red Sea coastal plain. The eastern part of the mountain belt comprises mainly a tableland at altitudes between approximate 1,000 and 2,500 m and descends to altitudes below 1,000 m asl at the western margin of the Rub al Khali desert (Figs. 7.2 and 7.3).

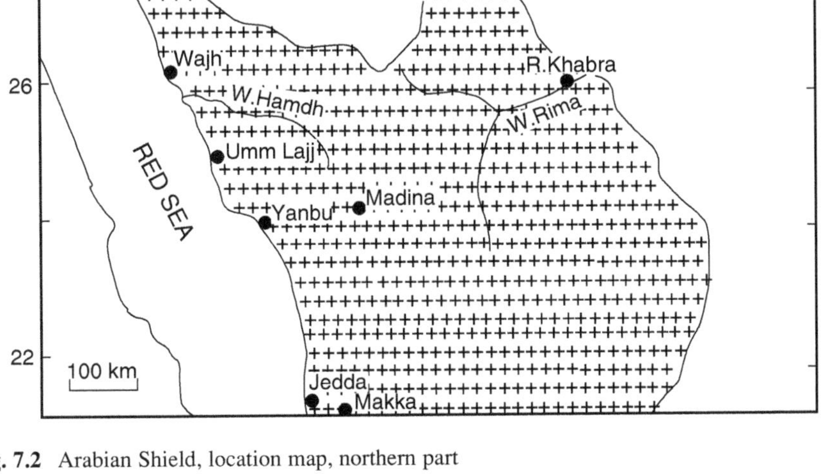

Fig. 7.2 Arabian Shield, location map, northern part

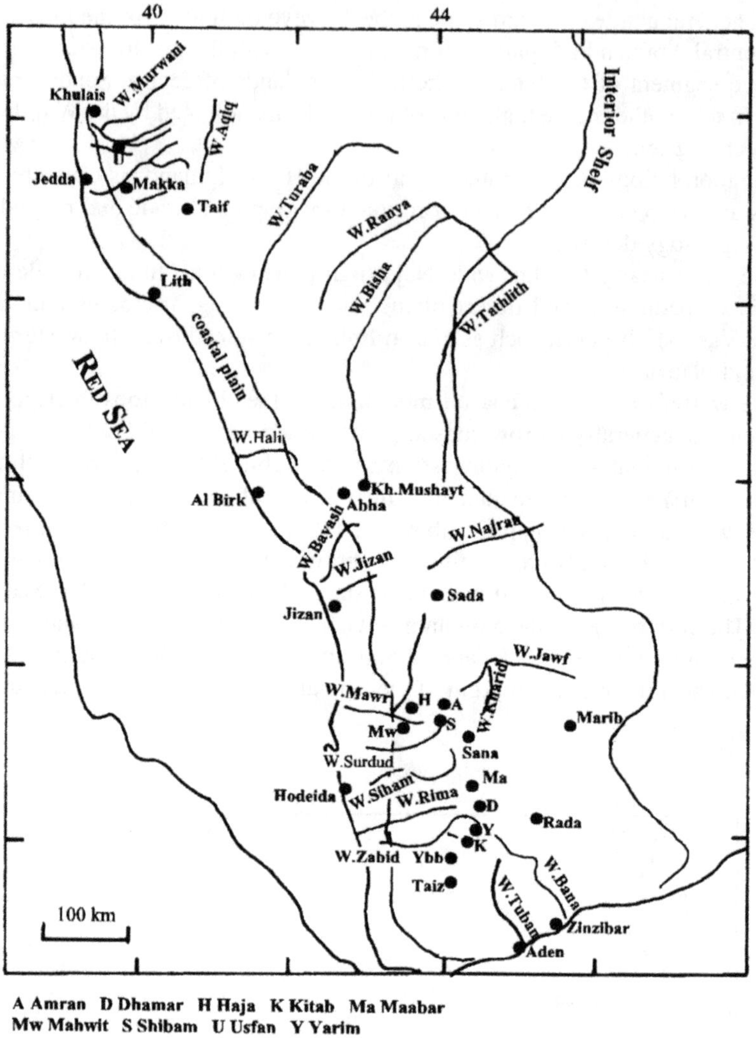

Fig. 7.3 Arabian Shield, location map, southern part

7.1.2.2 Basalt Fields

Vast basalt fields, such as Harrat Rahat, are spread over the western part of the Nejd plateau.

The Harrat Rahat fields of Oligocene to Quaternary basalts extend on the southeastern rim of the Hijaz mountains over 300 km between Madina and Taif in a width of about 50 km. The margins of the basalt massif are accompanied by steep wadis, which end in sabkha depressions, such as Wadi Aqiq in the east, or descend in steep slopes toward the Red Sea in the west.

Topographic elevations of Harrat Rahat range from 620 m asl at Madina to a mean altitude of 900 m asl and peaks of >1,500 m asl on the mountain crest.

The Tertiary Yemen volcanics form strongly dissected mountain massifs over an extensive area in the southwest of the Arabian Plate.

7.1.2.3 Wadi Systems of the Arabian Shield and Its Foreland

The crest of the coastal mountains runs near the western margin of the Arabian Shield and marks approximately the main water divide between the Red Sea and the hydrographic Gulf basin (Euphrates–middle Gulf–Rub al Khali sub-basins).

The highlands of the Arabian Shield are drained by a number of major wadis in east–northeast direction toward the interior of the Arabian Peninsula and by many relatively short and steep wadi systems with westward drainage to the Red Sea.

From the Asir mountains, a network of major wadis crosses the Nejd highlands, some wadis pass the Tuwayq mountains of the Interior Shelf before they vanish in the arid sand deserts of Nefud, Ad Dahna and Rub al Khali. Altitudes of the wadi catchments in the Asir mountains range from 2,000 to >3,000 m.

In the higher mountain ranges, the wadis have generally steep gradients; in parts of the eastern slopes of the Hijaz and Asir mountains, the eastward directed wadis are entrenched in 60–120 m deep canyons. Major wadi systems with eastward drainage on the Arabian Shield are Wadi al Jauf, Wadi Najran, Wadi Bisha which continues into the paleo-stream of Wadi ad Dawasir, Wadi Turaba and Wadi Rima.

Wadi al Jawf in Yemen drains into Ramlat as Sabatayn. The drainage system of Wadi Najran, covering a catchment of 4,250 km^2 in the highlands on the border of Saudi Arabia and Yemen, ends in the sand seas of Rub al Khali.

Wadi Bisha and its tributaries Wadi Ranya and Wadi Tathlith drain the east slope of the northern Asir mountains and the Hijaz highlands of southwestern Saudi Arabia continuing into Wadi ad Dawasir on the northern margin of the Rub al Khali.

Wadi Turaba with its headwaters in the Asir mountains near Taif has a catchment of 3,700 km^2.

Wadi Rima and its tributary wadis originate in the area east of Madina and constitute, with a length of the main wadi of 400 km and a catchment area of 71,000 km^2, the largest wadi system of Saudi Arabia. East of the Arabian Shield, Wadi Rima continues into the paleo-river course of Wadi Batin, which ends in the Euphrates depression near Basra.

About 90 wadi systems with relatively small catchment areas drain toward the Red Sea coast in Saudi Arabia. Combined average runoff of wadis in Saudi Arabia toward the Red Sea coast is estimated at 1.27×10^9 m^3 over a total cross-section of more than 1,000 km.

In the northern part of the Red Sea–Gulf of Aqaba coast between Aqaba and Jedda, volumes of surface flow are rather limited.

Wadis in the area between Jedda, Makka and Madina have their headwaters mainly in the Harrat Rahat volcanic massif. Mean annual rainfall on Harrat Rahat is 50–100 mm, and high seepage losses of wadi runoff occur into the volcanics and interbedded sediments.

On the western escarpment of the Yemen highlands, seven major wadis (wadis Mawr, Surdud, Siham, Rima, Zabid, Rasgan, Mauza) drain toward the Red Sea. Average annual stream flow varies between 20×10^6 and 180×10^6 m^3, the highest average flow volume being recorded for Wadi Mawr. The mean combined surface flow of the seven major wadis is around 565×10^6 m^3.

7.1.3 Climate and Vegetation

The climate on the Arabian Shield ranges from arid conditions with average annual rainfall of less than 100 mm in the north and east to more humid conditions in the Asir mountain zone of Saudi Arabia and Yemen with average annual rainfall of more than 500 mm.

Mean annual precipitation is e.g., 50–80 mm for Wadi Najran on the eastern slope of the Yemen–Saudi Arabian highlands, 250 mm for Wadi Turaba on the northeastern slope of the Hijaz, and 100 mm for Wadi Rima in the northeast of the Arabian Shield (Fig. 7.4).

Precipitation is distributed over three seasonal rain periods in winter, spring and summer. In winter between December and March, atmospheric depressions advancing from the Mediterranean Sea or the Red Sea produce heavy rain over a few days. These events occur, on average, once in the winter season. In spring during April and May, rainfall is caused from prevailing southeasterly winds and affects mainly the eastern slopes of the Asir mountains in Saudi Arabia and of the northern part of the Yemen highlands. From late July, the climate comes under the influence of moisture inflow from south and southwest with rainfall mainly on the high mountain areas and the western escarpment. The summer rains may continue into September with generally decreasing intensity.

Most of the northern shield areas are covered by desert landscapes, oases of generally small extent are found in some of the wadis and morphologic depressions.

The Asir mountains with relatively high rainfall are covered by terraced plantations up to altitudes of 2,400 m asl and by sparse conifer vegetation and by subtropical and temperate vegetation on the mountain slopes. On the western escarpment, intensive agricultural cultivation is practiced on a system of numerous narrow terraces.

References. Al-Sayari and Zötl (1978), Beydoun (1991), ITALCONSULT (1969), Jado and Zötl (1984), Kruck et al. (1996), Wohlfahrt (1980).

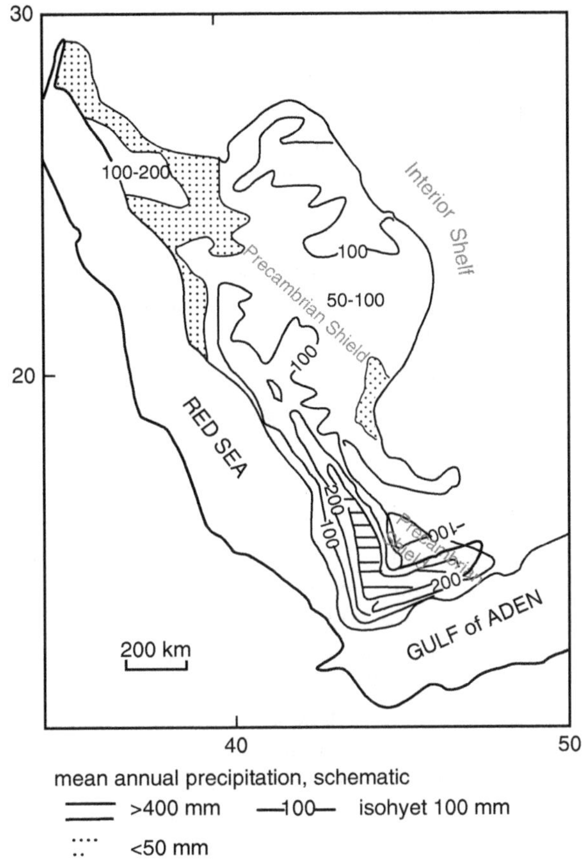

Fig. 7.4 Mean annual precipitation on the Arabian Shield. After Alsharhan et al. (2001), ESCWA (1999b)

7.2 Aquifers on the Crystalline Shield

7.2.1 Aquifers in Wadis and Morphologic Depressions

7.2.1.1 Hydrogeologic Characteristics

No aquifer systems with large horizontal and vertical extent occur on the outcrops of crystalline Precambrian rocks of the shield area. The Precambrian basement rocks contain groundwater in weathered surficial sections and fissured zones but do not generally provide aquifers with substantial productivity. Main groundwater resources on the crystalline shield areas are found in wadi courses and morphologic basins prevailingly in Quaternary sediments.

Productive alluvial aquifers are located, in particular, in the middle and upper courses of large wadi systems which extend far back into the highlands. The climatic conditions in the southwestern mountainous part of the shield create significant renewable groundwater resources in limited areas of the extensive wadi networks.

Major alluvial aquifers occur in the wadi systems of Wadi Bisha and Wadi Tathlith, which join at the eastern margin of the shield into Wadi ad Dawasir, and in Wadi Rima and its tributaries.

Outside of the main wadi systems, the highlands comprise numerous small local depressions with shallow wells used for irrigation. Small springs issue at many places in the upstream sections of the wadis.

Alluvial wadi aquifers are usually composed of unsorted coarse and uncemented deposits with relatively high permeability. The wadi aquifers are, in general, characterized by:

- Limited thickness, not exceeding a few tens of metres
- Narrow width varying from a few metres to a few hundred metres
- Small aquifer volumes
- High permeability
- Extent over long wadi stretches

Groundwater in wadi aquifers of the Arabian Shield originates prevailingly from infiltration of sporadic runoff. The Precambrian basement rocks generally constitute the base of the wadi aquifers.

Alluvial fans or deltas extend from the mouth of large wadi systems on the upper boundary of coastal or inland plains.

Historically, the wadi aquifers were the primary source of groundwater supply in the shield areas. Alluvial aquifers at medium altitudes provided water supply for major settlements, such as Madina, Taif and Bisha.

Various wadi systems with relatively short courses and steep slopes extend over the western parts of the Hijaz and Asir mountains, the Yemen highlands and the escarpment toward the Red Sea coast. The water resources of detrital aquifers in these wadis are exploited prevailingly in the coastal plains (Sect. 9.3.3.2).

7.2.1.2 Wadi Bisha

Wadi Bisha runs from the Hijaz mountain area into the landscapes of alluvial plains and low hills of An Nejd. In its upper course, the wadi is incised by 200–300 m into the mountain massif. The basement of the catchment of Wadi Bisha is composed of granite–gneiss, granodiorites, sericite schists and green schists. The granitic rocks are generally intensively weathered, permeability in the schists is restricted mainly to narrow joints.

The valley channel is hundreds of metres to some kilometres wide. The sedimentary filling of the wadi comprises, from bottom to top:

- 10–40 m thick silt, sand and gravel
- Around 8 m thick younger terrace deposits with sand and pebble layers
- 1–1.5 m thick channel alluvium of coarse sand, gravel and boulders
- A thin cover of eolian sand and silt on the central flood plain

Aquiferous zones are found in the 2–5 m thick weathering zone of granitic rocks and in the coarse wadi sediments. The thickness of the unconsolidated Quaternary wadi deposits reaches generally 20 m and, in some buried stream channels, several tens of metres. Depth to groundwater generally ranges from 4 to 18 m below land surface. The groundwater is unconfined in most of the wadi channel alluvium, but confined by impervious layers in the lower flood plain deposits.

7.2.1.3 Wadi Rima

The wadi channel of the several hundred kilometres long Wadi Rima opens, in its lower course, into a 5–8 km wide peneplain and enters, in the east, into the sandstone outcrops of the Interior Shelf. Wadi Rima and its tributaries intersect, in the west, prevailingly schists of the Mudama formation and, in the central part of its course, granitic rocks.

Accumulations of unconsolidated rocks are restricted to the main wadi and its tributaries. The thickness of the unconsolidated alluvial sediments does not exceed 10 m in Wadi Rima and is less than 5 m in some of the tributaries. Fanglomerates occupy the margins of the main wadi and the wadi terraces, reaching a width of up to 3 km in some areas. In the granite areas, coarse sands extend over most of the marginal sections of the main wadi; fine sand, silt and sabkha sediments cover the flat wadi floor. The fine grained wadi sediments interfinger, toward the borders of the wadi, with fanglomerates and talus fans.

Shallow wells, which yield a few l/s, are located mainly on the border of the main wadi channel. On the fine grained sediments of the flat wadi floor, infiltration rates of stream runoff are generally low.

In the area covered by schists, coarser detrital sediments fill only a few narrow wadi channels, which contain shallow groundwater lenses after ephemeral rainfall.

Outcropping granite and granitic gneiss on the peneplain is weathered down to depths of 20 m.

Mean annual rainfall in the Wadi Rima catchment is around 100 mm.

7.2.1.4 Wadi Wajj

The catchment of Wadi Wajj, covering a narrow drainage system of 1,600 km^2 in the Asir mountains around Taif, includes meta-volcanics, volcaniclastic rocks, granites, amphibolite and quartz-mica schists, dikes, and Quaternary sediments of up to 20 m thickness on the wadi floor.

7.2.1.5 Aquifers of the Harrat Rahat Basalt Area

The Harrat Rahat basalts rest on marls with sand and gravel beds: terrestrial deposits, which cover the Precambrian basement, and wadi sediments of a drainage system preceding the volcanic eruptions. Groundwater is extracted through shallow wells in morphologic depressions and wadi courses on the margins of the basalt field. Well yields of 1–50 l/s are obtained from wadi sediments or weathered basalt.

Near Madina, the spring Ain Zerqa issued with a discharge of more than 100 l/s on the northern, topographically lower end of the basalt field. The spring discharge has been replaced by groundwater exploitation through boreholes from alluvial deposits covered by Quaternary basalt. The combined well yield for the Madina water supply amounted to 120 l/s in 1980.

7.2.1.6 Highlands East of Jedda

The highlands east of Jedda in the southern Hijaz are crossed by a chain of east–west directed parallel drainage networks including the Murwani–Khulais, Shamiya, Fatima and Naaman wadi systems (Jado and Zötl 1984: 113). The length of individual wadis is around 35 km; none of the wadis reaches the coast.

The catchments areas of Wadi Fatima and its tributaries, extending over 4,650 km^2, are covered mainly by igneous rocks (granitic to dioritic plutons), metamorphic and volcaniclastic rocks and, in the upstream reaches, basalts of Harrat Rahat.

In Wadi Fatima, the wadi filling consists of a few metres to >30 m thick sands and gravels, grading into fine sand and clay on the lower reaches. Average annual rainfall is about 280 mm, decreasing downstream to 70 mm.

Many wells penetrate the fractured and weathered surficial rock layers, which may depleted completely during the dry period.

Groundwater of the shallow Wadi Fatima aquifer in wadi sediments and fractured and weathered crystalline rocks is exploited for water supply of Makka and Jedda. A considerable amount of rainfall is channelled into surface water and the groundwater is recharged from sporadic runoff events.

The catchment of Wadi Murwani–Khulais and its numerous tributaries covers 5,330 km^2. The catchment is situated mainly in Precambrian rocks, the headwaters reaching into the Harrat Rahat basalt field. Downstream, the wadi system enters into the tectonic depression of Khulais and finally into the Red Sea coastal plain.

The alluvial filling of Wadi Murwani–Khulais consists of sand, gravel, clay and shale, reaching a thickness of 20 m on the Khulais plain, where the Quaternary is underlain by Tertiary sediments of >160 m thickness.

The catchment of Wadi Yamaniya with a size of 620 km^2 is, to a large extent, covered by dikes, granite, diorite, amphibolite and gneiss.

Alluvial deposits consist of gravel, sand, silty sand and clay, the total thickness varying from a few metres to about 100 m. Saturated thickness of the wadi aquifer reaches 20–70 m in the downstream sections.

7.2.1.7 Wadi Fara

Wadi Fara extends for about 30 km in northeast–southwest direction on the western escarpment of the Hijaz highlands between Madina and the Red Sea coast at Yanbu al Bahr. The morphologic depression of the 2–6 km wide wadi covers 120 km^2 within a mountainous catchment in Precambrian rocks, mainly granite, gneiss and schists, which are crossed by dikes. The wadi filling is composed of up to 50 m thick sands, clay and gravels with conglomeratic layers at the bottom. Mean annual precipitation directly over the wadi is only 22 mm.

Fresh groundwater with EC values of around 900 μS/cm occurs in the alluvial wadi deposits. Groundwater on the fringes of the wadi, where a thin layer of unconsolidated sediments overlies the Precambrian rocks, is brackish with EC values of 3,000 to >5,000 μS/cm.

7.2.2 *Groundwater Regimes*

7.2.2.1 Hydraulic Parameters

Reported values of hydraulic conductivity for wadi aquifers of the Arabian Shield vary from 2×10^{-5} to 2×10^{-3} m/s in the coarse grained channel alluvium and are around 1×10^{-6} to 1×10^{-5} m/s in the flood plain deposits.

Hydraulic conductivities of the wadi sediments are reported as:

- 1×10^{-4} to 6×10^{-4} m/s for Wadi Wajj near Taif
- 2×10^{-5} to 2×10^{-4} m/s for the channel alluvium of Wadi Bisha
- 1×10^{-6} to 1×10^{-5} m/s for flood plain deposits of Wadi Bisha
- 5.6×10^{-7} to 1.85×10^{-5} m/s for Wadi Yamaniya

Transmissivities of the 10–70 m thick wadi aquifers vary from 2 to 1,000 m^2/d with average values around 200 m^2/d.

Transmissivity values of 200–1,152 m^2/d have been tested in Wadi as Safra (southwest of Madina) and of 2.4–165 m^2/d in Wadi Yamaniya.

7.2.2.2 Groundwater Recharge

Groundwater recharge to the wadi aquifers results prevailingly from infiltration of surface runoff. The contribution of transmission losses from surface runoff in the wadi course depends on runoff volume and duration, soil profile characteristics and depth to water table. Particularly important in the recharge mechanism are initial moisture content accumulated from sequences of flood events and soil heterogeneity in the centre of the wadi channel and on the edges of the wadi course.

In the areas covered by basalt flows, the rugged topographic relief with many hollows and local depressions appears to favour recharge in limited quantities outside the main wadi courses.

Significant indirect recharge may occur in particular in the semi-arid southern parts of the Asir mountains with mean annual rainfall above 250 mm and in wadis receiving runoff from the semi-arid high mountain zone.

Rough estimates of recharge volumes in wadis of the Hijaz–Asir mountains have been obtained from a comparative assessment of wadi catchments and areas cultivated from groundwater abstraction and from cumulative well yield in the individual catchment areas. Cultivated areas correspond to around 3–6% of the catchment areas and cumulative well yields may be in the order of 10–15% of average rainfall over the catchment areas. Estimates of recharge rates from these assessments indicate that wadi infiltration may concentrate an equivalent of 2–5 mm/a over the entire wadi catchments and up to 40 mm/a in the more humid mountain areas.

For the highlands east of Jedda, recharge rates of 5–7% of mean annual rainfall have been estimated.

7.2.2.3 Groundwater Discharge

The wadi aquifers discharge in springs in the wadi beds and through evapotranspiration in sabkha zones. On the eastern margin of the shield area, groundwater moves in various wadi courses as subsurface flow into wadi or sedimentary aquifers of the adjoining Interior Shelf. Groundwater from aquifers of the westward oriented wadis enters the coastal plain as subsurface flow or surface base flow.

Springs issue from alluvial aquifers in wadis Fatima and Fara west of Madina and in the Khulais and Taif areas. Mean spring discharge ranges up to 1 m^3/s in some areas. Discharge of springs issuing from basalt fields, such as Harrat Rahat, is generally low but can reach 100 l/s at a few discharge points. Springs from fractured granite or gneiss issue in the Lith and Jizan areas near the Red Sea coast.

In Wadi Fara, groundwater was previously exploited from the alluvial wadi filling through around 300 "ayoun", artificially induced springs (cf. Sect. 8.3.2.4). The shallow groundwater exploitation structures with free outflow have been replaced by deeper wells with a mean production rate of 93,550 m^3/d in the 1970s, which greatly exceeds the estimated groundwater flow volume of 45,000 m^3/d.

Natural and artificial discharge from wadi aquifers of the Arabian Shield in Saudi Arabia with catchments between 180 and 2,000 km^2 are in the order of 4 to 9×10^6 m^3/a for individual wadi systems.

The volume of groundwater flow in the highlands east of Jedda may be around $10^5 \times 10^6$ m^3/a (data of 1967). Groundwater flow in the Wadi Yamaniya aquifer was calculated as 52×10^6 m^3/a in 1978.

In the higher mountain zones, infiltration of surface runoff into wadi channel deposits can be relatively high. The deposits in the majority of the stream channels are, however, thin and do not allow a transmission of important volumes of groundwater flow to the lower reaches of the wadis or a significant groundwater storage.

7.2.3 *Groundwater Salinity and Hydrochemistry*

7.2.3.1 General Distribution of Groundwater Salinity

Groundwater with low to moderate salinity occurs in fissured igneous rocks and wadi fillings mainly in the more humid high altitude areas of the Arabian Shield. In most wadis, groundwater salinity increases down gradient. The general range of groundwater salinity is between 300 and 7,500 mg/l TDS.

The increase of groundwater salinity toward lower topographic elevations of the wadi courses is modified in some areas by local features:

- In upstream wadi sections, groundwater salinity can be elevated by evaporation, where the morphologic relief causes very shallow water tables, or where impermeable dikes cross the wadi course.
- The down-gradient increase of groundwater salinity can be interrupted by inflow of low salinity water from tributary wadis.

The low salinity groundwater is generally Cl water with Ca or Na as predominant cation. Ca–HCO_3 waters occur in some headwater areas, e.g., in Wadi Fatima. HCO_3 concentrations of most groundwaters are low, reflecting the prevailing siliceous lithology of the aquiferous rocks and the low CO_2 production in the poorly developed soil cover.

Rain water on the highlands of the shield is generally Ca–HCO_3 type water with salinities around 40 mg/l TDS. Flood waters on the highlands have salinities of 150–250 mg/l TDS and Ca–HCO_3 or Na–Cl character.

The hydrochemical composition of the groundwaters is dominated by evaporative enrichment of Cl and SO_4 concentrations and of equivalent cations and, to some extent, by dissolution of lithogenic components. "The chemical weathering reactions of the main rock-forming minerals can be an important long-term neutralization process of the chemical composition of groundwater for providing Mg, Ca, Na, K and HCO_3 ions and being significant for Mg and Ca ions" (Alyamani and Hussein 1995).

Sources of Na are, in particular, andesine, oligoclase and albite, of Ca and Mg hornblende and biotite.

The brackish groundwaters in the downstream wadi courses are generally Na–Cl waters. High SO_4 concentrations occur in some areas, where lenses of gypsum are intercalated in the wadi sediments.

7.2.3.2 Wadi Bisha

In the upper reaches of Wadi Bisha near Khamis Mushayt with mean annual precipitation of around 400 mm, fresh groundwater occurs in fissure-type aquifers: basalts, gabbro, diorite, granite. The fresh water with salinities between 480 and 770 mg/l TDS is HCO_3 water with major cations – Ca, Mg, Na – at approximately equal percentages. Brackish water with salinities between 1,800 and 3,000 mg/l TDS is found in areas with a cover of Quaternary sandy–silty deposits and loess above green schists, chlorite sericite slate, granite–gneiss. Predominant ions of the brackish water are Ca and Cl.

In Wadi Bishabsha, an upstream tributary of Wadi Bisha, groundwater in a shallow aquifer in the crystalline rocks is generally brackish with electrical conductivity values of 2,000–4,000 μS/cm. Groundwater salinity tends to increase in downstream direction, but local anomalies with elevated water salinity are found also upstream in irrigation areas and in areas where groundwater flow is influenced by geologic barriers, like faults and dikes. Main mechanisms controlling the groundwater hydrochemistry are:

- Dissolution of carbonate minerals, creating an oxidized HCO_3 type water
- Evaporative enrichment of dissolved salts from ponds, shallow water table and in irrigation areas

The general source of elevated groundwater salinity is attributed to evaporation processes affecting the shallow groundwater. During floods from low rainfall intensity, water is trapped in hollows and local morphologic depressions in the crystalline rocks and, during subsequent flood and recharge events, remaining pockets of mineralized water and salts dispersed on the surface are flushed into the shallow aquifer.

In the lower course of Wadi Bisha, groundwater salinity in the gravel aquifer of the central wadi filling ranges from 400 to 1,200 mg/l TDS. In the peripheral silty and sandy wadi deposits, the groundwater is generally brackish with salinities of 1,800–6,000 mg/l TDS.

According to Lloyd et al. (1980), the main wadi recharge in the central course of Wadi Bisha with relatively thick alluvial deposits is modern and originates from flood flow infiltration. The recent recharge produces Ca–Na–HCO_3 type groundwater with low to moderate salinity (electrical conductivity $<2{,}000$ μS/cm). Groundwater on the wadi edges has a higher salinity with electrical conductivity from 2,000 to $>6{,}000$ μS/cm and elevated concentrations of Na, Ca, Cl and SO_4. The origin of the brackish water on the wadi edges is attributed to recharge in low rates over a thin alluvial aquifer in an extensive area adjacent to the main wadi channel, where groundwater salinity is increased by evaporation and dissolution of gypsum. Shallow subsurface inflow of the brackish water reaches the wadi edges, where it is held back by the fast flowing fresh water of the main alluvial aquifer with minor mixing.

7.2.3.3 Wadi Rima

The unconsolidated sediments of the up to several kilometres wide Wadi Rima contain shallow groundwater with relatively low salinity above groundwater with higher mineralization. Generally, low salinity of the groundwater coincides with detectable tritium contents, indicating very recent recharge. The higher salinity in the somewhat older groundwater appears to be related to evaporative enrichment under the arid climate conditions.

The salinity in gravel aquifers of the central wadi filling is 410–1,220 mg/l TDS, and 1,820–5,750 mg/l TDS in the peripheral silty sands and gravels. In some areas, shallow wells extracting low salinity groundwater are located at the edges of the wadi channel, where coarse wadi sediments favour runoff infiltration.

Predominant cations in groundwaters of the wadi aquifer are Ca and Na, Cl is generally the predominant anion.

7.2.3.4 Wadi Wajj

Groundwater in Wadi Wajj, situated in the Asir mountains in the Taif area at altitudes around 1,600 m asl, varies from fresh water with average salinity of 530 mg/l TDS in1978 to 837 mg/l TDS in 2003 upstream of Taif to brackish water with salinities of >2,000 mg/l TDS downstream of Taif. The fresh groundwater is mainly Ca–HCO_3 type water, the brackish water downstream of Taiz is prevailingly Na–Cl water. HCO_3 concentrations are in ranges of 160–300 mg/l. "Given the fact that no carbonate rocks exist in the area, the possible sources of bicarbonate include presence of organic matter in the groundwater which is oxidized to produce carbon dioxide, which in turn promotes dissolution of minerals. This weathering enriches groundwater in Ca, Mg, and bicarbonates. Weathering of silicates may also be accounted as a source of bicarbonate" (Al-Shaibani 2008:160). The increased salinity of the brackish groundwater downstream of Taif results probably from effects of evapotranspiration, irrigation return flow and waste water infiltration. The quality of groundwater in the area downstream of Taiz improved between 1978 and 2003 after installation of drainage and waste water collector systems.

7.2.3.5 Harrat Rahat Basalt Area

Salinity of groundwater from wells tapping aquiferous basalts at the margin of Harrat Rahat ranges from 1,500 to 7,000 mg/l TDS. The groundwater with relatively low salinity is Na–HCO_3 to Na–SO_4 type water. Groundwaters with higher salinity are Na–SO_4 type waters, elevated sulfate concentrations being derived from evaporite layers in alluvial sediments above the basalt, or Na–Cl type waters, where the basalts contain intercalations of saliferous clay layers.

Groundwater from wells in wadis along the margins of the Harrat Rahat basalts has salinities of 320–2,720 mg/l TDS. Generally, groundwater salinity is relatively low at the wadi edges and increases toward the main wadi course and in direction of groundwater flow. Locally, the down gradient increase of groundwater salinity is interrupted by inflow of low salinity groundwater from tributaries. The groundwater in the wadi aquifers is Ca–HCO_3 to Na–Cl type water.

7.2.3.6 Highlands East of Jedda

In the aquifers of the main wadi systems on the highlands east of Jedda, fresh groundwater occurs in the steep upper and middle wadi courses. The general range of groundwater salinity in the wadi aquifers varies from 300 to 7,500 mg/l TDS.

The mineral content of the groundwaters depends mainly on the mineralogic composition of the crystalline rocks and wadi sediments in the catchment area. Evaporation processes favoured by natural conditions or irrigation, are responsible for an increase in mineralization, especially of Na and Cl. Salinity generally increases toward the sides of the valley.

Fresh groundwater with salinities of 400–800 mg/l TDS occurs in the topographically upper reaches of the wadi aquifers (e.g., Wadis Murwani, Shamiya, Fatima). The fresh water is found mainly within catchment areas, which are covered by igneous and metamorphic rocks and basalt flows. Predominant cations in the fresh water are Ca and Na, major anions HCO_3, Cl, SO_4 are in approximately equal percentages with a slight tendency to SO_4 predominance.

The groundwater salinity generally increases downstream to >1,000 mg/l and the water type changes to Na–Cl predominance; Ca, Na and Cl concentrations generally increase along the flow path.

Brackish to saline groundwater occurs in outcrop areas of marine sediments in local tectonic depressions. Elevated groundwater salinities are also found in wadi courses in schists and amphibolites crossed by dikes, which act as groundwater barriers and cause high evaporative enrichment of seasonally trapped groundwater. Agricultural activities may be an additional source of elevated groundwater salinity in some areas.

In the Khulais plain, groundwater salinity increases to 2,300 mg/l TDS with average concentrations of 820 mg/l Cl and 490 mg/l SO_4. In the Usfan plain, most groundwaters are brackish with average salinities of 1,400–1,500 mg/l TDS. Downstream of the Usfan area near the coastal plain, the average groundwater salinity increases to around 3,000 mg/l TDS and Cl concentrations to 1,700 mg/l.

SO_4 concentrations are elevated to around 700 mg/l, where lenses of gypsum are intercalated in the wadi filling, e.g., in Wadi Shamiya.

Examples of the hydrochemical composition of groundwaters from the Arabian Shield in western Saudi Arabia are shown as Piper diagrams in Figs. 7.5–7.7. Arithmetic means of brackish groundwater samples from five wadis of the western highlands between Madina and Taif (mean TDS values between 1,300 and 3,400 mg/l) indicate a predominance of Ca–Cl to Na–Cl type waters (Fig. 7.5). Groundwater samples from the recharge area of Wadi Fatima between Makka and Taif are mainly fresh waters (EC values 455–1,366 μS/cm) of Ca–HCO_3 to Ca–Na–Cl type (Fig. 7.6). Groundwater in various wadis of the Arabian Shield south of Makka (Damm area, Kotb et al. 1990) is prevailingly brackish Na–Cl type water (TDS values 1,940–6,400 mg/l). Samples of fresh groundwater in that area (TDS values 530–940 mg/l) are mainly Ca–Na–Cl–SO_4 waters (Fig. 7.7).

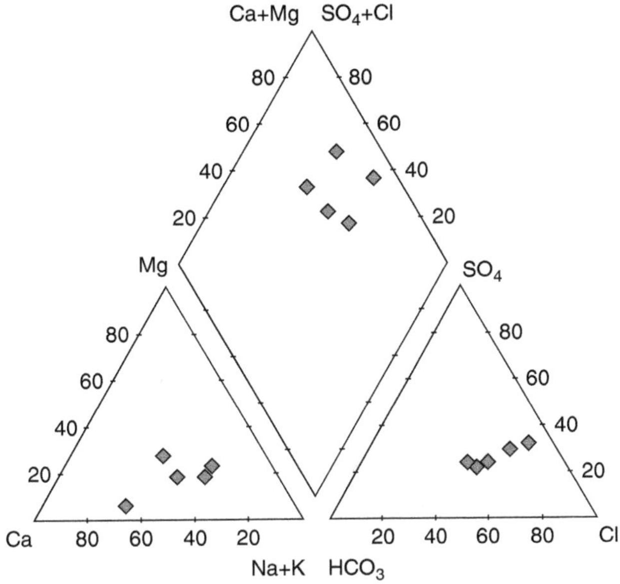

Fig. 7.5 Piper diagram: Groundwater samples from wadis on the western escarpment of the Arabian Shield between Madina and Taif. Data from Bazuhair et al. (1991)

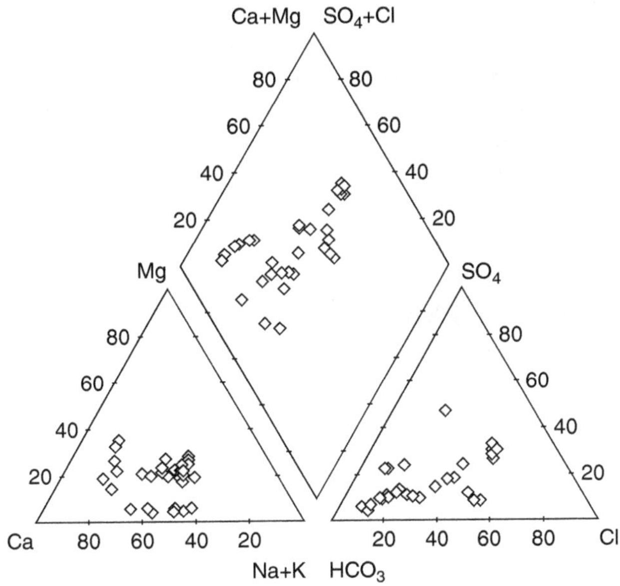

Fig. 7.6 Piper diagram: Groundwater samples from the Wadi Fatima catchment, Arabian Shield, western Saudi Arabia

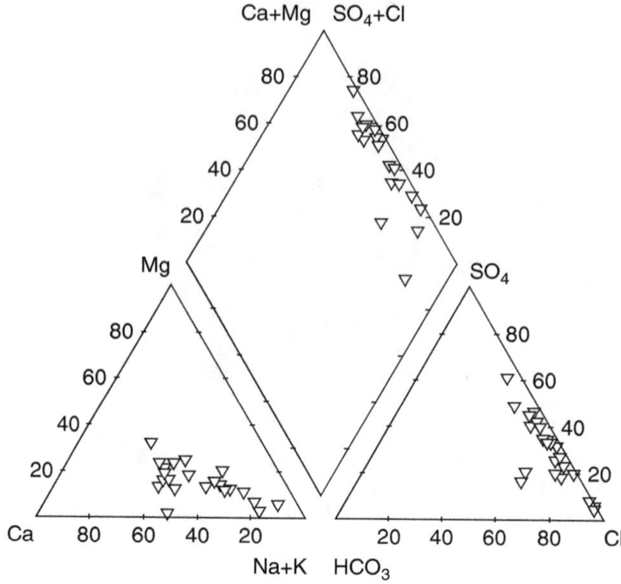

Fig. 7.7 Piper diagram: Groundwater samples from the Damm area, Arabian Shield, western Saudi Arabia. Data from Kotb et al. (1990)

References. Abdulrazzak et al. (1988), Al-Ahmedi and Sen (1989), Al-Shaibani (2008), Alyamani and Hussein (1995), Alyamani and Atkinson (1993), Basmaci and Al-Kabir (1988), Basmaci and Hussein (1988), Bazuhair and Hussein (1990), Durozoy (1972), Hammad (1978), ITALCONSULT (1969), Jado and Zötl (1984), Kotb et al. (1990), Lloyd et al. (1980), Memon et al. (1984), Robertson (1992), Sorman and Abdulrazzak (1993).

7.3 Sedimentary and Volcanic Aquifers Within the Arabian Shield in Yemen

7.3.1 Main Aquifers

Within the sequence of sedimentary and volcanic formations of the southern Arabian Shield, the Amran limestone and the Yemen volcanics constitute aquifers with wide lateral extent, but with generally low to moderate productivity. Aquifers with higher productivity are found in the tectonically induced intermountain basins of the western Yemen highlands, where Mesozoic–Quaternary sediments combine relatively high transmissivities with favourable recharge conditions (Table 7.1; Fig. 7.8).

Table 7.1 Hydrostratigraphic scheme of the shield area in Yemen after van der Gun and Ahmed (1995), Robertson (1992)

Stratigraphic age	Formation	Prevailing lithology	Aquiferous properties
Quaternary	Unconsolidated wadi and basin sediments		Aquiferous in intermountain basins and wadis
		Volcanics	Locally minor aquifers
Tertiary	Yemen volcanics	Flood basalts and various volcanics and volcaniclastics	Fissure type *aquifer* with low to moderate productivity
	Medj-Zir	Sandstone	Locally upper part of Tawila sandstone aquifer
Cretaceous	Tawila	Sandstone	*Major aquifer* in intermountain basins and on the western escarpment
Jurassic	Amran	Limestone	Fissure type *aquifer* over wide plateau areas with low to moderate productivity
	Kohlan	Sandstone	Local aquifer
Paleozoic	Akbra	Shale, glacial deposits	Mainly aquitard
	Wajid	Sandstone	*Major aquifer* in northeastern Yemen (Interior Shelf area)
Precambrian		Crystalline basement	General base of aquifer systems

7.3.1.1 Mesozoic–Quaternary Aquifers of the Highlands and the Western Escarpment

The Jurassic *Amran* formation, composed of massive limestones, shales and marls, forms extensive outcrops over the highland zone between Sanaa and Sada. The limestones provide a fissure-type aquifer with generally low to moderate productivity. Wells with relatively high yield are found at some locations, in particular near wadis. The alternation of permeable and low permeability layers creates various aquiferous sections within the formation; perched groundwater lenses occur in various areas and springs issue along outcrops of aquitard horizons. The Amran limestones probably constitute, in some areas, a joint aquifer system with underlying sandstones of the Kohlan and Wajid formations, which are, however, generally situated at considerable depth (probably 400–500 m below land surface near Sanaa and more than 1,000 m below land surface north of Sanaa).

The Cretaceous *Tawila* sandstones are exposed in rather narrow outcrop belts on the eastern rim of the Yemen volcanics lava flows, on top of a few mountain plateaus on the western margins of the highlands, and on some steps of the western escarpment. The sandstones contribute major components of the aquifer systems in the Sanaa and Rada groundwater basins. In the highlands west of Sanaa between Shibam and Al Mahwit, Tawila sandstone plateaus comprise aquiferous units of local extent with small springs issuing at the base of the sandstones on top of the underlying Amran series. A relatively extensive aquifer system is developed in the

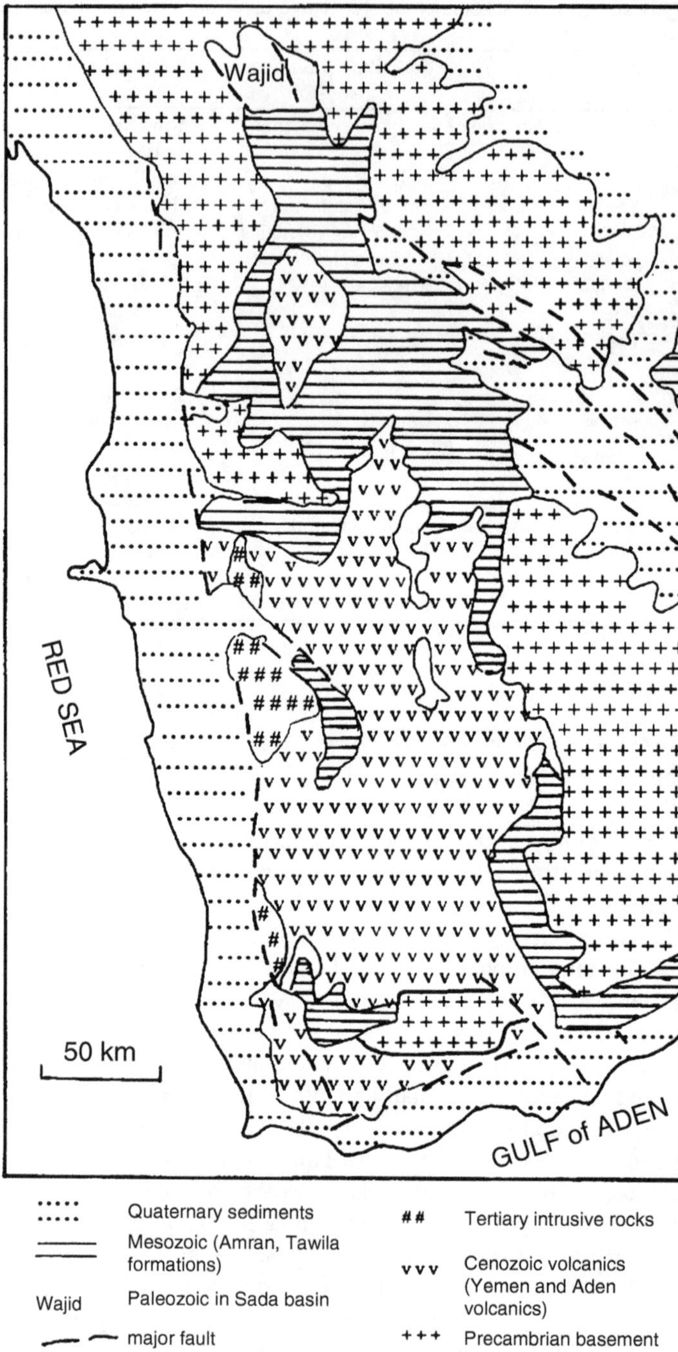

Fig. 7.8 Outcrops of geologic formations in the western highlands and escarpment of Yemen. Simplified sketch after Grolier and Overstreet (1978), van der Gun and Ahmed (1995)

Tawila sandstones jointly with the underlying Amran limestones in the upper catchment of Wadi Surdud on the western slope of the highlands west of Sanaa. Groundwater from the Tawila–Amran aquifer issues in a group of springs at the contact between the sedimentary aquifer with impermeable rocks of the basement, sustaining perennial surface flow in Wadi Surdud toward the coastal plain. Mean discharge of the Surdud springs is 750 l/s from a catchment of 325 km^2.

7.3.1.2 Volcanic Aquifers

The Tertiary Yemen volcanics cover wide areas of the western highlands of Yemen from north of Sanaa to south of Taiz and on the Sahara massif between Haja and Sada. The generally several hundred metres and up to 2,000 m thick Yemen volcanics constitute a fissure type aquifer with, in many areas, poor productivity. Relatively favourable conditions are found in intermountain plains around Maabar–Dhamar–Kitab (central highland plains), where the hydraulic properties of the volcanic rocks have been increased by tectonic fracturing.

In the Sanaa intermountain basin, the Yemen volcanics comprise three main aquiferous units:

- Basal basalt, consisting of 150 m to more than 300 m thick homogeneous dense basalts, which are intensely fractured and have relatively good aquiferous characteristics.
- Stratoid rocks, comprising pyroclastics, mainly rhyolites, bedded tuffs and ignimbrites with aquiclude or aquitard layers formed by decomposed argillaceous tuffs.
- Several hundred metres thick chaotic and stratoid rocks with basaltic and rhyolitic lava flows, which are aquiferous in fractured zones.

7.3.1.3 Aquifers in Intermountain Basins

Intermountain plains are scattered over the western highlands of Yemen. Most of these plains are located relatively near the main water divide that separates the Red Sea basin from the Rub al Khali drainage basin. Some of the plains provide relatively favourable conditions for groundwater development because of several factors (Van der Gun and Ahmed 1995):

- Groundwater levels are, under unexploited conditions, within a few tens of metres below the ground surface.
- The hydraulic conductivity of the rocks in the plains is higher than that of the surrounding rock units.
- Subhorizontal topography and the presence of alluvial deposits favour groundwater replenishment.

- Surface water leaves the plains only on rare occasions and in insignificant amounts; natural discharge from these areas is mainly by evaporation and subsurface outflow.

Many of the plains have a relatively small extent. More extensive aquiferous zones are found in the intermountain basins of the Amran valley north of Sanaa, the Sanaa plain and, further south, the central highland plains and the Rada plain. The Sada plain is situated on the margin of the shield area and, as it is underlain prevailingly by Paleozoic sandstones, may be considered part of the Interior Shelf (Sect. 5.2.1.2).

The *Amran valley* is located in the Yemen highlands about 50 km northwest of Sanaa. The 30 km long and 5–10 km wide valley corresponds to a southwest–northeast oriented tectonic graben, which subsided, along faults with throws of several hundred metres, within mountain plateaus covered by Jurassic Amran limestones. The valley floor is situated at 2,100–2,300 m asl, mountain peaks in the surroundings of the valley reach elevations of more than 3,000 m asl. The valley floor, including some lateral branches of the valley, extends over 230 km^2, the catchment area of the valley covers 1,280 km^2. Average annual rainfall in the Amran valley catchment is around 300 mm with main precipitation events in the months of March to May and August to September.

To the east, the Amran graben is accompanied by a line of Quaternary volcanoes.

The Amran graben is filled with Quaternary detrital deposits of up to 300 m thickness: sand, silt, clay, gravel, loam with intercalations of basaltic flows and dikes. The Quaternary sediments form a generally continuous aquifer complex with intercalated basalt layers and the underlying Amran limestones. The productivity of the aquifer is attributed mainly to the moderately high transmissivity of the unconsolidated Quaternary sediments with some additional contribution from the underlying limestones in fractured zones. Well yields average around 20 m^3/h, reaching up to 50 m^3/h.

Intensive exploitation of the Amran valley aquifer for irrigation has resulted in a heavy decline of groundwater levels of up to 2 m/a.

The *Sanaa plain* extends as an intermountain valley with irregular margins in approximately south–north direction over around 700 km^2. The plain is situated at an elevation of about 2,200 m asl, the surrounding mountains rise to 3,000 m asl. Mean annual precipitation in the Sanaa area is around 300 mm.

The Sanaa plain is covered by Quaternary alluvial deposits comprising silt, clay, sand, gravel and volcanic detritus. The geologic sequence in the subsurface of the plain is composed of:

- Sandstones and clays of the Kohlan series (Triassic to early Jurassic)
- Carbonates of the Amran series (Jurassic)
- Tawila sandstones (Cretaceous)
- Sandstones, clays and silts of the Medj-Zir series (Paleogene)
- Yemen volcanics (Tertiary)

The highlands surrounding the valley are covered prevailingly by the Yemen volcanics and, in limited areas, Quaternary basaltic flows.

Tectonic movements and erosion have created a complex geologic–hydrogeologic structure of the basin with the following general distribution of formations underlying the Quaternary basin sediments: In the northern part of the Sanaa plain, the Quaternary sediments rest mainly on the Tawila sandstone formation. On the northern tip of the plain, the Tawila formation has been eroded and the Quaternary is underlain by the Amran formation and the Kohlan sandstones. Toward south, the Tawila sandstones dip to depths of several hundred metres below land surface and are overlain by trap basalts of the Yemen volcanics.

The Tawila sandstones are the most important exploited aquifer unit together with aquiferous sections of the overlying Medj-Zir formation. The thickness of the Tawila aquifer ranges from 100 m in the northern part of the plain to about 400 m in the central and southern parts. Traditionally, the Quaternary sediments provided the main source for water supply and small scale irrigation through groundwater extraction from shallow dug wells. With increasing groundwater exploitation and decline of water levels, the Quaternary aquifer with its low to moderate productivity has lost its importance as major water resource.

Groundwater in the Tawila sandstone aquifer is phreatic, where the sandstones crop out or are located at shallow depth, and is confined in southern part of Sanaa plain. The Amran limestone and the Yemen volcanics are aquiferous in fracture and fissure zones, but well yields are generally low to moderate.

The "*central highland plains*" of Yemen around Maabar–Dhamar–Kitab consist of a number of plains of varying size. On some plains, such as Qa Jahran, more than 100 m thick alluvial deposits with relatively high permeability through tectonic fracturing constitute the main aquifer system. On other plains with only thin cover by alluvial material, groundwater is mainly found in the volcanic rocks.

The *Rada* basin comprises a zone of connected plains extending, in the southeast of the Yemen highlands, over 400 km^2 at altitudes between 2,000 and 2,700 m asl. The plains are surrounded by mountains and hills, which are covered by Tertiary–Quaternary volcanics in the west and by crystalline basement in the east. Mean annual rainfall is 165–250 mm.

The Cretaceous Tawila sandstones act as main aquifer in the Rada basin. The thickness of the sandstones increases from 150 m in the northeast to more than 600 m in the centre of the basin, where the sandstone formation is covered by Tertiary volcanics. Fractured basalt and tuffs of the 100 m to more than 1,000 m thick Tertiary Yemen volcanics form, together with interbedded alluvial deposits, an aquifer in the southern and western parts of the basin. In the northeast of the plain area, shallow groundwater is found in the weathered zone of outcropping Precambrian metamorphic rocks and in a thin cover of alluvial deposits. Holocene lava flows and ash layers, extending over the northwestern margin of the Rada basin, are generally unsaturated but provide relatively favourable conditions for enhancing the recharge.

Alluvial deposits in wadi courses constitute shallow aquifers in many areas of the Yemen highlands and the western escarpment. The wadi deposits are prevailingly unsorted and unconsolidated accumulations of sand, silt and gravel. The dimensions of the wadi aquifers are rather limited with a width of a few metres to a few hundred metres and a thickness not exceeding a few tens of metres.

7.3.2 Groundwater Regimes

7.3.2.1 Hydraulic Parameters

Values of transmissivity and hydraulic conductivity of some of the main sedimentary and volcanic aquifers within the Arabian Shield of Yemen are listed in Table 7.2.

Transmissivities are generally low to moderate. Relatively high transmissivities of some hundred m^2/d are found in the Tawila sandstone aquifer in some areas.

7.3.2.2 Groundwater Recharge

A high percentage of groundwater recharge of the sedimentary and volcanic aquifers in the Yemen highlands and the western escarpment as well as in aquifers of the intermountain basins occurs through infiltration of wadi runoff. The wadi aquifers provide particularly favourable recharge conditions: Flood flow from rainfall events is collected in the wadi systems and intercepted by infiltration into the permeable wadi deposits, which also drain runoff from springs and seepages. Infiltration rates are relatively high in recent lava flows, e.g., on the margins of the Sanaa and Rada basins, where the broken basalt surface and layers of volcanic ash favour infiltration of rainfall and flood flow. An additional contribution to

Table 7.2 Hydraulic parameter values of aquifers in Yemen

Area	Aquifer	Transmissivity (m^2/d)	Hydraulic conductivity (m/s)
Sanaa intermountain basin	Various aquifers	10–500	
	Tawila sandstone	26–550	
	Yemen volcanics	30–110	
	Alluvial deposits	8.6–194	
Rada basin and western escarpment	Yemen volcanics	0.5–15	
Rada basin	Alluvial deposits	90–290	10^{-7} to 3×10^{-5}
	Tawila sandstone	30–540	2×10^{-6} to 1×10^{-5}
Sada basin	Alluvial deposits	10–105	

Data from Italconsult (1972), Boehmer (1988), Aust (1993)

groundwater recharge is expected in irrigated areas, where precipitation together with the artificially created soil moisture can exceed the saturation capacity of the soil.

The aquifer system of the Sanaa plain receives main volumes of groundwater replenishment from deep percolation of recharge on the high plateaus and mountains surrounding the plain. Within the plain, surface runoff, which rarely continues for more than a few days, infiltrates into the shallow aquifer.

Mean recharge rates in the Sanaa groundwater basin have been estimated at 30 mm/a for the tributary wadi catchments, 40 mm/a for the sandstone outcrop in the northeast of the plain and 10 mm for the plain areas in general.

Groundwater recharge rates are estimated at 72.5 mm/a for the Ayoun Surdud catchment, 31.5 mm/a on average for the Mahwit province, and 6–20 mm/a for the Rada basin.

7.3.2.3 Groundwater Movement and Flow Volumes

Groundwater flow in the *Amran valley* roughly follows the slope of the land surface toward northeast with hydraulic gradients between 1 and 12 m/km. Subsurface outflow from the Amran basin aquifer occurs at the northern end of the valley into the Amran limestone aquifer. Downstream of the Amran valley, groundwater flows through the limestone aquifer in rather steep gradients toward northeast, contributing to the discharge of springs in the Wadi Kharid catchment.

Groundwater movement in the Tawila sandstone aquifer of the *Sanaa* plain appears to be directed, in general, from the western and eastern fringes of the plain toward the centre of the plain and then toward north. The groundwater flow pattern appears, however, to be complicated by hydraulic discontinuities through faulting and volcanic intrusions.

Groundwater balance estimates for the Sanaa plain indicate an inflow of $20–30 \times 10^6$ m^3/a from sandstone outcrops in the west and of 33×10^6 m^3/a from the east.

Groundwater discharges in the Sanaa basin in a number of generally small springs rising from the Tawila sandstone, Yemen volcanics, Quaternary basalt or Quaternary basin sediments. Discharge of most springs is less than 3 l/s, the discharge of a few springs on the mountain foothills in the west of the plain reaches 20–50 l/s. Springs rising from Quaternary basalt in northwest of the Sanaa plain have a discharge of around 100 l/s. At present, the groundwater regime in the Sanaa plain is, to a large extent, affected by groundwater extraction from boreholes.

The Sanaa basin comprises the upper hydrologic catchment of Wadi Kharid, which merges into Wadi Jawf about 50 km north of the Sanaa plain. It is assumed that some subsurface outflow from the Sanaa plain reaches the Wadi al Kharid area through the Amran limestones.

In the *Rada plain*, groundwater flow follows the general topographic slope from west toward a narrow outlet of the basin in the northeast. Under natural conditions, groundwater discharge occurred mainly through springs on the margins of the plain and evapotranspiration from shallow groundwater in the plain. Natural groundwater outflow in the northeast of the plain toward the Rub al Khali was 0.6×10^6 m^3/a. Groundwater flow is influenced by step faults through:

- High permeability in the fault zones
- Abrupt changes of aquifer thickness and related changes of transmissivity along the faults

Components of a groundwater balance for the Rada plains have been estimated as follows:

Recharge from rainfall on the plains	3.3×10^6 m^3/a
Inflow from a volcanic plateau in the northwest	7.2×10^6 m^3/a
Subsurface outflow in the north	-1.3×10^6 m^3/a
Discharge through springs, evapotranspiration and well abstraction	-9.2×10^6 m^3/a

Artificial groundwater abstraction increased from 2.4×10^6 m^3/a in 1980 to 18.7×10^6 m^3/a in 1988.

At present, the groundwater regime in the Amran, Sanaa and Rada plains is predominantly controlled by groundwater extraction from wells, exploitation exceeding significantly the renewable resources.

On the *western escarpment* of the Yemen highlands, groundwater flows in a general westward direction to the Tihama coastal plain, into which groundwater from the escarpment aquifers discharges as subsurface inflow, mainly in the sand and gravel aquifers of larger wadis, or as baseflow in the wadis. In Wadi Surdud, springs with a mean discharge of 750 l/s issue from the Tawila–Amran aquifer complex at an altitude of 1,100 m asl on the boundary of the aquiferous formations with outcropping rocks of the Precambrian basement. The catchment of Wadi Surdud receives a subsurface inflow of 2.4×10^6m^3/a from the highlands near Sanaa east of the surface water divide. Most of the spring discharge reaches the coastal plain as base flow.

The total groundwater balance of Mahwit province with catchment areas of 1,365 km^2 amounts to 70×10^6 m^3/a, out of which around 68% discharge from the Mesozoic–Tertiary sedimentary and volcanic rocks, and 32% constitute short-term circulation of shallow groundwater in wadi aquifers.

7.3.3 Groundwater Salinity and Hydrochemistry

Groundwater in the Yemen highlands and the western escarpment is prevailingly Ca–HCO_3 water with low to moderate salinity. The salinity of groundwaters in the unconsolidated sediments of the intermountain basins as well as in the Tawila sandstone and Yemen volcanics aquifers of the Yemen highlands ranges generally from 300 to 600 mg/l TDS. The relatively low groundwater salinity appears to be related to relatively short circulation periods and flushing of aquifers in a relief with high altitude differences and high hydraulic gradients.

Groundwater in recharge zones is mainly Ca–HCO_3 type water of low mineralization. Downgradient there is a gradual increase in dissolved solids, and sulfates and chlorides become more prominent and eventually dominant downstream of the highlands in groundwater evaporation zones of the coastal plains.

In the Sanaa plain, groundwater samples from the different aquifers are mainly fresh waters of Ca–HCO_3 type with HCO_3 concentrations of 80–305 mg/l. Contamination from waste water is indicated in considerable parts of the urban area of Sanaa.

Groundwater in the Wadi Kharid area downstream of the Sanaa plain is mainly Ca–HCO_3 type water with salinities between 450 and 950 mg/l TDS. Ca–Na–HCO_3 or Na–HCO_3 waters originate probably from volcanic aquifers.

The Quaternary aquifer of the Amran plain, contains fresh water with EC values of 450–700 μS/cm. Predominant ions are Ca and HCO_3 (Fig. 7.9), HCO_3 concentrations are in a range of 100–250 mg/l, Mg/Ca ratios generally 0.4–0.8. A few samples have somewhat elevated SO_4 concentrations of up to 175 mg/l and slightly elevated EC values. NO_3 concentrations are elevated in a few samples up to 80 mg/l.

Water discharging from the Mesozoic aquifer system in the relatively large springs in Wadi Surdud is Ca–HCO_3 water with a salinity around 600 mg/l TDS.

References. Aust (1993), Charalambous (1982), Italconsult (1972), Jungfer (1984), Min Agric Yemen (1977), Neumann-Redlin (1991), Robertson (1992), van der Gun and Ahmed (1995), Wagner and Nash (1978).

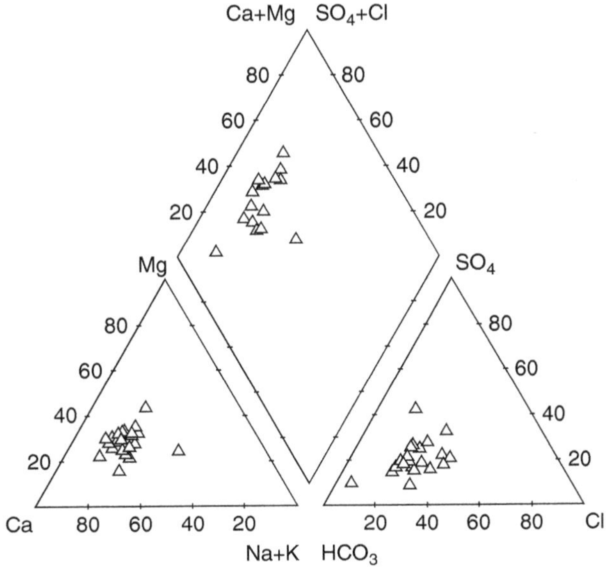

Fig. 7.9 Piper diagram: Groundwater samples from the Amran valley, northern Yemen. Data from Wagner and Nash (1978)

7.4 Information from Isotope Data

7.4.1 Groundwater Age

Tritium values above detection level indicate recent recharge in many groundwater samples from the highlands and the western escarpment of the Arabian Shield.

In wadi aquifers of the Jedda–Makka–Taif area, tritium values range from 15 to 38 TU.

In some wadi systems, a zonation of ^{3}H values appears to be related to specific features of the hydrogeologic regime:

In Wadi Rima, tritium data of samples from middle and downstream reaches are higher than values in the upper catchment area. The ^{3}H distribution indicates, that short term groundwater circulation predominates at lower altitudes sustained by seasonal flood flow infiltration, while aquiferous sections in at higher altitudes include higher components of longer term groundwater movement.

In Wadi Bisha, the ^{3}H values reflect, together with the hydrochemical data, a partition of the wadi aquifer into a central groundwater flow system with significant recharge and into marginal zones, which receive more limited replenishment. ^{3}H values are around 40–50 TU in the wadi centre with fast groundwater flow; in the "edge water" on the margins of the wadi course, ^{3}H values range from 2 to 14 TU.

On the escarpment west of Sanaa, recent groundwater recharge is indicated by ^{3}H values of 7–29 TU and ^{14}C values of 89–114 pmc.

The spring water at Ayoun Surdud, discharging from the Tawila–Amran aquifer complex at 1,100 m asl, has ^{14}C values of 25–30 pmc, corresponding to retention periods of several thousand years. ^{3}H values of 5–7 TU at Ayoun Surdud show, that the spring water contains components of recent flood infiltration.

In groundwater samples from the Sanaa high plain, ^{3}H values above 2.1 TU were found in 3 out of 15 samples (3.7–15.9 TU). ^{14}C data of three groundwater samples from the Sanaa plain correspond to groundwater ages of 8,200–19,200 years.

7.4.2 Stable Isotopes of Oxygen and Hydrogen

$\delta^{18}O$ values of groundwater samples from the Arabian Shield vary in a general range of −2.6 to −1.1‰ with d values scattering around the GMWL.

In the Hijaz highlands near Madina (wadis Shamiya, Fatima, Naaman, Murwani, Khulais plain) and in the highlands of Yemen around Sanaa, $\delta^{18}O$ values of −1.8 to −2.6‰ (Madina, Fig. 7.10) and −1.6 to −2.6‰ (Sanaa) are reported. On average less negative $\delta^{18}O$ values (−1.1 to −1.8‰) are found in groundwaters of wadi aquifers on the escarpment east of Jedda and on the eastern margin of the Arabian Shield in aquiferous deposits of the Wadi Rima stream system (Wadi Rima, Wadi Maraghan, Fig. 7.11). The data suggest an influence of an altitude effect with relatively depleted (more negative) $\delta^{18}O$ and $\delta^{2}H$ in the higher mountain areas. Relatively depleted $\delta^{18}O$ values of −1.9 to −2.6‰ found on some wadi stretches of the western escarpment may be attributed to recharge from flood flow originating at higher altitudes.

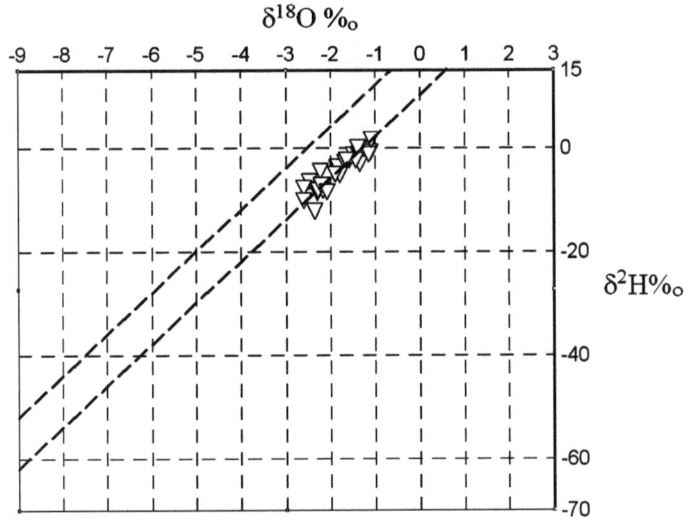

Fig. 7.10 $\delta^{18}O/\delta^{2}H$ diagram: Groundwater samples from wadi aquifers in the highlands and the escarpment between Jedda and Madina, western Saudi Arabia. Data from Jado and Zötl (1984)

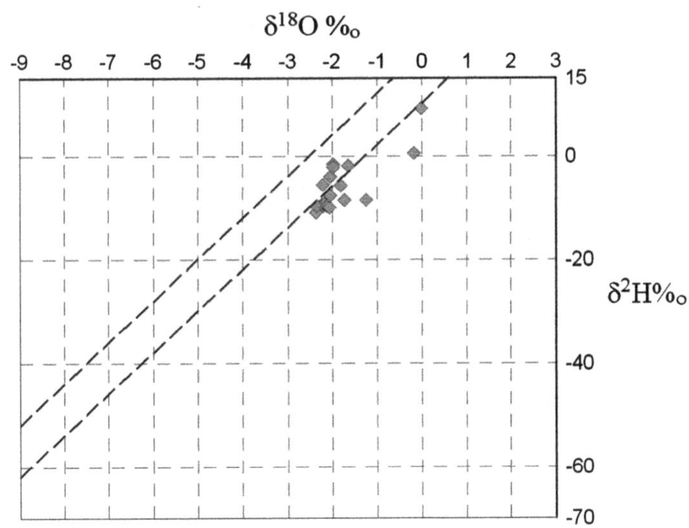

Fig. 7.11 $\delta^{18}O/\delta^{2}H$ diagram: Groundwater samples from Wadi Rima, Arabian Shield, western Saudi Arabia. Data from Jado and Zötl (1984)

d values of groundwater samples from most areas of the Arabian Shield are between +9 and +13‰.

Relatively enriched $\delta^{18}O$ values of up +0.1‰ with d values of +2.6‰ in the lower reaches of Wadi Murwani–Khulays indicate impacts of evaporation of infiltrating surface water or of irrigation water return flow.

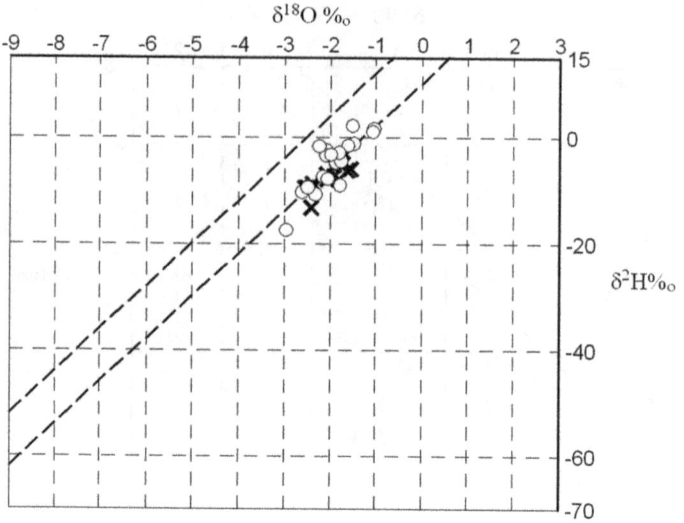

Fig. 7.12 $\delta^{18}O/\delta^2H$ diagram: Groundwater samples from the Sanaa-Mahwit area, Yemen x Sanaa high plain, O Mahwit escarpment. Data from Neumann-Redlin (1991), Jungfer (1984)

In Wadi Bisha on the eastern slope of the Hijaz highlands, the $\delta^{18}O$ data confirm the zonation of the aquifer into a central course with fast flowing groundwater and marginal zones, defined from hydrochemical and tritium data. $\delta^{18}O$ is -2.3 to -2.7‰ in the central zone and -1.4 to -1.8‰ on the wadi margins. The values in the wadi centre may be assumed to contain components of recharge at higher altitudes, the less negative values of the marginal groundwater appear to reflect local recharge affected by evaporative enrichment.

The $\delta^{18}O$ and δ^2H values of groundwater from Al Mahwit province on the western escarpment of the Yemen highlands scatter around the GMWL with a range of $\delta^{18}O$ from -2 to -1‰ similar to values of groundwater on the western escarpment of Saudi Arabia (Fig. 7.12).

The $\delta^{18}O$ values of the generally young groundwaters in the Mahwit province seem to depend mainly on the altitude of the catchment; accordingly it may be assumed that the groundwater extracted from boreholes or discharging in springs originates from catchment areas of local extent. $\delta^{18}O$ values of water of the springs Ayoun Surdud with retention periods of several thousand years suggest a mean altitude of the recharge area between 2,500 and 3,000 m asl. The main recharge area of the springs, which issue at an altitude of 1,100 m asl may be assumed in the outcrop area of the Tawila sandstone aquifer in the highlands northwest of Sanaa (Figs. 7.10–7.12).

References. Basmaci and Al-Kabir (1988), Jado and Zötl (1984), Jungfer (1984), Lloyd et al. (1980), Neumann-Redlin (1991), Wagner and Geyh (1999).

Chapter 8
Oman Mountains

8.1 Geographic and Geologic Setup

The southeastern margin of the Arabian Peninsula is accompanied by the Oman mountains, a sickle-shaped 700 km long and up to 140 km wide range, which reaches a peak elevation of more than 3,000 m asl. The core of the Oman mountains is formed by a sequence of autochthonous formations of Precambrian to Cretaceous age, which is overlain by allochthonous complexes: the partly metamorphic sedimentary rocks of the Hawasina unit and the Semail ophiolites.

8.1.1 Morphology and Climate

The Oman mountains separate the coastal belt along the Gulf of Oman from the Rub al Khali. The mountain ranges of Al Hajar al Gharbi and Al Hajar ash Sharqi form a nearly continuous morphologic barrier with altitudes of 1,000–2,000 m asl, which is interrupted by a few passes with lower elevation, in particular at Wadi Semail west of Muscat, and further north near Al Ain–Buraimi. The Jebel Hajar al Gharbi is occupied by high mountain massifs above 2,000 m asl at Jebel Akhdar, Jebel Nakhl and Jebel Khawr, cumulating in Jebel Akhdar with 3,009 m asl. The Jebel Hajar ash Sharqi south of Wadi Semail reaches altitudes of 2,251 m asl at Jebel Tawa and 2,223 m asl at Jebel Khadar.

In the north, the high mountains of Al Hajar al Gharbi are followed by a zone of prevailingly ophiolite hills and mountains with moderately high altitudes of 500–1,100 m asl, situated mainly on territory of the United Arab Emirates. On the northern tip of the Oman mountains, the Musandam peninsula with the up to 2,087 m high Ruus al Jibal limestone massif protrudes as an about 30 km wide spur into the Strait of Hormuz which connects the Gulf of Oman with the Arabian–Persian Gulf (Figs. 8.1 and 8.2).

The eastern slopes of the Oman mountains descend, along the northern and southern stretches, in partly abrupt escarpments directly to the coast of the Gulf of Oman. Between Fujayra and Seeb, an around 30 km wide coastal plain, Al Batina,

W. Wagner, *Groundwater in the Arab Middle East*,
DOI 10.1007/978-3-642-19351-4_8, © Springer-Verlag Berlin Heidelberg 2011

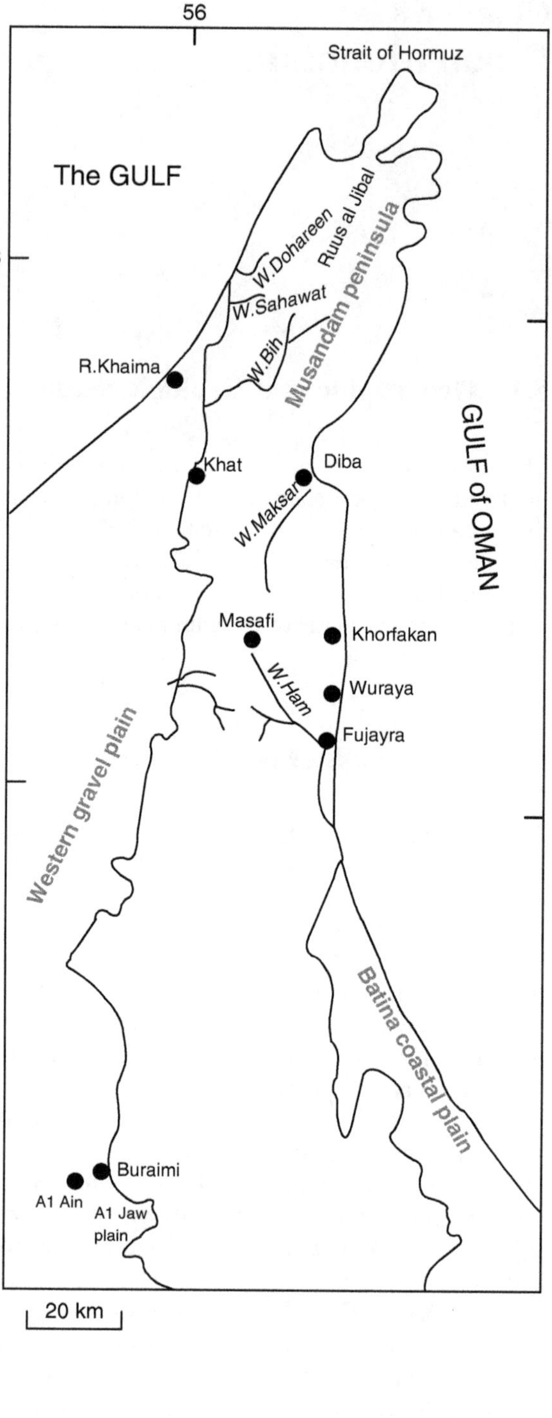

Fig. 8.1 Northern Oman mountains, location map

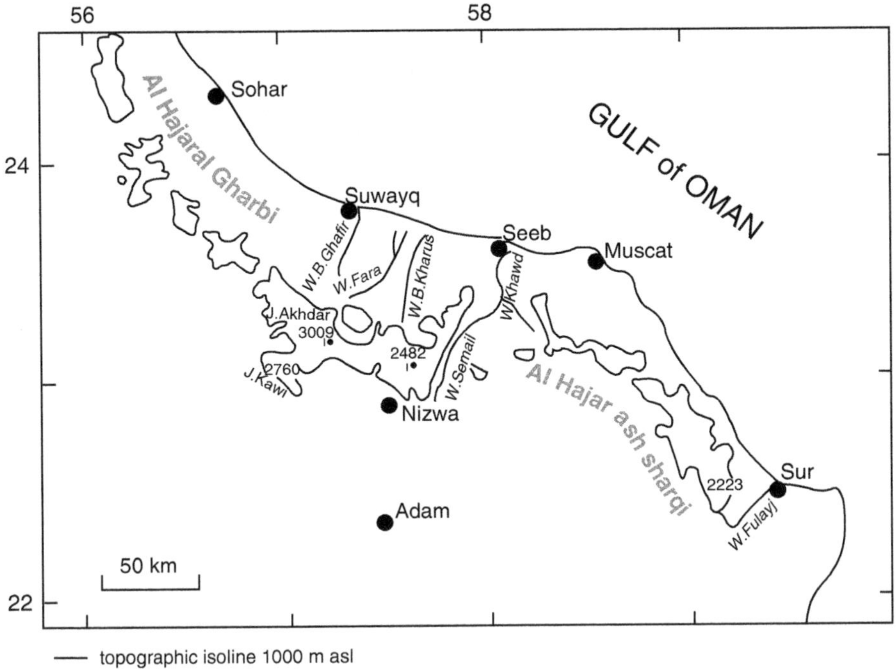

Fig. 8.2 Oman mountains, location map of Al Hajar al Sharqi and Al Hajar al Gharbi

extends for nearly 300 km between the mountains and the coast line. In the west, the Oman mountains are adjoined by a flat gravel plain which grades further east into the sand desert of Rub al Khali. In the southwest, the foothills of the Oman mountains descend into the Rub al Khali and the desert Jidda al Hararis.

The crest area of the Oman mountains forms approximately the main surface water divide between the Gulf of Oman in the east and the Rub al Khali and the Gulf coast in the west. The wadis draining toward the Batina coastal plain in the east are frequently deeply incised on the mountain slopes; wadis directed toward the Rub al Khali in the west have generally a more flat broad morphology.

The climate in the Oman mountains is mainly arid with mean annual precipitation ranging from generally 100–150 mm to more than 300 mm at elevations above 2,000 m asl in the high mountain zone of Jebel el Akhdar. Two types of rainstorms predominate:

- Storms related to frontal depression systems, which originate over the Mediterranean Sea and trace into Oman and the United Arab Emirates during the winter months
- Orographic or convectional rainfall in the mountains during summer

Depression systems occur a few times per season or may be absent for 2 or 3 years. Convectional rainfall is relatively frequent in Oman, but is an exceptional event in the northern part of the Oman mountains in the United Arab Emirates.

The moisture of winter as well as of summer rainfall is derived from vapour masses of the Gulf of Oman and of the Gulf between the United Arab Emirates and Iran.

Barren rocks with very limited soil cover dominate the landscapes of the arid Oman mountains. Scarce vegetation with shrubs and bushes is found on the ophiolite outcrops and on the higher mountain areas. The high mountain massif of Jebel el Akhdar is covered by a vegetation of grass, bushes and low trees. Small oases extend along some stretches of larger wadis with intensive cultivation in the wadi channels and broader valleys.

8.1.2 Geology

The Oman mountains are built up of three main geologic units (Fig. 8.3):

- An autochthonous sequence of Precambrian to Cretaceous rock units
- Two allochthonous sequences: a lower sequence, the Hawasina nappe, composed of sedimentary rocks and metasediments, and the ophiolites of the Semail nappe as upper sequence

The autochthonous units, the Hawasina sequence and the Semail sequence occupied initially consecutive belts: the sediments of the mainly Mesozoic Hajar group on the Arabian Shelf in the west, the Semail oceanic crust in the east, and the Hawasina sediments overlying older oceanic crust in between. During the Upper Cretaceous, the Hawasina and Semail sequences were thrust as tectonic nappes over the autochthonous units, the Hawasina being stacked above the Hajar sediments and the Semail nappe above the Hawasina with partly metamorphic alterations of the Hawasina sediments. The stacked nappes probably formed a chain of low-relief islands along the site of the present mountain chain. The thrust tectonics stagging the Hawasina and Semail series over the autochthonous may be attributed to horizontal crustal compression in response to relative movements of the Arabian Plate and adjacent micro-continents toward the southwest edge of the Asian continent (Glennie 1995: 68).

The orogenic uplift and consecutive erosion processes, creating the present Oman mountain chains, followed in the Oligocene–Miocene.

8.1.2.1 Autochthonous Sequence

The autochthonous sequence has been sub-divided into a lower and upper autochthonous, corresponding to Precambrian–Lower Permian and Upper Permian–Cretaceous rock units, respectively. The lower autochthonous comprises the Precambrian basement, Eocambrian to Ordovician limestones, dolomites, siltstones and quartzites, and sandstones of the Carboniferous–Lower Permian Haushi group. The upper autochthonus is made up prevailingly of limestone and dolomite formations of the Hajar group. The lower autochthonous is exposed in relatively limited areas of the Oman mountains in some domal structures of Jebel Akhdar and Saih Hatat in the Hajar al Gharbi massif west of Muscat.

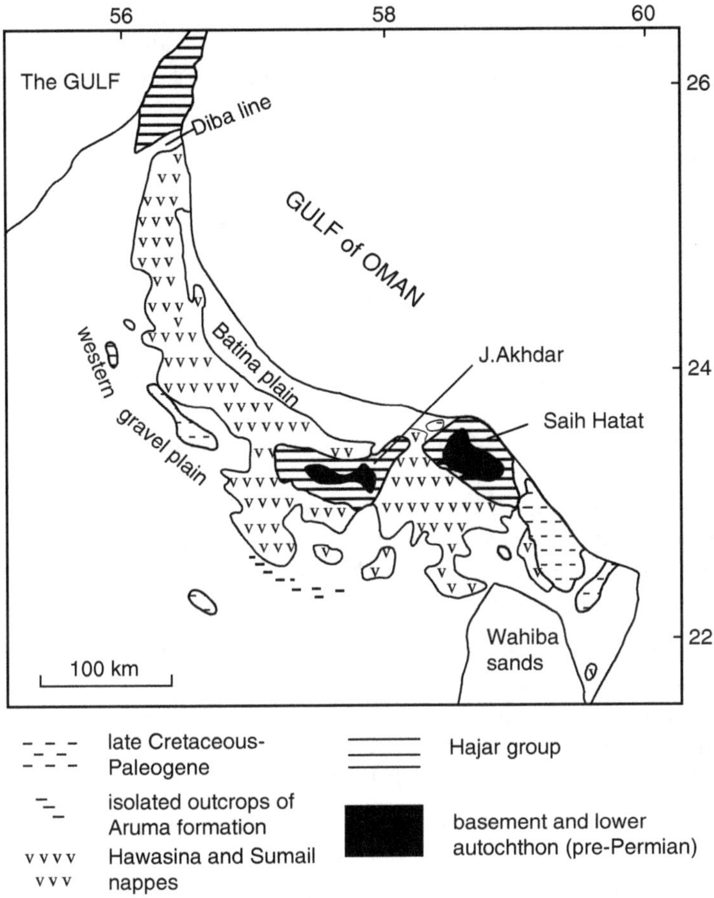

Fig. 8.3 Outcrops of main geologic units in the Oman mountains. After Glennie (1995), generalized

The Hajar group of the upper autochthonous is an up to 3,000 m thick sequence of limestones with some dolomite intercalations that was deposited in shallow marine waters on the Arabian continental shelf. The group is separated from the siliciclastics and dolomites of the underlying strata by a regional erosional unconformity and is followed stratigraphically by the late Cretaceous Aruma group. The parautochthonous Hajar limestone massif of the Musandam peninsula is, in the south, separated from the ophiolite mountains by a major fault zone, the Dibba zone.

The late Cretaceous–Tertiary sequence of the upper autochthonous comprises:

- Carbonates of the Campanian–Maastrichtian Aruma formation, which has been subdivided in Interior Oman into two units:
  - Fiqa formation below: shales and carbonates, with conglomerates and turbidities (Muti formation) at the base
  - Simsima formation above: mainly shallow marine limestones

- The late Cretaceous Juweiza formation on the northwestern boundary of the Oman mountains, composed of detrital deposits of the uplifted allochthonous units
- The Paleogene Hadramaut group with the Umm er Radhuma formation, the gypsiferous Rus formation and the Damam formation

The shallow marine limestones of the Maastrichtian Simsima formation and of the Paleogene Hadramaut group were deposited above the autochthonous–allochthonous units of the Oman mountains after the nappe emplacement. These late Cretaceous–Paleogene rocks were deformed during the mid-Tertiary, when the Oman mountains were uplifted into the present arched structure.

8.1.2.2 Allochthonous Sequence

The Hawasina consists, to a large percentage, of calcarenites that were deposited as turbidities in deep waters northeast of the Arabian continental shelf at roughly the same time span as the Hajar group. The thickness of the Hawasina reaches 1,600 m and decreases probably to around 200 m at distal points of the sedimentation area. The series comprises metasediments, rocks in chert–limestone facies and volcanic intercalations, including the following rock types:

- Metamorphics: quartzite, various types of schists with bands of crystalline marble and enclosures of amphibolites
- Chert–limestone facies: bedded chert and limestone with shale layer
- Volcanics: intermediate and basic lavas, partly vesicular, porphyritic or with pillow structures, agglomerates, tuff beds

The Semail sequence represents principally an 8–15 km thick slate of former oceanic crust of the Tethys sea floor and constitutes the largest and best preserved ophiolite complex of the world. The main rock units of the ophiolite sequence range from ultrabasics to gabbro, sheeted diabase dikes and volcanics. These units include the following lithological types:

- Ultrabasics: peridotite to serpentinite and harzburgite with magnesite and chrysotile veins and silicified alteration products
- Gabbros and ultrabasics: gabbro with intermixed ultrabasic rocks
- Gabbros: coarse grained leucocratic and melanocratic gabbro varieties, commonly layered and with zones of serpentinites, breccias and pegmatites
- Sheeted dikes: medium to fine grained diabase, which forms swarms of sub-parallel sub-vertical dikes
- Volcanis: basaltic lavas with pillow structure

The dolerite dikes, the basalt, and, to a lesser extent, the gabbro underwent secondary hydrothermal alteration (greenschist facies metamorphism) when they were part of the ocean ridge. The peridotite was affected by secondary hydrothermal alteration resulting in a conversion of main minerals to serpentinite.

References. Alsharhan (1989), Glennie (1995).

8.2 Hydrogeologic Conditions in the Hajar Limestone Areas

8.2.1 The Aquifer

Limestones and dolomites of the Permian–Cretaceous Hajar supergroup constitute an important aquifer in the outcrop areas of the Musandam peninsula in the northern Oman mountains, in Jebel Akhdar and in some areas of Jebel Hajar esh Sharqi (Table 8.1).

8.2.1.1 Musandam Peninsula

On the Musandam peninsula, the Hajar carbonate complex forms the Ruus al Jibal mountain range rising with steep morphologic relief from sea level to around 2,000 m asl. The aquifer is here composed of carbonate formations with a total thickness of more than 3,000 m, which are karstified to a depth of several hundred metres below surface. On sub-regional scale, the karstified carbonate complex possibly behaves as one continuous aquifer system, zones of preferential groundwater flow may, however, be concentrated to the deeply incised major wadi systems.

Major wadis in the Ruus al Jibal massif contain unconsolidated wadi sediments with around 30 m and up to more than 100 m thickness. The wadi sediments constitute, in some areas, an aquifer which is in contact with the underlying karst aquifer but with locally higher – perched – water levels.

The mountain massif is crossed by major fault systems with NNW–SSE to N–S and WSW–ESE orientation and by a few extensive west–east trending faults. The fault zones appear to be accompanied by particularly intensive karstification and are followed, along some stretches, by the courses of the westward oriented main wadis: Wadi Bih, Wadi Dohareen, Wadi Sahawat.

The karstic complex is bounded by:

- The sea coast in the northwest and in the east
- A regional fault zone, assumed to constitute a thrust fault, in the west
- The Dibba fault system in the southeast, delimiting the carbonate rocks against the allochthonous Hawasina and Semail units

Along the sub-regional fault line in the west, the karst aquifer is in lateral contact with unconsolidated Quaternary sediments ("gravel fans") which overlie marls and limestones of the Upper Cretaceous Juweiza formation. In some areas, the Hajar aquifer is adjoined to the west by outcrops of rocks with low permeability, e.g., in the Khat area, where hot springs discharge along the outcrop boundary of the karst aquifer.

8.2.1.2 Jebel Akhdar

The Hajar aquifer of Jebel Akhdar is composed mainly of shallow marine bioclastic and clayey limestones and dolomites with a total thickness of 2,000–3,000 m.

Table 8.1 Hydrostratigraphic sequence in the Oman mountains

Stratigraphic age		Autochthonous		Allochthonous		Western mountain foreland	
		Formation or group	Aquiferous properties	Unit	Aquiferous properties	Unit	Aquiferous properties
Tertiary	Neogene					Fars	Brackish water aquifer
	Paleogene					Hadramaut limestone, chalk, marl	*Aquifer*
Cretaceous	Maastrichtian–Coniacian			Semail ophiolite	*Shallow* fissured *aquifer*	Aruma limestone	Mainly aquitard
	Cenomanian–Lower Cretaceous	Hajar limestone and dolomite	*Major aquifer* in Musandam peninsula and Jebel Akhdar	Hawasina nappe, metasediments, metavolcanics	Generally aquitard		
Jurassic–Permian							

On the southern flank of Jebel Akhdar, the Hajar group dips steeply beneath the alluvial plain and re-emerges 80 km further south in the frontal Adam mountains.

In the synclinal structure between Jebel Akhdar and Jebel Adam, the Hajar aquifer is overlain by a marl aquitard of the Aruma formation.

8.2.2 Groundwater Regimes

8.2.2.1 Hydraulic Parameters

For the Hajar aquifer in Wadi Bih, a mean hydraulic conductivity of 7.7×10^{-4} m/s and a mean transmissivity of 2,000 m^2/d were deduced from pumping tests. In the overlying wadi sediments, horizontal hydraulic conductivities range from 4×10^{-4} to 8×10^{-4} m/s.

Transmissivities of fractured limestones and dolomites of the Hajar aquifer in the Jebel Akhdar area are in the order of 1.7–1,500 m^2/d.

8.2.2.2 Groundwater Recharge and Groundwater Flow

Groundwater recharge on the outcrop of the Hajar aquifer on the Musandam peninsula was estimated at 15 mm/a corresponding to 10% of the mean annual precipitation. Estimates of the recharge volume in the Wadi Bih catchment for the period 1996–2005 amount to a mean of 3.66×10^6 m^3/a, corresponding to 10% of the mean annual precipitation of 150 mm over a catchment of approximately 500 km^2. Annual groundwater withdrawal reaches around 11.2×10^6 m^3/a.

Fresh water replenishment in the Hajar aquifer of the Musandam peninsula may depend prevailingly on indirect recharge from flood flows in the deeply incised main wadis. Groundwater movement in most of the Hajar aquifer of the Ruus al Jibal massif appears to be directed toward west to northwest. Deep groundwater levels in the major wadi systems, the zones of preferential groundwater flow, may reflect a rather smooth sub-regional water table.

Groundwater from the southeastern part of the Hajar aquifer in the Ruus al Jibal massif discharges in springs in the Khat area at the contact of karst limestones with low permeability rocks. The spring water, which is considered to represent deep groundwater flow, rising probably from a depth of around 400 m below land surface. Groundwater discharge of $1.46–2.06 \times 10^6$ m^3/a in the Khat springs indicates a recharge rate of 12 mm/a over the catchment of 125 km^2, or 8–11% of annual rainfall.

North of Khat, groundwater probably moves as subsurface outflow from the karst aquifer into Quaternary sand and gravel deposits, which extend in the coastal area west of the limestone massif and from which large quantities of groundwater are extracted mainly for irrigation. Under undisturbed conditions, subsurface outflow from the northern part of the Ruus al Jibal limestone massif probably entered coastal sabkhas or the sea.

In the east of the Ruus al Jibal massif, some groundwater flow probably reaches the coast of the Gulf of Oman.

In the Hajar aquifer of Jebel Akhdar, groundwater flow is directed:

- To springs at intermediate altitude of the northern mountain slope
- To the Batina coastal plain
- To the piedmont zone adjacent to the southern mountain slope

8.2.3 Groundwater Salinity and Hydrochemistry

Groundwater in the Hajar aquifer of the Musandam peninsula is prevailingly brackish, but fresh water occurs at various locations in the larger wadis. EC values range from 550 to 15,000 μS/cm. The following salinity ranges and water types have been found in wadi courses and in well fields at the foothill zone of the mountains:

- Wadi Dohareen, Wadi Ghalila, Sahawat well field: fresh water with equal percentages of major anions or Cl type fresh water and brackish Na–Cl water with EC values generally between 660 and 6,000 μS/cm, increasing during exploration up to 15,000 μS/cm
- Wadi Bih: prevailingly brackish Na–Cl type water, fresh Cl type groundwater at some locations; salinity between 600 and 7,000 mg/l TDS with an average of 2,122 mg/l
- Burairat well field downstream of Wadi Bih: brackish Na–Cl water with salinities between 1,000 and 6,640 mg/l TDS, on average around 4,000 mg/l

In the Khat area, brackish Na–Cl type groundwater with EC values around 2,500 μS/cm discharges in springs and has been tapped in boreholes.

The fresh water occurrences are possibly restricted to relatively shallow water layers in major wadi courses, sustained by recent recharge mainly through infiltration of sporadic runoff, while groundwater under the mountain plateaus and at greater depth below the wadis is brackish to saline.

Groundwater in sand and gravel fans in the foothill zone west of the Ruus al Jibal mountains appears similar in its hydrochemical composition to brackish groundwater contained in the Hajar limestone aquifer with, on average, higher Na and Cl concentrations.

The hydrochemical composition of groundwater from the Hajar aquifer of the Musandam peninsula appears to reflect a mixture of fresh groundwater and brackish Na–Cl water with varying proportions of fresh and brackish groundwater in different areas (Fig. 8.4).

The major ion contents in the fresh groundwater (EC values 600–1,500 μS/cm) result probably from:

- Slightly enriched atmospheric inputs (Cl and SO_4 contents of 35–150 mg/l)
- Dissolution of soil and rock carbonate through interaction with biogenic CO_2 (HCO_3 contents of 130–230 mg/l and equivalent cation contents)
- Mixture with varying amounts of brackish water (Na and Cl contents above 150 mg/l)

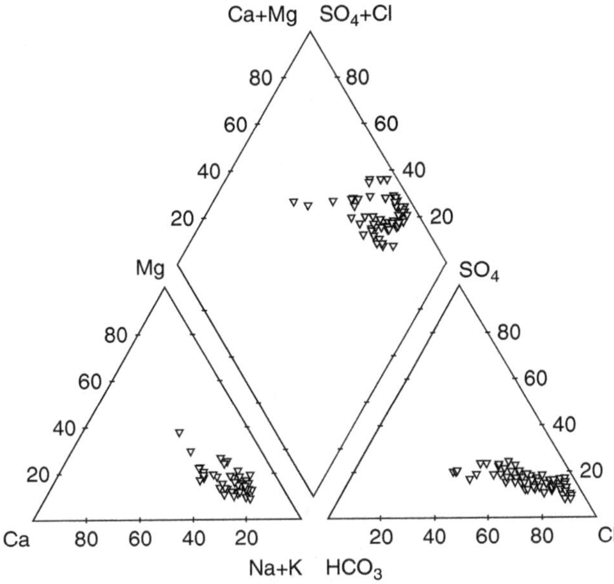

Fig. 8.4 Piper diagram: Groundwater samples from the Hajar aquifer of the Musandam peninsula, United Arab Emirates. Data from files of Ministry of Agriculture and Fisheries and Ministry of Agriculture and Water Resources, Dubai

According to the hydrochemical composition of the brackish to saline groundwater, recent sea water intrusion is apparently not the source of elevated groundwater salinity. "Evaluation of the chemistry of the water, in particular the Ca/Mg and the Cl/SO_4, suggests that the end member is not modern sea water, but rather a variable degree of water–rock reactions, with Na–Cl rich evaporite systems or brine". Hydrochemical modelling "confirms that sea water intrusion is not the source of the higher salinity, but suggests a mix of groundwater with an older and possibly deeper source of water and younger groundwater, both from the same recharge area" (Alsharhan et al. 2001: 298).

"Some Ca, HCO_3 and Mg is added by rock weathering, most of the solutes appear to be added from a regional brine that is lower in Mg and SO_4 and higher in Cl and K than sea water" (Rizk et al. 2007).

The ranges of major ion concentrations of fresh water samples and of a hypothetical brackish water component are listed in Table 8.2.

Ratios between total salinity (EC values) and Na and Cl concentrations of groundwater from the Hajar aquifer of the Musandam peninsula follow rather closely a simple mixing line, indicating that the elevated groundwater salinity originates from one major source.

In the recharge areas of the Hajar aquifer in Jebel Akhdar, Ca–HCO_3 type water prevails, Mg–HCO_3 water is found at a few locations. Groundwater salinity ranges from 315 to 550 mg/l TDS.

Table 8.2 Major ion concentrations (mg/l) of fresh and brackish components of groundwater in Hajar aquifer of the Musandam peninsula (from Wagner 1996c, data from UAE Min. Electricity and Water)

	EC (μS/cm)	HCO_3	SO_4	Cl	Ca	Mg	Na + K
		(mg/l)					
Fresh water component, range	550–1,500	128–232	35–135	40–333	30–116	12–63	40–233
Brackish water component, assumed	8,500	200	650	2,400	250	130	1,400

Higher groundwater salinity is observed in limestone aquifers of Jebel Akhdar outside the wadis or morphologic depressions (plateau section, mean groundwater salinity 720 mg/l TDS). The differences in groundwater salinity are "seen as reflecting different recharge conditions existing between the bare limestone surface of the plateau and the 20 m thick gravel deposits with its capacity to easily accept and store large volumes of flood water" (Macumber 1995: 484).

In the Adam mountains, the Hajar aquifer, emerging on an anticlinal structure 80 km south of Jebel Akhdar, contains Na–Cl type groundwater under reducing conditions with a salinity of 1,477–3,093 mg/l TDS.

References. Al-Sayari and Zötl (1978), Alsharhan et al. (2001), Harress (1975), IWACO (1986), Macumber (1995), Matter et al. (2006), Rizk et al. (1997, 2007), Wagner (1995a, 1996b).

8.3 Groundwater in the Ophiolite Mountains

8.3.1 Aquiferous Zones

The term "ophiolite mountains" is applied here to the areas in the Oman mountains, which are covered by rocks of the Semail and Hawasina nappes. Aquiferous layers are found in these rock complexes in near-surface weathering zones and in fracture zones. Shallow groundwater occurs, in particular, along wadi courses which follow stretches of intensive fracturing and/or are filled with coarse stream deposits (Table 8.1).

The metasediments and metavolcanics of the Hawasina nappe have generally low permeabilities and act as aquitards.

The Semail ophiolite complex comprises peridotites of the earth mantle and crustal rocks: basaltic pillow lavas, dolerite dikes, gabbro. The crustal rocks have been affected by hydrothermal alteration, creating veins with minerals which produce clayey material during weathering. "The peridotite is affected by secondary hydrothermal alteration, resulting in a conversion of the main minerals to serpentine. Subaerial meteoric weathering of peridotite led to the formation of oxides and clay minerals, commonly rendering the peridotite very friable and porous" (Dewandel et al. 2005: 712).

Hydraulic conductivities in the ophiolite complex are attributed to fissure conductivity and fracture conductivity. Significant fissure conductivity is generally restricted to a near surface zone down to a depth of around 50 m below surface. Apparently, most fissures are closed at a depth of a few tens of metres.

Fissure hydraulic conductivity is generally higher in gabbro and dolerite than in peridotite, probably because many of the fissure openings in the peridotite are partly or completely sealed by carbonates.

Fracture hydraulic conductivity is superimposed on the fissure conductivity and can reach much deeper than 50 m below surface, creating locally extensive secondary porosity. The hydraulic conductivity of tectonic fractures affects all types of ophiolitic rocks and the most productive wells are located in zones of tectonic fracturing.

8.3.2 Groundwater Regimes

8.3.2.1 Hydraulic Parameters

The following values for hydraulic conductivity and transmissivity of ophiolite rocks in the Oman mountains have been estimated:

Fissure hydraulic conductivity of gabbro and dolerite	10^{-5} to 10^{-7} m/s
Fissure hydraulic conductivity of peridotite	10^{-7} m/s
Fracture hydraulic conductivity	10^{-5} m/s
Fracture transmissivity values	9–8,600 m^2/d, geometric mean 90 m^2/d
Transmissivity of fractured and weathered peridotite	7–8 m^2/d

In Wadi Dibba, on the margin of the ophiolite mountains, transmissivities of Quaternary wadi bed gravels range from 19 to 13,700 m^2/d.

8.3.2.2 Groundwater Recharge

Recharge rates in the ophiolite mountains of Oman are about 7% of the mean annual precipitation of 250 mm or 18 mm/a. Absence of soil favours recharge rather than evapotranspiration.

The following estimates of groundwater recharge in the ophiolite mountains of Oman have been reported:

- Recharge to the shallow peridotite aquifer: 18 ± 8 mm/a or 7% of mean precipitation, in rainy years up to 10% of total precipitation
- Average annual recharge in wadis in the ophiolite mountains of the northern part of Oman in wadis Ghalaji, Kabir, Hajr, Asih: 9–12% of the rainfall of 73–149 mm/a
- Recharge to deep groundwater (thermal springs): less than 0.1% of total precipitation

Recharge to crustal rock aquifers was estimated at about 20 mm/a (8% of total precipitation).

Total – direct and indirect – recharge in the Wadi Ham catchment (Fujayra Emirate) upstream of Wadi Ham dam may be in the order of 7–9% of mean rainfall ($2.0–2.5 \times 10^6$ m^3/a over 190 km^2 or 10–13 mm/a).

For the Wadi Wuraya catchment, a mean recharge of 14 mm per year, equal to 10% of rainfall has been assumed. This recharge rate is supposed to sustain perennial surface flow of about 60 l/s in a particular section of the wadi, where perennial base flow sustains waterfalls within the wadi.

Recharge in the catchment of the Khat springs is estimated at 8–11% of mean annual rainfall of 150 mm, or 12–16 mm/a.

The estimates obviously imply high uncertainties and can indicate only that mean groundwater recharge may be in the order of 15 mm/a or 10% of rainfall. This includes direct recharge from rainfall as well as indirect recharge from flood flows in wadi sediments.

8.3.2.3 Groundwater Flow and Flow Volumes

Groundwater flow in the ophiolite mountains is directed toward east to the coastal plain on the Gulf of Oman and toward west or southwest to the gravel plain in the foreland of the Oman mountains. The courses of major wadis may act as zones of preferential groundwater flow in many cases. The groundwater divide between flow to the Gulf of Oman and the western gravel plain on the fringes of the Rub al Khali sub-basin probably coincides approximately with the surface water divide on the crest of the Oman mountains.

To some extent, groundwater discharges in springs within the mountain areas, prevailingly in stream channels, and is used for water supply, diverted for local irrigation or re-infiltrates further downstream in the wadi bed. A major part of groundwater flow in the ophiolite mountains reaches the plains adjoining the Oman mountains in the east and west through subsurface inflow along the main wadi courses. Some diffuse subsurface flow appears also to reach the plains over wider stretches of the boundary between ophiolite outcrops and Quaternary cover.

A tentative water balance of Wadi Ham with a catchment area of 190 km^2 upstream of the Fujayra coastal plain has the following components:

Rainfall	27.6×10^6 m^3/a
Surface runoff	5.35×10^6 m^3/a
Subsurface flow	2×10^6 m^3/a
Evapotranspiration	20.2×10^6 m^3/a
Subsurface outflow to the coastal plain	0.5×10^6 m^3/a
Consumptive use of shallow groundwater in the mountain area	1.5×10^6 m^3/a

Recharge 7% of 145 mm mean annual rainfall or 10 mm/a.

Baseflow in the Wadi Wuraya waterfall area is around 1.8×10^6 m^3/a from a catchment of 128 km^2.

In Jebel Hajar ash Sharqi, the peridotites are, in general, exposed in the upstream parts of the mountainous ophiolitic catchment, while crustal rocks (gabbro, dolerite, basalt) tend to occupy the downstream areas. Many of the wadis in the peridotite area have perennial flow; base flow in the wadis decreases abruptly, where the wadi reaches the gabbro.

The peridotite aquifer has a low hydraulic conductivity but a relatively high storage capacity, which can maintain base flow over large periods (0.1–0.4 $l/s/km^2$ over 6 months). The base flow drained from the peridotite aquifer infiltrates downstream into the gabbro aquifer, which has a higher hydraulic conductivity, providing an annual water input into the gabbro of 1.7×10^4 m^3/km^2.

Storm runoff infiltration into the gabbro reaches around $32 \times 10^3 m^3/km^2$.

8.3.2.4 Groundwater Exploitation Through Falaj Systems

In the ophiolite mountain area, many oases are irrigated by falaj systems. A falaj is a drainage gallery, which in its upstream section cuts below the groundwater surface and acts as drainage collector of groundwater. The downstream sections of the falaj conduct the collected water to the surface, where the water discharges in free outflow. The main tunnel of the falaj may reach a depth of 30 m, which decreases gradually as the falaj approaches the surface.

The gallery is connected to the surface by shafts at distances of 15–30 m. The shafts serve for aeration, for transport of the excavated material to the surface and for access during maintenance and repair of the gallery. Falaj systems in the ophiolite mountains generally extend over distances of 1–5 km.

A falaj can be deepened down to a few metres below the groundwater surface only, implying a limitation of the volume of collected groundwater. The discharge of a falaj is determined mainly by the volume of water drained in the uppermost section of the gallery.

Construction or maintaining of falaj systems can be particularly suitable in shallow aquifers with relatively low permeability, where over a moderately extended drainage section adequate volumes of groundwater can be collected. The traditional groundwater exploitation through a falaj has advantages in that:

- Costs of investment for construction, maintenance and operation are relatively low
- No devices and no energy for water lifting are required
- Water can be drawn from shallow aquifers with relatively low permeability

Disadvantages are that the water discharge is highly dependent on seasonal rainfall, it cannot be adapted to actual water demands and is strongly affected by natural or artificial decline of the water level.

Falaj irrigation systems – also known as qanat or karez – have been constructed since around 800 BC. During the sixth century BC, when Persian rule extended from the Indus to the Nile, technology of constructing falaj systems spread throughout the empire from Mesopotamia to the shores of the Mediterranean Sea, parts of Egypt, Afghanistan and central Asia.

The falaj systems of Oman and United Arab Emirates have provided communities with water for irrigation and domestic purposes for the last 1,500–2,000 years. In Oman in 1993, about 55% of the cropped land was irrigated by falaj water in 1990, and 4,800 active falaj systems delivered some 900×10^6 m^3/a of water representing 60% of the country's total water usage (Information of Ministry of Water Resources, Sultanate of Oman). In the United Arab Emirates, falaj systems are part of the agricultural heritage, but recently many have run dry because of low rates of recharge and excessive groundwater pumping.

8.3.3 Groundwater Salinity and Hydrochemistry

8.3.3.1 Shallow Groundwater

Groundwater salinity in the near surface aquiferous zones of the ophiolite complex is generally low to moderate with TDS values of 300–400 mg/l. Most wells in the ophiolite mountain area are located in wadis and tap low salinity groundwater which is recharged from infiltration of wadi runoff. Outside the main wadis, recharge rates are certainly much lower, the infiltrating water may therefore be affected to a higher degree by evaporative processes leading to an increase of major ion concentrations and in particular of Na and Cl contents.

Surface water and shallow groundwater within the fissured and fractured peridotites is Mg–HCO_3 type water with a groundwater salinity of 245–1,187 mg/l TDS and Mg/Ca ratios of 2.94–6.25 (Fig. 8.5).

Groundwater in Wadi Wuraya in the northern ophiolite mountains (United Arab Emirates) is Mg–Cl–HCO_3 water with a salinity of 280–340 mg/l TDS.

Aquiferous peridotites of Jebel Hajar ash Sharqi contain shallow groundwater of Mg–HCO_3 type with EC values of less than 850 μS/cm.

Groundwaters in the ophiolite mountains of Oman show wide ranges of Mg/Na and HCO_3/Cl ratios, Ca and SO_4 percentages are generally low (Fig. 8.6). Ranges of electrical conductivity are reported as 416–3,820 μS/cm in peridotites and 630–7,500 μS/cm in gabbro.

Groundwater salinity in ophiolite aquifers on the southern foothills of Jebel Akhdar range from 245 to 810 mg/l TDS, Cl concentrations from 51 to 202 mg/l. These ophiolite groundwaters, which were sampled from strongly serpentinized peridotites, are mainly Mg–HCO_3 type waters.

According to Neal and Stanger (1984), the Mg–(Na)–HCO_3–(Cl) type waters are the result of weathering and evaporative processes in a system open to the atmosphere. The high magnesium and bicarbonate concentrations are attributed to the dissolution of near surface brucite according to

$$Mg(OH)_2 + 2\ CO_2 \rightarrow Mg^{2+} + 2HCO_3^-.$$

The wide spread occurrence of Mg–HCO_3 type surface and shallow groundwater in serpentinized rocks results from a high solubility of brucite. Brucite dissolution in the

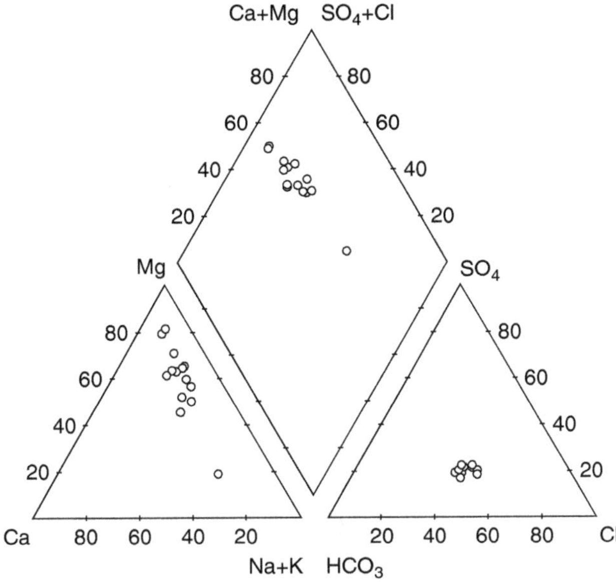

Fig. 8.5 Piper diagram: Groundwater samples from the ophiolite mountains of the United Arab Emirates. Well fields Haray, Lulaya and Wuraya, data from files of Ministry of Agriculture and Fisheries and Ministry of Agriculture and Water Resources, Dubai

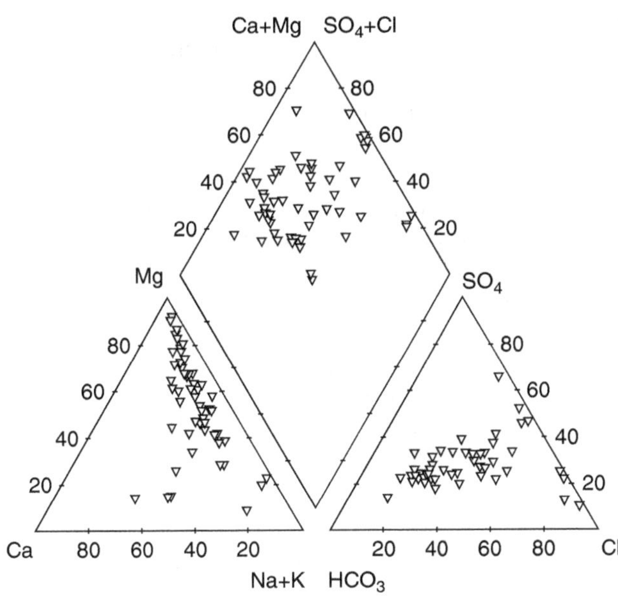

Fig. 8.6 Piper diagram: Groundwater samples from the ophiolite mountains of Oman (dolerite, gabbro, peridotite). Data from Dewandel et al. (2005)

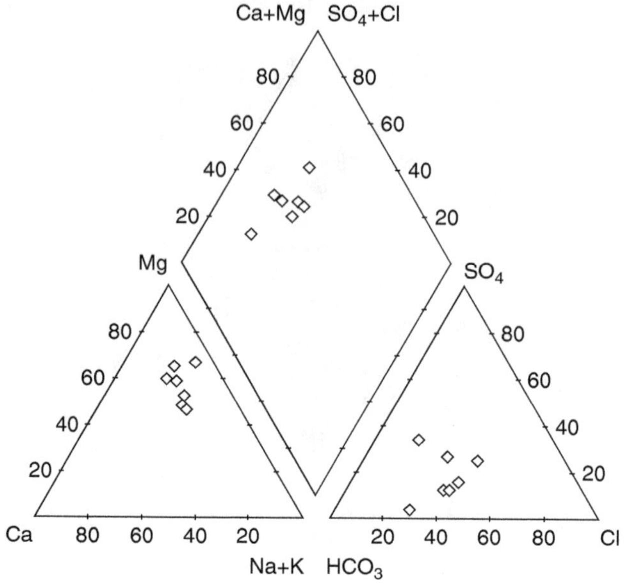

Fig. 8.7 Piper diagram: Falaj water from ophiolites in the United Arab Emirates. Data from Alsharhan et al. (2001)

shallow aquifer leads to the formation of secondary magnesite and abundance of magnesite veins of some millimetres to decimetres in thickness in the ultramafites.

Water discharging from falaj systems in the ophiolite mountains is prevailingly fresh water to slightly brackish water in some areas.

Falaj water in the United Arab Emirates draining fissured ophiolite rocks or overlying gravels is generally fresh water of Mg–HCO_3 type (Fig. 8.7) with EC values between 380 and 750 μS/cm. Components of alkaline groundwater from serpentinites are indicated in falaj water at several locations through pH values of up to 10.9 and up to 1,700 μS/cm elevated EC values.

Salinity of falaj water is generally low near the water divide in the mountain area and increases toward east and west with distance from the recharge area.

In the gravel plains at the coast and in the western foreland of the Oman mountains, EC values of falaj water increase to 1,250 and 1,850 μS/cm, respectively. Falaj water from the Hajar aquifer of the Khat area has EC values of 2,200–2,800 μS/cm. The Falaj Ain Soukhna near Al Ain, draining a Paleogene carbonate aquifer, has an EC value of 11,000 μS/cm.

8.3.3.2 Alkaline Deeper Groundwater

At many locations in the ophiolite mountains, highly alkaline brackish water springs emerge from ultramafic rocks. The spring water is Na–Cl water with elevated concentrations of Ca and OH and pH values of >9, up to 12.2. Most of the hyperalkaline springs have a low discharge and are located in the outcrop zone of ultramafic rocks between the partially serpentinized harzburgites and overlying ophiolite units.

The temperatures of 21–41°C of the brackish spring waters with EC values around 1,600 μS/cm suggest that the groundwater circulation ranges from near surface to around 700 m depth.

The hydrochemical characteristics of the hyperalkaline water are attributed to the impact of low temperature serpentinization of ophiolite rocks on circulating groundwater of meteoric origin. Several specific hydrogeologic conditions are responsible for the evolution of the hydrochemical composition of these waters.

Groundwater circulation is related to infrequent and irregular recharge pulses in an arid environment and slow movement in rocks with relatively low permeability. Accordingly, groundwater reacts during retention periods of several years to decades under closed conditions with the aquifer matrix. The initially acid to neutral Mg–HCO_3 water of the shallow ophiolite aquifer is altered through serpentinization/hydration processes at the contact of the water with the ophiolitic rocks: through silicate dissolution, Mg, Ca and OH are released into solution, pH rises and HCO_3 is converted to CO_3. With increasing alkalinity, dissolution of olivine and pyroxene phases result in the formation of serpentinite and an increase of Ca and OH in the water. Ca is derived from clinopyroxene.

During the serpentinization process, the water becomes progressively more reducing as ferrous iron is released from silicate minerals, and the oxidized dissolved solids SO_4 and NO_3 are reduced to HS and NO_2. Finally, the water attains an alkaline, highly reducing hydrochemical composition with high Ca and OH content.

The brackish nature of the water with elevated Na and Cl concentrations is attributed to the leaching of marine salts, which are dispersed in the ophiolitic rocks as remnants from the submarine phase before the uplift of the Oman mountains.

The discharge points of the hyperalkaline groundwaters are frequently accompanied by by-products of the serpentinization processes:

- Emerging hydrogen gas with minor amounts of methane and hydrocarbons
- $Ca(OH)_2$ and $Mg(OH)_2$ deposits as portlandite and amorphous gel

References. Abdulla and Durabi (1997), Al-Sayari and Zötl (1978), Dewandel et al. (2005), Faig (1990), Fritz et al. (1992), Matter et al. (2006), Neal and Stanger (1984), Rizk and El-Etr (1997), Rizk et al. (1997), Wagner (1996b), Wagner and Karanjac (1997).

8.4 Information from Isotope Data

8.4.1 Groundwater Age

A high percentage of groundwater extracted from the Hajar limestone aquifer of the Musandam peninsula appears to be old with varying admixtures of recently recharged groundwater. ^3H values show an impact of recent recharge at several locations in Wadi Dohareen and Sahawat well field (3.6–6.3 TU), Burairat well field (2.8–4.2 TU) and Wadi Bih (3.5–8.8 TU, tritium data from Gonfiantini 1992).

Indications of recent recharge according to ^3H data have been found in fresh water as well as in brackish water.

^{14}C values in the Hajar aquifer of the Musandam peninsula indicate generally high groundwater ages with 7.1 pmc in brackish water and 18.5 pmc in fresh water of Wadi Bih, and 9.5 pmc in brackish water of the Burairat well field. The age of these waters is more than 10,000 years, possibly several 10,000 years, but admixture of recent recharge has to be assumed from the ^3H values.

The ^3H values from the Khat area are generally low (0.8–2.6 TU), the groundwater appears to contain only minor components of recent recharge. ^3H values from Tawiyen boreholes south of Khat (6.7–7.2 TU) indicate significant recent recharge. From a corresponding ^{14}C value of 54 pmc, mixtures between old and recently recharged groundwater may be assumed.

The groundwaters of Jebel Akhdar are considered modern although the low ^3H values indicate retention periods of at least decades in extensive catchment areas.

8.4.2 *Stable Isotopes of Oxygen and Hydrogen*

8.4.2.1 Stable Isotopes in Precipitation and Runoff

Meteoric water lines for rainfall at Bahrain are stated by different authors as

$$\delta^2H = 6.3 \times \delta^{18}O + 11.4‰ \quad \text{and} \quad \delta^2H = 4.9 \times \delta^{18}O + 9.7‰.$$

Similar meteoric water lines have been deduced from data of stable isotopes in rainfall samples in Oman and the United Arab Emirates with gradients between 4.26 and 5.1 and d values of +8 to +9.7‰. For samples from rainfall events with more than 20 mm, the meteoric water line of northern Oman is

$$\delta^2H = 7.5 \times \delta^{18}O + 16.1‰.$$

In the United Arab Emirates, a trend has been observed that data of winter precipitation scatter around the Mediterranean Meteoric Water Line (d = +20‰), while stable isotopes of summer rains are more enriched with values plotting close to the Global Meteoric Water Line (d = +10‰).

The $\delta^{18}O/\delta^2H$ relationship of rainfall in northern Oman is non-linear: samples with $\delta^{18}O$ values between −3 and −5‰ scatter around a meteoric water line with a slope of around 8, while samples with more enriched $\delta^{18}O$ between −3 and +8‰ tend to follow an evaporation line with a slope of around 5. The slope of around 5 found in sample sets of precipitation of the Oman mountains as well as in Bahrain is explained by an impact of evaporation during rainout, i.e. the rain drops are affected by evaporation during the fall of the droplets.

The variation of stable isotope values may be related to the seasonally varying origin of precipitation from two sources: the Mediterranean Sea and the Indian Ocean.

Samples of runoff in northern Oman scatter mainly around the more depleted part of the meteoric water line

$$\delta^2\mathrm{H} = 7.5 \times \delta^{18}\mathrm{O} + 16.1‰.$$

8.4.2.2 Stable Isotopes in Groundwater

Groundwater samples from the Hajar aquifer of the Musandam peninsula show a rather homogeneous cluster of $\delta^{18}O/\delta^2H$ values scattering close to the MMWL. $\delta^{18}O$ values of most samples are in a range between −3.3 and −3.8‰, the majority of d values is between +17 and +20‰. More negative $\delta^{18}O$ values reaching −4.26‰ and higher d values of up to +23‰ are found in some samples of Wadi Bih (Fig. 8.8).

The high deuterium excess in the groundwater may be attributed to recharge from moisture originating from nearby seas (Arabian Gulf, Gulf of Oman), and the scatter close to the Mediterranean Meteoric Water Line suggests that the recharge took place without evaporation of water drops.

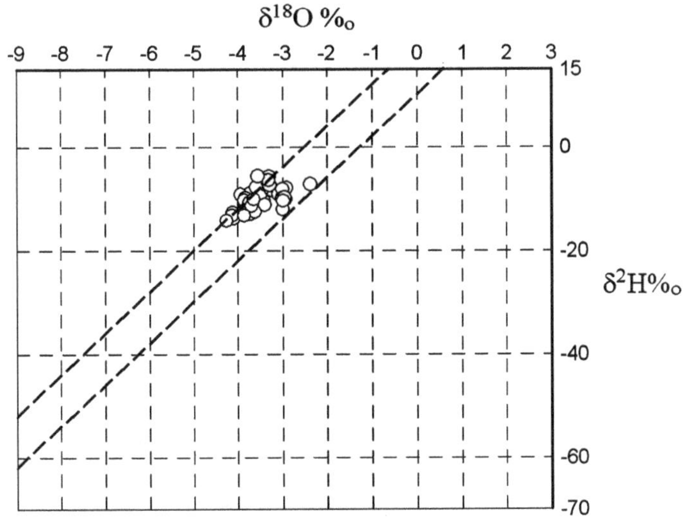

Fig. 8.8 $\delta^{18}O/\delta^2H$ diagram: Groundwater samples from the Hajar aquifer of the Musandam Peninsula in the United Arab Emirates. Data from Akiti et al. (1992), Gonfiantini and Akiti (1985), Kulaib (1991), Rizk et al. (2007), Wagner (1996b)

The distribution of stable isotope values may, in general, reflect the morphologic and recharge conditions. The generally depleted isotope composition of the groundwaters in the Musandam peninsula suggest that recharge occurs mainly at higher altitudes. For Wadi Bih, an average elevation of recharge zones of 1,450 m asl has been calculated, recharge probably occurs prevailingly in the higher ranges of the 1,050–2,090 m high mountain zones surrounding Wadi Bih with rapid recharge and infiltration in wadis steeply incised in the Hajar limestone massif.

The $\delta^{18}O/\delta^{2}H$ values of groundwater in the Khat area in the southwest of the Musandam peninsula are less negative than the values of groundwater in the main wadis of the Musandam mountain area. $\delta^{18}O$ values of groundwater in the Khat area vary from −2.9 to −3.1‰, d values from +12.1 to +15.8‰. The groundwater in the Khat area, discharging in thermal springs, appears to be related to a confined aquifer with recharge at relatively distant parts of the catchment at moderately high altitudes.

The $\delta^{18}O/\delta^{2}H$ values of the brackish groundwater of the Hajar aquifer of the Musandam peninsula indicate, that sea water is not the source of elevated salinity. "Only brine with salinity significantly greater than sea water could provide the observed solute distribution without perturbing the groundwater isotope signature. It appears there is a small water flux but large solute flux associated with the source of solutes. Most of the solutes appear to be derived from regional brine". "The solutes and the water are largely from different sources" (Rizk et al. 2007).

$\delta^{18}O$ values of most samples from the ophiolite mountains in the United Arab Emirates form a rather homogeneous cluster between −1.9 and −2.5‰ with d values scattering near the GMWL and varying from +9 to +14.5‰ (Fig. 8.9).

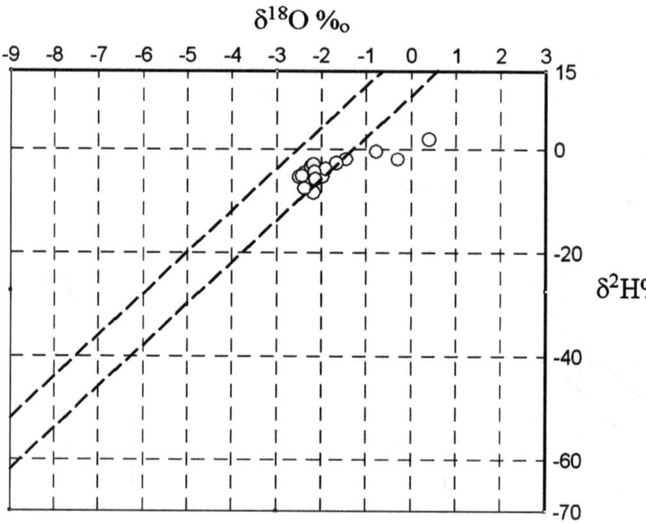

Fig. 8.9 $\delta^{18}O/\delta^{2}H$ diagram: Groundwater samples from the ophiolite mountains in the United Arab Emirates. Data from files of Ministry of Agriculture and Fisheries and Ministry of Agriculture and Water Resources, Dubai

The $\delta^{18}O/\delta^{2}H$ values are significantly less depleted than values in the Hajar carbonate aquifer of the Musandam peninsula. The data reflect the recharge conditions in the ophiolite mountains at moderately high altitudes and in a not very pronounced topography, where evaporation effects are more intensive than in the high mountain areas. $\delta^{18}O/\delta^{2}H$ data at some locations of the ophiolite mountains are apparently influenced by local evaporation with $\delta^{18}O$ values enriched up to +0.4‰ and low d values between +5.8 and −0.5‰.

Groundwater with relatively depleted $\delta^{18}O$ and $\delta^{2}H$ values occurs in the Hajar carbonate aquifer in Jebel Hajar al Gharbi in a west–east belt across the northern catchments of Jebel Akhdar, Jebel Nakhl and the Saiq plateau with $\delta^{18}O$ values between −2 and −4.5‰ and d values around +14‰. The distribution of data shows a clear altitude effect: the most depleted $\delta^{18}O$ values of <−3‰ are restricted to the eastern part of Jebel Akhdar and to the upper catchments of the northward flowing wadis. $\delta^{18}O/\delta^{2}H$ values of groundwater samples from Jebel Hajar al Gharbi scatter around a meteoric water line

$\delta^{2}H = 5.1 \times \delta^{18}O + 3‰$ (Akhdar water line, Macumber et al. 1997),

indicating a consistent evaporation effect. The evaporative effect appears, however, less pronounced in the data set with $\delta^{18}O$ values of <−3‰, which with a d value around +14‰ resembles the composition of groundwater in parts of the Oman mountains in the United Arab Emirates; d values are, however, on average lower than d values of the Hajar aquifer of the Musandam peninsula.

Groundwater with depleted isotope values originating from high altitude recharge is also found in some springs and falaj msystems emerging at the mountain foot in Wadi Semail, but is not observed in springs along the southern flanks of Jebel Akhdar.

A plume of isotopically depleted (higher altitude) groundwater ($\delta^{2}H$ >10: $\delta^{18}O$ >3) emerges from the eastern parts of Jebel Akhdar and passes, through a gap in the ophiolitic "barrier", northwards across the lower catchment of the Wadi Al Maawil towards the coast near Barka. Apart from this stable isotope content, the presence of the plume is reflected in relatively high Ca/Mg ratios. The data indicate a pathway of groundwater from the limestone catchments of Jebel Akhdar onto the coastal plain.

On the southern flanks of the high mountain area of Jebel Akhdar toward interior Oman, stable isotope values of oxygen and hydrogen are less depleted with $\delta^{18}O$ values between −2.5 and −0.8‰ and show a significant evaporation effect.

In the ophiolite aquifer of the southern foothills of Jebel Akhdar, situated at 440–596 m asl, $\delta^{18}O$ values range from −0.6 to −1.4‰, $\delta^{18}O/\delta^{2}H$ values scatter around or below the GMWL with d values between +1.9 and +10.2‰. The stable isotope values, which are more enriched in comparison to values in the more northern ophiolite areas in the United Arab Emirates, indicate that recharge may be influenced by southern moisture sources.

$\delta^{18}O/\delta^{2}H$ values of alkaline springs in the Oman mountains tend to deviate from the ophiolite fresh water toward more enriched $\delta^{18}O$ values reaching up to +0.43‰ and relatively depleted $\delta^{2}H$ values. d values can be as low as −8.5‰ (Fig. 8.10). The deviations are probably caused by isotopic exchange between groundwater and

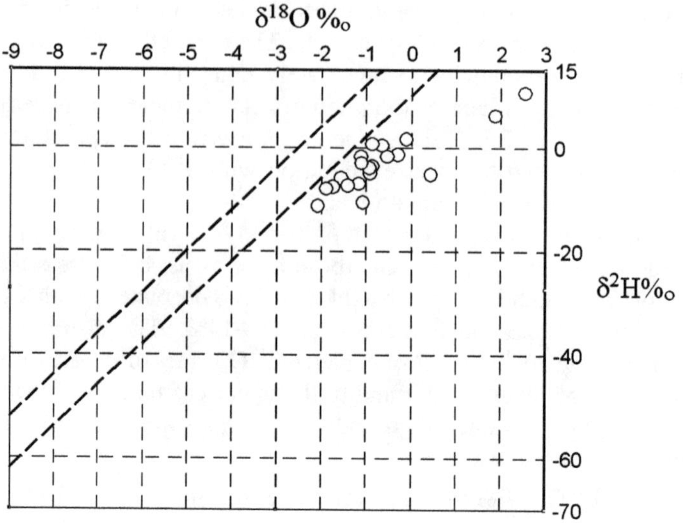

Fig. 8.10 $\delta^{18}O/\delta^2H$ diagram: Water samples from alkaline springs in the ophiolite mountains of Oman. Data from Neal and Stanger (without year)

hydrous phases of the serpentinite sequence. Ionic exchange with hydrous phases of the serpentinites, which have a δ^2H value of around −65‰ and $\delta^{18}O$ values of +7.84‰, leads to very negative δ^2H values in the groundwater until −11.2‰ and, accordingly, to low or even negative d values (Fig. 8.10).

References. Akiti et al. (1992), Al-Sayari and Zötl (1978), Gonfiantini (1992), Gonfiantini and Akiti (1985), Kulaib (1991), Macumber et al. (1997), Matter et al. (2006), Neal and Stanger (1984), Rizk et al. (1997, 2007), Wagner (1996b), Wagner and Geyh (1999).

Chapter 9
Aquifers in Coastal Plains of the Arabian Plate

9.1 General Features of the Coastal Plains

9.1.1 *Distribution of Coastal Aquifers*

The western, southern and eastern margins of the Arabian Plate are delimited by sea coasts. Along considerable stretches of the coasts, productive aquifers are found, which are composed in particular of

- Unconsolidated fluvial, littoral or marine sediments of prevailingly Quaternary age
- Carbonate rocks of various ages (Paleozoic to Cretaceous, Upper Cretaceous, Tertiary)

Many of the coastal aquifers extend over relatively narrow plains, which comprise important agricultural areas. Plains with irrigated agriculture adjoin, in particular, the coast of:

- The Mediterranean Sea in Syria, Lebanon and the Gaza Strip
- The Red Sea in Saudi Arabia and Yemen
- The Gulf of Oman in Oman and the United Arab Emirates

The total area irrigated through groundwater extraction in coastal plains along the margins of the Arabian Plate was estimated at 170,000 ha (Al-Rifai 1993).

Intensive exploitation of groundwater from coastal aquifers, in particular for irrigation, has created wide-spread problems of declining groundwater levels, depletion of aquifers and salt water intrusion. The abstraction of groundwater in many coastal plains is several times higher than the present recharge and a large part of the groundwater resources in coastal plains is heavily overexploited.

9.1.2 *Groundwater Regimes*

Depending on the sub-regional or local hydrogeologic conditions, different types of groundwater regimes in the coastal aquifers are found in different areas along the sea coasts of the Arabian Plate:

W. Wagner, *Groundwater in the Arab Middle East*,
DOI 10.1007/978-3-642-19351-4_9, © Springer-Verlag Berlin Heidelberg 2011

- Sand and gravel aquifers of prevailingly Quaternary age extend over various strips along the Mediterranean Sea, Red Sea, the Gulf of Aden and Arabian Sea, and the Gulf of Oman. The groundwater regime in these coastal sand and gravel aquifers is determined mainly by:
  - The pattern of deposition of Pleistocene–Holocene fluvial and alluvial sediments
  - Inflow of surface water through streams and subsurface inflow through aquifers
  - Groundwater discharge in springs and sabkhas
  - Hydraulic interactions between sea water and groundwater.
- The groundwater regimes in coastal carbonate aquifers of the Lebanon and Ansariye mountains are integrated parts of the regime of the corresponding extensive karst aquifers in the mountain areas. The coastal areas are, in some parts, major discharge zones, such as at Nahr Sinn and Nahr Banias in Syria and at the submarine discharge area of Chekka in Lebanon. Similar conditions may exist in some parts of the Hajar limestone aquifer on the eastern coast of the Musandam peninsula.
- Wide parts of the coastal areas along the Gulf in Saudi Arabia and the United Arab Emirates are situated downstream of the main sabkha discharge zone of the multi-aquifer system of the eastern Arabian platform. Main components of the groundwater regime in the coastal flats are assumed to be very slow movement of brackish and saline water, infiltration of tidal sea water inundations, vertical salt water movement under sabkhas, limitation of groundwater circulation on barriers of stagnant brines.
- Special groundwater flow conditions are observed in the Gulf coastal area, in particular in Qatar and Bahrain: Brackish groundwater moves above saline water directly to the coast and local lenses of fresh to slightly brackish water are sustained by rainfall infiltration on karstic surfaces.

The distribution of fresh water aquifers in coastal sand and gravel deposits is, in general, inhomogeneous. Groundwater recharge and occurrence of fresh groundwater is mainly concentrated to the inflow zones of wadis from the adjoining highlands. The aquifers containing fresh water are composed, in most plains, of a number of semi-independent "groundwater flow domains". The groundwater flow domains are more or less fan shaped and include wadis as recharge zones, one or more corresponding discharge zones and a connecting aquifer zone. Groundwater discharge occurs in the topographically lower parts of the plains by evaporation in coastal salt flats or by submarine groundwater outflow. Outside the groundwater flow domains, groundwater can be nearly stagnant (van der Gun et al. 1992).

9.1.3 Hydrochemical Aspects

9.1.3.1 Fresh Water/Salt Water Equilibrium

In many coastal aquifers, fresh water is adjoined by salt water on a rather steep interface. A typical salinity distribution in coastal sand and gravel aquifers consists of shallow fresh water lenses in the groundwater flow domains around main wadi courses, brackish water in the areas between the groundwater flow domains, and a wedge of saline water extending from the sea coast under the fresh and brackish water aquifer.

Exploitation of fresh water resources in the coastal aquifers is, in particular in the semi-arid and arid zones, frequently accompanied by an increase in water salinity mainly through sea water intrusion. Apart from direct sea water intrusion, various sources can contribute to an increase of groundwater salinity in exploited coastal aquifers, such as ascending brackish or saline groundwater, irrigation return flow, evaporation in sabkhas or leaching of salts from evaporites.

In coastal aquifers, the hydraulic gradient is generally directed toward the sea coast. If sea water is present in an aquiferous formation on the coast, a contact zone is formed between the lighter fresh water flowing to the sea and the heavier underlying sea water. Hydrochemical dispersion causes a transition zone along the contact between sea water and fresh water with varying salinity and density. The width of the contact zone can be small in comparison to the thickness of the aquifer. In such cases, a stationary interface develops and the fresh groundwater flows to the sea above the interface.

By abstracting groundwater from a coastal aquifer in excess of replenishment, the piezometric head in the fresh water body becomes lower than the head in the adjacent sea water wedge, and the interface starts to advance landward until a new equilibrium is reached.

9.1.3.2 Sea Water Intrusion and Ion Exchange

The chemical composition of sea water changes, as it intrudes a fresh water aquifer, through mixing and chemical reactions. These changes are particularly evident within the sea water front of an initial intrusion that mixes with the fresh water. Subsequent intrusions cause only minor additional changes of the water composition.

In the transition zone of fresh water and sea water, chloride concentrations are in a range between the background concentration in groundwater and the sea water concentration. The front of this transition zone is characterized by ion exchange: In clay minerals, especially montmorillonite, free surfaces with negative charges are occupied by cations. In a typical fresh water aquifer these mineral surfaces are saturated mainly with calcium ions, whereas in a typical salt water aquifer, the surfaces are occupied mainly by sodium ions. During sea water intrusion into a fresh water aquifer, sodium from the aqueous solution is fixed and calcium is released from mineral surfaces. Magnesium and potassium may also be exchanged for calcium but the Na–Ca exchange is the most significant ion exchange process. This exchange is assumed to be instantaneous (Kafri and Arad 1979).

During intrusion of fresh water into a salt water aquifer, Na is released from mineral surfaces while Ca from the aqueous solution is fixed on the surfaces. Ion exchange processes from recent sea water intrusion are generally indicated in the hydrochemical composition of the affected groundwater. In saline or brackish

Table 9.1 Changes in ionic ratios in groundwater affected by salt water/fresh water fronts (after Richter et al. 1993)

Ionic ratios	Recent salt water intrusion		Fresh water intrusion
	Advancing front	Behind advancing front	
Ca/Na	Increase	No change	Decrease
(Ca + Mg)/Na	Constant	No change	Constant
Na/Cl	Decrease	No change	Increase

water related to fossil sea water intrusion, an impact of ion exchange processes may not recognizable.

Table 9.1 summarizes major changes on ionic ratios as a result of these changes in water facies.

Behind the ion exchange front, simple dilution controls the composition of the brackish water and the chloride concentration is directly proportional to mixing rates. If sea water intrudes a fresh water aquifer, Na/Cl ratios decrease from ratios, which may be greater than 1.0, to ratios, which often are lower than the value in sea water (<0.86). If fresh water replaces marine water or washes out marine sediments, very high Na/Cl ratios can result.

9.1.3.3 Carbonate Dissolution and Sulfate Reduction

Sea water intrusion can be accompanied by processes of carbonate dissolution and of sulfate reduction: An undersaturation with respect to calcium carbonate can occur in mixtures of fresh water and sea water, and the mixed water can dissolve carbonates causing an increase of HCO_3 and Ca concentrations.

Sea water has a relatively high SO_4 concentration. If aquifers with reducing conditions are intruded by sea water, the SO_4 concentration of the intruding sea water may decrease through redox reactions with organic matter. The resulting changes of pH and CO_2 contents can lead to an additional carbonate dissolution.

References. Al-Rifai (1993), Appelo and Geirnaert (1991), Custodio (1987), Freeze and Cherry (1979), Hanshaw et al. (1971), Kafri and Arad (1979), Richter et al. (1993), van der Gun et al. (1992).

9.2 Mediterranean Sea Coast

9.2.1 Coast Along the Ansariye and Lebanon Mountains

On the western slopes of the Ansariye and Lebanon mountains, outcrops of Mesozoic formations reach, over long stretches, directly to the Mediterranean Sea coast. Quaternary sand and gravel plains of several kilometres width accompany the coast along the northern part of the Ansariye mountains between Nahr el Kebir esh Shimali and Banias, on the Akar plain between the Ansariye and Lebanon mountains, and between Saida and Sur on the southwestern slope of the Lebanon mountains (locations on Fig. 2.4). The main groundwater resources of these plains are contained generally in Mesozoic carbonate aquifers, which are replenished from recharge in the adjoining mountains.

In the Nahr el Kabir esh Shimali–Lataqiye area, Paleogene–Miocene–Quaternary sediments provide a shallow aquifer. In the Akar plain, Neogene–Quaternary sediments constitute the upper section of a multi-aquifer system.

In the coastal plain of Sarafand, situated on the Mediterranean Sea coast between Saida and Tyr in southwestern Lebanon, main groundwater resources are found in the Cenomanian–Turonian carbonate aquifer underlying Quaternary detrital sediments. Fresh groundwater with a salinity of 500–700 mg/l TDS, which is recharged mainly in the adjoining mountain areas, is exploited from the Upper Cretaceous aquifer at 100–300 m depth below the plain surface.

The groundwater quality in the plain appears not significantly affected by sea water intrusion, but the impact of intensive irrigation is recognized in increasing salinity values in the shallow aquifer as well as in the hydraulically connected deeper carbonate aquifer.

9.2.2 Gaza Strip

The Gaza Strip extends over an about 42 km long and, in west–east direction, 6–13 km wide coastal area between the Mediterranean Sea and the Negev desert. Topographic elevations ascend from the sea shore to about 200 m asl in the east.

Mean annual precipitation ranges from 450 mm in the north to 200 mm in the south. The strip is crossed in east–west direction by two wadis with low and irregular flow during winter: Wadi Gaza and Wadi Salqa (Fig. 9.1).

The Gaza plain is underlain by an 80–180 m thick sequence of Pleistocene marine sands and sandstones with intercalated clay and silt layers. The Pleistocene deposits rest on the Neogene Saqiye formation, composed mainly of shale and clay. A 1–2 km wide belt along the Mediterranean Sea coast is covered by 20–40 m high sand dunes.

The Pleistocene sands and sandstones constitute the main aquifer of the Gaza Strip. Along the coast, the 90 m thick aquifer is divided by clay and silt layers into a shallow water table sub-aquifer and several confined sub-aquifers. Inland, the clay lenses wedge out toward east and the Pleistocene sediments form a single aquifer. Toward south and east, the saturated thickness of the Pleistocene aquifer decreases to about 30 m. The shales and clays of the Saqiye formation constitute an aquiclude at the base of the Pleistocene aquifer.

In the east, the Pleistocene is underlain by aquiferous Eocene chalks and limestones.

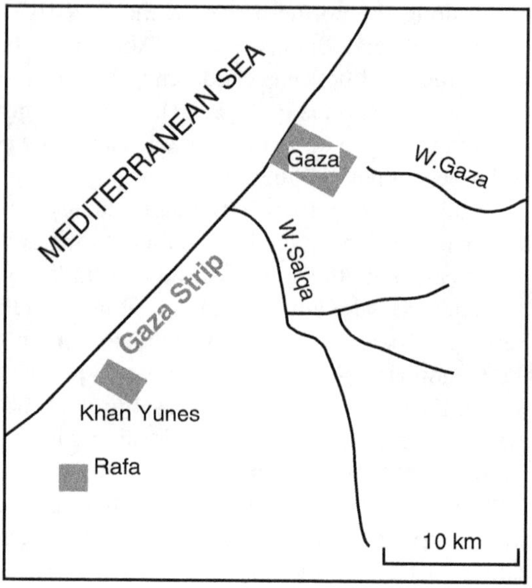

Fig. 9.1 Gaza coastal area, location map

The Gaza coastal aquifer is replenished through subsurface inflow from the east, infiltration of sporadic runoff in wadis Gaza and Salqa, and from rainfall infiltration in the sand dunes near the coast.

The lateral inflow into the Gaza Strip is estimated at 10–15 $\times$ 10^6 m^3/a. Recharge from Wadi Gaza amounts to around 1.5–2 $\times$ 10^6 m3/a, total fresh water recharge to approximately 52–62 $\times$ 10^6 m^3/a.

Pumping has significantly altered the natural flow pattern. The groundwater balance in the Gaza Strip is strongly negative: groundwater extraction from wells of 70 $\times$ 10^6 to 100 $\times$ 10^6 m^3/a exceeds the sustainable yield by 15 $\times$ 10^6 to 45 $\times$ 10^6 m^3/a (data of 1970–1990). The hydrologic deficit resulted in lowering of groundwater levels in many areas with groundwater levels in depression cones at approximately 2 m below mean sea level. In these areas, groundwater flows towards the centre of the depression cones, inhibiting the flushing of salts towards the sea and creating zones of salt accumulation in the groundwater.

Groundwater salinity is elevated in many parts of the Gaza Strip. Groundwater with moderate salinity and Cl concentrations of less than 250 mg/l occurs on the northern border of the Gaza Strip and in the southwestern coastal plain. Direct recharge in sand dunes appears to be a source of relatively low groundwater salinity.

The main sources of poor groundwater quality, indicated by elevated Cl and NO_3 concentrations, are:

- Sea water which has intruded into the aquifer in areas of a few hundred metres to 2.5 km from the coast line
- Lateral inflow of brackish water (Cl 500–2,000 mg/l) from the west in the middle and southern areas of the Gaza Strip
- Contamination from domestic sewage and agricultural activities

- Saline water in deeper aquifers and in the Eocene aquifer adjoining the coastal plain in the southeast
- Deep brines at the base of the coastal aquifer (Cl 4,000–6,000 mg/l) ascending during groundwater exploitation

In some areas, groundwater salinity in the deeper sub-aquifers exceeds sea water salinity, reaching up to 40–60 g/kg TDS.

Most municipal wells in the Gaza Strip show nitrate levels in excess of 50 mg/l. The main sources of NO_3 are fertilizers and domestic sewage.

References. Al-Agha (1995), Al-Jamal et al. (1997), El-Nakhal and Himida (1980), Hashash and Aranyossy (1995), Melloul and Collin (1994), Mogheir et al. (2006), Qahman and Larabi (2006), Sunna (1995), Yakirevich et al. (1998).

9.3 Red Sea Coast

9.3.1 Main Geographic and Geologic Features

The Red Sea coast forms, together with the coast of the Gulf of Aqaba, the western boundary of the Arabian Shield over a length of around 2,200 km. The western margins of the highlands of the shield, which are covered by rocks of the crystalline basement and, in some areas in the south, by sedimentary rocks and volcanics, generally form steep escarpments descending to coastal plains or directly to the sea shore.

The crest zone of the highlands with the main surface water divide at around 100–180 km distance from the Red Sea is situated at approximately 1,000 m asl in the Hijaz highlands in the north and above 2,000 m in the Asir mountains in the south.

On the foot of the Hijaz highlands between Aqaba and Jedda, coastal plains occupy limited areas between rocky sections of the coast. South of Jedda, the Tihama forms a nearly continuous coastal plain over 1,200 km until the southern tip of the Arabian Peninsula at Bab el Mandeb.

The coastal plains generally ascend from the sea shore to the mountain foothills at around 100–200 m asl.

The Red Sea coastal plains occupy the eastern margin of the Red Sea graben structure. The graben, formed mainly during the Tertiary, is filled with fluvial, marine and littoral Cenozoic sediments with a total thickness of several thousand metres, which overlie the Precambrian basement. Subsidence of the Red Sea graben was initiated in the Oligocene and is continuing until present. The graben deposits comprise mainly shallow marine Tertiary sediments and prevailingly fluviatile and alluvial Quaternary sediments. The graben tectonics were accompanied by Tertiary dike intrusions and Pleistocene–Holocene basaltic volcanism.

The surface of the coastal plains is generally covered by Quaternary sediments: alluvial deposits, gravel fans, sabkha deposits and occasional sand dunes. About 90 wadis with catchments of small to moderately large size enter the coastal plain from the mountains in the east (locations of major wadis on Fig. 7.3). The wadi courses cross part of the coastal plain and generally disappear in sabkha zones without reaching the coast.

The climate within the Red Sea coastal zone is arid with mean annual rainfall of 100 to less than 50 mm and main precipitation during the winter months. Rainfall in many of the catchment areas of wadis entering the coastal plains is, however, considerably higher reaching 300–400 mm/a on the escarpment of the Tihama and 400–700 mm/a on the main water divide of the Asir mountains of southwestern Saudi Arabia and Yemen. In the Asir mountains, seasonal rain periods are distributed to three seasons: December–January (Mediterranean winter rains), April, and August–September.

Surface water flow at the entrance into the coastal plains is restricted, in most wadis, to a few days after storm events in the catchment areas.

Main agricultural areas are found in the vicinity of the wadi courses, as far as water resources for irrigation are available. The areas between the wadis are used mainly as nomadic pasture land. The salt encrusted sabkha zone is generally barren of vegetation.

The total average runoff in wadis draining toward the Red Sea coast has been estimated at 1.3×10^9 m^3/a for the Saudi Arabian part of the coast and 565×10^6 m^3/a for the Tihama in Yemen.

9.3.2 Coastal Plains Between Aqaba and Jedda

The coastal area along the northern part of the Red Sea, including the Gulf of Aqaba, comprises some plains of relatively small extent.

The Ifal depression on the Gulf of Aqaba at the northwestern foot of the Midyan highlands contains accumulations of Quaternary sediments of up to 90 m thickness and 2,000 m thick Tertiary sediments. Groundwater contained in these sediments is generally brackish; in particular dissolution of anhydrite layers in the Tertiary deposits contributes to the elevated groundwater salinity.

In the up to 40 km wide coastal plain at the Wadi al Hamdh delta between Al Wajh and Umm Lajj, mainly brackish groundwater is found in Quaternary deposits, except for the area around Umm Lajj in the south of the plain, where a barrier of basement rocks protects a fresh water aquifer from sea water intrusion (locations on Fig. 7.2).

The coastal plain around Jedda comprises the gravel fans of three major wadi systems: Wadi Khulais, Wadi Usfan and the Wadi Fatima–Wadi Naaman system, in addition to detrital deposits of several small wadis (locations on Fig. 7.3). The main wadis constitute a system of parallel drainage networks within the southern Hijaz highlands with a general drainage trend from east to west toward the Red Sea.

The catchment areas of the individual main wadi systems extend over 2,000–5,330 km^2 and are covered by Precambrian basement rocks and by Tertiary volcanics of the Harrat Rahat plateau.

Tertiary sediments of up to 150 m thickness in the coastal plain comprise mainly sandy clay, fine sand and gravel and are, to some part, semi-consolidated. Deposition of the Pliocene–Pleistocene sediments occurred simultaneously with the sinking of the Red Sea graben.

Quaternary sediments have an average thickness of 20 m; stream channels can contain Quaternary deposits of several tens of metres thickness. The coastal plain of Jedda comprises a lower strip along the coast at 2–5 m asl, which is covered by Quaternary detrital deposits, and a higher eastern part rising to around 20 m asl with outcrops of Tertiary and Precambrian rocks. The Quaternary deposits rest on consolidated Miocene rocks.

Around Wadi Usfan, the top of the Precambrian basement is situated in a tectonic depression at more than 140 m below sea level.

The Khulais plain, 110 km northeast of Jedda, extends over a tectonic depression within the foothill area of the Hijaz highlands. The stream systems of Wadi Murwani, Wadi Abu Hanifa, Wadi Ghuran drain into the plain.

In the Khulais depression, a sequence of Cretaceous–Quaternary sediments overlies the Precambrian basement:

- 250 m thick Cretaceous detrital deposits: conglomerate, breccia, sandstone, marl, shale
- Eocene–Oligocene sandstone and siltstones of 20–200 m thickness with iron layers
- Up to 60 m thick basalt flows
- Up to 30 m thick Quaternary gravels, sand, silt and clay

Unconfined groundwater is found in the alluvium of the wadi courses at 5–20 m below surface. The Cretaceous–Tertiary sediments contain partly confined groundwater with piezometric levels at 20–40 m below land surface.

9.3.3 Tihama Plain in Saudi Arabia and Yemen

9.3.3.1 Geographic–Geologic Setup

The Tihama extends between Jedda and the southwestern tip of the Arabian Peninsula at Bab el Mandeb over a 1,200 km long strip along the southern Red Sea coast in the southwest of the Arabian Peninsula. The Al Birk basalt plateau divides that coastal strip into a northern segment (Tihama ash Sham) and a southern segment, which comprises the Tihama Asir or Jizan plain in Saudi Arabia and the Tihama of Yemen.

The topography of the 30–40 km wide coastal plain of the Tihama ascends from the Red Sea to around 200 m asl at the margin of a foothill zone, which rises up to 600 m asl. In the east, a more than 1,000 m high escarpment demarcates the Asir mountain massif with peak elevations of more than 3,000 m asl. The steeply ascending mountains consist of crystalline rocks of the Arabian Shield and, in Yemen, also of Tertiary volcanics and Mesozoic sedimentary rocks (Sect. 7.1.1).

The Tihama coastal plain comprises different morphologic areas:

- The upper wadi courses with about 4 km wide 10–20 m deep channels and a 100–200 m wide wadi bed with varying zones of accumulation and erosion
- The middle and lower wadi courses, which widen about 10–15 km downstream of the mountain rim into deltaic accumulation plains with numerous stream arms
- A 2–4 km wide sabkha zone along the coast
- Extensive plains between the main wadi courses

In the coastal plain, Quaternary unconsolidated sediments overlie a thick sequence of Tertiary–Pleistocene deposits. The Tertiary graben sediments comprise early Miocene sandstones, marls, shales, and a middle Miocene lagoonal series of halite, anhydrite, clay and sand.

Near Jizan the Tertiary sediments reach a thickness of 3,900 m with a 1,200 m thick halite succession, overlain by marls, clays and evaporites of the Baydh formation.

The Red Sea graben in Yemen is filled with a total thickness of several thousand metres of fluvial, marine and coastal sediments.

Subsidence during the Quaternary resulted in accumulation of >150 m thick deposits near the coast. In deeper graben zones near the coast, the thickness reaches >250 m. Toward the mountain escarpment, the thickness of Quaternary decreases to around 30 m along the wadi courses and <20 m in the areas between the wadis.

The Quaternary of the Tihama comprises a succession of fluviatile sediments of prevailingly Pleistocene age, consisting of loam, sandy loam and gravel. Gravel horizons compose around 20–35% of the wadi sediments. The Pleistocene wadi courses coincide more or less with the present wadis.

At or near the surface, the Quaternary consists mainly of alluvial fan and alluvial plain sediments in the eastern – upper – part of the plain and of eolian deposits in the west. In the north of the Tihama of Yemen, salt marshes constitute an evaporitic zone near the coast as. Further west – under the Red Sea – the Quaternary is represented mainly by coral reefs, sandy limestones and evaporites.

Volcanic activities during the Pleistocene formed the basalt cover of the Al Birk plateau.

9.3.3.2 Main Wadis

Main wadis entering the Tihama plain south of Jedda are the wadis Fatima, Yiba, Hali, Jizan, Dhamad in Saudi Arabia, the wadis Mawr, Surdud, Siham, Rima and Zabid in Yemen (Fig. 7.3).

The size of individual catchments of main wadis entering the Tihama in Saudi Arabia is 1,280–4,500 km^2.

Surface runoff in the wadis is fed mainly by rainfall in the relatively humid higher parts of the catchments. Mean precipitation increases from 100–150 mm/a in the coastal plain to 300–400 mm/a on the mountain foothills and 400–700 mm/a in the highlands.

Table 9.2 Average annual streamflow in major wadis of the Tihama of Yemen (after Khouri 1991, ESCWA-UNEP 1996)

Wadi Mawr	180×10^6 m^3
Wadi Surdud	60×10^6 m^3
Wadi Siham	90×10^6 m^3
Wadi Rima	65×10^6 m^3
Wadi Zabid	120×10^6 m^3
Wadi Rasgan	30×10^6 m^3
Wadi Mauza	20×10^6 m^3
Total	565×10^6 m^3

Floods in wadis of the western escarpment reach the Tihama plain about 3–5 times per year. Mean runoff coefficients (% of annual rainfall) are 3.8 for Wadi Hali, 3.7 for Wadi Yiba, 13 for the upper catchment of Wadi Jizan, 11 for Wadi Dhamad. Mean runoff in Wadi Jizan is 80×10^6 m^3/a, and 59.7×10^6 m^3/a in Wadi Dhamad.

In the plain, the wadi courses generally divert into several channels on the recent gravel fans and stream flow diminishes through infiltration and evaporation. Only rarely surface runoff discharges into the sea.

The total annual streamflow of seven major wadis entering the Tihama plain in Yemen amounts to 565×10^6 m^3 (Table 9.2). The wadi runoff has a considerable base flow component in the lower courses, ranging from 20 to 40% of total flow. Streamflow is characterized by high flood volumes of short duration and relatively constant base flow with seasonal fluctuations.

9.3.3.3 Aquiferous Zones and Groundwater Regimes

Fresh water aquifers of the Tihama plain are concentrated in zones surrounding major wadis, which enter the plain from the adjacent highlands, the "groundwater flow domains" (Sect. 9.1.2).

Gravel fans with relatively high permeability extend from the mouth of larger wadis into the upper reaches of the gravel plain, and infiltration of surface runoff in the wadis provides the main source of groundwater recharge in the plain.

Because of the very limited amounts of precipitation and a very high evaporation rate in the plain, direct groundwater recharge from rain water within the coastal plain is generally negligible. The groundwater is replenished mainly from infiltration of wadi runoff in the upper sections of the coastal plain. After flood events, a rise of groundwater levels by 0.5–4 m is observed in aquifers near the wadi courses.

In the Jizan plain, the following distribution of hydraulic conductivity values has been observed:

- 1×10^{-3} to 8×10^{-4} m/s around the wadi channels in the upper and middle reaches of the plain
- 1×10^{-4} to 8×10^{-4} m/s at the margins of the wadi and delta areas
- 1×10^{-6} to 1×10^{-4} m/s in areas with low thickness of Quaternary deposits

Hydraulic conductivity decreases with an increasing proportion of fine grained sediments in the interfluvial areas and near the coast.

Transmissivities are 300–13,000 m^2/d in the wadi domains and 500–3,500 m^2/d on the margins of the coarse wadi sediments. In the Tihama of Yemen, transmissivities tend to range from 500 to 3,000 m^2/d.

In the Jizan plain, 30–50% of the flow of minor and intermediate floods infiltrate into the underground and 20–30% of the flow of flood events with high discharge. Mean annual recharge amounts to 30–35% of the surface runoff. A high percentage of the groundwater flow evaporates in the coastal sabkha zone.

Groundwater flow in a 20 km wide section of the Jizan plain, including the flow domains of Wadi Yiba, Wadi Hali and Wadi Jizan, has been estimated at $35 \times 10^6\ m^3/a$. For Wadi Dhamad, the estimated groundwater flow is $10 \times 10^6\ m^3/a$.

Surface runoff in six wadis crossing the southern Jizan plain is, on average, $251 \times 10^6\ m^3/a$, of which $44 \times 10^6\ m^3$ (1.4 l/s) or 17% contribute to the groundwater recharge. The recharge rates vary from 9 to 36% of runoff in different wadi domains.

Groundwater flows in east–west to northeast–southwest direction approximately parallel to the wadi channels. Hydraulic gradients are 0.5% in the upper parts of the plain and decrease to 0.07% in the coastal deltas.

In the coastal area around Wadi Surdud in Yemen, the intensive use of groundwater for irrigation has negative impacts:

- Steadily declining groundwater levels, particularly in the eastern part of the Tihama
- Progressive sea water intrusion in the western part of the Tihama

Under intensive exploitation, the fresh water/salt water interface may progress annually by some tens of metres and groundwater levels may fall to depths below which groundwater pumping is no longer economic.

9.3.3.4 Groundwater Salinity and Hydrochemistry

Occurrence of fresh groundwater in the Tihama coastal plain of Saudi Arabia is restricted mainly to Quaternary gravel deposits in channel like systems corresponding to the present drainage pattern and to former wadi systems. Groundwater in the underlying Tertiary sediments (Baydh formation) is generally saline. Quaternary sediments in the areas between the major groundwater flow domains contain brackish to saline waters.

Water of flood flow in the wadis has, in general, an electrical conductivity of 300–400 μS/cm; electrical conductivity of baseflow varies between 500 and 900 μS/cm. Water from surface runoff in wadis is mainly Ca–HCO_3 type water with SO_4 percentages of 25–35 meq%.

Groundwater with low electrical conductivity and with a hydrochemical composition similar to surface runoff is found near the wadis and in the foothill areas. Groundwater with electrical conductivity of <1,000 μS/cm generally forms narrow tongues along the major wadi courses in the upper and middle reaches of the coastal plain. With distance from the recharge areas near the foothills, groundwater salinity increases up to 4,800 mg/l TDS and the hydrochemical type of the groundwater changes from Ca–HCO_3 to Na–HCO_3 and Na–Cl.

In groundwater with low to moderate salinity, percentages of the major anions HCO_3, SO_4, Cl are in approximately equal ranges with a tendency of increasing Cl percentages with increasing salinity.

Fresh groundwater in channel systems of the Jizan plain overlies the generally brackish to saline groundwater. Six major fresh water bearing drainage channel systems have been explored, which correspond to ancient east–west to northeast–southwest directed wadi courses. Migrations of the Quaternary drainage pattern have been caused by modification of the land surface through tectonic movements and basaltic volcanism. The thickness of the main fresh water bearing channels ranges from 70 m to more than 100 m.

Tongues of brackish groundwater with electrical conductivity values of <5,000 μS/cm extend along the major wadi courses to a distance of 2–3 km from the sea shore.

In the Tihama of Yemen, groundwater salinity generally increases in direction of groundwater flow from east to west. Along the coast, shallow groundwater is generally brackish. In some zones with concentrated submarine groundwater outflow, fresh groundwater occurrences can reach near the coast to depths of 100–200 m.

Elevated groundwater salinity in the Tihama of Yemen is caused by irrigation return flow in the agricultural areas and by influences of saline water in the coastal sabkha zone, by sea water intrusion and from mixtures with groundwater from the saline Baydh aquifer. Ascending brines may be a major source of the high groundwater salinity in the Jizan coastal aquifer.

Saline waters of the sea and the sabkha zone are Na–Cl waters, the groundwater of the Baydh formation is Na–Cl water with, in comparison to sea water, relatively high Ca concentration. Total salinity of Red Sea water is 46 g/kg, of Baydh groundwater 23 g/kg; The Ca/Na ratio in groundwater of the Baydh formation is with 0.59 much higher than the ration of 0.04 in sea water.

References. Al-Sayari and Zötl (1978), GEOCON (1995), Hammad (1978), Hussein et al. (1993), Jado and Zötl (1984), Klinge (1982), van der Gun and Ahmed (1995), van der Gun et al. (1992), Van Overmeeren (1989).

9.4 Coast of the Gulf of Aden and the Arabian Sea

9.4.1 Geologic and Morphologic Structure

9.4.1.1 Morphology and Climate

The southern coast of the Arabian Peninsula is accompanied for about 1,400 km from Bab el Mandeb toward east by a belt of mountains and highlands, extending from the Asir mountains around Taiz with peak elevations above 3,000 m asl to the highlands of Hadramaut and southern Dhofar, where the higher mountain peaks reach altitudes between 1,000 and 2,000 m asl. The mountain landscape of the southern Yemen highlands descends to the Gulf of Aden in a step-wise broken escarpment, grading about 900 km east of Aden into the morphologic Al Ghayda depression. A nearly 300 km long belt of the Dhofar mountains extends parallel to the coast of the Arabian Sea, including the massifs of the Jebel al Qamar (peak 1,911 m asl), Jebel Qara (999 m asl), and Jebel Samhan (1,812 m asl), and passes toward northeast into the flat desert landscape of Jidda al Harasis.

The climate on the coastal plains along the Gulf of Aden–Arabian Sea is very dry, somewhat less arid conditions prevail in the catchment areas of some of the wadis of southern Hadramaut and in particular in the mountain zone around Taiz–Ibb.

In the Dhofar area in southern Oman, rainfall originates from two principal forms: infrequent storms of high intensity (every 2–3 years) and the annual monsoon which affects the Dhofar mountains and the Salala plain. Rainstorms originate from either low pressure systems moving across the Arabian Peninsula or from the less frequent but more intensive cyclonic systems moving westward from the Arabian Sea. The annual monsoon during the months of June to August brings mist and low clouds over the Salala plain and the Dhofar mountains. Precipitation takes the form of light rain and condensation from the mist on vegetation (occult precipitation). The monsoon precipitation is characterized by long duration but low intensity creating very limited runoff. The native vegetation intercepts a proportion of the mist and fog causing it to drip onto the ground.

Mean annual precipitation is 100 mm on the Salala plain and 100–150 mm on Jebel al Qara.

9.4.1.2 Main Geologic Features

During the Tertiary, the rift graben of the present Gulf of Aden opened between the Arabian sector of the Afro–Arabian Plate and northeastern Africa. Intensive block faulting caused the mountains of southern Yemen to break into numerous blocks, separated by faults running in ENE direction approximately parallel to the Gulf of Aden.

The highlands accompanying the southern coast of the Arabian Peninsula belong to the Arabian Shield in the west and the Hadramaut segment of the Arabian Platform in the east. Main geologic structures of the platform, which run in an acute angled trend along the coast of the Arabian Sea, are the southern Hadramaut arch, the Al Ghayda synclinal depression, and the uplift of the Dhofar mountains which may be viewed as an eastern continuation of the northern Hadramaut arch. In the east, the Arabian Sea coast is accompanied by a 4–20 km wide and about 250 km long structurally complex coastal belt, which is bounded in the north by the monoclinal limestone plateaus of Jebel al Qara and Najd.

The mountain and highland zone adjoining the Gulf of Aden–Arabian Sea coast include outcrops of Precambrian rocks and late Miocene–Pliocene volcanics

(Aden volcanics) and, on the Hadramaut–Dhofar uplift on the southern margin of the Arabian Platform, Cretaceous–Quaternary sedimentary rocks.

The plateaus of the southern Yemen highlands and the Dhofar mountains have a slight monoclinal ENE dip. The escarpment and the highlands of these southern plateau zone are covered prevailingly by Paleogene carbonate formations (Hadramaut group), with few exposures of the underlying Cretaceous Mukalla sandstones.

The escarpment reaches over long stretches to the shores of the Gulf of Aden and the Arabian Sea. Coastal plains of generally moderate size extend over a few stretches on the downstream end of larger wadi systems.

9.4.1.3 Main Wadis and Coastal Plains

Main wadis and coastal plains on the southern coast of the Arabian Peninsula comprise the Tuban delta, Abyan delta, Ahwar delta, Maifaa plain and the Salala plain (Fig. 9.2).

The Tuban delta is situated next to Aden on the foot of the 627 m high volcano Jebel Shamsan, 170 km east of Bab el Mandeb. The headwaters of Wadi Tuban are located in the Yemen highlands around Taiz at altitudes of more than 2,000 m asl. The catchment covers 7,150 km^2, average annual surface flow is 125×10^6 m^3.

The Abyan delta at Zinjibar, around 50 km east of Aden, is crossed by two major wadi channels: Wadi Bana and Wadi Hasan. Wadi Bana runs for 250 km from its headwaters in the Yemen highlands at altitudes of 2,000 m asl until the coast. Average annual surface flow in Wadi Bana, with a catchment of 12,000 km^2, is 170×10^6 m^3. Wadi Hasan rises in the south of the Yemen highlands about 80 km from the coast; average annual surface flow is 30×10^6 m^3.

The delta of Wadi Ahwar, about 200 km east of Aden, is relatively small in comparison to the Tuban and Abyan deltas. Wadi Ahwar rises in the Hadramaut highlands some 80 km north of the coast.

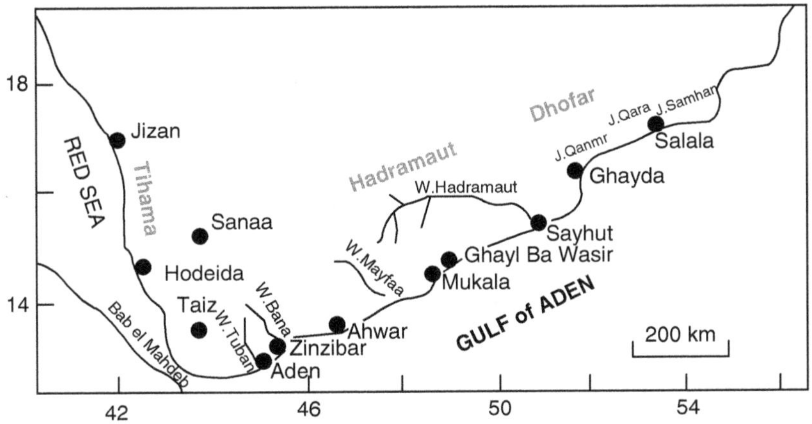

Fig. 9.2 Coastal areas in the south of the Arabian Peninsula, location map

Wadi Maifaa rises in the Hadramaut highlands and reaches the coast after an about 120 km long coarse 360 km east of Aden. Near the coast, the wadi opens into a vast plain.

Wadi Hadramaut with its headwaters at around 1,000 m asl in the Hadramaut highlands has a total length of around 360 km. The wadi runs in approximately west–east direction in a 400–450 m deep channel between the northern and southern Hadramaut highlands. In the east, Wadi Masila which constitutes the lower course of Wadi Hadramaut turns toward southeast and widens into a 20 km wide plain which slopes to the coast of the Arabian Sea. Average annual surface flow in Wadi Hadramaut is around 380×10^6 m^3.

9.4.2 Hydrogeologic Features

9.4.2.1 Aquiferous Zones

Fresh water occurs on the coastal plains of the Gulf of Aden and the Arabian Sea in gravel fans and delta areas of the major wadis. Fresh water flow domains are found, in particular, in the deltas of Wadi Tuban, Wadi Abyan and Wadi Ahwar, the Maifaa plain, and the Salala plain in Oman.

The thickness of the aquiferous Quaternary deposits of the Tuban delta increases from around 30 m near the mountain foothills to at least 200 m in the lower delta.

In the Abyan delta, a narrow northern zone of Quaternary deposits is separated from the main delta by outcropping or near-surface ridges of the basement. Near the coast, the up to 100 m thick unconsolidated Quaternary sediments are underlain by sandstones. Near the coast, the unconsolidated Quaternary is underlain by clay layers and a several hundred metres thick sandstone sequence.

Transmissivities of the Quaternary aquifer are probably in a range of 300–10,000 m^2/d; permeability of the underlying sandstones appears to be rather low.

The delta deposits of Wadi Ahwar, composed of Quaternary sands, gravels and loams, constitute a generally some tens of metres thick aquifer, which is underlain by up to 50 m thick Pliocene conglomerates. Sand dunes cover the eastern part of the plain.

In the rather small coastal plain of Ghayl Ba Wasir, fresh to brackish groundwater occurs in karstified Paleogene rocks, which include Middle Eocene clays and marls rocks and an Oligocene shale, limestone, evaporite sequence. The karstification appears to be related mainly to the dissolution of gypsum. The groundwater probably leaks from the underlying Cretaceous sandstones into the karstic aquifer along a fault zone, where groundwater flow in the sandstone meets a barrier of Lower Eocene marls. The thickness of the karst aquifer is about 20 m and transmissivities vary from 400 to 18,000 m^2/d. The groundwater is extracted for water supply of Mukalla.

At Salala a 12–40 km wide triangular plain extends between the Arabian Sea and Jebel el Qara over 240 km^2. The plain is underlain by the $>$1,000 m thick

Oligocene–Miocene Taqa formation, a carbonate sequence with chalky and porous marine limestones. The Taqa formation is covered by Quaternary alluvial deposits, calcarenites and eolian sands. The thickness of the alluvium reaches up to 40 m under main wadi channels, and a paleo-channel system has been incised up to about 130 m into the Tertiary deposits.

Major faults constitute the boundary between the Salala plain and the adjoining outcrops of Jurassic to early Tertiary limestones of Jebel el Qara.

Main aquiferous zones are found in the Salala plain in fissured and karstified sandy limestones of the upper part of the Taqa formation and in alluvial deposits of wadi channels. The aquifer base is formed by low permeability turbidite layers in the Taqa formation.

The wadis crossing the Salala plain are mostly dry except during short-term floods from convectional storms or cyclonic events.

9.4.2.2 Groundwater Recharge and Groundwater Flow

Groundwater in the coastal plains of the Gulf of Aden and the Arabian Sea is recharged mainly through infiltration of runoff in the wadis entering the plains from the adjoining highlands. To some extent, replenishment of shallow coastal aquifers may occur through subsurface inflow and upward leakage from underlying sandstones, such as in the plain of Ghayl Ba Wasir.

At Salala groundwater moves from the limestone aquifer of Jebel el Qara into the coastal plain toward the Arabian Sea coast. The Jebel el Qara aquifer receives significant recharge during the monsoon season. Most of the fresh groundwater is channelled from Jebel el Qara into the centre of Salala plain. The total subsurface inflow into the plain is in the order of 32×10^6 m^3/a. Some spring discharge occurs at the foot of the Jebel el Qara escarpment, forming small pools. The largest spring is Ain Razat with a discharge of around 200 l/s.

Main groundwater flow in the plain is directed along fault zones, incised wadis and paleo-channel systems in north–south direction toward discharge zones near the coast, in particular in "Khawrs", shallow open water surfaces.

9.4.2.3 Groundwater Salinity

In the delta areas on the southern coast of the Arabian Peninsula, fresh groundwater is restricted prevailingly to the main flow domains accompanying the upper reaches of wadis in the coastal plains. In the lateral fringes of the plains, groundwater is generally brackish.

Groundwater is brackish in part of the Tuban delta and most of the Abyan delta. Groundwater salinity in the Wadi Ahwar coastal plain is highly varying, reaching locally nearly 10,000 mg/l TDS.

In the Paleogene aquifer of the Ghayl Ba Wasir plain, elevated sulfate concentrations of 300–1,800 mg/l restrict the usability of the groundwater.

The Salala plain aquifer comprises a central zone with relatively fresh groundwater with EC values of <2,000 μS/cm and brackish water zones with EC values of >2,000 μS/cm, reaching up to 15,000 μS/cm in the west and east. At the northern boundary of the plain along the foot of Jebel el Qara, fresh water overlies brackish water. The greatest thickness of fresh groundwater is associated with karstic wadi channels. A tongue of fresh groundwater extends through the centre of the plain to the coastal margin. Outside the main fresh water tongue, the fresh water layer in the central aquifer zone is only a few metres thick.

Water in the open water bodies of the Khawrs is brackish to saline.

References. Clark et al. (1987), ESCWA-UNEP (1996), Farid (1993), GEOCON (1995), Hawkins et al. (1981), Jado and Zötl (1984), Khouri (1991), Milligan and Gharbi (1995), O'Boy (1995).

9.5 Batina Coastal Plain in the United Arab Emirates and Oman

9.5.1 Geographic and Geologic Set-Up

Along the eastern coast of the United Arab Emirates and of Oman, relatively narrow strips of coastal plains extend between the Oman mountains and the Gulf of Oman. On the northeastern margin of the ophiolite mountains, alluvial flats of limited extent occupy the lower ends of wadis between promontories of the ophiolite massif, mainly Wadi Dibba and Wadi Wuraya. South of Khorfakan, coastal flats and wadi fans form a 10–30 km wide almost continuous plain, Al Batina, which extends over a length of about 300 km until the area of Muscat. The altitude of the plain rises from sea level to somewhat less than 100 m asl on the mountain foothills in the west.

In many areas, the upper (western) part of the Batina plain is characterized by a morphology with stream channels, gravel fans, fluvial terraces and, frequently, low hills or ridges with outcrops of ophiolite rocks. Downstream near the coast, the terrain becomes flat with sabkhas and occasional low sand dunes.

The Batina plain is crossed by a number of wadis, which drain the eastern slopes of the Oman mountains toward the Gulf of Oman. Major wadis are Wadi Dibba, Wadi Wuraya and Wadi Ham in the United Arab Emirates, Wadi Ahin, Wadi Bani Ghafir, Wadi Al Fulayj, Wadi Semail in Oman (Figs. 9.3 and 8.2).

Average annual rainfall in the Batina plain and in the adjoining ophiolite mountains generally ranges from 120 to 165 mm, but may be as low as 60 mm in the coastal plains of Oman. Rainfall events are experienced primarily in February and March from depressions moving southeastward from the Mediterranean Sea. Summer rainfall is more sporadic and more localized in the United Arab Emirates; further south in Oman, precipitation from low pressure systems during summer becomes more important.

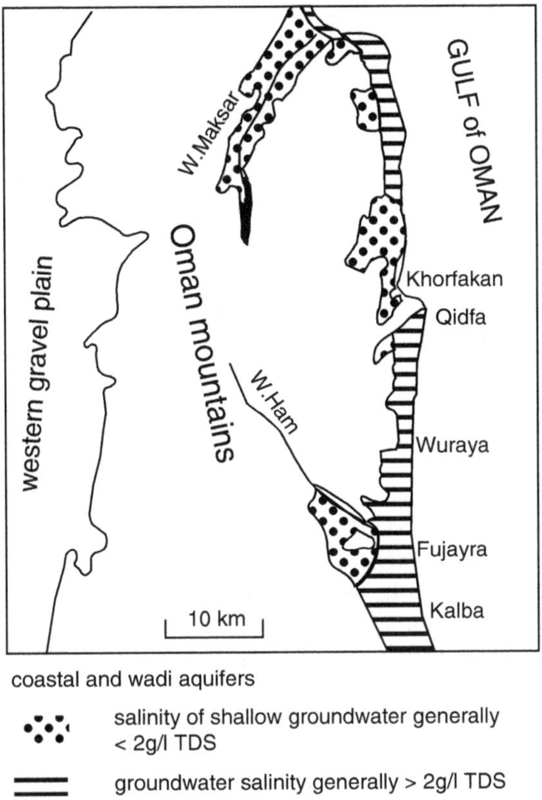

Fig. 9.3 Batina coastal plain in the United Arab Emirates, salt water intrusion and locations. After Wagner (1995a)

The coastal plains of Al Batina are covered by Quaternary unconsolidated to semi-consolidated sands and gravels with interbedded clay lenses or layers, which rest on clayey conglomerates or directly upon fissured ophiolites. The Quaternary cover has generally a wedge shape, thickening from a few metres close to the mountains to more than 100 m along the coast.

The Oman mountains west of the Batina coastal plain are composed mainly of basic igneous rocks of the Semail ophiolite complex and metamorphic rocks of the Hawasina nappe. Carbonate rocks of the Hajar group occur in the higher reaches of some wadi catchments in particular of Jebel Akhdar.

9.5.2 Aquiferous Zones and Groundwater Regimes

9.5.2.1 Aquifers

The main aquiferous zones in the Batina coastal plain are constituted by Quaternary sands and gravels. The Semail ophiolite complex, adjoining and underlying the Quaternary aquifer, provides a fissure type aquifer with generally low productivity.

The sand and gravel aquifers generally have a saturated thickness of less than 50 m, which may, however, increase to more than 100 m near the coast. A thick sequence of marls and clays underlies the coarser Quaternary sediments in some areas near the coast. On the northeastern coast of Oman, the thickness of the lower clayey Quaternary reaches up to 600 m.

Shallow alluvial aquifers extend, in particular, along active wadi channels and grade laterally into an irregular network of semi-confined aquifers in the interfluvial areas. Deeper fresh groundwater under confined conditions is found in some areas of the Batina in Oman.

The water table is situated at 10–15 m below ground surface near the coast and at greater depth further inland.

9.5.2.2 Groundwater Recharge

The Quaternary coastal aquifers are recharged primarily by sporadic flood flow infiltration. Main recharge is confined to the active wadi systems, which have a limited storage capacity and mean circulation times of a few years to decades. Recharge to the more extensive shallow interfluvial aquifers is more limited, mean circulation times are in the order of some tens to hundreds of years.

Main recharge in the northern part of the Batina is related to rainfall events in winter, while toward south in the Batina of Oman, recharge from summer rains become more dominant.

Gravel and sand aquifers in the plains receive replenishment from:

- Recharge from flood flow reaching the plains
- Direct recharge from rainfall over the plains
- Subsurface inflow from the mountain area

Recharge from local rainfall appears to constitute an important component of groundwater replenishment in the areas, where coarse gravel deposits provide favourable infiltration conditions. Mean recharge may reach 10–20% of local rainfall, around 19–30 mm/a.

Recharge rates of 2 mm/a are estimated for the coastal plains of northeastern Oman and of up to 54 mm/a for the catchment areas in the adjoining mountains.

Under irrigated conditions, rainfall infiltration added to a high moisture content in the unsaturated zone can produce a pulse which is rapidly transmitted to the water table by piston flow. Small infiltrations of 10 mm and less can activate the piston flow process. Periodic rainfall produces, however, a small flux component compared to the essential ongoing contribution of leaching. Consequently, leaching largely determines the chemical composition of the shallow groundwater, except near major wadis.

Under non irrigated conditions, drainage from the soil profile into the groundwater appears to be very small.

The central and southern plains of Al Batina in Oman mostly comprise gently undulating surfaces with extensive drainage networks, on which surface flow is periodically activated by rainfall on the plain. Infiltration within this morphologic setting is concentrated at locations along active drainage lines, where infiltration rates may be relatively high by ponding of runoff.

Deeper confined groundwater is recharged at higher elevations in the mountain or foothill zone, following buried channels and permeable routes through the clastic sediments. Mean circulation times of the deeper groundwater range up to several thousands of years.

9.5.2.3 Hydraulic Parameters

Hydraulic conductivities in the shallow Quaternary aquifer of the coastal plains of northeastern Oman are, on average, 2×10^{-4} m/s, transmissivities 1–500 m^2/d. Hydraulic conductivity in the underlying clayey Quaternary deposits is in the order of 10^{-5} m/s.

Transmissivity values of the coastal aquifers in the Batina plain in the United Arab Emirates have been reported as:

- 8.5–8,630 m^2/d for the Fujayra coastal plain, with <100 m^2/d to around 200 m^2/d in the Wadi Ham channel on the upstream part of the plain and up to >1,000 m^2/d in lower parts of the plain
- 14–13,770 m^2/d for the coastal plain of Wadi Dibba

9.5.2.4 Northeastern Coast of Oman

About 20 larger wadis enter the Batina coastal plain in northeastern Oman, the headwaters of which are situated at distances of 40–60 km from the coast in the Oman mountains (Fig. 8.2).

The upper reaches of the coastal area are generally occupied by a piedmont zone with stream terraces. The coastal plain between the piedmont zone and the sea shore includes alluvial deposits of the present wadis, coastal dunes and sabkhas. Tertiary deposits crop out in the piedmont zone as a series of hills and ridges.

Surface runoff in the wadi channels follows periods of heavy rainfall. Many streams originating in the mountains may flow for an extended period in downstream reaches within the piedmont zone. Generally, flood flows entering the plain from the mountains disperse into numerous tributaries. 15–40% of the volume of flood flow from of the mountains may discharge into the Gulf of Oman.

Wadi Bani Ghafir originates in Jabal Akhdar at an altitude of around 2,600 m asl and reaches the coast of the Gulf of Oman at Suwayq 100 km west of Muscat. The upper catchment is situated in outcrops of carbonate rocks of the Hajar group, in the middle catchment the wadi runs in a narrow stream valley through the Semail ophiolites. Downstream of the mountain area, the wadi opens into a broad gently sloping coastal plain with shallow braided channels. Near the mountain front, the coastal plain is divided into incised channels and ancient terraces.

The size of the catchment area at the entrance of the wadi into the coastal plain is 655 km^2. Mean annual runoff is 1.5×10^6 m^3, corresponding to 1.7% of mean annual precipitation of 139 mm. Previously, about 0.7×10^6 m^3 discharged on average into the sea. Construction of a recharge dam with a capacity of 0.78×10^6 m^3 reduced the volume of flood flow to coastal lagoon areas and into the sea.

Wadi Semail drains a large catchment of relatively low altitude hills between the mountain massifs of Jebel Hajar el Gharbi and Jebel Hajar el Sharqi and merges downstream into the Al Khawd fan in the coastal area of Seeb. The catchment of the wadi is covered mainly by ophiolites and extends, in the west and east, into outcrops of limestone and shale sequences of the Hajar group.

Before reaching the coastal plain, Wadi Semail flows through a narrow gorge in ophiolites, which form a barrier separating the large upstream catchment from the relatively small coastal plain formed by the Al Khawd fan.

Groundwater replenishment in the Al Khawd fan occurs from flood runoff originating in the upper catchment and probably by groundwater flow through the ophiolite barrier. The Al Khawd fan has been intruded by sea water, salt water extending at depth to about 6 km inland from the coast.

Wadi Al Fulayj with a total drainage area of over 750 km^2 reaches the Gulf of Oman south of Jebel Hajar al Sharqi about 150 km south of Muscat. The catchment area in the mountains, with peaks rising to an elevation of 2,327 m asl, are covered by sedimentary rocks of Tertiary age: marl, marly limestone, with calcarenitic, sandstone and siltstone intercalations, breccia, conglomerates and bioclastic limestones. Unconsolidated deposits along the wadi reach a thickness of up to 80 m.

The wadi widens into a coastal plain covered by unconsolidated coastal deposits of the active wadi channel, terrace deposits and isolated outcrops of marly rocks. The coastal deposits, composed of silt, gravel and sand with clay, constitute the principal aquifer of the plain. The thickness of the aquiferous unconsolidated sediments varies from a few metres to >30 m, with a maximum of 80 m along the main wadi course. Near the coast, coarser material of variable thickness between 10 and 40 m overlies thick marly clays.

9.5.2.5 Eastern Plains of the United Arab Emirates

The up to 4 km wide *Wadi Dibba*, with a length of 30 km in the northeast of the United Arab Emirates, runs along a major fault zone in the northeast of the United Arab Emirates, which separates the autochthonous Hajar mountains of the Musandam peninsula from the allochthonous ophiolite mountains.

The Quaternary wadi sediments of Wadi Dibba comprise:

- Up to 55 m thick coarse gravel and sand of the recent wadi bed
- Up to 30 m thick unconsolidated to semi-consolidated gravels of a lower terrace
- Up to 50 m thick conglomeratic gravels of an upper terrace

Relatively high transmissivities of up to around 14,000 m^2/d are found in well sorted coarse wadi bed gravels.

The wadi aquifer is replenished from infiltration of stream flow in Wadi Dibba and from subsurface inflow from the ophiolite aquifer of the highlands adjoining the wadi in the southeast.

Wadi Wuraya has a catchment of 129 km^2 mainly in the ophiolite mountains; average annual streamflow is 0.93×10^6 m^3. Along one stretch within the ophiolite mountains, characterized by small waterfalls, the wadi has a perennial base flow of around 60 l/s.

At the downstream end of Wadi Wuraya, a coastal plain extends over about 25 km^2. The coastal plain comprises a sand and gravel aquifer, the thickness of which may reach several tens of metres. Fresh water in the aquifer appears to be underlain by brackish groundwater in, at least, part of the plain.

Near the coast, the sand and gravel cover of the plain appears to be split into several around 500–1,000 m wide channels separated by outcrops or near-surface occurrence of ophiolite rocks.

Wadi Ham extends from the ophiolite mountains south of Masafi in the west to the Gulf of Oman in the east. In its downstream part, the wadi crosses the coastal plain around Fujayra. The catchment area at the entrance into the coastal plain is 190 km^2; mean annual flow is 0.82×10^6 m^3 corresponding to a runoff coefficient of 8%.

The sand and gravel aquifer of Wadi Ham extends over two sections:

- An about 2 km wide and 3.5 km long section directly downstream of the entrance of Wadi Ham into the coastal area (wadi section)
- The 4.5 km long plain section, widening to more than 8 km toward the coast

A recharge dam (Wadi Ham dam) is situated at the upstream limit of the wadi section.

In the wadi section, the saturated aquifer thickness is 10–40 m, groundwater salinity is relatively low, and the gradient of the groundwater surface is relatively steep. In the plain section, the saturated aquifer thickness is generally >50 m and up to >90 m near the coast, the hydraulic groundwater gradient is low; groundwater salinity increases toward the coast.

Groundwater throughflow through the wadi section of the aquifer has been estimated at 2.5×10^6 m^3/a. The throughflow is dominantly influenced by the infiltration of flood flow enhanced through the Wadi Ham recharge dam.

Direct recharge from rainfall in the coastal plain is estimated at 20% of mean annual rainfall or 20,000 m^3/a/km^2.

Average groundwater replenishment in the Fujayra plain is around 2.6×10^6 m^3/a, composed of:

Subsurface inflow	0.5×10^6 m^3/a
Direct recharge from local precipitation	0.5×10^6 m^3/a
Flood flow infiltration from Wadi Ham	1.6×10^6 m^3/a

9.5.3 Groundwater Salinity and Hydrochemistry

9.5.3.1 General Distribution of Groundwater Salinity

In the unconsolidated sand and gravel aquifers, recharge from wadi runoff (natural or dam recharge) can produce fresh water lenses along wadi courses and in parts of the coastal plains. Groundwater with somewhat higher salinity, surrounding or underlying the fresh water lenses, possibly originate from recharge outside the wadi courses, where infiltration of rainwater may produce recharge at low rates but over large areas.

In the unconsolidated aquifers of the coastal plain, groundwater salinity increases in direction towards the sea, and brackish groundwater prevails in the plain along the sea coast. Salt water underlies fresh water or brackish water in considerable parts of the coastal aquifers.

In many parts of the coastal plains, the groundwater regime and the distribution of groundwater salinity is significantly affected by artificial measures, in particular groundwater abstraction and, in some wadi courses, by runoff retention through wadi dams.

The intensive groundwater abstraction has caused a wide-spread increase of water salinity in irrigation and water supply wells.

9.5.3.2 Northeastern Coastal Plain of Oman

In the 90 km long strip of the coastal alluvial aquifer around Sohar on the northern end of the Batina of Oman, groundwaters with low to moderate salinity are found in alluvial aquifers associated with narrow wadi channels (EC $<1{,}000$ µS/cm). Within the coastal plain, EC values increase up to 16,000 µS/cm.

In the Al Khawd fan (Seeb area, 40 km west of Muscat, Fig. 8.2), the alluvial deposits contain fresh groundwater with EC values of less than 1,000 µS/cm to a depth of approximately 100 m below surface within and near the wadi channels. Groundwater from the interfluvial areas is more saline with EC values up to 2,000 µS/cm.

The fresh to brackish groundwater is underlain by saline groundwater; the fresh water/salt water boundary is situated at a depth of 150–200 m at a distance of 4.5 km from the coast.

In some areas, highly saline groundwater is found with EC values in the order of 70,000 µS/cm and Cl concentrations of up to 30,000 mg/l. The saline water possibly originates from a sabkha type brine. Below the saline groundwater, a deeper aquifer, which has an artesian head about 7 m asl, contains mainly brackish groundwater with EC values between 1,000 and 3,000 µS/cm.

9.5.3.3 Eastern Plains of the United Arab Emirates

Groundwater is exploited in the eastern coastal plans of the United Arab Emirates from various well fields for domestic supply and from many irrigation wells.

Fresh groundwater is extracted in well fields in the upper parts of the Fujayra and Khorfakan coastal plains with salinities of 350–900 mg/l in the Shaara well field (Fujayra) and EC values of 670–1,360 μS/cm in Khorfakan well field. The well fields Dibba, Qidfa and Sur Kalba produce fresh to brackish groundwater with EC values of 540–14,500 μS/cm (Fig. 9.4). High EC values and Cl concentrations of up to 5,000 mg/l indicate a strong impact of sea water intrusion in some wells (locations on Fig. 9.3).

In the Fujayra coastal plain, groundwater salinity is low to moderate in the wadi section of the sand and gravel aquifer downstream of Wadi Ham dam with TDS values between 350 and 900 mg/l.

In the Shaara well field, situated on the Fujayra coastal plain immediately downstream of the Wadi Ham recharge dam, fluctuations of groundwater salinity are clearly related to recharge during flood events. Salinity decreases sharply after rainfall in all observed boreholes of the well field but rises to previous levels of around 700 mg/l TDS within a few months.

The observations reflect the dominant but brief impact of recharge through Wadi Ham dam on the groundwater regime in the wadi section. Recharge enhanced through the dam has an almost immediate effect on groundwater levels and groundwater salinity. After several weeks, conditions start to return to the pre-recharge state, as the recharge wave moves down-gradient and groundwater extraction in Shaara well field removes part of the fresh recharge.

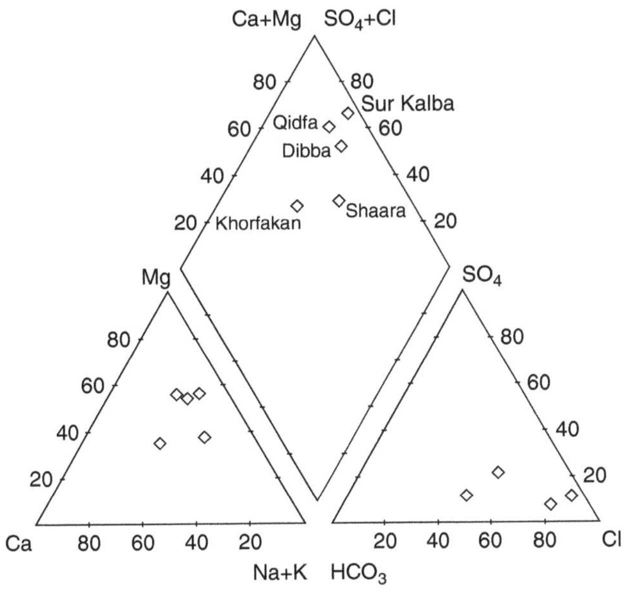

Fig. 9.4 Piper diagram: Mean values of groundwater samples from well fields on the eastern coastal plains of the United Arab Emirates. Data from files of Ministry of Agriculture and Fisheries and Ministry of Agriculture and Water Resources, Dubai

A great number of shallow irrigation wells is operated in the eastern coastal plain of the United Arab Emirates. EC values of the extracted water vary in a wide range from 300 to 12,500 μS/cm. Predominant ions are generally Na and Cl.

Brackish Cl water extends over wide parts of the sand and gravel aquifer of the Fujayra coastal plain: wells from Fujayra and Kalba well fields, most samples of Sur Kalba well field, samples from many irrigation boreholes. The samples may be classified as Ca–Cl, Mg–Cl or Na–Cl type water. Saline groundwater of Ca–Cl type with EC values of 35–38 mS/cm was found in 100–300 m deep investigation boreholes.

9.5.3.4 Hydrochemical Characteristics

Predominant ions in fresh water of the Quaternary sand and gravel aquifers of the Batina plain are Mg, Na, Cl and, in some samples, HCO_3. Mean concentrations of major ions are higher than in shallow fresh water of the adjoining ophiolite aquifer. Brackish groundwater in the lower reaches of the coastal plain are Cl waters with varying predominance of major cations Na, Mg, Ca.

In the sand and gravel aquifer of Wadi Ham in the upper part of the Fujayra coastal plain, two different hydrochemical types of groundwater are found (Fig. 9.5):

- Groundwater with low salinity originating from recharge of flood runoff, which is enhanced by a recharge dam
- Groundwater with salinities of around 1,000 mg/l and Na–Cl dominated composition

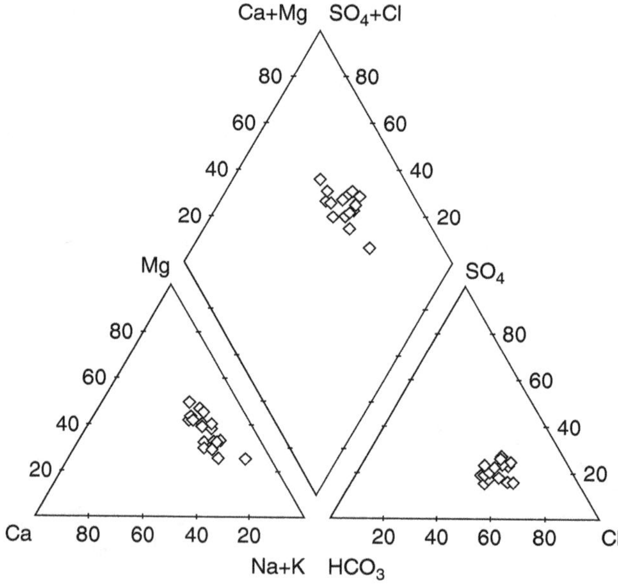

Fig. 9.5 Piper diagram: Groundwater samples from boreholes of Shaara well field on the Fujayra coastal plain, United Arab Emirates. Data from files of Ministry of Agriculture and Fisheries and Ministry of Agriculture and Water Resources, Dubai

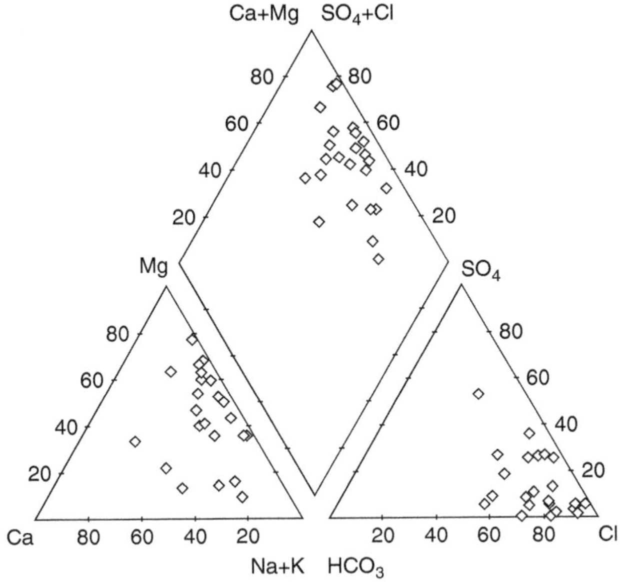

Fig. 9.6 Piper diagram: Groundwater samples from irrigation wells on the Fujayra coastal plain, United Arab Emirates. Data from files of Ministry of Agriculture and Fisheries and Ministry of Agriculture and Water Resources, Dubai

Fluctuations of salinity and hydrochemical composition of the groundwater in that section after flood events are particularly evident for Cl and Na, and can also be seen for SO_4, Ca and Mg. HCO_3 concentrations remain more constant at levels of 150–200 mg/l.

The hydrochemical composition of the fresh groundwater in the Fujayra plain appears to be controlled by the following processes:

- HCO_3 contents of rain water of around 20–60 mg/l increase in the flood recharge water through interaction between biogenic soil CO_2 with carbonate or silicate minerals to mean concentrations of up to 200 mg/l.
- Cl concentrations of around 5–30 mg/l in rainwater are enriched to a varying degree in the infiltrating water to an average of 70 mg/l.
- In water with moderate salinity underlying the fresh water lens in Shaara well field, Cl concentrations are enriched to about 200 mg/l. This water possibly originates from rainfall recharge in more extensive parts of the catchment area.

Mg–HCO_3 type waters, which are carried into the coastal plains in runoff from the ophiolite mountains, are apparently converted largely to water with Cl predominance during the recharge process through evaporative ion enrichment. Enrichment of the Cl concentration is probably relatively low in fast recharge from flood flow and higher in dispersed recharge from rainfall. Considering average recharge rates of 10–25% of precipitation in the catchment, Cl concentrations may be enriched by a factor of 4–10 in comparison to rain water. HCO_3 concentrations remain generally at a rather low level because of limited CO_2 production in a poor soil cover under the arid climate conditions.

In some boreholes in the Fujayra coastal plain, groundwater with high salinity and elevated pH is encountered (TDS up to 4,750 μS/cm, pH up to 10). These groundwaters appear to be related to slow movement of deep groundwater from the ophiolite mountains.

9.5.3.5 Sea Water Intrusion

Salt water intrusion from the sea is experienced in many coastal stretches of Al Batina, where groundwater is exploitated intensively.

Sea water intrusion of coastal aquifers in northern Oman has been caused by heavy abstraction from shallow wells for irrigation and by large municipal well fields, which has e.g., induced intrusion of a saline wedge within the Al Khawd fan.

Apart from sea water intrusion, brackish or saline water from coastal sabkhas and irrigation return flow may contribute to the groundwater salinity. In detail, a complex horizontal and vertical distribution of various hydrochemical sub-types of groundwater may be expected in the eastern coastal plains.

In Wadi Fara in northeastern Oman, EC values in the Quaternary aquifer near the coast increase from 1,600 μS/cm at shallow depth to 50,000 μS/cm at depths of 30–120 m below surface.

In the Fujayra coastal plain, a fresh water tongue extends along Wadi Ham from the margin of the ophiolite mountains about 6 km into the plain area. The lower – eastern – end of the fresh water tongue marks the position of a fresh water/salt water interface. Water samples in the range of the fresh water tongue are similar in their hydrochemical composition to samples from Shaara well field.

Groundwater with a salinity of >2,000 mg/l extended on the Fujayra plain to a distance of up to 4 km from the sea coast during 1984. Subsequently, increase of water salinity has affected many irrigation wells and also boreholes of the Fujayra water supply well field. In particular in the Kalba area south of Fujayra, high groundwater salinity is found in wells over wide parts of the coastal plain (Fig. 9.6).

Brackish to saline groundwater near the coast is prevailingly Na–Cl or Mg–Cl type water.

The hydrochemical composition of brackish to saline groundwaters found in the Fujayra coastal plain is influenced by cation exchange processes, typical for coastal sand aquifers with progressing sea water intrusion. Na of the intruding sea water is exchanged with cations adsorbed on the surfaces of the aquifer material. Adsorbed exchangeable cations are in particular Mg, the dominant cation in most fresh groundwaters of the Fujayra coastal plain, and Ca. The exchange process can be formulated schematically as

$$Na^+ + 1/2(Mg \text{ or } Ca) - X_2 \rightarrow Na - X + 1/2(Mg \text{ or } Ca)^{2+},$$

where X indicates the soil exchanger.

References. Abdulla and Durabi (1997), Akiti et al. (1992), Al-Asam and Wagner (1997), Al-Battashi and Ali (1997), Appelo and Geirnaert (1991), Appelo and Postma (1993: 143), Clark et al. (1989), Faig (1990), Kacimov et al. (2009), Kulaib (1991), Lakey and Al Hinai (1997), Lakey et al. (1995), Macumber et al. (1997), Rizk and El-Etr (1997), Rizk et al. (1997), Wagner (1995a, 1996b, c), Wagner and Karanjac (1997).

9.6 Information from Isotope Data

9.6.1 Indications of Recent Recharge

^{3}H and ^{14}C values reflect the general groundwater regime in coastal plains of the Arabian Peninsula. ^{3}H values and relatively high ^{14}C values indicate recent recharge along major wadi courses; ^{3}H values are generally low in groundwater near the coast and in interfluvial areas with less frequent recent recharge and longer residence periods of the groundwater.

In the Tihama Asir in Saudi Arabia, about one third of groundwater samples have ^{3}H values close to or below detection limit, indicating residence periods of more than two decades.

Detectable tritium, up to 84 TU, was found in many samples from the upper clastic layers of wadi aquifers, reflecting active present-day recharge and relatively fast groundwater circulation. ^{3}H values below detection level prevail in the deeper parts of the wadi aquifers and in downstream sections of the wadis near the coast.

In samples extracted from 14 boreholes in the Tihama Asir from depths between 25 and 124 m, ^{14}C values vary between 62.9 and 88.2 pmc corresponding to ages between a few years and 2,500 years B.P. ^{3}H values in these samples are generally low.

From tritium levels in groundwater of the Salala coastal plain, ages of less than 35 years have been estimated for most samples.

Groundwaters in the coastal plains of northeastern Oman show three principal categories:

- Shallow alluvial aquifers along active wadi channels have tritium values of 6–15 TU and high ^{14}C values. The aquifers contain recently recharged groundwater with a turnover time of a few years.
- In shallow groundwater in interfluvial areas, tritium values are low. These groundwaters receive less frequent recharge and have mean residence times in the order of tens to hundreds of years.
- Confined fresh water at greater depth is free of detectable tritium and mean residence times range up to several thousand years.

Tritium levels of 6–15 TU in shallow groundwater of the upper ranges of the coastal plains in Oman indicate short circulation times of a few years. Near the coast, tritium levels vary over ranges of <1–20 TU, reflecting the different frequency of recharge near wadis and in interfluvial areas.

Tritium levels of shallow groundwater in the coastal plain of Wadi Maawil downstream of Jebel Akhdar (80 km west of Muscat) are between 7 and 13 TU corresponding to recent recharge events. In deeper confined groundwater in the piedmont area, ^{3}H is below detection level; from ^{14}C values groundwater ages of around 5,000 years have been estimated for the confined groundwater.

In the Batina plain of the United Arab Emirates, tritium data indicate significant recent recharge in the Fujayra coastal plain (9–20 TU in Sur Kalba well field, up to 12.2 TU in Fujayra well field) and at Khorfakan (6.6–11.3 TU, data of 1984–1989). ^{3}H values below 3 TU prevail in the lower parts of the coastal plains.

The distribution of tritium levels in groundwater of coastal alluvial aquifers in the United Arab Emirates suggest a mean residence time of 5 years. Groundwater in shallow interfluvial aquifers along the coast are sub-modern in age (tritium free but with modern ^{14}C activities). Piezometric heads in these aquifers show long term fluctuations, not directly related to the active groundwater circulation near wadis.

The occurrence of modern recharge is also indicated by ^{14}C values of >50 pmc, up to 100 pmc, in the Fujayra coastal plain.

Relatively low values of ^{14}C (36.9 pmc) and ^{3}H (1.5–4.2 TU) are found in wells of the Dibba well field, which may receive groundwater flow from recharge in more distant parts of the relatively extensive low altitude catchment of Wadi Dibba.

9.6.2 *Stable Isotopes of Oxygen and Hydrogen*

δ^{18}O values of shallow groundwaters in the coastal plains at the Red Sea and the Gulf of Oman vary around −0.5 and −3‰, as far as the groundwaters are not influenced by evaporative enrichment or sea water intrusion. d values scatter around the GMWL, with, however, significant differences in mean d values for different coastal areas.

9.6.2.1 Coastal Aquifers at the Red Sea Coast

^{18}O values of a large group of groundwater samples from the Jizan and Jedda coastal areas in Saudi Arabia are between −0.5 and −2‰ with d values around +10‰. In some parts of the Jizan and Jedda coastal plains, data points tend to deviate from the GMWL to more positive δ^{18}O values (Figs. 9.7 and 9.8).

Local variations of the stable isotope composition in groundwater of the Tihama plain in Saudi Arabia appear to be related to the groundwater recharge at different altitudes of the catchment areas and evaporative isotope enrichment in the wadis and coastal plains, where groundwater levels are situated near surface. The δ^{18}O and δ^2H values of the deeper groundwater from 45 to 125 m deep boreholes are in the same range as values from the shallow groundwater but do not show an impact of evaporation. Apparently, the shallow and the deep groundwater originate from the same recharge areas.

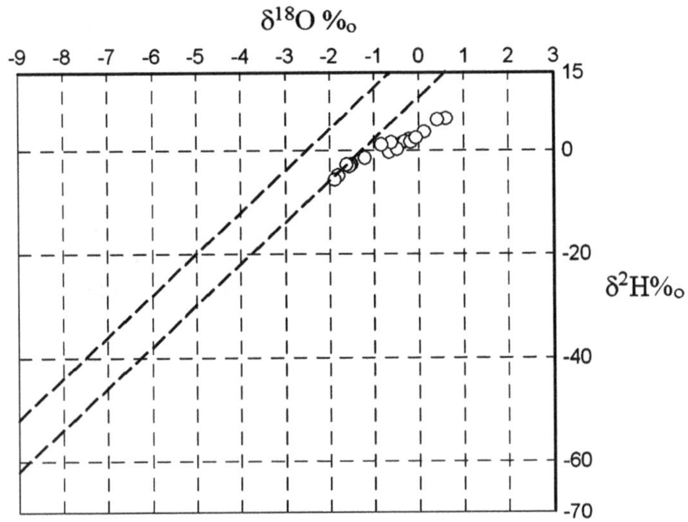

Fig. 9.7 $\delta^{18}O/\delta^2H$ diagram: Groundwater samples from the Jedda coastal plain. Data from Jado and Zötl (1984)

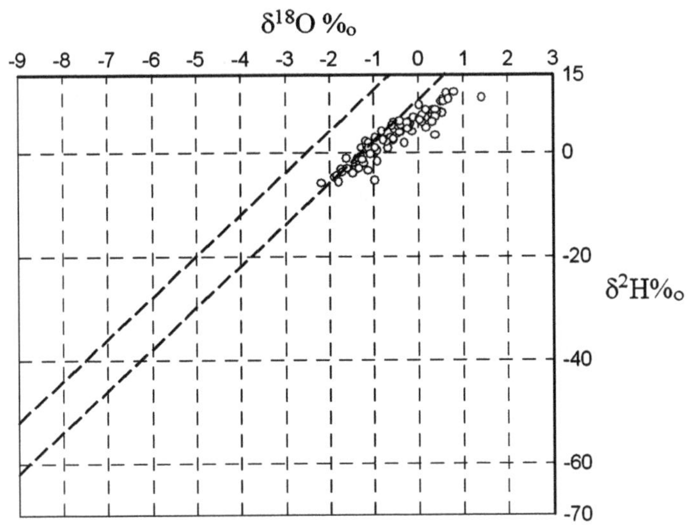

Fig. 9.8 $\delta^{18}O/\delta^2H$ diagram: Groundwater samples from the Tihama coastal plain in Saudi Arabia. Data from Jado and Zötl (1984)

9.6.2.2 Precipitation in Oman

Precipitation in northern Oman is dominated by two types of meteorologic events:

- Barometric depression systems originating over the Mediterranean Sea or Red Sea in winter.
- Orographic or convectional rainfall occurring during the summer months in the mountain areas, when high temperatures create convection cells which fill with moist air from the Gulf of Oman.

Both systems precipitate vapour masses mainly from the Gulf of Oman with initial ^{18}O values between −2.8 and −4‰. The convectional rains during summer are affected by high evaporation causing a shift of $\delta^{18}O$ values from the GMWL to more positive values. Winter rains are not affected significantly by evaporation and $\delta^{18}O/\delta^{2}H$ values plot above the GMWL near the MMWL.

On the Salala plain at the Arabian Sea coast and the adjoining the Dhofar mountains, the monsoon during the months of June to August provides a major source of the annual precipitation. $\delta^{18}O/\delta^{2}H$ values in monsoon precipitation deviate only slightly from sea water and tritium values are generally low with 2–6 TU. "This is interpreted as being a function of a low altitude single stage weather system with rapid evaporation and almost immediate precipitation" (Clark et al. 1987).

9.6.2.3 Salala Plain

Most groundwater samples from the Salala plain have $\delta^{18}O/\delta^{2}H$ values in the typical range of the monsoon precipitation with $\delta^{18}O$ between −0.5 and +0.3‰ and $\delta^{2}H$ between +0.3 and +9‰.

The $\delta^{18}O/\delta^{2}H$ and tritium values show, that the groundwater in the Salala plain originates prevailingly from present-day monsoon recharge on the plain and the adjoining Dhofar mountains. The $\delta^{18}O/\delta^{2}H$ values of most groundwaters sampled in the Salala plain ($\delta^{18}O$ −0.49 to +0.39‰, $\delta^{2}H$ +3 to +10‰) are identical to values of monsoon precipitation, which vary only slightly from sea water (Fig. 9.9). The occult precipitation from the monsoon in the mountains is apparently the principal source of recharge.

A few groundwater samples from the Salala plain indicate recharge from more depleted rains from cyclonic meteorologic systems, which reach the coast as infrequent events from the Arabian Sea.

9.6.2.4 Batina Coast in Oman

$\delta^{18}O$ values of the shallow groundwater of the Batina coastal plain in Oman are generally in a range of −0.2 to −2.9‰ with a pattern of d values, which indicates a deviation from the GMWL toward an evaporation slope.

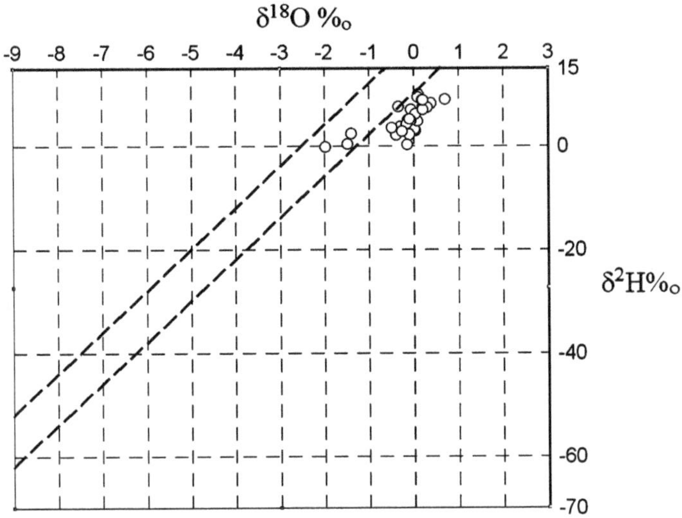

Fig. 9.9 $\delta^{18}O/\delta^{2}H$ diagram: Groundwater samples from the Salala coastal plain, Oman. Data from Clark et al. (1989)

The stable isotope composition of the recent groundwater along active wadi channels indicates the dominance of recharge from convective rainfall in the mountains during summer and varying evaporative isotope enrichment. Samples from sites along the wadi course show fluctuations of up to 1.3‰ in $\delta^{18}O$ within 1 year, probably related to isotopic signatures of individual storms.

The water of deeper confined groundwater has more depleted $\delta^{18}O/\delta^{2}H$ values than the shallow fresh groundwater, indicating recharge from non-evaporated winter rainfall.

A relatively enriched isotopic composition and $\delta^{18}O$ values of up to +2‰ is found in groundwater samples near the coast, indicating influences of evaporitic enrichment or sea water intrusion.

In a 90 km long strip around Sohar in the north of the Batina coastal plain of Oman, d values show a trend of deviation from the GMWL to an evaporation slope. $\delta^{18}O/\delta^{2}H$ values of the groundwater in the extensive coastal aquifers are in the same general range as values of groundwaters along wadi courses. Infiltration of wadi runoff is probably the principal mechanism of recharge to the coastal aquifers. Summer rainfall in the mountains may contribute significantly to the groundwater recharge of the plain aquifers.

In the coastal aquifers of Wadi Maawil and Wadi Bani Kharus (70–90 km west of Muscat), $\delta^{18}O/\delta^{2}H$ values of shallow wells indicate recharge by infiltration of evaporated wadi runoff originating from summer rains in the mountains.

A plume of isotopically depleted (higher altitude) groundwater ($\delta^{18}O < -3‰$, $\delta^{2}H < -10‰$) emerges from the eastern parts of the Jebel Akhdar and passes through a gap in the ophiolite "barrier" northwards across the lower catchment of Wadi al Maawil toward the coast near Barka. The plume, which is also recognizable in hydrochemical parameters, indicates a pathway of groundwater from the limestone catchments of Jebel Akhdar into the coastal plain.

In the Al Khawd fan (40 km west of Muscat), $\delta^{18}O/\delta^{2}H$ values follow an evaporation line, which deviates from the GMWL to values of −1‰ to +0.5‰. The groundwater is replenished mainly from a catchment in the piedmont area at less than 450 m asl. $\delta^{18}O/\delta^{2}H$ values of the deeper groundwater are more depleted with $\delta^{18}O$ between −2 and −1‰.

Local variations of ranges of $\delta^{18}O/\delta^{2}H$ values on the Al Khawd fan appear to be related to varying influences of recharge from flood flow originating on the high Jebel Hajar el Gharbi mountain zone and on the topographically lower ophiolite mountains around Wadi Semail. Probably, the uppermost areas of the fan are recharged mainly by flood runoff with relatively enriched $\delta^{18}O/\delta^{2}H$ values from the extensive Semail catchment. During more intensive rainfall events, floods from the higher mountain areas with isotopically more depleted water reach the middle and lower areas of the fan.

The presence of intruded sea water is evident in groundwaters with positive values of $\delta^{18}O$ and $\delta^{2}H$ near the coast.

Deeper confined groundwater in the Al Khawd fan has $\delta^{18}O$ values around −2.8 to −3.5‰, showing no signs of evaporative enrichment. Recharge is assumed to occur from winter rains at higher elevations.

9.6.2.5 Batina Plain in the United Arab Emirates

On the eastern coastal plain of the United Arab Emirates, $\delta^{18}O$ values are, on average more negative and d values higher than values of shallow groundwater from the Batina plain of Oman. $\delta^{18}O$ values extracted from well fields in the coastal area of Fujayra–Khorfakan range generally between −2.7 and −2.1‰, d values between +10.8 and +17.3‰ (Fig. 9.10 and Table 9.3). The differences in stable isotope values between the coastal plains of Oman and the United Arab Emirates are possibly related to the decrease of the frequency of summer rains from south to north.

In some samples from production wells on the Fujayra coastal plain, impacts of isotopic enrichment are recognizable ($\delta^{18}O$ −2‰, d +9.4‰).

Stable isotope values of groundwater in the Dibba well field are more negative than in other parts of the eastern coastal plains of the United Arab Emirates ($\delta^{18}O$ −3 to −2.7‰) without deviations of d values from the general range. The trend to more negative $\delta^{18}O$ values is possibly related to an inflow of groundwater recharged at higher altitudes in relatively distant parts of Wadi al Maksar.

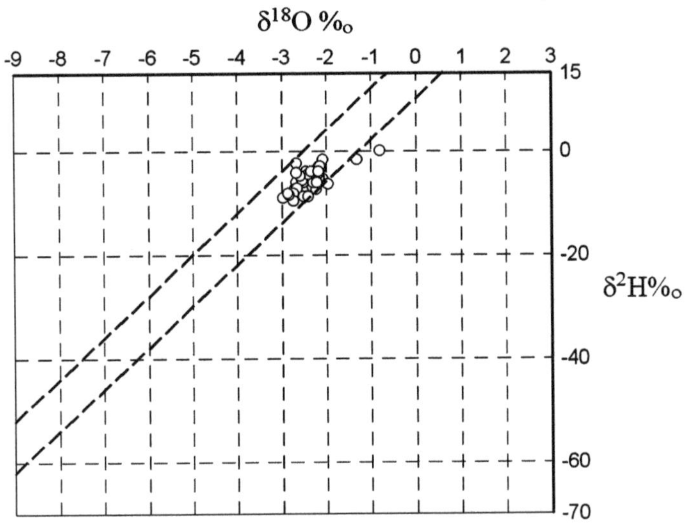

Fig. 9.10 $\delta^{18}O/\delta^2H$ diagram: Groundwater samples from the Batina coastal plain in the United Arab Emirates. Data from files of Ministry of Agriculture and Fisheries and Ministry of Agriculture and Water Resources, Dubai

Table 9.3 Isotope values of groundwater from the eastern coastal plain of the United Arab Emirates

Area	$\delta^{18}O$	δ^2H	d	n	^3H	^{14}C	Sampling period
	Mean values (‰)				TU	pmc	
Dibba–Khorfakan	−2.6	−6.1	+13.3	11	<1.5–11.3	92.6	1984–1989
Fujayra–Kalba	−2.1	−4.6	+11.0	42	<1.2–20.3	56.2–91.3	

Data from Akiti et al. (1992), Gonfiantini (1992), Kulaib (1991), Isotope laboratory WAJ, Amman

References. Al-Sayari and Zötl (1978), Alsharhan et al. (2001: 240), Clark et al. (1987, 1989), Jado and Zötl (1984), Kulaib (1991), Macumber et al. (1997), Wagner (1996c), Wagner and Geyh.

References

Abderrahman WA (1989) Effect of groundwater use on the chemistry of spring water in Al-Hassa Oasis. J King Abdulaziz Univ Earth Sci 3:259–265, Jeddah

Abderrahman WA, Rasheeduddin M (1994) Future groundwater conditions under long-term water stresses in an arid urban area. Water Resour Manag 8:245–264

Abderrahman WA, Rasheeduddin M, Al-Harazin IM, Esuflebbe JM, Eqnaibi BS (1995) Impacts of management practices on groundwater conditions in the Eastern Province, Saudi Arabia. Hydrogeol J 3(4):32–41

Abdulla MS, Durabi AA (1997) Survey on groundwater recharge and flow in Wadi Wurrayah. In: Third Gulf Water Conference, vol 1, Muscat, p 123

Abdulrazzak MJ, Sorman AU, Abu Riraiza O (1988) Estimation of natural groundwater recharge under Saudi Arabian arid climate conditions. In: Simmers I (ed) Estimation of natural groundwater recharge. NATO ASI Series, pp 125–138

Abu-Ajamiieh MM (1967) A quantitative assessment of the groundwater potential of the Rijam Formation aquifer in the Jafr Basin. Sandstone aquifers of East Jordan, Report of FAO, NRA, Amman

Abumaizer MS (1996) Hydrochemical properties and environmental isotopes of groundwaters of the Middle Aquifer in the Yarmouk Basin, Jordan. Isotope Field Applications for Groundwater Studies in the Middle East. IAEA-TECDOC, 098: 301-621, Vienna

Abu-Sharar TM, Rimawi O (1993) Water chemistry of the Dhuleil Aquifer (Jordan) as influenced by long-term pumpage. J Hydrol 149(1–4):49–66

ACSAD (1981) Geological map of the Hamad basin project, 1:500, 000. ACSAD, GTZ, BGR, Damascus, Hannover

ACSAD (1983) Studies of the Hamad Basin, Atlas and Part 1, App.4 Groundwater Resources. Rep. ACSAD, Damascus (English and Arabic)

Agrocomplect (1984–85) Study and complete design works for the integral development of the Syrian Akkar plain and Bekaa area. Rep. Agrocomplect – Syrian A.R. Dir. Land Reclamation, Sofia

Ahmed AM, Kraft W (1972) A contribution to the hydrogeology of the western desert in Iraq. J Geol Soc Iraq 5:135–148, Baghdad

Akiti T (1988) Aspect of isotope hydrology of the United Arab Emirates. Internal Report. IAEA, Vienna

Akiti T, Gonfiantini R, Kulaib A (1992) Aspects of environmental isotope hydrogeology of the United Arab Emirates. First Gulf Water Conference, vol 1, Dubai

Al Ansari NA, Hijab BR, Al Shammaa AM (1995) Utilization of groundwater for agricultural activities, South Nukhaib area, Western Desert, Iraq. In: Sultanate of Oman international conference on water resources management in arid countries, Muscat, pp 450–456

Al Awadi E, Mukhopadhyay A (1995) Hydrogeology of the Dammam aquifer of the Gudair area, Kuwait. Sultanate of Oman Internat. In: Conference on Water Resources Management in Arid Countries, Muscat, pp 428–435

W. Wagner, *Groundwater in the Arab Middle East*,
DOI 10.1007/978-3-642-19351-4, © Springer-Verlag Berlin Heidelberg 2011

Al Charideh AR (2007) Environmental isotopic and hydrochemical study of water in the karst aquifer and submarine springs of the Syrian coast. Hydrogeol J 15(2):346–351

Al Ejel F, Abderahim AH (1974) Syrian geology. Damascus (Arabic)

Al Mashadani AM (1984) Dynamic evolution of the Iraqi sedimentary basins: consequences on the distribution of fluids. Thesis, University of Pau

Al Mashadani AM (1995) Survey and evaluation of available data on water resources in Iraq. Unpubished Report, ESCWA, Amman

Al Sulaimi J, Al-Rabaa S, Al-Muhanna A, Amer A (1992) Assessment of groundwater resources in Kuwait using remote sensing technology, vol 3. Geology Report, KISR 4038, Kuwait

Al-Agha MR (1995) Environmental contamination of groundwater in the Gaza Strip. Environ Geol 25:109–113

Al-Ahmedi ME, Sen Z (1989) The depletion of water resources in Wadi As-Safra, Saudi Arabia. J King Abdulaziz Univ Earth Sci 2:1–15, Jeddah

Al-Asam MS, Wagner W (1997) Investigations for development of groundwater management strategies in the eastern coastal plain of the United Arab Emirates. In: Third Gulf Water Conference Proceedings, Muscat, pp 329–339

Al-Awadi E, Mukhopadhyay A, Al-Haddad A (1995) Compatibility of desalinated water with the Dammam formation at the northwest Shigaya water-well field, Kuwait. Hydrogeol J 3(4):56–73

Al-Awadi E, Mukhopadhyay A, Al-Senafi MN (1998) Geology and hydrogeology of the Dammam Formation in Kuwait. Hydrogeol J 6(2):302–314

Al-Battashi MB, Ali SR (1997) Groundwater recharge dam and hydrogeology of Wadi Al-Fulayj, Wilayat Sur, Sultanate of Oman. In: Third Gulf Water Conference, vol 1, Muscat, pp 295–311

Al-Hajari SA (1990) Geology of the Tertiary and its influence on the aquifer system of Qatar and eastern Arabia. Thesis Univ. South Carolina

Al-Hajari SA, Kendall ChG (1992) The sedimentology of the Lower Eocene Rus Formation of Qatar and neighboring countries. J Univ Kuwait 19:153–173

Al-Harthy S, Farbridge RA, Wyness AJ (1995) Irrigation development in Nejd, southern Oman. Sultanate of Oman Internat. In: Conference on Water Resources Management in Arid Countries, Muscat, pp 86–147

Al-Homoud AS, Allison RJ, Sunna B, White K (1995) Geology, geomorphology, hydrology, groundwater and physical resources of the desertified Badia environment in Jordan. GeoJournal 37(1):51–67

Al-Jamal K, Al-Yaqubi A, Shoblak M (1997) Non conventional water resources in Palestine (Gaza Governerates). Expert Group Meeting use of non-conventional water resources and application of appropriate technologies for groundwater management in the ESCWA countries. E/ESCWA/ENR/1997/WG.3/CP.8, Manama

Al-Kharabsheh A (1995) Possibilities of artificial groundwater recharge in the Azraq Basin: potential surface water utilization of five representative catchment areas (Jordan). Thesis, University of Würzburg

Allam MN (1995) Central well field for agricultural development in the Nejd. In: Sultanate of Oman international conference on water resources management in arid countries, Muscat, pp 220–227

Al-Mahmood AMY (1992) Management and development of water resources in the State of Qatar. In: Proceedings of the First Gulf Water Conference, Water Science and Technology Association, Bahrain (Arabic)

Almasamir A, Sarcis I (1992) Peculiarities in the chemical composition of ground waters in the northern part of the valley surrounding river Oronto. AR Syria Eng Geol Hydrogeol 22:36–45, Sofia (in Bulgarian)

Almomani M (1993) Environmental isotope study of the upper ground water system in Azraq Basin, Jordan. In: IAEA Workshop in the Middle East, Damascus

Almomani M (1994) Summary report about the hydrochemistry and isotope hydrology in the basalt aquifer systems in Jordan, Amman

Almomani M (1996) Environmental isotope and hydrochemical study of the shallow and deep ground water in the Azraq Basin, Jordan. In: Isotope Field Applications for Groundwater Studies in the Middle East. IAEA-TECDOC, 098, Vienna, pp 57–101

Al-Momani M (1991) Relationship between A.W.S.A. well field and North East desert ground water (upper aquifer) and sources of salinity in deep B2/A7 aquifer, Azraq Basin, Jordan. Presented in the workshop of "Isotope Hydrology in the Middle East Countries" held in Ankara, October 1991

Almomani M, Seiler K-P (1994) Environmental isotope study of the shallow and deep ground water in the Azraq Basin, Jordan. In: Isotopes in water resources management: SM-633/4P. IAEA, Vienna

Al-Murad MA, Zubari WK (1992) The status of the Dammam aquifer in Kuwait, a quantitative and qualitative assessment in the period 1970–1990. In: Proceedings of First Gulf Water Conference Dubai, Water Science and Technology Association, Bahrain (Arabic)

Al-Murad MA (1994) Evaluation of Kuwait aquifer system and assessment of future wellfields abstracion using a numerical 3D flow model. Thesis Arabian Gulf Univ. Bahrain

Al-Mutawa AM (1993) Water resources in the United Arab Emirates. Regional Symposium. Water use and conservation. ESCWA, Amman

Al-Rashed MF (1994) Modelling of the Shegaya, Sulaibiya and Umm Gudair fields in Kuwait. Int J Water Resour Dev 10(1):39–54

Al-Rifai N (1993) Problem of seawater intrusion into groundwater coastal aquifers in ten selected countries of the Near (and Middle) East. FAO Expert consultation seawater intrusion into coastal aquifers in the Mediterranean Basin and the Near East, Cairo

Al-Ruwaih F (1984) Ground-water chemistry of Dibdiba Formation, North Kuwait. Groundwater 22(4):412–417

Al-Ruwaih F (1985) Hydrochemical classification of the groundwater of Umm Al-Aish, Kuwait. J Univ Kuwait (Sci) 12:287–296

Al-Ruwaih FM (1993) Studies of groundwater chemistry of the Al-Shagaya field-D, Kuwait. J Univ Kuwait (Sci) 20:127–143

Al-Ruwaih FM (1995) Chemistry of groundwater in the Dammam aquifer, Kuwait. Hydrogeol J 3(4):42–55

Al-Ruwaih FM (1996a) Hydrogeochemical variation of carbonate aquifer, Al-Sulaibiya, Kuwait. Water Air Soil Pollut 90(3–4):489–505

Al-Ruwaih FM (1996b) Hydrogeology and groundwater modeling of the carbonate aquifer of Al-Salaybiya, Kuwait. J Sci Eng 23:89–100

Al-Ruwaih FM, Al-Haddad A (1994) Hydrogeological and management studies of Al-Shagaya, Field D, Kuwait. Int J Environ Hydrol 2(2):11–20

Al-Saafin AK, Bader TA, Shehata W, Hötzl H, Wohnlich S, Zötl JG (1990) Groundwater recharge in an arid karst area in Saudi Arabia. Selected papers on hydrogeology from the 82th international geological congress. Washington DC, USA, pp 29–41

Al-Sagabi IA (1978) Ground water potentility of Tabuk and Saq aquifers in Tabuk area. Inst Appl Geol Res Ser 7. M.Sc. Thesis, Institute of Applied Geology, King Abdullaziz University

Al-Sagaby I, Moallim MA (1996) Use of isotopes to study groundwater recharge through sand dunes in Qasim area, Saudi Arabia. Isotope field applications for groundwater studies in the Middle East. IAEA-TECDOC-890, Vienna, pp 11–32

Al-Sayari SS, Zötl JG (eds) (1978) Quaternary period in Saudi Arabia, 1: Sedimentological, hydrogeological, hydrochemical, geomorphological and climatological investigations in central and eastern Saudi Arabia. New York, Vienna

Al-Shaibani AM (2008) Hydrogeology and hydrochemistry of a shallow alluvial aquifer, western Saudi Arabia. Hydrogeol J 16(1):155–165

Al-Turbak AS, Al-Hassoun SA, Al-Brahim AM, Al-Dhowalia KH (1996) Natural groundwater recharge through sand dunes to underlying aquifers. KACST project AR-11-53 rep King Saud univ, Riyadh

Alsharhan AS (1989) Petroleum geology of the United Arab Emirates. J Petrol Geol 12(3):253–288

Alsharhan AS, Nairn AEM (1997) Sedimentary basins and petroleum geology of the middle East. Amsterdam

Alsharhan AS, Rizk ZA, Nairn AEM, Bakhit DW, Alhajari SA (2001) Hydrogeology of an arid region: the Arabian Gulf and adjoining areas. Amsterdam

Alyamani MS, Atkinson TC (1993) Identification of chemical processes using primary component analysis. J King Abdulaziz Univ Earth Sci 6:135–146, Jeddah

Alyamani MS, Hussein MT (1995) Hydrochemical study of groundwater in recharge area, Wafi Fatimah basin, Saudi Arabia. GeoJournal 37(1):81–89

Al-Youssef M (1994) Geomorphological study of surface drainage patterns in Qatar Peninsula, Arabian Gulf. Thesis, Arabian Gulf University of Bahrain (cited after Zubari 1997)

Amer A, Barrat JM, Mukhopadhyay A (1990) Assessment of groundwater resources in Kuwait using a finite difference model. Water Resour Dev 6(2):104–114

Andrews IJ (1992) Cretaceous and Paleogene lithostratigraphy in the subsurface of Jordan. Natural Resources Authority Subsurface Geol Bull 5, Amman

Appelo CAJ, Geirnaert W (1991) Processes accompanying the intrusion of salt water. In: Hydrogeology of salt water intrusion. Int Contrib Hydrogeol 11:291–304, Hannover

Appelo CAJ, Postma D (1993) Geochemistry, groundwater and pollution. Rotterdam

Arad A (1988) B, F, and Sr as tracers in carbonate aquifers and in karstic geothermal systems in Israel. In: IAH 21st Congress on Karst hydrogeology and karst environmental protection, Guilin, pp 922–934

Arsalan FA (1976) Geologie und Hydrologie der Azraq-Depression (Ost-Jordanien). Thesis Technische Hochschule Aachen

Aust H (1993) Grundwasser aus Vulkaniten: Erschließungs- und Qualitätsprobleme, dargestellt anhand eines Beispiels aus dem Jemen. Geol Jahrb A 142:217–233

Bajbouj MK (1982) Le bassin du Yarmouk. Étude hydrologogique et hydrogéologique. Thesis, Inst. Nat. Polytechnique Lorraine, Nancy

Bajjali W (2006) Recharge mechanism and hydrochemistry evaluation of groundwater in the Nuaimeh area, Jordan, using environmental isotope techniques. Hydrogeol J 14(1–2):180–191

Bajjali W (2008) Evaluation of groundwater in a three-aquifer system in Ramtha area, Jordan: recharge mechanisms, hydraulic relationship and geochemical evaluation. Hydrogeol J 16:1193–1205

Bajjali W, Clark ID, Fritz P (1997) The artesian thermal groundwaters of northern Jordan: insights into their recharge history and age. J Hydrol 192:355–382

Bakiewicz W, Milne DM, Noori M (1982) Hydrogeology of the Umm Er Radhuma aquifer, Saudi Arabia, with reference to fossil gradients. Q J Eng Geol Lond 15:105–126

Ba-Monem AM (1995) Water resources and its applications in agricultural development in Yemen: a special study on irrigation in Wadi Hadramout. Sultanate of Oman Internat. In: Conference on Water Resources Management in Arid Countries, Muscat, pp 303–320

Basmaci Y, Al-Kabir M (1988) Recharge characteristics of the aquifers of Jeddah-Makkah-Taif region. In: Simmers I (ed) Estimation of natural groundwater recharge. NATO Series, pp 367–375

Basmaci Y, Hussein JAA (1988) Groundwater recharge over western Saudi Arabia. In: Simmers I (ed) Estimation of natural groundwater recharge. NATO Series, pp 395–403

Bassam AM (1998) Determination of hydrochemical processes and classification of hydrochemical facies for the Sakaka Aquifer, northeastern Saudi Arabia. J Afr Earth Sci 27(1):27–38

Batayneh AT (2006) Use of electrical resistivity methods for detecting subsurface fresh and saline water and delineating their interfacial configuration: a case study of the eastern Dead Sea coastal aquifers. Jordan Hydrogeol J 14(7):1277–1283

Bazuhair ASA (1989) Optimum aquifer yield of for aquifers in Al-Kharj area, Saudi Arabia. J King Abdulaziz Univ Earth Sci 2:37–49, Jeddah

Bazuhair AS, Hussein MT (1990) Springs in Saudi Arabia. J King Abdulaziz Univ Earth Sci 3:251–258, Jeddah

Bazuhair AS, Al Leheibe AH, Hussein MT (1991) Hydrochemival co-parameters of groundwaters in the Arabian Shield, Saudi Arabia. J King Abdulaziz Univ Earth Sci 4:149–160, Jeddah

Bender F (1974a) Geology of Jordan. Contributions to the Regional Geology of the Earth, Berlin-Stuttgart

Bender F (1974b) Explanatory notes on the geologic map of the Wadi Araba, Jordan. Geol Jahrb B10:62

Bender F (1975) Geologic map of Jordan, 1:500.000. U.S. Dept. Interior Geol. Survey Prof. Paper, 560-I: Plate 1, Reston

Bender F (1982) On the evolution of the Wadi Araba – Jordan rift. Geol Jahrb B45:3–20

Bender F et al (1968) Geologische Karte von Jordanien 1:250.000, 5 sheets, Hannover

Beydoun ZR (1991) Arabian Plate hydrocarbon geology and potential - a plate tectonic approach, vol 33. American Association of Petroleum Geologists, Studies in Geology, Canada

BGR – WAJ (1994) Groundwater resources of northern Jordan, vol 3, Structural features of the main hydrogeological units in northern Jordan. Report of Water Authority of Jordan. Federal Institute of Geosciences and Natural Resources, Amman

Boeckh E et al (1970) German geological mission in Syria. Report. Rep BGR, Hannover

Boehmer WK (1988) Groundwater programme Rada integrated rural development project. Rep Ilaco, Arnhem

Bramkamp RA, Ramirez LF, Steinecke M, Reiss WH (1963) Geological map of the Jawf-Sakakah quadrangle, Kingdom of Saudi Arabia, U.S. Geological Survey of Miscellaneous Geological Investigation Map I-210A, 1:500,000, Washington DC

BRGM (1976) Hydrogeological investigation of Al-Wasia aquifer in the eastern province of Saudi Arabia. Regional study. Min Agric Water, Riyadh (cited after Subyani and Sen 1991)

BRGM (1977) Al Hassa development project. Groundwater resources study and management progamme. Final Report, vol 3, App. II-7, Results of isotopic analyses. BRGM, Paris

Buday T (1980) The regional geology of Iraq, stratigraphy and paleogeography. State Organisation for Minerals, Baghdad

Burdon DJ (1954) Infiltration rates in the Yarmouk Basin of Syria – Jordan. Publ Assoc Int Hydrol Sci 37:343–355, Rome

Burdon DJ, Safadi Ch (1963) The karst groundwaters of Syria. FAO rep, Athens

Burdon DJ (1982) Environmental conditions favouring deposition of potential aquifers. In: ACSAD workshop on regional correlation of geologic formations in the Hamad basin, Damascus

Burdon DJ, Al-Sharhan A (1968) The problem of the paleokarstic Dammam limestone aquifer in Kuwait. J Hydrol 6:385–404

Burdon DJ, Al-Sharhan A (1986) The problem of the paleokarstic Dammam limestone aquifer in Kuwait. J Hydrol 6:385–404

Carmi I, Gat JR (1978) Changes in the isotope composition of precipitation of the Eastern Mediterranean Sea area: a monitor for climate change? Israel Meteorol Res Pap 2:124–135

Caro R, Eagleson PS (1981) Estimating aquifer recharge due to rainfall. J Hydrol 53:158–211

Cavelier C, Salatt A, Heuze Y (1970) Geological description of the Qatar Peninsula (Arabian Gulf). Qatar University, Qatar

Charalambous AN (1982) Problems of groundwater development in the Sanaa basin, Yemen Arab Republic. Int Assoc Hydrol Publ 136:265–274

Chebaane M, El-Naser H, Fitch J, Hijazi A, Jabbarin A (2004) Participatory groundwater management in Jordan: Development and analysis of options. Hydrogeol J 12(1):14–32

Cherif OH, El Deeb WMZ (1984) The Midle Eocene-Oligocene of the northern Hafit area, south of Al Ain City (United Arab Emirates). Géologie Méditeranéenne 11(2):207–217

Clark ID, Fritz P, Quinn OP, Rippon PW, Nash H, Sayyid Barghash bin Ghalib al Said (1987) Modern and fossil groundwater in an arid environment: a look at the hydrogeology of southern Oman. Isotope Techniques in Water Resources Development, IAEA, Vienna, pp 167–188

Clark ID, Ravenscroft P, Fritz P (1989) Origin and age of coastal groundwaters in Northern Oman. In: Proceedings of 10th Salt Water Intrusion Meeting (SWIM), Ghent, pp 16–20

Clark ID, Bajjali WT, Phipps GC (1995) Constraining ^{14}C ages in sulfate reducing groundwaters: two case studies from arid regions. In: Isotopes in Water Resources Management: SM-336/10; IAEA, Vienna

Custodio E (1987) Hydrogeochemistry and tracers. In: Custodio E: Ground-water problems in coastal areas. Studies and reports in hydrology 45:213–269, UNESCO

Dabbagh AE, Hötzl H, Zötl JG (1988) Karst features of the Arabian platform and their influence on aquifers. In: IAH 21st congress on Karst hydrogeology and karst environment protection, Guilin, pp 452–459

Dansgaard W (1964) Stable isotopes in precipitation. Tellus 16:436–468

Davidson ES, Hirzallah B (1966) Hydrogeology of the Jericho area, Jordan. Rep Central Water Authority

Dewandel B, Lachasagne P, Boudier F, Al-Hattali S, Ladouche B, Pinault JL, Al-Suleimani Z (2005) A coneptual hydrogeological model of ophiolite hard-rock aquifers in Oman based on a multiscale and a multidisciplinary approach. Hydrogeol J 13(5–6):708–726

Dincer T, Al-Mugrin A, Zimmermann U (1974) Study of the infiltration and recharge through the sand dunes in arid zones with special reference to the stable isotopes and thermonuclear tritium. J Hydrol 23(1–2):79–109

Downing RA, Price M, Jones GP (1993) The hydrogeology of the chalk of north-west Europe. Oxford

Domenico PA, Schwartz FW (1990) Physical and chemical hydrogeology. Wiley, New York

Droubi A (1983) Study of water resources in the Daw Basin, part 2, chap 6. Hydrochemistry. Rep ACSAD, Damascus (Arabic)

Dubertret L (1955) Carte géologique du Liban au 1/200.000e, carte et notice explicative. Beirut

Dubertret L (1966) Liban, Syrie et bordure des pays voisins. Paris

Durozoy G (1972) Hyrogéologie des basalts du Harrat Rahat. Bull BRGM 2(III):37–50

ECWA (1981) Assessment of the water resources situation in the ECWA region. E/ECWA/NR/L/1/Rev, 1, Beirut

Edgell HS (1997) Aquifers of Saudi Arabia and their geological framework. Arab J Sci Eng 22(1C):47–64, Dhahran

El Hames AS (2005) Determination of groundwater availability in shallow arid region aquifers utilizing GIS technology: a case study in Hada Al-Sham, Western Saudi Arabia. Hydrogeol J 13(5):640–648

El Naser H, Gedeon R (1996) Hydrochemistry and isotopic composition of the Nubian Sandstone aquifers of Disi-Mudawwara area, south Jordan. In: Isotope field applications for groundwater studies in the Middle East, vol 890. IAEA-TECDOC, Vienna, pp 16–47

El-Nakhal HAH, Himida IH (1980) Evaluation of the groundwater quality in Gaza Strip, Palestine. In: The First Arab Symposium Water Resources, 1, part 2, ACSAD, Damascus, pp 29–42

ESCWA (1995) Assessment of water quality in the ESCWA region. E/ESCWA/ENR/1995/41, Amman

ESCWA (1996) Investigation of the regional basalt aquifer system in Jordan and Syria. Reporrt of UN Economic and Social Commission for Western Asia, Ministry of Water and Irrigation Jordan, Ministry of Irrigation Syrian A.R., BGR, E/ESCWA/ENR/1996/11, Amman

ESCWA (1999a) Groundwater quality control and conservation in the ESCWA region, E/ESCWA/ENR/1999/1, New York

ESCWA (1999b) Groundwater resources in Paleogene carbonate aquifers in the ESCWA region, preliminary evaluation, E/ESCWA/ENR/1999/6, New York

ESCWA (1999c) Application of satellite remote-sensing methods for hydrogeology in the ESCWA region, E/ESCWA/ENR/1999/2, New York

ESCWA-UNEP (1996) Water resources assessment in the ESCWA region using remote sensing and GIS techniques. ESCWA, UNEP, Islamic Development Bank, ESCWA/ENR/1996/7, Amman

Faig ES (1990) Contribution to the study of the hydrogeology and hydrochemistry of Wadi Dibba in the United Arab Emirates. In: Etudes recentes sur la geologie de l'Afrique. CIFEG Publ. Occasionelle 1980/22: 397-399, Orleans

Falkner RD (1994) Fossil water or renewable resource: the case for one Arabian aquifer. Proc Inst Civil Eng Water Maritime Energy 106(4):325–331 (cited after Zubari 1997)

FAO (1979) Survey and evaluation of available data on shared water resources in the Gulf States and the Arabian Peninsula, vol 1–3. Food and Agricultural Organization of the United Nations, Rome

Faradzhev VA (1966) The geological map of Syria, scale 1:200, 000, sheet I-37-XIV (Al-Qaryatein), Explanatory notes. Min. Industry Syrian A.R., Technoexport, Damascus

Farid MS (1993) Groundwater and water security in the future of Salalah region. Geoscientific Research in Northeast Africa, Rotterdam, pp 735–739

Faroog AS, Rogers JJ (1980) Sedimentary facies of the Sakaka sandstone, north western Saudi Arabia. Newslett Stratigr 8(3):171–179

Freeze A, Cherry JA (1979) Groundwater. Prentice Hall, Englewood Cliffs

Fritz P, Fontes JCh (eds) (1980) Handbook of environmental isotope geochemistry, vol 1, The terrestrial environment. Elsevier, Amsterdam

Fritz P, Clark ID, Fontes J-Ch, Whiticar MJ, Faber E (1992) Deuterium and ^{13}C evidence for low temperature production of hydrogen and methane in a highly alkaline groundwater environment in Oman. In: Proceedings of water-rock interaction, vol 7. IAH, Rotterdam, pp 793–796

Froehlich K, Yurtsever Y (1995) Isotope techniques for water resources in arid and semiarid regions. In: Adar E, Leibundgut Ch (eds) Application of tracers in arid zone hydrology, vol 232. Vienna, IAHS, pp 3–12

Froehlich K, Geyh MA, Verhagen Th, Wirth K (1987) Isotopenhydrologische Methoden zur Begutachtung von Grundwasser in Trockengebieten. Weltforum, Cologne

Gat JR (1981) Paleo-climatic conditions in the Levant as revealed by the isotopic composition of palaeowaters. Israel Meteorol Res Pap 3:31–82

Gat JR, Carmi I (1970) Evolution of the isotopic composition of atmospheric waters in the Mediterranean Sea area. J Geophys Res 75(15):3039–3048

Gat JR, Dansgaard W (1972) Stable isotope survey of the fresh water occurrences in Israel and the northern Jordan Rift valley. J Hydrol 16(3):177–211

Gat JR, Mazor E, Tzur Y (1969) The stable isotope composition of mineral waters in the Jordan rift valley. Israel J Hydrol 9(3):334–352

Gavrieli I, Burg A, Guttman J (2002) Transition from confined to phreatic conditions as the factor controlling salinization and change in redox state, Upper sub-aquifer of the Judea Group, Israel. Hydrogeol J 10:483–494

GDC (1980) Umm Er Radhuma study, Bahrain Assignment, vol 3, Groundwater resources. Groundwater Development Consultants, Ministry of Agriculture and Water, Kingdom of Saudi Arabia

Gedeon R, Amro H, Jawawdeh J, Kilani S, Smith B (1995a) Natural isotopes in groundwater from the Amman-Zarka basin, Jordan: hydrochemical and regulatory implications. In: Adar E, Leibundgut Ch (eds) Application of tracers in arid zone hydrology, vol 232. Vienna, IAHS, pp 81–97

Gedeon R, Amro H, Jawawdeh J, Smith B, Cook JM (1995b) Uran rich mineralized groundwater from central Amman, Jordan. In: Adar E, Leibundgut Ch (eds) Application of tracers in arid zone hydrology, vol 232. Vienna, IAHS, pp 225–232

GEOCON (1995) Republic of Yemen, groundwater resources available for development. Geological consulting company, International Academy for Sciences on Nature and Society Russia, Ministry of oil and mineral resources, Sanaa, Moscow

Geyh MA (1996) Isotope hydrological study of the Basalt Aquifer system in Syria and Jordan. Rep BGR, Hannover

Geyh MA, Wagner W (1982) Environmental isotope study of groundwater in the Hamad region. Report on Cooperation in Hydrogeological and Geological Studies in the Hamad and Ad-Dawwa Projects, 1978–1982, vol 7. BGR, Hannover

Geyh M, Khouri J, Rajab R, Wagner W (1985) Environmental isotope study in the Hamad region. Geol Jahrb C38:3–15

Gibbs B (1993) The hydrogeology of the Azraq Basin, NE Badia, Jordan. Thesis, University College, London

GITEC and HSI (1995) Groundwater investigations in the Hammad and Sirhan basins. Report of Minerals, Water and Irrigation, GITEC, HSI, GEMT, Amman

Glennie KW (1995) The geology of the Oman Mountains. NHBS, Beaconsfield

Gonfiantini R (1992) Investigation of groundwater resources of the United Arab Emirates by using isotope techniques. In: International Atomic Energy Agency Report. UAE/8/002, Vienna

Gonfiantini R, Akiti T (1985) Isotope investigations as a tool for regional hydrological studies in the United Arab Emirates. Progress report. International Atomic Energy Agency, Vienna

Grolier MJ, Overstreet WC (1978) Preliminary geologic map of region east of Sadah, Yemen Arab Republic. US Geol Survey open-file rep 76–740

GTZ and NRA (1977) National Water Master Plan of Jordan. Rep. Agrar- und Hydrotechnik and BGR, Hannover, Frankfurt, Amman

Gvirtzman H (1994) Groundwater allocation in Judea and Samaria. In: Isaac J and Shuval H (eds) Water and peace in the Middle East. Stud Environ Sci 58:205–218

Handy AH, Alomani FF (1984) Chemical characteristics of principal and secondary aquifers in Saudi Arabia. Water Research Stdies Div Publ 2 vol 7

Hanshaw BB, Back W, Deike RG (1971) A geochemical hypothesis for dolomitization by groundwater. Economic Geol 66:710–724

Haddad M, Jayyousi A, Mizyed N, Nuseibeh M (1996) An overview of deep groundwater in the Palestinian territory. In: Proceedings of regional seminar on integration of information between oil drilling and hydrogeology of deep aquifers. INWRDAM, Amman, pp 173–186

Hammad FA (1978) Hydrogeological aspects of Yanbu an Nakhl area, Saudi Arabian Kingdom. In: The first Arab symposium on water resources, vol 1, pt 2. ACSAD, Damascus, pp 43–53

Harhash IE, Yousif AM (1985) Groundwater in Qatar, summary of hydrological studies and results. Department of Agriculture and Water Research, Ministry of Industry and Agriculture, State of Qatar

Harress HM (1975) Hydrological expert opinion on thermal springs as planning basis for a health center. Khat, Ras el Khaima, UAE

Hashash A, Aranyossy JF (1995) Mise en évidence de l'origine de la salinité des eaux de l'aquifère de Sarafand (sud-Liban). In: Adar E, Leibundgut Ch (eds) Application of tracers in arid zone hydrology, vol 232. IAHS, Vienna, pp 35–42

Hawkins TRW, Hindle D, Stagnell R (1981) Outlines of the stratigraphy and structural frame work of the southern Dhofar (Sultanate of Oman). Geol Mijnbouw 60(2):247–256

Heim A (1928) Die artesischen Quellen der Bharain- Inseln im Persischen Golf. Ecol Geol Helv 21:1–6

Helal AH (1965) General geology and litho-stratigraphic subdivision of the Devonian rocks of the Jauf area, Saudi Arabia. N Jahrb Geol Paläont Monatshefte 1965:527–551

Hirzalla B (1973) Groundwater resources of the Jordan Valley. Rep NRA, Amman

Hobler M et al (1991) Groundwater resources of southern Jordan, vol 1. Report of Federal Institute for Geosciences and Natural Resources, Hannover

Hötzl H (1995) Groundwater recharge in an arid karst area (Saudi Arabia). In: Adar E, Leibundgut Ch (eds) Application of tracers in arid zone hydrology, vol 232. IAHS, Vienna, pp 195–207

Hötzl H, Wohnlich S, Zötl G, Benischke R (1993) Verkarstung und Grundwasser im As Summan Plateau (Saudi Arabien). Steirische Beiträge zur Hydrogeologie 5-158

Hughes Clarke MW (1988) Stratigraphy and rock unit nomenclature in the oil-producing area of interior Oman. J Petrol Geol 11(1):5–60

Hughes AG, Mansour MM, Robins NS (2008) Evaluation of distributed recharge in an upland semi-arid karst system: the West Bank Mountain Aquifer, Middle East. Hydrogeol J 16(5): 845–854

Hussein MT, Bazuhair A, Al-Yamani MST (1993) Groundwater availabilty in the Khulais plain, western Saudi Arabia. Hydrol Sci J 38(3):203–213

Imes JL, Wood WW (2007) Solute and isotope constraint of groundwater recharge simulation in an arid environment, Abu Dhabi Emirate, United Arab Emirates. Hydrogeol J 15(7): 1307–1316

Issar A (1990) Water shall flow from the rock. Springer, Heidelberg

Italconsult (1971) Water and agricultural studies in Bahrain. FAO, Min. Agriculture Water, Rome

Italconsult (1972) Water supply for Sanaa and Hodeida. Sanaa basin hydrogeological investigation. Interim report. UNDP-WHO-Yemen Arab Report Min Public Works, Rome

Italconsult (1973) Climate in area VI south, Hydrological Spec. Pap. Italconsult Hydrol. Serv., Kindom of Saudi Arabia Min. Agric. Water, Rome

Italconsult (1973) Water supply for Sanaa and Hodeida. Rep WHO UNDP, Rome

IWACO (1986) Ground water study, project 21/81. Drilling of deep water wells at various locations in the UAE. Min. Agriculture Fisheries, IWACO, Bin Ham Est., Dubai

Jado AR, Zötl JG (eds) (1984) Quaternary period in Saudi Arabia, vol 2: Sedimentological, hydrogeological, hydrochemical, geomorphological, and climatological investigations in western Saudi Arabia. Wien, New York

Jordan H, Hebert D, Saker I, Fröhlich K (1980) Tritiumuntersuchungen an Grundwässern Kuweits. Z Angew Geol 26(3):130–132

Jungfer EV (1984) Das Wasserproblem im Becken von Sana, anthropogene und physische Ursachen einer zunehmenden Austrocknung. In: Kopp H, Schneider G (eds) Entwicklungsprozesse in der Arabischen Republik Jemen, Wiesbaden, pp 171–194

Kacimov AR, Sherif HM, Perret JS, Al-Mushiki A (2009) Control of sea-water intrusion by salt-water pumping: coast of Oman. Hydrogeol J 17(3):541–558

Kafri U, Arad A (1979) Current subsurface intrusion of Mediterranean seawater: a possible source of groundwater salinity in the rift-valley system. Israel J Hydrol 44:267–287

Karanjac J (1995) Mathematical model of ground water system of Wadi Ham, Fujayrah Emirate. United Arab Emirates, Amman

Kareh R (1967, 1968) Les source sous-marines de Chekka (Liban-Nord) - Exploitation d´une nappe karstique a exutoires sous marins. Hannon 2:35–59, 3:93–121

Kattan Z (1996a) Environmental isotope study of the major karst springs in Damascus limestone aquifer systems: case of the Fijeh and Barada springs. In: Isotope field applications for groundwater studies in the Middle East. IAEA-TECDOC, Vienna, pp 127–150

Kattan Z (1996b) Chemical and environmental isotope study of the fissured basaltic aquifer systems of Yarmouk Basin, Syria. In: Isotope field applications for groundwater studies in the Middle East, vol 980. IAEA-TECDOC, Vienna, pp 151–183

Kattan Z (1996c) Chemical and environmental isotope study of precipitation in Syria. In: Isotope field applications for groundwater studies in the Middle East, vol 890. IAEA-TECDOC, Vienna, pp 185–202

Kattan Z (2002) Effects of sulphate reduction and geogenic CO_2 incorporation on the determination of ^{14}C groundwater ages – a case study of the Paleogene groundwater system in north-eastern Syria. Hydrogeol J 10(4):495–508

Kemper E (1980) Stratigraphy and facies analysis of the Hamad desert in Syria, Iraq, Jordan. Rep BGR, ACSAD, Hannover

Kew GA (1995) Irrigation in central Oman. In: Sultanate of Oman international conference on water resources management in arid countries, Muscat,pp 129–136

Khair K, Haddad F (1993) The fractured carbonate rocks of Lebanon and their seasonal springs. IAH Mem 24:1135–1144, Oslo

Khair K, Haddad F, Fattouh S (1991) The effects of overexploitation on coastal aquifers in Lebanon, with special reference to saline intrusion. Selected papers on aquifer overexploitation. Select Pap Hydrogeol 3:359–362

Khair K, Aker N, Zahruddine K (1992) Hydrogeologic units of Lebanon. Appl Hydrogeol 2:39–49

Khalifa MA (1997) Hydrogeology of the geothermal fractured-rock well field at Jabal Hafit, Abu Dhabi Emirate. In: Third Gulf Water Conference Proceedings, vol 1, Muscat, pp 125–140

Khayat S, Hötzl H, Geyer S, Ali W (2006) Hydrochemical investigation of water from the Pleistocene wells and springs, Jericho area, Palestine. Hydrogeol J 14(1–2):192–202

Khouri J (1982) Hydrogeology of the Syrian steppe and adjoining areas. Q J Eng Geol Lond 15:135–154, London

Khouri J (1991) Hydrology and hydogeology of the major basin and aquifer systems in the Mashrek, Arabian Peninsula and Nile Valley. ACSAD, GTZ, BGR, Damascus

Khouri J, Agha WR (1979) Hydrogeological and geophysical study of salt water intrusion phenomenon in the central area of the Dawwa Basin in the Syrian arid zones. ACSAD, Damascus

Klinge H (1982) Untersuchungen zum Grundwasserhaushalt der Tihama (Saudi Arabien). Giessener Geol Schriften, 34, Giessen

König RW (1994) Grundwasserneubildung in semiariden Gebieten Jordaniens berechnet mit Hilfe der Halogenverteilung in der ungesättigten Zone. Thesis, University of Würzburg

Kotb H, Hussein MT, Zaidi S (1990) Groundwater quality of Damm area, Saudi Arabia. J King Abdulaziz Univ Earth Sci 3:241–250, Jeddah

Kozlov VV, Artyemov AV, Kalis AF (1966) The geological map of Syria sclae 1:200, 000, sheets I-36-XVIII, I-37-XIII (Trablous, Homs), explanatory notes. Min. Industry Syrian A.R, Technoexport, Damascus

Krasheninikov VA, Golovin DI, Mourayov VI (1996) The Paleogene of Syria - Stratigraphy, lithology, geochronology. Geol Jahrb B86

Krasnov AA, Kazmin VG, Kulakov VV (1966) The geological map of Syria 1:200,000, sheets I-36-VI, I-37-I,II, explanatory notes, Moscow

Kroitorou L, Mazor E, Issar A (1992) Flow regimes in karstic systems: the Judean anticlinorium, central Israel. IAH Mem 13:339–354

Kroitoru L, Mazor E, Gilad D (1985) Hydrological characteristics of the Wadi Kelt and Elisha springs. In: Diskin M (ed) Scientific basis for water resources management. IAHS-AISH Publ 153:207–218

Kruck W, Rajab R, Wagner W (1981) Geological map of the Hamad basin project, 1:500,000. Damascus, Hannover

Kruck W, Schaeffer U, Thiele J (1996) Explanatory notes on the geological map of the Republic of Yemen – western part. Geol Jahrb B87

Kulaib AA (1991) Study of the hydrology of the United Arab Emirates by using isotope techniques. Ministry of Electricity and Water, Dubai

Lakey R, Al Hinai H (1997) Application of one dimensional flow modelling to estimate groundwater recharge – Al Batinah, Sultanate of Oman. Third Gulf Water Conference, Muscat, vol 1, pp 357–372

Lakey R, Easton P, Hinai A (1995) Groundwater recharge processes, eastern Batinah, Oman. In: Sultanate of Oman international conference on water resources management in arid countries: 511–520, Muscat

Lenz R (1969) Eine Strukturkarte von Westjordanien im Maßstab 1:250 000. Beih Geol Jahrb 81:93–112

Lerner DN, Issar AS, Simmers I (1990) Groundwater recharge, vol 8. International Contributions on Hydrogeology. IAH, Hannover

Litak RK, Barazangi M, Brew G, Sawaf T, Al-Imam A, Al-Youssef W (1998) Structure and evolution of the petroliferous Euphrates graben system, southeast Syria. Am Assoc Petrol Geol Bull 82(6):1173–1190

Lloyd JW (1965) The hydrochemistry of the aquifers of north-eastern Jordan. J Hydrol 3:319–330

Lloyd JW (1980) Aspects of environmental isotope chemistry in groundwaters in eastern Jordan. In: Arid-zone hydrology: investigations with isotope techniques. IAEA, Vienna, pp 193–204

Lloyd JW (1981) Case-studies in groundwater resources evaluation. Oxford Science, Oxford, pp 122–132

Lloyd JW (1986) A review of aridity and groundwater. J Hydrol Process 1(1):63–78

Lloyd JW, Farag MH (1978) Fossil ground-water gradients in arid regional sedimentary basins. Ground Water 16(6):388–393

Lloyd JW, Heathcote JA (1985) Natural inorganic hydrochemistry in relation to ground water. Oxford Science, Oxford

Lloyd JW, Pim RH (1990) The hydrogeology and groundwater resources development of the Cambro-Ordovician sandstone aquifer in Saudi Arabia and Jordan. J Hydrol 121:1–20

Lloyd JW, Fritz P, Charlesworth D (1980) A conceptual hydrochemical model for alluvial aquifers on the Saudi Arabian basement shield. International Atomic Energy Agency, Arid-zone hydrology investigations with isotope techniques, Vienna

Lloyd JW, Pike JG, Eccleston BL, Chidley TRE (1987) The hydrogeology of complex lens conditions in Qatar. J Hydrol 89:239–258

Lloyd JW (1995) The reality of recharge in semi-arid and arid areas. In: Sultanate of Oman international conference on water resources management in arid countries: 466–479, Muscat

Lovelock PER (1984) A review of the tectonics of the northern Middle East region. Shell International Petroleum, Maatschappij

Luijendijk E, Bruggemann A (2008) Groundwater resources in the Jabal Al Hass region, northwest Syria: an assessment of past use and future potential. Hydrogeol J 16(3):511–530

Macumber P (1995) Freshwater lenses in the hyper-arid region of central Oman. In: Sultanate of Oman international conference on water resources management in arid countries, Muscat, pp 480–487

Macumber PG, Bargash bin Ghalib Al-Said, Kew GA, Tennakoon TB (1995) Hydrogeologic implications of a cyclonic rainfall event in central Oman. In: Nash H and McCall GJH (eds) Groundwater quality. Chapman and Hall, London, pp 87–97

Macumber PG, Niwas JM, Alin Al Abadi, Seneviratue (1997) A new isotopic water line for northern Oman. In: Proceedings of 3rd Gulf Water Conference, Muscat, pp 141–163

Margane A, Hobler M, Almomani M, Subah A (2002) Contributions to the hydogeology of northern and central Jordan. Geol Jahrb C68, Hannover

Martin G, Krapp L (1977) Die Hydrogeologie des Umm Er Radhuma Dolomit- Wasserträgers im "King Faisal Bedouin Settlement"- Projekt bei Haradh, Saudi-Arabien. Geol Jahrb C17:37–55

Matter JM, Waber HN, Loew S, Matter A (2006) Recharge areas and geochemical evolution of groundwater in an alluvial aquifer system in the Sultanate of Oman. Hydrogeol J 14(1–2): 203–224

Matthess G (1990) Die Beschaffenheit des Grundwassers, 2nd edn. Borntraeger, Berlin-Suttgart

Matthess D, Ubell K (1983) Allgemeine Hydrogeologie – Grundwasserhaushalt. Borntraeger, Berlin-Stuttgart

MAW (1984) Water atlas of Saudi Arabia. Ministry of Agriculture and Water, Riyadh

Mazor E, Mero F (1969) The origin of the Tiberias-Noit mineral water association in the Tiberias – Dead Sea Valley, Israel. J Hydrol 8:91–106

Mazor E, Molcho M (1972) Geochemical studies on the Feshcha springs, Dead Sea Basin. J Hydrol 15:37–47

Mazor E, Kaufman A, Carmi I (1973) Hammat Gader (Israel): geochemistry of a mixed thermal spring complex. J Hydrol 18:289–303

McBride JH, Barazangi M, Best J, A-Saad D, Sawaf T, Al-Otri M, Gebran A (1990) Seismic reflection structure of intracratonic Palmyride fold-thrust belt and surrounding Arabian Platform, Syria. Am Assoc Petrol Geol Bull 74(3):238–259

McKenzie JA, Hsu KJ, Schneider JF (1980) Movement of subsurface waters under the sabkha, Abu Dhabi, UAE, and its relation to evaporative dolomite genesis. Concepts and models of dolomitization (eds. Zenger DH, Dunham JB, Ethington RL), Special Publ. Soc Econ Paleontol Mineral 28:11–30

Medvedev VYa (1966) The geological map of Syria, scale 1:200, 000, sheet I-37-XXII (Ar-Raqqa). Explanatory notes. Min. Industry Syrian A.R, Technoexport, Damascus

Melloul AJ, Collin M (1994) The hydrologic malaise of the Gaza Strip. Israel J Earth Sci 43: 105–116

Memon BA, Kazi A, Bazuhair AS (1984) Hydrogeology of Wadi Al-Yammaniyah, Saudi Arabia. Ground Water 22(4):406–411

Metcalf, Eddy (1980) Water resources in Qatar for agricultural development. Technical Center for Industrial Development, Doha (cited after Zubari 1997)

Michelson H, Lipson-Benitah S (1986) The litho- and biostratigraphy of the southern Golan heights. Israel J earth sci 35:221–240

Mijatovic B, Bakic M (1966) Le karst du Liban, Etude de son evolution d'apres les recherches hydrogeologiques

Mikhailov I (1964) Schematic map of the first-below-the-surface water-bearing formation of Syria. Min. Industry Syrian A.R, Technoexport, Moscow

Mikhailov IYa (1966) The geological map of Syria, scale 1:200, 000, sheet I-37-XX (Salamiyeh). Explanatory notes. Min. Industry Syrian A.R, Technoexport, Damascus

Milligan NH, Gharbi A (1995) Groundwater management in the Salalah plain. In: Sultanate of Oman international conference on water resources management in arid countries, Muscat, pp 530–538

Min Agric Yemen (1977) Development of Wadi al Kharid, Stage I. Rep. Min. Agric., Bureau central d'etudes pour les equipments d'outre-mer, Sir A. Gibbs & Partners

Mogheir Y, Singh VP, de Lima JLM (2006) Spatial assessment and redesign of a groundwater quality network using entropy theory, Gaza Strip. Hydrogeol J 14(5):700–712

Mukhopadhyay A (1995) Distribution of transmissivity in the Dammam limestone formation, Kuwait. Ground water 33(5):801–805

Mukhopadhyay A (1999) Spatial estimation of transmissivity using artificial neural network. Gorund Water 37(3):458–464

Mukhopadhyay A (2003) Allocation of visual, statistical and artificial neural network methods in the differentiation of water from the exploited aquifers in Kuwait. Hydogeol J 11(3):343–356

Mukhopadhyay A, Szekely F, Senay Y (1994a) Artificial groundwater recharge experiments in carbonate and clastic aquifers of Kuwait. Water Resour Bull 30(6):1091–1107

Mukhopadhyay A, Al-Sulaimi J, Barrat JM (1994b) Numerical modeling of ground-water resources management options in Kuwait. Ground Water 32(6):917–928

Murris RJ (1984) Middle East, stratigraphic evolution and oil habitat. Am Assoc Petrol Geol Mem 35:353–372

Naimi AI (1965) The ground water of northeastern Saudi Arabia. Fifth Arab Petroleum Congress, Cairo, pp 16–23

Nations Unies (1967) Carte hydrogeologique du Liban, 1: 200,000

Nativ R, Adar E, Dahan O, Geyh MA (1995) Water recharge and solute transport through the vadose zone of fractured chalk under desert conditions. Water Resour Res 31:253–2612

Neal C, Stanger G (lecture without year) Past and present serpentinization of ultramafic rocks, an example from the Semail ophiolite nappe of northern Oman

Neal C, Stanger G (1984) Calcium and magnesium hydroxide precipitation from alkaline groundwaters in Oman, and their significance to the process of serpentinization. Mineral Mag 48(2): 237–241

Neumann-Redlin Ch (1991) Groundwater resources study, Al Mahwit Province. Yemen A.R. Investigations for the assessment of a groundwater budget. Final report. Rep BGR, Hannover

Nitsch M (1990) Soils salinization in the Wadi Dhuleil and Wadi Arja irrigation projects. Rep BGR, Hannover

O'Boy CA (1995) Salalah Plain water resources. In: Sultanate of Oman international conference on water resources management in arid countries, Muscat, pp 723–726

Otkun G (1972) Observations on Mesozoic sandstone aquifers in Saudi Arabia. 24 International Geological Congress, Montreal, Section 11, pp 28–35

Otkun G (1973) General aspects of Paleocene carbonate aquifer in Saudi Arabia, vol 5. International symposium on development of groundwater resources, Madras, pp 29–38

Otkun G (1974) Outlines of hydrogeology of Al Hassa springs and the results of some recent observations. Rep. Min Agriculture and Water, Riyadh

Otkun G (1975) More about Paleogene karst aquifer in Saudi Arabia. Mem IAH 12:25–38

Oufland AK (1966a) The geological map of Syria, scale 1:200, 000, sheet I-37-XXI (Ar Rasafeh). Explanatory notes. Min. Industry Syrian A.R, Technoexport, Damascus

Oufland AK (1966b) The geological map of Syria, scale 1:200, 000, sheet J-37-III (Jrablus). Explanatory notes. Min. Industry Syrian A.R, Technoexport, Damascus

Parker OH (1969) The hydrogeology of the Mesozoic-Cainozoic aquifers of the western highlands and plateau of East Jordan. UNDP-FAO Techn Rep 212,2

Parker DH, Pike JG (1976) Groundwater resources in Qatar and their potential for development. Integrated water and land use project, Technical Note No. 34, FAO/UNDP and Qatar Ministry of Industry and Agriculture, Qatar

Parsons RM Co (1963) Groundwater Resources in Kuwait. State of Kuwait
Pike JG (1985) Groundwater resources and development in the central region of the Arabian Gulf. In: Memoirs of 18th Congress. IAH Hydrogeology in the Service of Man, Cambridge, pp 46–55
Piper AM (1944) A graphic procedure in the geochemical interpretation of water-analyses. Am Geophys Union Trans 25:914–923
Ponikarov VP et al (1964) Geological map of Syria 1:200, 000, 19 sheets. Min Industry Syrian A. R, Technoexport, Moscow
Ponikarov VP et al (1967a) The geology of Syria. Explanatory notes on the geological map of Syria 1:500,000. Part I: stratigraphy, igneous rocks and rectonics. Damascus
Ponikarov VP et al (1967b) The geology of Syria. Explanatory notes on the geological map of Syria 1:500,000. Part II. Damascus
Powers RW, Ramirez LF, Redmond CD, Elberg EL (1966) Geology of the Arabian Peninsula. Sedimentary geology of Saudi Arabia. US Geological Survey Professional Paper, Washington DC, pp 560
Prizgognow V, Nitashov V, Sterikovich PN (1988) Determination of recharge zone of Damascus Basin by the use of oxygen-18 in groundwater. Moscow
Protasevich LN, Maksimov AA (1966) The geological map of Syria, scale 1:200, 000, sheets J-37-I, II (Halab, Antakya). Explanatory notes. Min. Industry Syrian A.R, Technoexport, Damascus
Qahman K, Larabi A (2006) Evaluation and numerical modeling of seawater intrusion in the Gaza aquifer (Palestine). Hydrogeol J 14(5):713–728
Qutob Z (1978) Hydrogeological investigations of the Al Ain oases and near by lying areas, U.A. E. In: The first Arab symposium on water resources, vol 1, pt 2. ACSAD, Damascus, pp 1–28
Rasheedudin M, Yazigil H, Al-Layla RI (1989) Numerical modeling of a multi-aquifer system in eastern Saudi Arabia. J Hydrol 107:193–222
Razvalayev AV (1966) The geological map of Syria, scale 1:200, 000, sheet I-37-VIII (Saba Abar). Explanatory notes. Min. Industry Syrian A.R, Technoexport, Damascus
Read R (1995) Wadi Safwan investigation Buraimi region. In: Sultanate of Oman international conference on water resources management in arid countries: 711–714, Muscat
Richter BC, Kreitler CW, Bledsoe BE (1993) Geochemical techniques for identifying sources of ground-water salinization
Rimawi O (1985) Hydrogeochemistry and isotope hydrology of the ground- and surface water in North Jordan (North-northeast of Mafraq, Dhuleil-Halabat, Azraq-Basin). Thesis, Technical University Munich, Munich
Rimawi O, Udluft P (1985) Natural Water groups and their origin of the shallow aquifers complex in Azraq-depression/Jordan. Geol Jahrb C33:17–38
Rizk ZS, El-Etr HA (1997) Hydrogeology and hydrochemistry of some springs in the United Arab Emirates. Arab J Sci Eng 22(1C):97–111, Dhahran
Rizk ZS, Al-Sharhan AS, Shindo S (1997) Evaluation of groundwater resources of United Arab Emirates. In: Third Gulf water conference, vol 1, Muscat, pp 95–122
Rizk ZS, Alsharhan AS, Wood WW (2007) Sources of dissolved solids and water in the Wadi Al Bih aquifer, Ras Al Khaimah Emirate, United Arab Emirates. Hydrogeol J 15(8).1453–1463
Robertson (1992) Republic of Yemen, The natural resources project – Hydrogeological map, 1:1,000,000, 2 sheets. Min.Oil Mineral Resources, Robertson group, Sanaa
Robinson BW, Al Ruwaih F (1985) The stable isotopic composition of water and sulfate from the Raudhatain and Umm Al Aish freshwater fields. Kuwait Chem Geol (Isot Geosci Sect) 58: 129–136
Rosenthal E (1987) Chemical composition of rainfall and groundwater in recharge areas of the Bet Shean – Harod multiple aquifer system. Israel J Hydrol 89:329–352
Rosenthal E (1988) Hydrochemical changes induced by overexploitation of groundwater at common outlets of the Bet Shean-Harod multiple aquifer system. Israel J Hydrol 97:107–128
Safadi Ch (1955) Aperçu sur l'hydrogeologie des terrains volcaniques de la Syrie Meridionale. Int Assoc Hydrol Sympos Ankara, pp 287–294

Salameh E (1996) Water quality degradation in Jordan (Impacts on Environment, Economy and Future Generations Resources Base). Friedrich Ebert Stiftung & Royal Society for Conservation of Nature, The Higher Council of Science and Technology, Amman

Salameh E (2004) Using environmental isotopes in the study of the recharge-discharge mechanisms of the Yarmouk catchment area in Jordan. Hydrogeol J 12(4):451–463

Salameh E, Khdier K (1985) Groundwater qualities in Jordan, vol 4. News Letter Water Research Stuy Center, Amman

Salameh E, Rimawi O (1984) Isotopic analyses and hydrochemistry of the thermal springs along the eastern side of the Jordan Dead Sea-Wadi Araba rift valley. J Hydrol 73(1–2):129–145

Salameh E, Rimawi O (1987a) Hydrochemistry of precipitation of Northern Jordan, vol 8. Bulletin of Water Research and Study Center, University of Jordan, Amman, pp 1–56

Salameh E, Rimawi O (1987b) Natural radioactivity and hydrochemistry of some Jordanian groundwater resources, vol 9. Bulletin of Water Research Study Center, University of Jordan, Amman

Salameh E, Shaqur T (1981) Geology and hydrology of the area west of Kerak City/Jordan. Dirasat 8(2):29–55, Amman

Salameh E, Udluft P (1985) The hydrodynamic pattern of the central part of Jordan. Geol Jahrb C38:39–53

Salameh E, Rimawi O, Abu Moghli I (1991) Precipitation water quality in Jordan, vol 16. Water Research and Study Center, University of Jordan, Amman

Salameh E, Alawi M, Batarseh M, Jiries A (2002) Determination of trihalomethanes and the ionic composition of groundwater at Amman City, Jordan. Hydrogeol J 10(2):332–339

Sanford WE, Wood WW (2001) Hydrology of the coastal sabkhas of Abu Dhabi, United Arab Emirates. Hydrogeol J 9(4):358–366

Sayed SAS, Al-Ruwaih FM (1995) Relationship among hydraulic characteristics of the Dammam aquifer and wells in Kuwait. Hydrogeol J 3(1):57–70

Scarpa DJ (1994) Eastward groundwater flow from the mountain aquifer. In: Issac J and Shuval H (eds) Water and peace in the Middle East. Elsevier, Amsterdam, pp 193–202

Schaeffer U (1997) Application of satellite remote-sensing methods for hydrogeology in the ESCWA region. Rep BGR, Hannover

Schelkes K (1997) Groundwater balance of the Jordan-Syrian basalt aquifer. Rep BGR, Hannover

Selkhozpromexport (1974) Hydrogeological and hydrological surveys and investigations in four areas of Syrian Arab Republic. Rep. Selkhozpromexport – Gruzgiprovodhkoz, Tbilisi

Selkhozpromexport (1986) Water Resources Use in Barada and Auvage Basins for irrigation and crops, Syrian Arab Republic, vol 2: Natural conditions. Book 2: Hydrogeology. Selkhozpromexport, Moscow

Sen Z, Al-Dakhel A (1986) Hydrochemical facies evolution in Umm Er Radhuma limestone, eastern Saudi Arabia. Ground Water 24(5):626–635

Shaban A, Khawlie M, Abdallah C (2006) Use of remote sensing and GIS to determine recharge potential zones: the case of Occidental Lebanon. Hydrogeol J 14(4):433–443

Shahab SJK (1997) Hydrochemical and environmental aspects of the upper aquifer system in Tulkarm-Qalqeelya area/Palestine. Thesis, University of Jordan, Amman

Shampine WJ, Dincer T, Noory M (1979) An evaluation of isotope concentrations in the groundwater of Saudi Arabia. Int Sympos Isotope Hydrol 2:443–463

Shatsky MA, Husein I (1996) Groundwater quality in the Saq aquifer, Saudi Arabia. Hydrol Sci J 41(5):683–696

Shatsky VN, Kazmin VG, Kulakov VV (1966) The geological map of Syria, scale 1:200, 000, sheet I-37-XIX, I-36-XXIV. Explanatory notes. Min. Industry Syrian A.R, Technoexport, Damascus

SMMP (1975) Riyadh additional water resources study. Rep Min Agriculture Water Saudi Arabia, Riyah (cited after Subyani and Sen 1991)

Sogreah (1968) Water and agricultural development studies area V. Rep Min Agriculture Water Saudi Arabia, Riyah (cited after Subyani and Sen 1991)

Sorman AU, Abdulrazzak MJ (1993) Infiltration recharge through wadi beds in arid regions. Hydrol Sci J 38(3):173–187

Soulidi-Kondratiev ED (1966) The geological map of Syria, scale 1:200, 000, sheet I-37-XV (Tudmor). Explanatory notes. Min. Industry Syrian A.R, Technoexport, Damascus

Sowayan AM, Allayla R (1989) Origin of the saline groundwater in Wadi Ar-Rumah, Saudi Arabia. Ground water 27(4):481–490

Subyani A, Sen Z (1991) Study of recharge outcrop relation of the Wasia aquifer in central Saudi Arabia. J King Abdulaziz Univ Earth Sci 4:137–147, Jeddah

Sunna BF (1995) Geology and hydrogeology of Jordan, Syria. West Bank and Gaza Strip. Rep ESCWA, Amman

Tóth J (1995) Hydraulic continuity in large sedimentary basins. Hydrogeol J 3(4):4–16

UNDP (1970) Liban. Étude des eaux souterraines, rapport technique. United Nations Development Programme, New York

United Nations (1982) Groundwater in the eastern Mediterranean and Western Asia, vol 9. Natural resources/water series, New York

Van der Gun JAM, Ahmed AA (1995) The water resources of Yemen. Report of WRAY-35, Sanaa, Delft

Van der Gun J, Elderhorst W, Kruseman G (1992) Predicting the long-term impacts of groundwater abstraction from an intensively exploited coastal aquifer. Selected papers on aquifer overexploitation. Select Pap Hydrogeol 3:303–318

Van Overmeeren RA (1989) Aquifer boundaries explored by geoelectrical measurements in the coastal plain of Yemen: a case of equivalence. Geophysics 54:38–48

Verhagen BT, Geyh MA, Fröhlich K, Wirth K (1991) Isotope hydrological methods for the quantitative evaluation of ground water resources in arid and semi-arid areas. Development of a methodology. Weltforum, Cologne

Vierhuff H, Trippler K (1977) National water master plan of Jordan, vol 4, Groundwater resources. GTZ and NRA, Hannover, Frankfurt, Amman

Wagner W (1995a) Report on mission to the United Arab Emirates, data compilation and evaluation for groundwater studies in the eastern coastal area and in the Hajar aquifer, northern Oman Mountains. Rep. ESCWA E/ESCWA/ENR/1995/6, Amman

Wagner W (1995b) Groundwater with low salinity in the dry regions of Western Asia, hydrochemical aspects. In: Sultanate of Oman international conference on water resources management in arid countries, Muscat, pp 443–449

Wagner W (1996a) Report on mission to Lebanon: Advice on a program for exploration of fresh water sources in the Chekka area. Rep. ESCWA, E/ESCWA/ENR/1996/4, Amman

Wagner W (1996b) Advice on groundwater investigations in Wadi Ham, Wadi Wurrayyah and Wadi Bih. Rep. ESCWA/ENR/1996/01, Amman

Wagner W (1996c) Groundwater quality control in the ESCWA region. In: Proc. Expert Group Meeting Implication Agenda 21. ESCWA/ENR/1996/5. UNEP, New York, pp 128–162

Wagner W (1997) Report on mission to the Syrian Arab Republic: hydrogeologic advice to ICARDA on sustainable agricultural groundwater management in dry areas. E/ESCWA/TCD/1997/32, Amman

Wagner W, Geyh M (1999) Application of environmental isotope methods for groundwater studies in the ESCWA region. Geol Jahrb C67

Wagner W, Karanjac J (1997) Mathematical model of the Wadi Ham aquifer, Fujayrah coastal plain, United Arab Emirates. Advisory services to the Ministry of Agriculture and Fisheries, United Arab Emirates. Rep. ESCWA, E/ESCWA/TCD/1997/6, Amman

Wagner W, Kruck W (1982) Report on Activities and Results in the Field of Geology and Hydrogeology, 1978-1982, Part II-1. Geology of the Hamad Region. Rep BGR, ACSAD, Hannover, Damascus

Wagner W, Nash HG (1978) Report on a preliminary appraisal of ground water occurrance in the Amran valley, Yemen Arab Republic. Rep BGR, Hannover

Wagner W, Khouri J, Rashdan J, Rofail N, Zahra S (1982) Report on activities and results in the field of geology and hydrogeology, 1978–1982. Part III, Hydrogeological investigations in the Hamad region. Rep. BGR, ACSAD, Hannover, Damascus

WAJ-ODA (1994) Qa Disi aquifer study, Jordan. Interim Rep. WAJ, UK ODA Technical Cooperation Programme

Wakshal E, Nielsen H (1982) Variations of $\delta^{34}S(SO_4)$, $\delta^{18}O(H_2O)$ and Cl/SO_4 ratio in rainwater over northern Israel, from the Mediterranean coast to the Jordan rift valley and the Golan heights. Earth and Planetary Sci Letters 61:272–282

Wakil M (1994) Underground water for supplemental irrigation in Syria: quantity and quality. ICARDA farm resource management program. Annual report for 1993, Aleppo, pp 42–58

Ward K, Smith P (1994) An investigation into the shallow groundwater resources of part of north-west Syria (Aleppo-Atareb-Idleb area). ICARDA farm resource management program. Annual report for 1993, Aleppo, pp 29–42

Wilson GW, Wozab DH (1954) Chemical quality of waters occurring in the Jordan valley area. Publ Internat Assoc Hydrol 37(2):170–178, Rome

Weinberger G, Rosenthal E, Ben-Zvi A, Zeitoun DG (1994) The Yarkin - Taninim groundwater basin, Israel, hyrogeology: case study and critical review. J Hydrol 161:227–255

Wiesemann G (1969) Zur Tektonik des Gebietes östlich des Grabenabschnittes Totes Meer-Jordantal. Beih Geol Jahrb 81:215–247

Wohlfahrt E (1980) Die Arabische Halbinsel. Berlin

Wolfart R (1961) Geological–hydrogeological research for the utilization of groundwater of the Wadi Nisah area, water to Riyadh, Saudi Arabia. Rep BGR, Hannover

Wolfart R (1966) Zur Geologie und Hydrogeologie von Syrien. Beihefte Geol Jahrb 68

Yakirevich A, Melloul A, Sorek S, Shaath S, Borisov V (1998) Simulation of seawater intrusion into the Khan Yunis area of the Gaza Strip coastal aquifer. Hydrogeol J 6(4):549–559

Yechieli Y, Kafri U, Goldman M, Voss CI (2001) Factors controlling the configuration of the fresh-saline water interface in the Dead Sea coastal aquifer system: synthesis of TDEM surveys and numerical groundwater modeling. Hydrogeol J 9(4):367–377

Yurtsever Y (1992) Interim report on preliminary evaluation of isotope results from reconnaissance samples collected in Kuwait. IAEA Technical Cooperation Project isotope Hydrology in the Middle East, Vienna

Yurtsever Y (1996) Isotope methodologies in water resources and examples of applications in the Escwa region. In: Proceedings of Expert Group Meeting Implication Agenda 21. ESCWA/ENR/1996/5, UNEP: 451-471, New York

Yurtsever Y, Gat J (1981) Atmospheric waters. In: Gat J, Gonfiantini U (eds) Stable isotope hydrology: Deuterium and Oxygen-81 in the water cycle. Technical Reports Series, vol 210. IAEA, Vienna, pp 103–142

Yurtsever Y, Payne BR (1979) Application of environmental isotopes to groundwater investigations in Qatar. International symposium on isotope hydrology 1978, vol 2, Vienna, pp 465–490

Zubari WK (1997) Regional study of the Paleogene carbonate aquifers in ESCWA countries. Umm er Radhuma and Dammam aquifers. Rep, ESCWA, Amman

Zubari WK (1999) The Dammam aquifer in Bahrain – Hydrochemical characterization and alternatives for management of groundwater quality. Hydrogeol J 7(2):197–208

Zubari WK, Khater AR, Al-Noaimi MA, Al-Junaid SS (1997) Spatial and temporal trends in groundwater salinity in Bahrain. Arab J Sci Eng 22(1C):81–94, Dhahran

Zuppi GM (1986) Environmental isotope study of groundwater systems in the Azraq Area, Jordan. Rep IAEA, Vienna

Abbreviations

a	year
ACSAD	Arab Center for the Studies of Arid Zones and Dry Lands, Damascus (Arab League)
asl	above sea level
BGR	Bundesanstalt für Geowissenschaften und Rohstoffe, Federal Institute for Geosciences and Natural Resources, Hannover
bls	below land surface
BP	years before present (before 1950)
d	day
EC	electrical conductivity (in Microsiemens/cm, μS/cm)
ESCWA	UN Economic and Social Commission for Western Asia, Beirut
FAO	UN Food and Agricultural Otganization, Rome
GTZ	Gesellschaft für Technische Zusammenarbeit, German Agency for Technical Cooperation, Eschborn
IAEA	UN International Atomic Energy Agency, Vienna
IAH	International Association of Hydrogeologists
IAHS	International Association of Hydrological Sciences
K	hydraulic conductivity (in metres per second)
KISR	Kuwait Institute for Scientific Research
l/s	litre per second
m	metre
m^3/a	m^3 per year
MAW	Ministry of Agriculture and Water (Saudi Arabia), Riyadh
n	number of samples
pmc	percent modern carbon
Q/s	specific well yield ($m^3/h/m$, well yield per m of drawdown)
T	transmissivity (m^3/d)
TDS	total dissolved solids (in mg/l)
TU	tritium unit
WAJ	Water Authority of Jordan, Aman

Index

G

H

I

J

K

GPSR Compliance
The European Union's (EU) General Product Safety Regulation (GPSR) is a set of rules that requires consumer products to be safe and our obligations to ensure this.

If you have any concerns about our products, you can contact us on

ProductSafety@springernature.com

In case Publisher is established outside the EU, the EU authorized representative is:

Springer Nature Customer Service Center GmbH
Europaplatz 3
69115 Heidelberg, Germany

www.ingramcontent.com/pod-product-compliance
Lightning Source LLC
Chambersburg PA
CBHW060748150926
51501CB00003B/125
9783642193507